Springer Geology

The book series Springer Geology comprises a broad portfolio of scientific books, aiming at researchers, students, and everyone interested in geology. The series includes peer-reviewed monographs, edited volumes, textbooks, and conference proceedings. It covers the entire research area of geology including, but not limited to, economic geology, mineral resources, historical geology, quantitative geology, structural geology, geomorphology, paleontology, and sedimentology.

More information about this series at http://www.springer.com/series/10172

Márton Veress

Covered Karsts

Springer

Márton Veress
Institute of Geography and Environmental Sciences
University of West Hungary
Szombathely, Hungary

ISSN 2197-9545 ISSN 2197-9553 (electronic)
Springer Geology
ISBN 978-94-017-7516-8 ISBN 978-94-017-7518-2 (eBook)
DOI 10.1007/978-94-017-7518-2

Library of Congress Control Number: 2016932407

Printed on acid-free paper

This Springer imprint is published by SpringerNature
The registered company is Springer Science+Business Media B.V. Dordrecht.

Preface

I had a double aim while writing this book. On the one hand, I have been dealing with covered karsts for about 40 years. During my investigations, I have collected a lot of experience and data that needed to be systematised. On the other hand, a number of books have been written on the topic of karst during the last few decades; many of them deal with covered karsts (mainly with covered karst features). However, in all of these books, the covered karsts are interpreted as a part of karst. In this book, the central topic is the covered karst itself. Non-covered karst topics are only mentioned and only to a certain extent if they contribute to the evolution of covered karst or to covered karstification.

The book is dissected into eight chapters. In the first chapter there is an introduction of karst (karst water, karst types, karst features). The second chapter presents those karst places where our researches on collecting data on covered karstification were carried out. Chapter 3 is about the methods applied during our covered karst researches. Chapter 4 contains the classification and description of covered karsts. In Chap. 5 we describe covered karst features. Chapter 6 is an overview on the processes taking place in the covered karst features and on the changes of the features. The evolution of covered karst features can be found in Chap. 7. Finally, in Chap. 8 we deal with the surface development related to the different covered karst types.

Szombathely, Hungary
Márton Veress

Acknowledgements

Many people contributed to the preparation of this book as well as to the necessary data collection. I would like to say thanks for their efforts here too. These people are the following: Deák Gy., Dombi L., Döbröntei L., Gárdonyi I., Györéné Koperecz É., Futó J., Jakab I., Lóczy D., Nacsa T., Németh H., Schläffer R., Szemes M., Széles Gy., Veressné Herczegh K. and Zentai Z.

Translated by: Lóczy D.

Contents

Chapter 1
General Description of Karst

Abstract The chapter presents the characteristics of karst, karst hydrology, mor-phology and typology. In the hydrological overview, karst water zones are identified and described; surface karst features (karren, dolines, poljes, etc.) are defined broken down to their main varieties with their brief characterisation. The main genetic types of caves are presented. Karst types are identified according to their geological conditions (such as covering, structure), and then the types by hydrology and cover are described. In the classification by climatic environment, tundra, temperate, mediterranean, tropical and high-mountain (glacio)karst are identified and described.

Keywords Karst water • Karst water zone • Karst morphology • Surface karst features • Karst cave • Karst type

1.1 Introduction

Research in the karst regions of the Earth dates back to the nineteenth century. Since then, the properties of karst have been investigated by a range of disciplines. For this reason and also because the karst is a complex system, there are many ways of characterisation. We regard hydrology, morphology and type the most important properties of karst. In this chapter, karst will be presented from these aspects.

There have been several basic works (Sweeting 1973; Jennings 1985; Trudgill 1985; White 1988; Jakucs 1977; Ford and Williams 1989, 2007) written on karst, which provide the foundations for our discussions below.

1.2 Karst Hydrology

The water infiltrating into the karst fills up rock joints until the impounding effect of the rock underlying the karst raises the water level to the elevation of the local base level. From that point, the water stored in the karst flows out from there (karst spring) and karst zones develop (Fig. 1.1). In the vadose zone between the ground

M. Veress, *Covered Karsts*, Springer Geology,
DOI 10.1007/978-94-017-7518-2_1

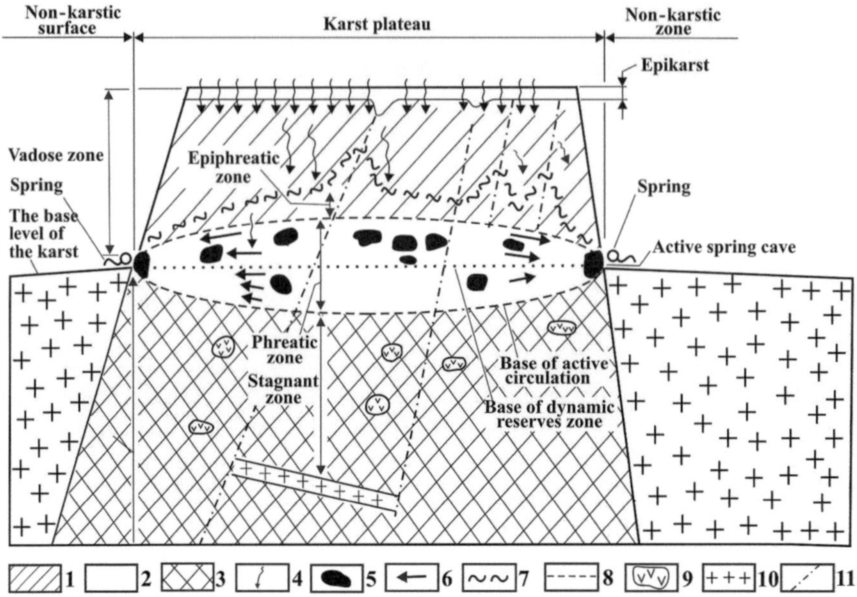

Fig. 1.1 The karst hydrological system (Veress 2004) *1* vadose zone, *2* phreatic zone, *3* stagnant karst water zone, *4* water percolation, *5* cavity filled with water, *6* water flow, *7* high karst water table, *8* low karst water table, *9* cavity filled with calcite, *10* non-karstic rock, *11* fault

surface and the karst water table, water percolates vertically downwards and only temporarily fills the joints (in rainy periods). In this zone, vertical solution and cavity formation take place. Beyond the karst water table follows the phreatic zone, where the water flows horizontally filling up joints continuously. The phreatic zone reaches down below the karst springs. Thus, its lower surface is concave, while its upper surface (the karst water table) is convex. The fluctuation of the karst water table can be very remarkable, occasionally reaching several hundred metres. In periods of high precipitation, it rises (high water table), while it sinks in dry periods (low water table). Between the two surfaces, we find the epiphreatic zone and below the lower surface the phreatic zone. Rock solution with horizontal cavity formation is mostly concentrated in these zones. The deep karst lies below the static karst water level (spring level). The part of this zone without water flow is called stagnant karst water zone. In the cavities of this zone, precipitation occurs.

1.3 Karst Features

1.3.1 Surface Karst Features

Surface karst features include karren, dolines, ponors, poljes, karst mounds, relict karst features, karst lakes and karst plains. Surface karst features primarily result from solution in the epikarst zone. The most important and one of the most common non-karstic landforms of karst regions is the valley.

With the exception of megakarren, karren are small-scale features, which develop in high density on karst terrains. According to size, Ginés (2004, 2009) identifies nanokarren (smaller than 1 mm), microkarren (less than 1 cm in width), mesokarren (the width and depth of features of tens of centimetres) and megakarren (which have sizes of tens of metres or even more).

Karren develop on bare surfaces under soil patches, below the soil and cover sediments (Bögli 1960; Zseni 2004, 2009; Slabe and Liu 2009; Veress 2010). Their most common formation environment is on high-mountain and tropical karsts, but they also occur on coasts (Lundberg 2009) and in caves (Bögli 1960). Their development is due to water flow and seepage (White 1988; Ford and Williams 2007; Veress 2010).

Water flow karren develop on bare surfaces. They include rillenkarren (1–2 cm wide, some tens of centimetre-long channels wedging out downslope), rinnenkarren (channels without outflow, several tens of metres long, some centimetres or tens of centimetres wide), trittkarren (heel-like depressions), meanderkarren (asymmetric channels) and wall karren (several metre-long channels on steep slopes). Water seepage karren can develop on both bare and soil-covered terrain. They include grikes (grike karren, depressions of 10–20 cm wide and several tens of metres long between parallel walls), pitkarren (pipe-like features in rocks) and kamenitzas (bowl-shaped depressions of several tens of centimetres diameter). Other karren features are rainpits, root karren and relict karren (spitzkarren, karren tables, karren inselbergs and monadnocks). Megakarren include pinnacle karst, tsingy, stone forests and giant chasms.

Giant chasms (bogaz) are solution forms of several metres width, several tens of metres depth and hundreds of metres length. They are, however, not clearly distinguishable from grikes. Giant chasms have several varieties: in high mountains, they often develop with shafts and widen to form karst streets.

In karst regions, dolines are the most frequently occurring features. Their size (diameter and depth) ranges from several metres to hundreds of metres. There are numerous types of dolines with different morphological characters. According to origin and the environment of formation, the following main varieties are identified (Cramer 1941; Williams 2004, 1983; Ford and Williams 2007; Waltham et al. 2005; Waltham and Fookes 2003; Gunn 1981; Fig. 1.2). Solution dolines (Figs. 1.2 and 1.3), the most typical features of karst regions, emerge on bare or soil-covered surfaces. No erosion valley or ravine is attached to solution dolines. Although solution dolines develop by surface solution (Jakucs 1977; Veress and Péntek 1996) or solution in the epikarst (Williams 1983), pipes and shafts are common on their floors. The varieties distinguished according to geological conditions, climate or the cover of the bearing surface are stepped doline (in horizontally bedded rocks) or half-doline (on the contact line between karstic and non-karstic rocks). It is called karst fenster if during deepening the doline floor reaches down to impermeable rocks. The solution dolines of the temperate belt are usually of bowl shape. On tropical karsts, cockpits, complex dolines, while in high mountains schachtdolines or schneedolines are found. Cockpits are large-scale landforms, divided into several small depressions (connected by karst streets) with drainage of several grades and possibly ponors on their floors. The interior of the 'doline rainforest karst' is

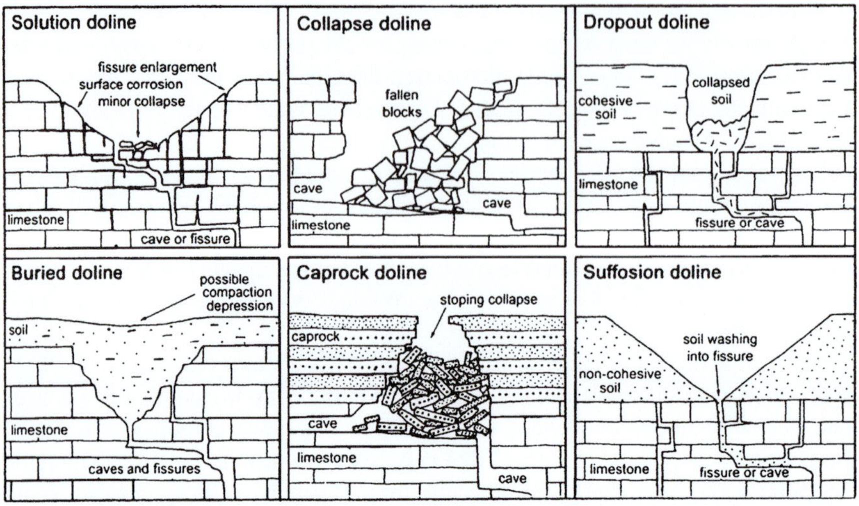

Fig. 1.2 Doline types (After Waltham and Fookes 2003, 2005)

Fig. 1.3 Solution doline (Pádis Plateau, Romania)

dissected by cracks and pipes with multiple crests and cones among them. In South China, a doline type, closed depression among the karst towers (inselbergs), is lined with cover sediments at the floor (Balázs 1991). This doline type corresponds to the depression of superficial deposit developed in tropical environment (see below). Schachtdolines are depressions with steep, often vertical, walls.

In the high mountains of Europe, paleokarsts were discovered: in the Alps Upper Triassic dolines (Bini and Pellegrin 1998) and Miocene caves (Audra et al. 2006; Frisch et al. 2002). The Würm glaciers of the Durmitor Mountains developed in dolines (Cvijič 1899, 1913); therefore, these features are older than Würm age. According to Bauer and Zötl (1972), the dolines on the karst of the Northern Limestone Alps have formed since the Late Tertiary. In various high mountains, like the Durmitor, Proklétije, Maglič and Orjel Mountains, glacier valleys developed from dolines (Cvijič 1899, 1911, 1913; Menkovič 1994). In some regions of the Northern Limestone Alps (in the Totes Gebirge), rows of large-scale dolines (of several hundred metres diameter) are found on the floors of glacial troughs (Veress 2012). The large size and the till partially filling numerous dolines indicate that they are paleodolines. As it has already been mentioned, they are dated to the Late Tertiary (Bauer and Zötl 1972). In the lower elevations of high mountains, steep-walled and deep dolines also occur, which were free of burial under ice and, thus, have not been reshaped by ice.

Dolines may develop into flat-bottomed forms with short slopes (bowl doline) or open forms without side slopes (ruined doline) (Hevesi 1984). Solution dolines may be also developed if a ponor loses its catchment, fills up and transforms into a doline (Jakucs 1977; Hevesi 1978) and if the expanding neighbouring dolines coalesce and form uvalas. Uvalas can be dissected by thresholds, half thresholds and hums; their rims are arcuate. On the uvala floor, new dolines often develop and merge into complex uvalas.

Collapse dolines (Fig. 1.2) derive from caved-in cavities and caves. They are generally features of steep walls with mounds of variable size in the centre, consisting of fallen rock debris.

Covered karst dolines also have several varieties. Caprock dolines (Waltham and Fookes 2003) are also of collapse origin, but the collapse affects consolidated non karstic cover sediments (Fig. 1.2). If the cover sediments are unconsolidated, subsidence dolines form. Gullies and ravines may lead to subsidence dolines or may be connected to them. Such dolines can be steep-walled, collapsed dropout dolines (Fig. 1.2) or suffosion dolines with gentle slopes, resulting from compaction or suffosion (Fig. 1.2). A compaction doline is a feature of small depth which develops by the compaction of the fill of the buried doline (Fig. 1.2). Cramer (1941) and Jennings (1985) identify alluvial streamsink dolines, typical at the katavothra of poljes and form through the intensive erosion of cover deposits by runoff. This type is regarded a depression of superficial deposit (DSD). Depressions of superficial deposit exclusively form in cover sediments, where the latter is locally eroded and transported into the karst (Veress 2009).

Ponors are depressions of inflow of surface waters into the karst, and, therefore, they invariably have a conducting passage or a ponor cave. Ponors develop on rock boundaries where non-karstic impermeable rocks contact with limestone. According to their position, ponors may be karst interior (Jakucs 1977; Hevesi 1980) or karst marginal (Jakucs 1956).

Karst marginal ponors are found where the cover sediment or cover rock wedges out at the terminal points of blind valleys on the covered karst terrain. Karst interior ponors form if a valley on the karst cover sediment incises and its floor reaches down to the limestone (Fig. 1.4b) and also at several other sites in the karst: below glaciers and in poljes, paleodolines, cockpits and intermountain plains of inselberg karsts. A special ponor type of poljes is the katavothron (estavelle). At high karst water levels, katavothra function as springs, while at low karst water levels as ponors. Katavothron-like functioning is observed for some foot caves of tropical karsts (Balázs 1990). Below glaciers, shafts functioning as ponors occur (Corbel 1957; Kunaver 1965; Ford et al. 1970).

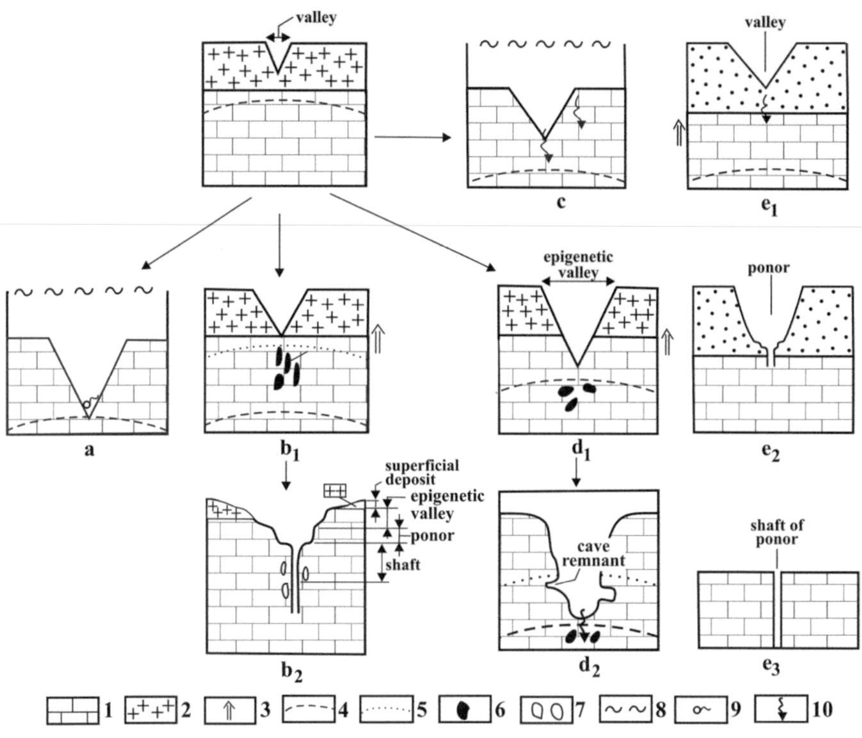

Fig. 1.4 Epigenetic valleys on karst (Modified after Veress 2004) (**a**) deepening of the epigenetic valley as the karst water table lies close to the valley floor and there is no water seepage from the watercourse, (**b**) formation of ponor passages from the inactive karst water cavities, (**c**) transformation of the epigenetic valley into dry valley as waters from the valley and its environs seep into the karst, (**d**) exposure of inactive karst water cavities by the incision of the epigenetic valley and (**e**) ponor develops on the floor of the valley, *1* limestone, *2* non-karstic rock, *3* uplift, *4* karst water table, *5* former karst water table, *6* expanding cavity, *7* non-expanding cavity, *8* former surface, *9* spring, *10* water seepage

Poljes are large, closed (undrained) forms with flat bottoms. On their floors, non-karstic rocks outcrop or the limestone is under non-karstic cover. Poljes are typical primarily in mediterranean karst (in the Dinaric Mountains) but also occur under tropical climate, on the margins of inselberg karsts (Sweeting 1973). Their rims are straight and the longer axis falls in the strike of the mountains where they are found. According to position, they are karst marginal or karst interior poljes, the latter being further classified into several varieties (Gams 1977, 1978; Delannoy 1997): structural, peripheral, border, piedmont or base level poljes. According to their hydrological properties, poljes belong to three types (Cvijič 1893; Cholnoky 1940):

- The dry polje has its floor above the high karst water table.
- The intermittently flooded polje has its floor below the high karst water table. Such poljes contain intermittent lakes of variable extension and katavothra.
- The bottom of the permanently flooded polje lies even below the level of the low karst water table. In such poljes, not only intermittent lakes but also lakes of permanent water surface develop, the levels of the latter ones are lower than the low karst water table and therefore there is no water seepage or flow towards the karst.

In karst regions, valleys are the most frequently occurring non-karstic landforms. In karst areas, valley formation always has a particular reason since rainwater infiltrates into the karstic rock. In surfaces of steep dip, on the margins of glacial troughs, short erosion valleys can emerge since water from intensive rainfalls – before infiltration – flows downslope and is capable of incision in combination with additional effects (e.g. cryofraction). A watercourse of abundant discharge, particularly if it crosses the karst only over a short distance, although its water seeps away, also can keep open (and incise) its valley. On permafrost surfaces, the water from watercourses cannot percolate deep and, therefore, is capable of valley formation.

On covered karst terrain (in the case of impermeable cover), cutting through the cover sediment, the watercourse continues to incise into the limestone until there remains an impermeable cover in the catchment (epigenetic valley), as runoff from the (patches of the) cover sediment flow into the valley. After inheritance, the valley has a permanent water flow if the valley floor reaches the karst water table. In spite of the removal of cover sediment, this situation can be sustained until valley incision keeps pace with the subsidence of karst water level (Fig. 1.4a). If the rate of karst water level subsidence is higher, the valley becomes a dry valley (Fig. 1.4c). If during inheritance the watercourse contacts a chimney in the limestone body, a ponor forms and the watercourse is beheaded (Fig. 1.4b). If during inheritance no beheading (capture) is possible, the cavities formed in the rock mass promote valley formation even in the case the valley floor lies above the karst water table. Part of the cavities is obliterated by valley formation, while another part is preserved in the valley side (Veress 2000, Fig. 1.4d).

The transformation of the valley from a glacial trough is not an inheritance process if the glacial trough developed in limestone reaches the non-karstic base (Fig. 1.5). A valley can form on the truncated anticline where non-karstic rocks are exposed (Fig. 1.5II) or through the caving in of the ceiling of an erosion cave

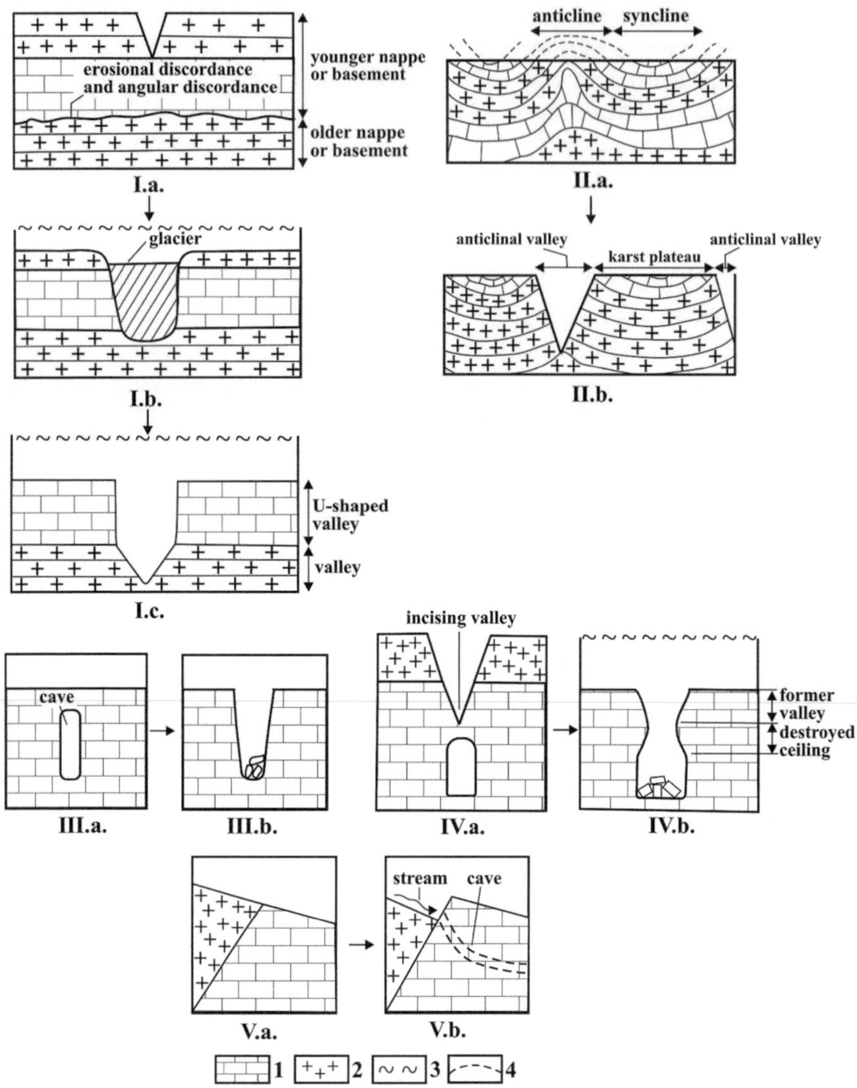

Fig. 1.5 Non-epigenetic valleys on karst (After Veress 2004) *I* inheritance of glacial trough over the non-karstic bedrock underlying the karst, *II* valley formation in the non-karstic strata exposed in the anticline, *III* opening of cave, *IV* thinning out of cave ceiling by epigenetic valley, *V* blind valley formation, *1* limestone, *2* non-karstic rock, *3* former surface, *4* karst water table

(Fig. 1.5III, IV). On the non-karstic terrains bordering the karst, blind valleys terminating on the karst margin and continuing on ponor passage develop. Pocket valleys begin on the margins of karst areas and are fed by karst springs. Their uppermost or other sections of variable length are often still deepened in carbonates.

Relict features, like hums, karst inselbergs, karst mounds and karst thresholds, are common on karst. Karstic mounds, typical of mediterranean karst (Veress and Péntek 2010) and tropical karst (Day 1978), are some metres to some tens of metres high. On tropical karst, they originate from the denudation of inselbergs (Balázs 1990). They are primary features on polygonal karst, where mounds of various shapes are found on the dividing walls between dolines (Balázs 1973; Day 1978).

The height of karst inselbergs of variable shapes ranges from several hundreds of metres to about 300 m: there are sigmoid, conical, steep cones, tall tower, pinnacle and flat-topped mogote types (White 1988). In the inselbergs, cave remnants occur at various levels, while at their base, foot caves can be found (Balázs 1962; Wilford and Wall 1965) and shaft caves may also be present in their area (Balázs 1990). The meandering rivers of the intermountain lowland carve overhangs at the footslopes.

Karstic summits and elevations of occasionally more than 100 m relative height are typical of the temperate karst. Along some side slopes, they are only separated from their environs by valleys. Karstic summits have a flat top, while karstic elevations are rounded on the top. On the side slopes of elevations and summits, different types of solution dolines (ruined or bowl dolines) are observed.

In karst depressions (dolines, ponors, in some portions of poljes), intermittent or permanent lakes appear. In covered karst dolines, both intermittent and permanent lakes are common.

Karst plains develop through solution primarily in tropical areas (in smaller extension also on mediterranean karst). On tropical karsts, intensive solution and the absence of uplift result in the formation of hardly dissected carbonate surfaces (Wilford and Wall 1965; Sweeting 1995), where weathering products or fluvial deposits can accumulate. The surface is completely planated by lateral and pluvial erosion. As it has been already mentioned, karst plains of various extensions can also originate on other types of karst to different effects, including the retreat of karst margins, the coalescence of dolines and then the removal of thresholds between dolines (Grund 1914), on intensively karstifying rock (e.g. gypsum) or if the horizontally bedded thin rock strata liable to karstification are eroded and the underlying non-karstic horizontal rock beds are exposed (Cvijič 1918).

1.3.2 *Caves*

Karst caves are underground karst features which can form by solution or the joint action of solution and erosion of karst watercourses. Vertically developed and vertically aligned caves are chimneys, shafts and giant chasms (bogaz), all formed in the vadose zone.

Chimneys are smaller in diameter (at most some metres) than shafts – the two types of form, however, cannot be sharply separated. There are at least two varieties of shafts: the shallower shafts are several tens of metres deep, while the deep shaft systems can be hundreds of metres deep. The latter are built up of several separate partial shafts, often connected by short horizontal passages. For both varieties, but particularly for the latter, partial shafts or chimneys developed parallel with the main shaft are typical. The shafts communicate through horizontal cave corridors or cut across such corridors. From the chimneys or shafts (groups of), blind chimneys branch out. Grouped chimneys are found in wreath-like arrangement. Chimneys and shafts occur on all types of karst; they are particularly widespread on covered karst. In high mountains, shafts of great size are characteristic.

In the phreatic zone, mostly horizontal cavities of circular cross-section form. Cave formation can follow a different path. Looped passages develop along the intersecting cracks and bedding planes (Ford 1998). The closer they are to the karst water table, the lower is the dissection of loops in passages and also the more irregular the circular cross-section becomes. If the dissolving water flowed or seeped along a crack, vertical elongation is observed. If it happened along a bedding plane, horizontally elongated cavity cross-sections are generated (Jennings 1985). Naturally, the variable spatial positions of cracks and bedding planes result in highly variable directions of elongation.

In or along the phreatic zone, spring caves, relict caves, katavothra and foot caves originate. Spring caves are outflow sites of karst water. Relict caves are remnants of cavities exposed by valleys. Katavothra are produced by the alternating flow of water in the epiphreatic zone in essentially opposite directions.

Erosion caves result from the corrasional activity of water flowing in cavities generated by solution. One type, the ponor cave, still mostly forms in the vadose zone. The continuing cave passage develops from cavities which were created below the karst water table but by the time of the erosional development acquired a position above the karst water table temporarily (epiphreatic zone) or finally.

1.4 Karst Types

Karst regions can be classified according to their rock composition, geological conditions, covered character, elevation, hydrological conditions and climate (Gvozdetskiy 1965).

According to composing rocks, the karst can be limestone, marble, dolomite, halite or gypsum karst. In recent karsts, karstification takes place at present and in paleokarsts, it occurred in earlier times.

According to geological conditions, we speak about holokarst, merokarst and transitional karst (Cvijič 1924). Holokarsts are built up of uniformly developed limestone; the rocks of merokarsts show interruptions of non-karstic sequences; on transitional karst, the karstic rock mass is bordered by non-karstic (impermeable) rocks from below and on the sides.

According to structure, the karst can be geosyncline and platform karst (Komatina 1982) or horst-type karst (Jakucs 1977; Veress 2000; Hevesi 1991).

On platform-type karsts, relying on studies from the Russian Platform, syneclise and anteclise karsts are distinguished. For both varieties, covered karst areas can occur in considerable areal extension. The character and development of the non-karstic overlying rock cover are often controlled by geological history. Platform karsts have a variety in lower position (Russian Platform) and another in higher position (Middle Lena region, karst areas in South China).

On platform karsts – because of the lack of compression – bedding is horizontal, and as a result of the large amounts of terrestrial sediments, the contamination of karstic rocks is relatively high, and they are quite often interrupted by non-karstic interbeddings. Morphological dissection is negligible. On platform karst, the rate of water flow is low because the height of the base level is low compared to the karst surface. (The highly uplifted platform karsts in China are exceptions in this respect.) For this reason and because of the lithology, cavity formation is moderate, or if present, cavities are developed horizontally. Platform-type karsts occur in the United States (Kentucky, Indiana and other states) and on the coastal zones of North Africa.

Tectonically deformed karsts include karst in the folded belts and geosyncline (miogeosyncline and eugeosyncline) karsts. In the same mountains, mio- and eugeosyncline karsts usually cannot be distinguished from each other. On geosyncline karsts, the extension and character of covered karst areas are not controlled by the geological evolution but the geomorphological history (first of all glaciation).

After deformation, geosyncline karsts underwent uplift and the cover sediments were removed. (The sediments of the nappes accumulated in the miogeosyncline section, where limestones of chemical origin were formed in great thicknesses, particularly in lagoon environments. The karst regions here are generally the survived portions of nappes, such as the Northern Limestone Alps.) The nappe remnants represent plateaus transformed by glaciation. The nappe remnants are underlain by older nappes of non-karstic rocks or in situ accumulated also non-karstic rocks.

Horst-type karsts are widespread in Hungary, e.g. in the Transdanubian Mountains (Pécsi 1980). On this type of karst, an older planated karst surface has been dismembered into blocks of different elevation along faults. The uplifting, subsiding or oscillating movements of blocks generated diverse carbonaceous or non-carbonaceous sequences on the planated block surfaces.

According to elevation, there are coastal, plain, hill, low, medium and high-mountain (see geosyncline karst) karsts.

Coastal karst has numerous special properties: on the abrasional platform, abrasional karren and solutional abrasional benches form, and the role of bioerosion is also considerable (Jennings 1985; Gómez–Pajol and Fornós 2009). Coastal karst features (dolines, caves) are filled in fully or partially by brackish or saltwater where solution terminates. In the intertidal zone, solution and deposition alternate happens.

Karst can be open or covered karsts. Open karsts have a soil-free (bare karst) and a soil-covered variety (Gvozdetskiy 1965), while covered karst falls into the categories of cryptokarst (with impermeable cover sediment) and concealed karst (with permeable cover sediment) (Hevesi 1986) (Fig. 1.6). Buried karst is a relief where the bedrock is a karst rock, but there is no manifestation of karstification on the cover.

According to covered character and hydrological conditions, Jakucs (1977) distinguished between allogenic and autogenic karsts. This typology was further refined by Ford and Williams (1989). The autogenic karst is an open karst, and since it is higher than its environment on all sides, it only receives water from precipitation. As conceived by Jakucs (1977), allogenic is the open karst which receives water from the bordering higher non-karstic surfaces. In a more recent interpretation, the allogenic karst is a cryptokarst on which the exposure of the cover sediments causes ponor formation, and then as a consequence of valley incision, the rock boundary retreats and the location of ponor formation is shifted. On mixed karsts (essentially Jakucs' allogenic karst), non-karstic rocks in higher position alternate with karstic rocks in lower position (Fig. 1.6).

According to climate, tundra, temperate (fluvial), mediterranean, tropical and high-mountain karsts are distinguished. On tundra karst, as well as on the taiga or frost karst (in the temperate belt), the solution period is short, permafrost prevents solution or, if it is active, the solutes are not transported away. There are no solution dolines (or very subordinate and only form under special conditions), but covered

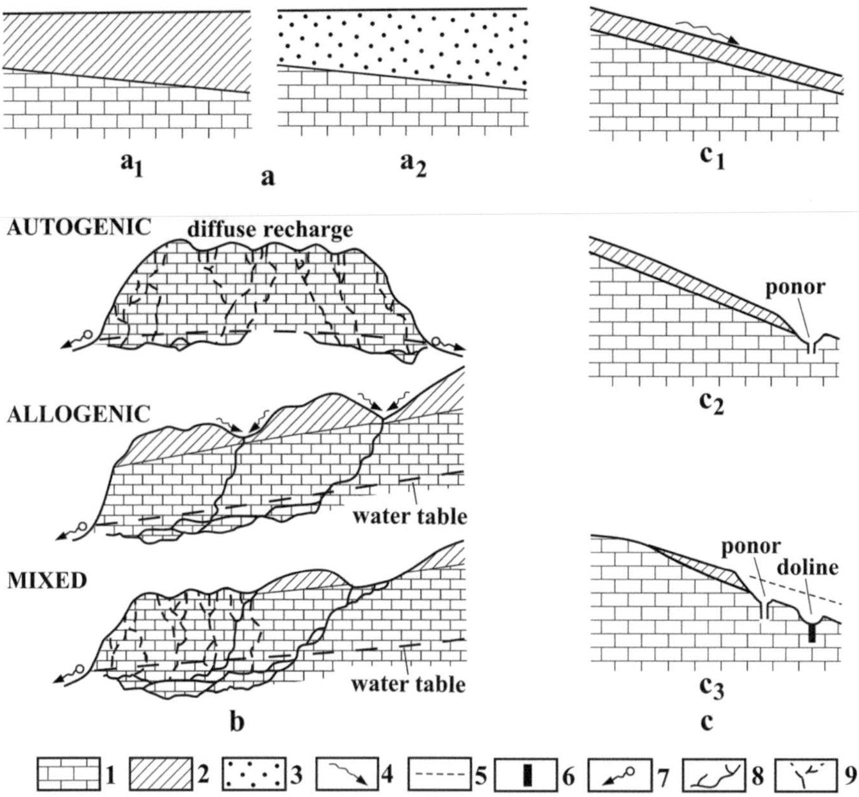

Fig. 1.6 Types of karst (After Jakucs 1977, using figures from Ford and Williams 2007): cryptokarst (**a₁**), concealed karst (**a₂**), (**b**) autogenic, allogenic and mixed karst (Ford and Williams 2007), (**c**) valley inheritance and ponor formation on cryptokarst (Jakucs 1977), *1* limestone, *2* impermeable cover, *3* permeable cover, *4* watercourse, *5* former surface, *6* chimney filled with sediment, *7* spring, *8* stream subsystem, *9* percolation subsystem

karst dolines (particularly on gypsum and halite) exist. On bare surfaces, karren occur and the cavities are of erosional origin.

On temperate karst, fluvial action is considerable, the period of solution is longer and CO_2 production is high in the soil. On open karsts, solution dolines (on inherited valley floors, ponor dolines developed from ponors) are predominant, but uvalas are also widespread. In addition to erosional caves, solution caves are also typical.

On mediterranean karsts, solution is active throughout the year. Dolines, uvalas, poljes and karstic mounds are widespread and cave formation is also intensive.

On tropical karst, solution is continuous and intensive and produces inselberg karst in varieties: fengcong, fenglin (Ford and Williams 2007; Yuan 1981, 1985; Lu 1985; Zhu 1988; Sweeting 1995) and, a third variety, the gufeng with typical karstic mounds (Balázs 1990). On tropical karst, polygonal and pinnacle karst can also be found. The polygonal karst (the cockpit karst) is fengcong karst or a variety of fengcong karst. The stone forest karst, the arête and pinnacle karst and the tsingy are varieties of pinnacle. Fengcong is composed of hills on a joint base, and the hills are divided by depressions (dolines, cockpits) of variable shape and size. Fenglin karst is composed of isolated hills with lowlands between them (Fig. 1.7). Both fenglin

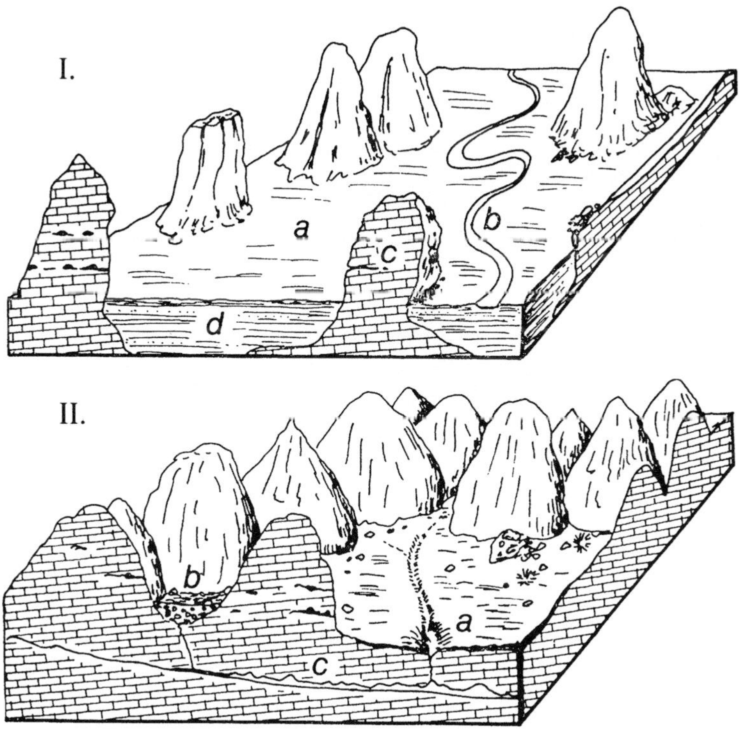

Fig. 1.7 Fenglin and fengcong in South China (After Balázs 1986): *I* fenglin, (**a**) intermountain plain, (**b**) river, (**c**) karst inselberg, (**d**) alluvium; *II* fengcong, (**a**) ponor, (**b**) doline, (**c**) underground stream (active cave)

and fengcong have low- and high-elevation varieties (Balázs 1990). The polygonal karst type is composed of a multitude of solution dolines formed close to each other. Dolines are bordered by polygonal crests (remnants of the original terrain) with karstic mounds (Williams 1971, 1972a, b; Balázs 1973; Day 1978). Fengcong has evolved from polygonal karst (Ford and Williams 2007), but it may happen that, as a consequence of the subsidence of the karst water table, fenglin transforms into polygonal karst (Ford and Williams 2007). Both types can be either autogenic or allogenic karst (Williams 1987) (Fig. 1.8).

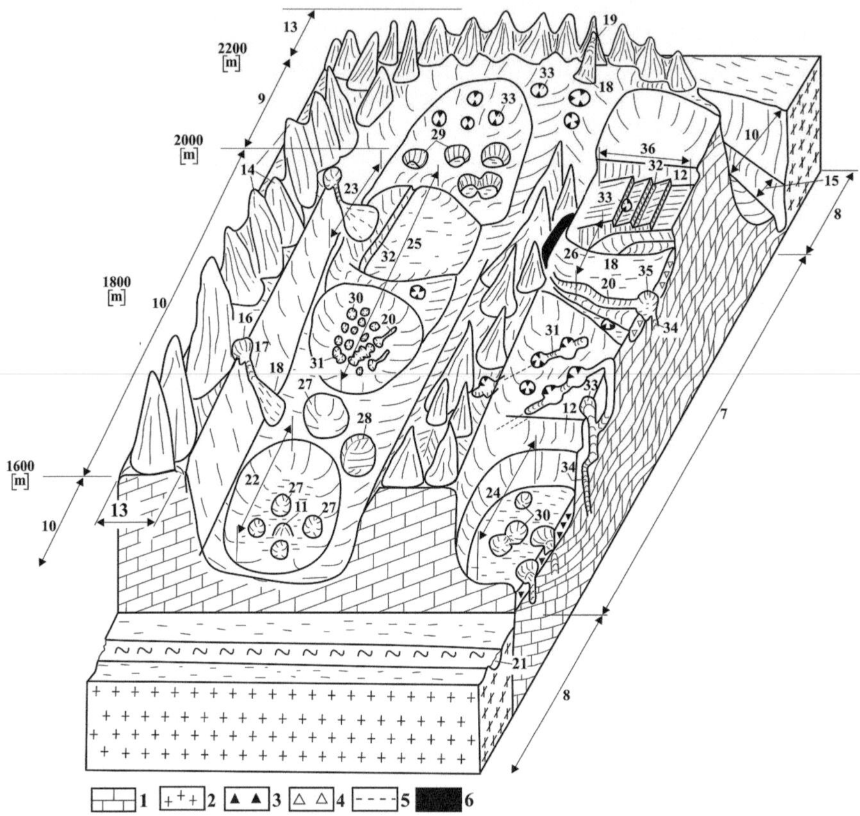

Fig. 1.8 Main features in the glacially transformed karstic high mountains (Modified after Veress 2010): *1* limestone, *2* older metamorphic basement, *3* moraine, *4* colluvial debris, *5* fault, *6* siliceous interbedding in limestone, *7* Klippen (plateau), *8* eroded portion of nappe, *9* cirque, *10* trough, *11* roche moutonnée, *12* tarn backwall, *13* arête, *14* horn, *15* river valley, *16* niche with circular rim, *17* niche of frost shattering, *18* talus cone, *19* path of rock avalanche, *20* furrow of covered karst depression, *21* river, *22* paleodoline, *23* asymmetric paleodoline, *24* paleodoline with moraine, *25* paleouvala, *26* partially filled paleodoline, *27* recent solution doline, *28* asymmetric solution doline, *29* schachtdoline, *30* subsidence doline, *31* subsidence uvala, *32* giant chasm (bogaz), *33* shaft system, *34* passage, chimney, shaft system in cross-section, *35* ponor, *36* schichtrippenkarst

In the high-mountain karst areas, surface karst features show zonation. High-mountain karst or its portion is also called glaciokarst, which is in a narrow sense only the glacially shaped portion of the mountains, while in a broader sense the bare, vegetation-free parts of the mountains. Typical glaciokarsts are free of vegetation and dissected by karren and dry valleys produced by snowmelt runoff (Smart 2004; Trimmel and Waltham 2004; Kunaver 2009; Sauro 2009; Monbaron and Wildberger 2009). On glaciokarst, both paleokarst and glacial features are present and one often determines the presence of the other. On glaciokarsts, Veress (2012) distinguishes between terrains with roches moutonnées; flat terrain; terrain with dolines and uvalas; terrain slightly dissected with glacial troughs; large-scale glacial troughs with bedrock basins; glacial trough with giant dolines and uvalas; and trough and terrain with arêtes. Paleodolines are the predominant features of glaciokarsts and typical representatives of karstification and glaciation (regarding their size, they are giant dolines). Such forms have hundreds of metres in diameter,

Fig. 1.9 Preglacial solution giant dolines (Durmitor Mountains, lower part of the Lokvice Valley). *1* preglacial giant doline, *2* ridge, *3* alluvial fan and apron, *4* glacial trough, *5* lower surface with moraines which borders the mountain

retained in spite of the glacial erosion, but their development was interrupted or modified (in their interior, open or covered karst formation also took place). Certain paleodolines were transformed, others were truncated and still others were filled to some degree. The paleodolines could have formed before the Pleistocene (Bauer 1952; Bauer and Zötl 1972). Two generations of giant paleodolines are distinguished by Veress (2012): preglacial (Fig. 1.9) and interglacial (Fig. 1.10) giant dolines. The former predate the Pleistocene glaciations, while the latter formed in one of the interglacials.

The Pleistocene could form in preglacial dolines (the dolines were transformed into cirques). In case of a doline row, this could happen in a single doline (that in the highest position) or simultaneously in several dolines. In the former case, the ice moved down from the uppermost doline and filled up the rest of the dolines. Glacial erosion was considerable both in the dolines and on inter-doline terrain. In the latter case, the ice flowing out from the members of the doline row initially moved over the terrain between the dolines. For this reason, glacial erosion probably reshaped the lower-lying members of the doline row to a lesser degree. Glacial erosion created a glacial trough from the doline row. Such a trough shows an arcuate planform, and the slopes of the doline and those of the trough are not distinct, while the floor of the trough has a deeper section (doline floor) and a less deep portion (threshold between dolines). The interglacial paleodolines formed on the floor of already existing troughs. Therefore, they dissect the trough floor but do not change its morphology. At the margins of major dolines, even parts of the trough floor are often preserved.

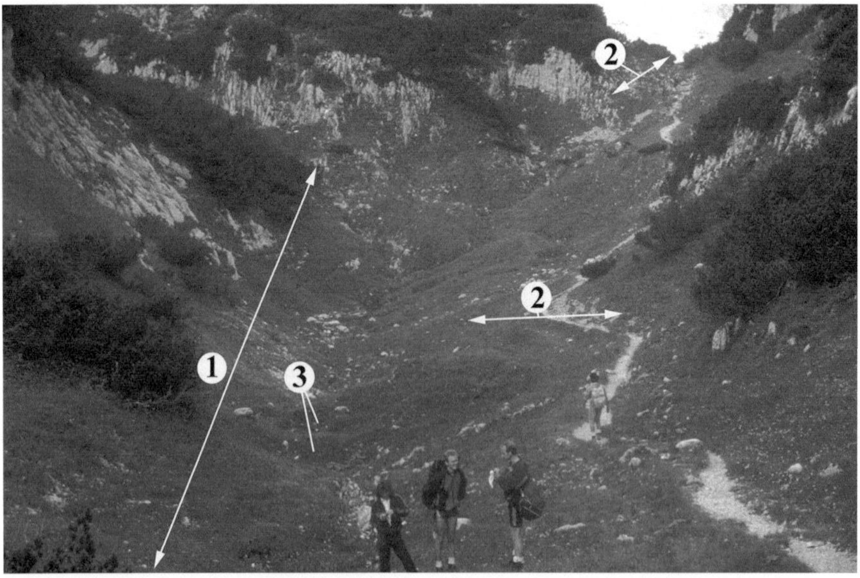

Fig. 1.10 Interglacial giant solution doline (Totes Gebirge). *1* interglacial giant doline, *2* floor remnant of glacier trough, *3* suffosion doline

In the course of glacial erosion, steps are carved on the surface of well-stratified limestones (Veress 2012). These steps are similar to cuestas that formed by selective denudation. Therefore, we use the term 'cuesta' for them in the text. Thus, paleokarst features are dissected by cuestas or recent karst features develop on such surfaces. If the dip direction of the valley is close to the dip direction of strata, the glacier has formed a valley of symmetrical cross-section (Fig. 1.11a$_1$), and the cuestas are found on the valley floor. If glacier movement was opposite to the dip direction, strata dip in opposite direction compared to the valley gradient (Fig. 1.11a$_2$). If it coincided with the dip direction, strata dip in the direction of the

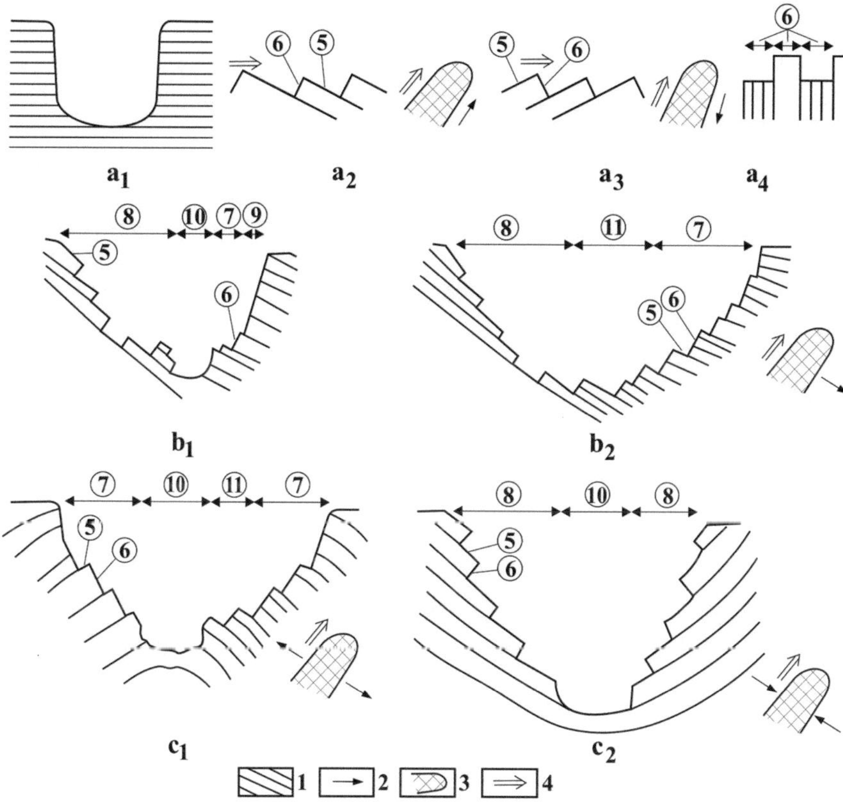

Fig. 1.11 Glacial troughs with cuestas and hogbacks. (**a**) Glacier direction identical with or opposite to the dip direction of strata; (**a$_1$**) valley in cross-section; (**a$_2$**) in longitudinal section, if the dip of strata and the direction of ice flow are identical; (**a$_3$**) in longitudinal section, if both directions are opposite; (**a$_4$**) vertical strata; (**b**) direction of glacier flow different from the dip direction of strata (at ca 90°); (**b$_1$**) a steep valley side sloping in opposite direction to rock dip; (**b$_2$**) stepped valley side sloping in opposite direction to rock dip; (**c$_1$**) glacier trough formed on anticline; (**c$_2$**) glacier trough formed on syncline, *1* strata, *2* dip direction of strata, *3* glacier, *4* direction of glacier flow, *5* bedding plane, *6* head of bed (scarp), *7* valley side with heads of beds, *8* valley side section with bedding planes, *9* steep valley side (cliff), *10* rock basin, karst depression, *11* valley floor with cuestas

valley slope (Fig. 1.11a_3). If the strata are vertical, hogbacks form (Fig. 1.11a_4). If the direction of glacier movement was different from the dip of strata (at an almost 90° angle), glacial erosion resulted in a valley of asymmetric cross-section, and the direction of cuestas points into the direction of the valley. The valley slope opposite to the dip of strata is often vertical (Fig. 1.11b_1). However, it could be dissected by cuestas (Fig. 1.11b_2). The scarp front (with heads of beds) dips towards the valley floor, while the bedding planes dip in the opposite direction. Cuestas of similar position are found on such valley floors and in the karstic (dolines) and non-karstic (bedrock basin) depressions of the karst (Fig. 1.11b_1). The slope of such valleys coinciding with the dip of strata is constituted by a series of bedding planes. The bedding planes at different heights are divided by heads of beds, which are steep or even overhanging independently from the dip (Fig. 1.11b).

If the glacial trough had been formed on an anticline and the ice moved in the direction of the strike of folding, its walls are constituted by cuestas. In this case, both valley sides (and also the valley floor) present a series of heads of beds sloping towards the interior of the valley and bedding planes dipping towards the valley margin (Fig. 1.11c_1). If the valley had been formed in a syncline and the ice moved along the strike of folding, both valley sides expose bedding planes dipping towards the valley floor. The head of bed outcrops either dips towards the valley margin or they are vertical (Fig. 1.11c_2).

High-mountain karsts show a zonal arrangement. In the lower regions (e.g. in the Alps between about 500 and 1,600 m), middle-mountain karst features (dolines of temperate climate) are characteristic, while at higher elevations (in the Alps above ca 1,600 m), glacially transformed paleodolines are common. Recent solution dolines are more typical in the interior of uncovered paleodolines, while covered karst dolines are more typical in paleodolines with covered floors. Above about 1,800–2,000 m elevation, shaft dolines, shaft systems, giant solutional chasms (bogaz) and karren take over predominance (Fig. 1.8).

References

Audra P, Bini A, Gabrovšek F, Häuselmann P, Hobléa F, Jeannin P-Y, Kunaver J, Monbaron M, Šušteršič F, Tognini P, Trimmel H, Wildberger A (2006) Cave genesis in the Alps between the Miocene and today: a review. Z Geomorphol NF 50(2):153–176

Balázs D (1962) Beiträge zur Speläologie des südchinesischen Karstgebietes. Karszt- és Barlang 2:3–82

Balázs D (1973) Relief types of tropical karst areas. In: Jakucs L (ed) IGU Symposium on Karst Morphogenesis. Attila Jozsef University, Szeged, pp 16–32

Balázs D (1986) Kína karsztvidékei (Karst regions in China). Karszt- és Barlang II:123–132 (in Hungarian)

Balázs D (1990) A dél-kínai karsztvidék főbb barlangtípusai (Main cave types of the South-China karst region). Karszt- és Barlang I:53–60 (in Hungarian)

Balázs D (1991) A zárt karsztos mélyedések globális rendszerezése, Dolinák-dolinaegyüttesek (Global systemization of closed karst depressions: Dolines and doline assemblages). Karszt- és Barlang I–II:35–44 (in Hungarian)

Bauer F (1952) Zur Verkarstung des Sengsengebirges in Oberösterreich. Mitt. Höhlenkomm, pp 7–14
Bauer F, Zötl J (1972) Karst of Austria. In: Herak M, Stringfield VT (eds) Karst, important Karst of the Northern Hemisphere. Elsevier, Amsterdam, pp 225–265
Bini A, Pellegrini A (1998) Il carsimo del Moncodeno. Geol Insubrica 3(2):296
Bögli A (1960) Kalklösung und Karrenbildung. Z Geomorphol NE 2:4–21
Cholnoky J (1940) A csillagoktól a tengerfenékig (From the stars to the sea bottom). Franklin Társulat, Budapest, 496 p
Corbel J (1957) Karsts hauts. Alpine Rev Geogr 32:135–158
Cramer H (1941) Die Systematik der Karstdolinen. N Jahrb Mineral Geol Paläont 85:293–382
Cvijič J (1893) Das Karstphaenomen. Versuch einer morphologischen Monographie. Geogr Abh Wien 5:218–329
Cvijič J (1899) Glacial and morphological studies of the mountains in Bosnia, Herzegovina and Monte Negro (in Serbian). Glas Srpske Kraljevske Akademije Nauka, Belgrade, LVII, 196 p
Cvijič J (1911) Base for the geography and geology of Macedonia and Old Serbia (in Serbian), vol III. Posebno izdanje Srpske Kraljevske Akademije Nauka, Belgrade, pp 1074–1094
Cvijič J (1913) The ice age in the Prokletije and surrounding mountains (in Serbian). Glas Srpske Kraljevske Akademije Nauka, Belgrade, XCI, p 149
Cvijič J (1918) L'hydrographie souterraine et l'évolution morphologique du karst. Trav Inst Geogr Alpine 6:375–426
Cvijič J (1924) Types morphologiques de terrains calcaires. Glas Geografshoe Drustvo 10:1–7
Day MJ (1978) Morphology and distribution of residual limestone hills (mogotes) in the karst of northern Puerto Rico. Bull Geol Soc Am 89:426–432
Delannoy JJ (1997) Recherches géomorphologiques sur les massifs karstiques du Vercors et de la transversale de Ronda (Andalousie). Les apports morphogiques du karst, tése de doctorat. Presses Universitaires du Septentrion, Lille, p 678
Ford DC (1998) Perspectives in karst hydrogeology and cavern genesis. Bull Hydrogeol 36:9–29
Ford DC, Williams PW (1989) Karst geomorphology and hydrology. Unwin Hyman, London, 601 p
Ford DC, Williams PW (2007) Karst hydrogeology and geomorphology. Wiley, Chichester, p 562
Ford DC, Fuller PG, Drake JJ (1970) Calcite precipitates at the soles of temperate glaciers. Nature 226:441–442
Frisch H, Kuhlemann J, Dunkl I, Székely B, Vinnemann T, Rettenbacher A (2002) Dachstein-Altfläche, Augenstein-Formation und Höhlenentwicklung – die Geschichte der letzten 35 Millionen Jahre in den zentralen Nördlichen Kalkalpen. Die Höhle 53(1):1–36
Gams I (1977) Towards the terminology of the polje. In: Proceedings of the 7th international speleological congress, Sheffield, England, September, pp 201–203
Gams I (1978) The polje: the problem of its definition. Z Geomorphol 22:170–181
Ginés A (2004) Karren. In: Gunn J (ed) Encyclopedia of caves and karst science. Fitzroy Dearborn, New York, pp 470–473
Ginés A (2009) Karrenfield landscapes and karren landforms. In: Ginés A, Knez M, Slabe T, Dreybrodt W (eds) Karst rock features, vol Carsologica, 9, Karren Sculpturing Zalozba ZRC. Institut za Raziskovanje krasa ZRC SAZU, Postojna-Ljubljana, pp 13–24
Gomez-Pujol L, Fornós JJ (2009) Coastal Karren in the Balearic Islands. In: Ginés A, Knez M, Slabe T, Dreybrodt W (eds) Karst rock features, vol Carsologica, 9, Karren Sculpturing Zalozba ZRC. Institut za raziskovanje krasa ZRC SAZU, Postojna-Ljubljana, pp 487–582
Grund A (1914) Der geographische Zyklus im Karst. Z Ges Erdkunde 52:621–640
Gunn J (1981) Hydrological process in karst depression. Z Geomorphol 25(3):315–331
Gvozdetskiy NA (1965) Types of karst in the U.S.S.R. Problems of speleological research, Prague, pp 47–54
Hevesi A (1978) A Bükk szerkezet- és felszínfejlődésének vázlata (Sketch of the structural and geomorphic evolution of the Bükk Mountains). Földr Ért XXVII(2):169–203 (in Hungarian)

Hevesi A (1980) Adatok a Bükk-hegység negyedidőszaki ősföldrajzi képéhez (Data on the Quaternary paleogeography of the Bükk Mountains). Földtani Közlemények 110(3–4):540–550 (in Hungarian)

Hevesi A (1984) Karsztformák kormeghatározásáról és mészkőhegységeink újharmadidőszak végi – jégkori arculatának megrajzolásában játszott szerepükről, a Bükk hegység példáján (On the dating of karst features and their role in the Late Tertiary – Pleistocene evolution: example of the Bükk Mountains). Földr Ért XXXIII:1–2, 25–36 (in Hungarian)

Hevesi A (1986) Hidegvizek létrehozta karsztok osztályozása (Classification of cold-water karsts). Földr Ért 35:231–254 (in Hungarian)

Hevesi A (1991) Magyarország karsztvidékeinek kialakulása és formakincse II. (Development and landforms of karst regions in Hungary II). Földtani Közlemények 115:99–120 (in Hungarian)

Jakucs l (1956) Adatok az Aggteleki-hegység és barlangjainak morfogenetikájához (Data on the morphogenesis of the Aggtelek Mountains and their caves). Földtani Közlemények 80(1):25–38 (in Hungarian)

Jakucs L (1977) Morphogenetics of karst regions. Adam Hilgar, Bristol, 284 p

Jennings JN (1985) Karst Geomorphology. Basil Blackwell, New York, 293 p

Komatina N (1982) A fejlődés feltételei és a karsztos területek feloszlása (Conditions of evolution and classification of karst areas). In: Burger A, Dubertret A (eds) Karsztterületek hidrogeológiája (Hydrogeology of karst regions). MKBT, Budapest, pp 23–25 (in Hungarian)

Kunaver J (1965) Guide through the high-mountainous Karst of the Julian Alps. 4th – international congress of speleology Postojna – Ljubljana – Dubrovnik, 40 p

Kunaver J (2009) The nature of limestone pavements in the central part of the southern Kanin plateau (Kaninski podi) Western Julian Alps. In: Ginés A, Knez M, Slabe T, Dreybrodt W (eds) Karst rock features – karren sculpturing zalozba ZRC, vol Carsologica, 9. Institut za raziskovanje krasa ZRC SAZU, Postojna-Ljubljana, pp 299–312

Lu Y (1985) Karst in China: landscapes, types. Rules Geological Publishing House, Beijing, 288 p

Lundberg J (2009) Coastal karren. In: Ginés A, Knez M, Slabe T, Dreybrodt W (eds) Karst rock features, vol Carsologica, 9, Karren Sculpturing Zalozba ZRC. Institut za raziskovanje krasa ZRC SAZU, Postojna-Ljubljana, pp 249–264

Menkovič L (1994) Glacial traces in the Djeravica area. Prokletije Mountains (in Serbian). Geogr Gadisnjak 30:139–146

Monbaron M, Wildberger A (2009) The karrenfields of the Muota valley: type localities of the main karren types after the nomenclature by Alfred Bögli. In: Gines A, Knez M, Slabe T, Dreybrodt W (eds) Karst rock features, vol Carsologica, 9, Karren Sculpturing Zalozba ZRC. Institut za raziskovanje krasa ZRC SAZU, Postojna-Ljubljana, pp 291–298

Pécsi M (1980) A Pannóniai-medence morfogenetikája (Morphogenesis of the Pannonian Basin). Földr Ért 29(1):105–127 (in Hungarian)

Sauro U (2009) Glaciokarst landforms of the lower Adige and sarca valleys. In: Gines A, Kner M, Slabe T, Dreybrodt W (eds) Karst rock features, vol Carsologica, 9, Karren Sculpturing Zalozba ZRC. Institut za raziskovanje krasa ZRC SAZU, Postojna-Ljubljana, pp 323–328

Slabe T, Liu H (2009) Significant subsoil rock forms. In: Gines A, Knez M, Slabe T, Dreybrodt W (eds) Karst rock features, vol Carsologica, 9, Karren Sculpturing, Zalozba ZRC. Institut za raziskovanje krasa ZRC SAZU, Postojna-Ljubljana, pp 123–137

Smart C (2004) Glacierized and glaciated karst. In: Gunn J (ed) Encyclopedia of caves and karst science. Fitzroy-Dearborn, New York, pp 389–390

Sweeting MM (1973) Karst landforms. Columbia University Press, New York, 362 p

Sweeting MM (1995) Karst in China: its geomorphology and environment. Springer-Verlag, Berlin, 265 p

Trimmel H, Waltham T (2004) Europe, alpine. In: Gunn J (ed) Encyclopedia of caves and karst science. Fitzroy Dearborn, New York, pp 325–328

Trudgill ST (1985) Limestone geomorphology. Longman, New York, 196 p

Veress M (2000) Covered karst evolution Northern Bakony Mountains, W-Hungary. A Bakony Természettud. Kut. Eredményei, 23, Bakonyi Természettudományi Múzeum, Zirc 167 p
Veress M (2004) A karszt (Karst). BDF Természetföldrajzi Tanszék, Szombathely, 215 p (in Hungarian)
Veress M (2009) Investigation of covered karst form development using geophysical measurements. Z Geomorphol 53(4):469–486
Veress M (2010) Karst environments. Karren formation in high mountains. Springer, Dordrecht, 230 p
Veress M (2012) Glacial erosion and karst evolution (Karren formation on the surface formed by glaciers). In: Veress B, Szigethy I (eds) Horizons in earth science research, vol 7. Nova Science Publishers, New York, pp 1–94
Veress M, Péntek K (1996) Theoretical model of surface karstic processes. Z Geomorphol 40(4):461–476
Veress M, Péntek K (2010) The development of hum slopes due to karren formation. Karst Dev 1(2):23–36
Waltham AC, Fookes PG (2003) Engineering classification of karst ground conditions. Q J Eng Geol Hydrogeol 36:101–118
Waltham AC, Fookes PG (2005) Engineering classification of karst ground conditions. Speleogenesis and Evolution of Karst Aquifers. Virtual Sci J www.speleogenesis.info
Waltham T, Bell F, Culshaw M (2005) Sinkholes and subsidence. Springer, Berlin, 382 p
White WB (1988) Geomorphology and hydrology of karst terrains. Oxford University Press, Oxford, 463 p
Wilford GE, Wall JRD (1965) Karst topography in Sarawak. J Trop Geogr 21:44–70
Williams PW (1971) Illustrating morphometric analysis of karst with examples from New Guinea. Z Geomorphol 15(1):40–61
Williams PW (1972a) Morphometric analysis of polygonal karst in New Guinea. Geol Soc Am Bull 83:761–796
Williams PW (1972b) The analysis of spatial characteristics of karst terrains. In: Chorley RJ (ed) Spatial analysis, Geomorphology. Methuen, London, pp 136–163
Williams PW (1983) The role of the subcutaneous zone in karst hydrology. J Hydrol (Neth) 61:45–67
Williams PW (1987) Geomorphic inheritance and the development of tower karst. Earth Surf Process Landf 12(5):453–465
Williams PW (2004) Dolines. In: Gunn J (ed) Encyclopedia of caves and karst science. Fitzroy Dearborn New York, London, pp 304–310
Yuan D (1981) A brief introduction to China's research in karst. Institute of Karst Geology, Guilin, 35 p
Yuan D (1985) New observation on tower karst. In: Proceedings of the 1st international conference on geomorphology, Manchester (also published by Institute of Karst Geology, Guilin, China), 14 p
Zhu X (1988) Guilin karst. Shanghai Scientific and Technical Publishers, Beijing, 188 p
Zseni A (2004) Talaj alatti karrformák (Subsoil karren features). Karsztfejlődés IX:157–175 (in Hungarian)
Zseni A (2009) Subsoil shaping. In: Ginés A, Knez M, Slabe T, Dreybrodt W (eds) Karst rock features, vol Carsologica, 9, Karren Sculpturing Zalozba ZRC. Institut za raziskovanje krasa ZRC SAZU, Postojna-Ljubljana, pp 103–121

Chapter 2
Study Areas

Abstract This chapter presents the locations where investigations and observations were made. The following karst regions studied in detail are described: Aggtelek Karst, Asiago Plateau, Bakony Mountains, Bükk Mountains, Durmitor, Northern Limestone Alps, Madagascar, Mecsek Mountains and Pádis plateau. The areas from where observations derive are the Atacama Desert, Biokovo mountains, Cerkniško polje, Crimean Peninsula, Dolomites Iceland, Middle Lena, the Lunan area and Parajd.

Keywords Karst type • Karst structure • Geology • Karst features • Aggtelek Karst • Asiago Plateau • Bakony Mountains • Bükk Mountains • Durmitor • Northern Limestone Alps • Madagascar • Mecsek Mountains • Pádis plateau • Atacama Desert • Biokovo mountains • Cerkniško polje • Crimean Peninsula • Dolomites • Iceland • Middle Lena • The Lunan area • Parajd

2.1 Introduction

The study areas (Fig. 2.1) from where the findings presented in this volume directly derive fall into two groups: in the first group of karsts, field research (measurements, mapping) were performed, while in the other group observations were predominant. The description of the karst regions includes the location, type, extension and geology of the karst (the bearing rocks, sediment cover). The description is followed by the subdivisions, morphology and evolution history of the karst as well as its surface landforms. Where it was possible, the individual study sites are also listed. The summary of the areas of detailed research is shown in Table 2.1 and the characteristics of observation areas in Table 2.2.

M. Veress, *Covered Karsts*, Springer Geology,
DOI 10.1007/978-94-017-7518-2_2

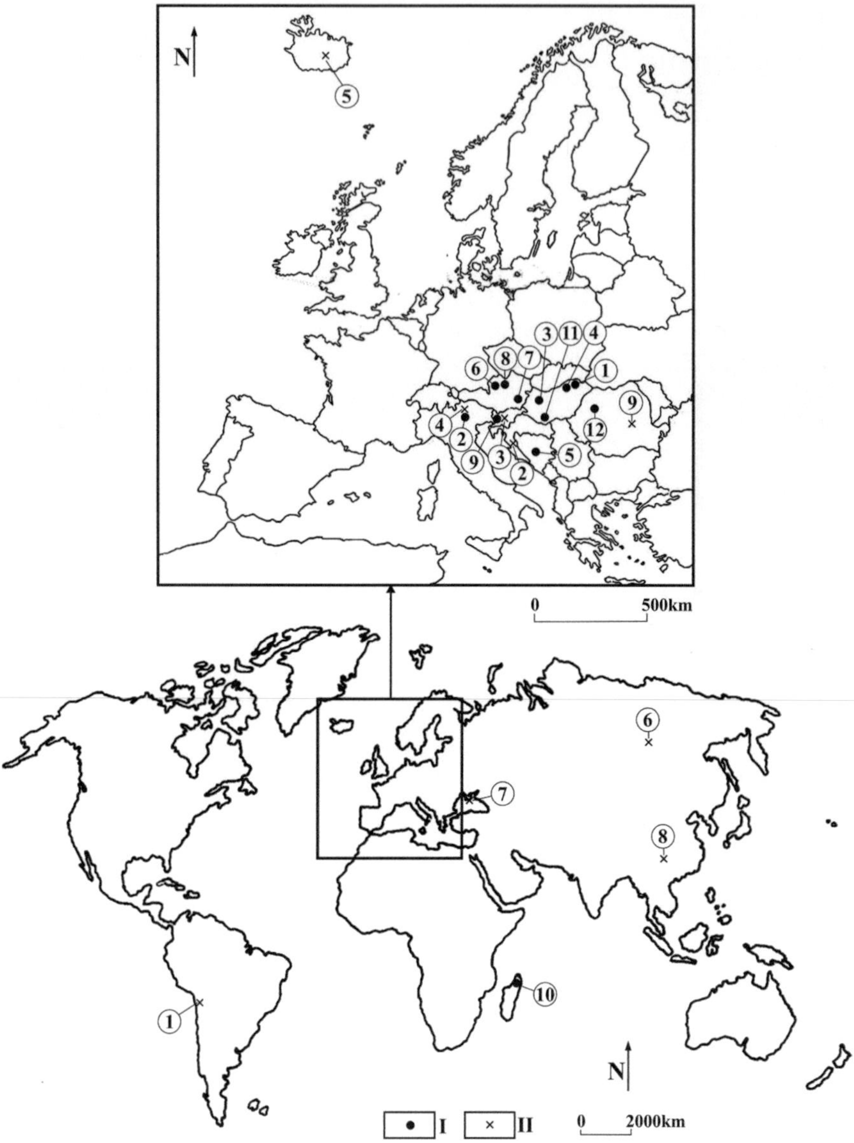

Fig. 2.1 Sites of research (*I*) and observation (*II*). *I* Research site; *II* observation site; *I 1* Aggtelek Karst, *2* Asiago Plateau, *3* Bakony Mountains, *4* Bükk Mountains, *5* Durmitor, *6* Dachstein, *7* Hochschwab Plateau, *8* Totes Gebirge, *9* Julian Alps, *10* Madagascar; *11* Mecsek Mountains, *12* Padis Plateau. *II 1* Atacama Desert, *2* Biokovo Mountains, *3* Cerkniško polje, *4* Dolomiti, *5* Iceland, *6* Middle Lena region, *7* Crimean Peninsula, *8* Lunan region, *9* Parajd

Table 2.1 The main characteristics of the investigated areas

Mountains	Location (country)	Location (physical geographical unit)	Elevation (top and average)	Geology (main rock types)	Tectonic position	References
Aggtelek Mts.	Hungary	Pannonian Basin	Nagyoldal: 604 m Av.: 350–500 m	Carbonate, marl (T)	Alcapa Megaunit, Aggtelek Unit	Haas (2001, 2004), Csontos and Vörös (2004), and Kövér et al. (2006)
Asiago Plateau	Italy	Southern Alps, Dolomiti	Cima: 2,208 m Av.: 1,600–2,000 m	Rhyolite, sandstone (P), limestone, dolomite, vulcanite (T)	Autochthonous – parautochthonous part of the Apulian Plate	Mauer (2000)
Bakony Mts.	Hungary	Pannonian Basin	Kőris hill: 709 m Av.: 400 m	Anchimetamorphic slate (O-C) Conglomerate–sandstone (P) Carbonate, marl, chert (T-E)	Alcapa Megaunit, Transdanubian Mountain Range Unit	Haas (2001, 2004) and Csontos and Vörös (2004)
Bükk Mts.	Hungary	Pannonian Basin	Istállós-kő: 959 m Av: 700–900 m	Anchimetamorphic slate (C) Carbonate, marl, chert (P-E) Ophiolite (T)	Alcapa Megaunit, Bükkium Unit	Haas (2001, 2004), Csontos and Vörös (2004), and Pelikán (2005)
Durmitor Mts.	Bosnia–Herzegovina, Crna Gora	Dinarides	Bobotov Kuk: 2,523 m Av.: 1,500 m	Anchimetamorphic clastics and metavolcanites (C-P), limestone, volcanites (T), ophiolite (J)	E-Bosnia-Durmitor block	Miljush (1978) and Ustaszewki et al. (2009)
Dachstein Mts.	Austria	Northern Limestone Alps (Eastern Alps)	Hoher Dachstein: 2,995 m Av.: 2,100 m	Dachstein Limestone (T_3)	Upper Austroalpine Nappe, Juvavic Facies Unit	Schwarzacher (2005)

(continued)

Table 2.1 (continued)

Mountains	Location (country)	Location (physical geographical unit)	Elevation (top and average)	Geology (main rock types)	Tectonic position	References
Hochschwab Plateau	Austria	Northern Limestone Alps (Eastern Alps)	Hochschwab: 2,277 m	Gutenstein Limestone Fm. (T_2), Steinalm Limestone (T_2), Wetterstein Limestone (T_2)	Upper Austroalpine Nappe, Juvavic Facies Unit	Plan and Decker (2006)
			Av.: 2,100 m			
Totes Gebirge (Dead Mountains)	Austria	Northern Limestone Alps (Eastern Alps)	Grosser Priel: 2,515 m	Werfen beds (T_1),	Upper Austroalpine Nappe, Tirolic facies unit	Plan et al. (2009)
			Av.: 2,000 m	Dachstein Limestone (T_3)		
Julian Alps	Slovenia	Southern Alps	Triglav: 2,864 m	Triassic limestone and dolostone (T)	Autochthonous, semi-autochthonous part of the Southern Alps	Obenholzner (1991)
			Av.: 2,300 m			
Madagascar	Madagascar	Africa	Maromokotro 2,876 m	Metamorphic rocks (Precambrian)	Parautochthon	Schlüter (2008)
			Av: 750–1,350	Carbonaceous rocks (Mesozoic)		
				Carbonaceous and argillaceous rocks (tertiary)		
Mecsek Mts.	Hungary	Pannonian Basin	Zengő: 682 m	Granite (C), conglomerate–sandstone (P-T)	Tisza Megaunit, Mecsek Zone	Fülöp (1990), Haas (2001, 2004), and Csontos and Vörös (2004)
			Av.: 300–500 m	Carbonate, marl, chert (T-K)		
				Basalt (K)		
Padis plateau (Bihar Mts.)	Roumania	Pannonian Basin (Transylvanian sub-basin)	Nagy-Bihar: 1,849 m	Limestone, dolomite, marl sandstone (T-K)	Tisza Megaunit, Villány–Bihar Unit	Patrulius et al. 1971
			Av.:1,300 m			

Table 2.2 Main characteristics of observation areas

Mountains	Location (country)	Location (geographically)	Elevation (top and average)	Geology (main rock types)	Tectonic position	References
Atacama Desert	Chile/Peru	Andes	Av.: 800 m	Eolic sediments (M-Q)	Andes	Scheuber and Andriessen (1990) and Reutter et al. (1996, 2007)
Biokovo Mts.	Croatia	Dinarides	Sveti Jure: 1,762 m Av.: 1,000 m	Limestone, bauxite (J_3-K_1), flysch (K-Ol)	Apulian-Periadriatic Unit, South Adriatic zone	Vlahović et al. (2005) and Korbar (2009)
Cirknitz polje (Javornik Hills)	Slovenia	Dinarides (SE Europe)	Av: 1,250 m	Limestone (Cretaceous)	Allochthonous	Šebela (2012)
Crimean peninsula	Ukraine	Crimean Mountains	Roman-Kosh:1,545 m Av.: 500–600 m	Sandstone (T-J), limestone, (J), limestone, marl, slate (K)	Meganticlinorium	Vakhrushev (2009)
Dolomitok	Italy	Southern Alps	Marmolada:3,344 m Av: 1,500–3,000 m	Dolomite, limestone, volcanics (T)	Parautochthonous	Mckenzie and Vasconcelos (2009)
Iceland	Iceland	N Atlantic Ocean	Hvannadalshnukur: 2,119	Basalt (M-Q), glacial sediments (Pl-Q)	Mid-Atlantic ridge (rise)	Thordarson and Hoskuldsson (2002) and Gudmundsson (2012)

(continued)

Table 2.2 (continued)

Mountains	Location (country)	Location (geographically)	Elevation (top and average)	Geology (main rock types)	Tectonic position	References
Middle Lena	Russia	Prilenskoe Plateau	Av.: 300–500	Limestone, dolomites, slate (Lower/Middle Cambrian)	Platform	Korzhuev (1961) and Solomonov et al. (2010)
The surroundings of the Lunan	China	South China	Av.: 1,600–1,400	Limestone (P)	Parautochthon	
Parajd	Roumania	Transylvanian Basin	Av.: 500–600 m	Evaporite (M)	E-Carpathians/Transylvanian Basin	Irimus et al. (2011), Sanders et al. (2002), and Krézsek and Bally (2006)

2.2 Areas of Detailed Field Research

2.2.1 *Aggtelek Karst*

The Aggtelek Karst (the Hungarian section of the Gömör-Torna [Slovenský kras]) belongs to the North Hungarian Mountains and is a temperate, middle-mountain karst. Its area is 200 km^2 and its main parts are the Aggtelek Mountains and the Alsó (lower) Hill area, which are further subdivided into plateaus: the Alsó Hill N of the Ménes Valley; the Haragistya and Szilas plateaus between the Ménes Valley and the Jósva Valley; and the Aggtelek Plateau and the Galyaság S of the Jósva (Fig. 2.2). The plateaus are blocks uplifted along faults or tilted, divided by valleys (Jósva, Ménes) and incised along tectonic lines or anticlines.

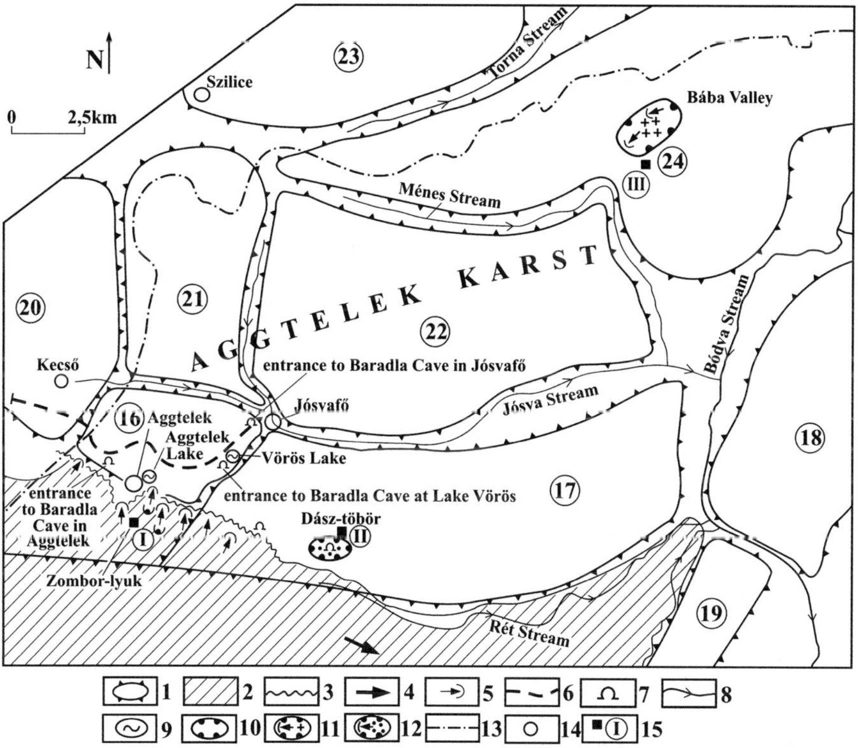

Fig. 2.2 Parts of the Aggtelek Karst. *1* Karst plateau, *2* cryptokarst, *3* rock boundary, *4* dip of cryptokarst, *5* ponor, *6* Baradla Cave, *7* cave entrance, *8* watercourse, *9* Dolina lake, *10* depression of superficial deposit, *11* depression of superficial deposit on interbedded sandstone, *12* depression in Pannonian sediment fill, *13* national border, *14* built-up area, *15* research sites, *16* Aggtelek Plateau, *17* Galyaság, *18* Szalonna Hill, *19* Rudabánya Mountains, *20* Kecső Plateau, *21* Haragistya, *22* Szinpetri Plateau, *23* Szilice Plateau, *24* Alsó Hill, *I* Zombor-lyuk, *II* Dász-töbör, *III* Bába Valley

The Karst is built up of Upper Permian–Lower Triassic gypsum/anhydrite and Triassic carbonates. The rocks of the Aggtelek Karst were deposited on the northern coast (Carpathian environment) of the Tethys Ocean, while the rock of the broader environs (Borsod Macrostructural Unit) on the southern coast of the Tethys sea arm (Less 1998). The rocks of the Karst are part of the Szilice Nappe, which is thrust southwards over the Borsod Unit (Kovács 1984) and underwent repeated compression in the Cretaceous and was transformed into secondary nappes (Klippen). They were covered by Oligocene–Miocene marine deposits (Bretka Limestone Conglomerate, Putnok Schlier) and Late Sarmatian volcanic tuff (Sásdi 1990). Patches of 'Pannonian' (Late Miocene–Early Pliocene) sediments occur in the southern and southeastern part of the Karst (Sásdi 1990). On the Pliocene–Pleistocene boundary, the mountains were tilted in SSE direction, and this surface became covered by a mantle of gravels transported from the N to the S (Borsod Gravel Formation). This cover was destroyed by watercourses running to the S which formed epigenetic valleys. Subsequently, the Galyaság Plateau was tilted in an opposite direction and, as a consequence, watercourses and valleys of northern direction developed there (Veress 2010a, 2012c).

Zámbó (1998) identified ponors with blind valleys on the floor of the uvala rows and dry valleys with doline rows as distinct karst landforms. Their floors are covered by weathering products (terra rossa, red clay).

According to their karstification properties, the mountains can be divided into four parts. The northern part, the eastern half of the Alsó Hill, is an autogenic karst, which may have been covered, but its karstification was not influenced by the possible former cover (Móga 2002). In the middle part the plateaus are autogenic karst blocks deprived of their cover sediments dissected by epigenetic valleys with doline rows (Zámbó 1998; Móga 2001, 2002). The southern part of the mountains is the Aggtelek Plateau and the Galyaság, the latter with uncovered elevated sections and covered lower-lying sections. There are both buried karst and concealed karst patches in the covered karst area. The cryptokarst terrains of the Galyaság, along with the Aggtelek Plateau, represent an allogenic karst. To the S and SE of this zone, the buried karst zone of the mountains is found (Fig. 2.2).

Karst features are diverse, large sized and of high density. On the elevations of the plateaus, dissolution and ruined dolines and doline rows on the epigenetic valley floors are widespread. Particularly in the Alsó Hill, large uvalas and uvala systems occur with deep shafts in the side walls of the dolines. A ponor row with blind valleys developed along the contact of the Galyaság and the Aggtelek Plateau. The blind valleys formed in Pannonian deposits, in the Borsod Gravel Formation and in their reworked sediments. The covered karst is not uniform but developed in patches of various sizes on the fills of the paleodolines. Covered karst patches are most common in the Galyaság, but, with the exception of the Aggtelek Plateau, they occur on all plateaus and their slopes.

Part of the blind valleys developed in depressions of the superficial deposit (DSD), which are typical in the paleodolines, paleouvalas and on valley floors with doline rows also in the area of the Galyaság (Veress 2010b). Depressions of the superficial deposit, however, also occur in uvala rows and epigenetic dry valleys.

The study sites here were the environs of the Zsombor ponor (I); the Dász doline (II), in the Galyaság area, the strip W of the Teresztenye Plateau, which includes the Keserű-tó basin (III) and the Vizetes (IV); in the Alsó Hill area, the Bába Valley (V) and the environs of the former village Derenk (VI).

2.2.2 The Asiago Plateau

The Asiago Plateau (Italy), between the Val d'Astico, Valsugana and Brenta Valleys, is a high-mountain karst (glaciokarst) in the Southern Alps. Its area is 600 km^2. The plateau is built up of Upper Triassic dolomite and Upper Triassic, Jurassic and Cretaceous limestones. Tectonically, it is an autochthonous, partially overthrust part of the Apulian Plate (Mauer 2000). The plateau is subdivided into a northern and a southern section separated by the Central Basin. The strata are slightly folded and of moderate dip. The southern and northern parts are composed of asymmetric anticlines, while the central part is a syncline (Sauro 1995). In the southern and northern parts, networks of epigenetic valleys developed. The preglacial surface with solution dolines in the N was transformed by glacial action, which created shallow glacier troughs from the erosional valleys. In the paleodolines transformed by ice, roches moutonnées occur together with recent solution or covered karst dolines.

On the strata of low dip glacial action carved cuestas with heads of beds nearly vertical and bedding planes almost horizontal. On the cuesta terrain a wealth of diverse karren developed. On the plateau numerous well-developed large shaft caves are found.

2.2.3 The Bakony Mountains

The Bakony Mountains belong to the Transdanubian Mountain Range in Hungary. It is a temperate middle-mountain karst, which is composed of the North and South Bakony, the former being further subdivided into the High and East Bakony (Fig. 2.3). Most heavily affected by karstification is the North Bakony (area: 1,070 km^2). As a tectonic unit, the Transdanubian Mountains are part of the Alpaca Macrostructural Unit (Csontos and Vörös 2004), previously called Pelso Macrostructural Unit (Fülöp 1989). The Transdanubian Mountains acquired its present position through a NE-directed shift from a South Alpine environment during the Miocene (Stegena et al. 1975). Its Triassic basement was transformed into an asymmetric syncline during the Mesozoic. In the southeastern part of the syncline, older (Permian) rocks are exposed, while in the northwestern section, Paleozoic rocks are subsided into the depth. Following the pelagic conditions in the Jurassic, chalk deposited in the central part during the Cretaceous and neritic deposits in the Eocene. Consequently, the main mass of the mountains is primarily constituted of Triassic limestones and dolomites, overlain by Jurassic, Cretaceous and

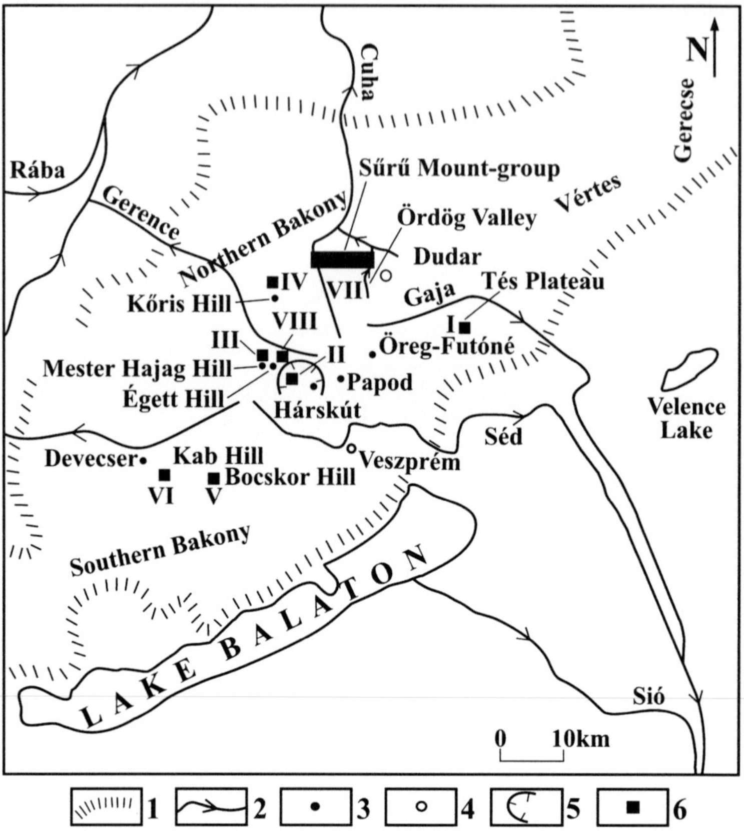

Fig. 2.3 The Bakony Mountains. *1* Boundary of mountains; *2* watercourse; *3* peak; *4* built-up area; *5* basin; *6* main research sites: *I* Tés Plateau; *II* Hárskúti Basin; *III* Mester-Hajag; *IV* Kőris Hill (Eleven-Förtés and Márvány-árok area), *V* Kab Hill; *VI* Pusztamiske and Devecser area, *VII* Sűrű hill group; *VIII* Égett Hill

Eocene limestones of small thickness. In the mountains karstification took place in various terrestrial periods: in the Upper Triassic (Raincsák 1980), in the Jurassic (Konda 1970) and the Cretaceous (Noszky 1964). Karst processes resulted in diverse landforms: dolines (Pataki 1983), poljes (Szabó 1956), karst mounds (Szabó 1966) and karst plains (Szabó 1956). The karst mounds have been truncated at the present. Paleokarst is indicated by manganese and bauxite lenses (Bárdossy 1977). During tropical karstification in the Late Cretaceous, the mountains became a karstic tropical peneplain (Bulla 1968a).

The peneplain was dismembered into blocks along faults. The area of the low but differently elevated mountains was covered by a deltaic gravel mantle in the Late Oligocene and Early Miocene (Korpás 1981), except for the highest elevated blocks (Pécsi 1980). By the present the gravel mantle has been partially removed. In the Pleistocene, at least on one occasion, the mountains were buried under loess. The loess mantle has also been eroded from the highest blocks and the steepest slopes.

With regard to their evolution, the blocks fall into different classes (Pécsi 1980). The blocks in higher position (above 550 m above sea level) are called horsts in summit position. In their areas Cainozoic rocks are missing, and patches of unconsolidated deposits (loess, clay, etc.) may occur in some paleokarstic depressions. The blocks in lower position (below ca 300–400 m) have mostly lost their gravel mantles (horsts in threshold position), but it could have also been preserved (horsts of cryptopeneplain type). The blocks of medium elevation (400–550 m) showed oscillating motion during the Cainozoic. Consequently, they were covered by Eocene limestones and gravels. This type of blocks is called horsts elevated to summit position and buried. Varieties of this type are exhumed horsts (only traces of cover deposits are preserved) and semiexhumed horsts (the truncated cone karst is buried under Miocene gravels).

Induced by the thinning of the crust under the Carpathian Basin (Stegena et al. 1975), the blocks of the South Bakony were affected by basaltic volcanism in the Pliocene (Lóczy 1913; Jugovics 1954), which produced the basalt mantles of Kab Hill and other hills (Tátika, Fekete Hill and others). The absolute age of basaltic volcanic activity in the South Bakony is established at 5–7 ka (Balogh et al. 1982). The basaltic eruptions from several centres were preceded by local pyroclast ejection (Lóczy 1913). The basal lava overflowed a carbonate surface composed of mounds and valleys and built up of Triassic dolomites and Triassic, Jurassic and Eocene limestones (Fig. 2.4). In the quiet intervals between eruptions, red clays developed on the basalt surfaces under warm and humid climate (Jámbor 1980). Sections of the basalt terrain and its environs were also mantled by loess.

The Bakony Mountains are constituted of blocks of flat summits and steep slopes divided by intramontane basins and grabens. In the areas of horsts of cryptopeneplains, valleys in the process of inheritance occur (if the valley is incised into gravel), while inheritance is completed (the valley floor is in limestone) on horsts elevated to summit position and on semiexhumed horsts. If the semiexhumed horsts are enclosed by horsts of cryptopeneplains, the inherited valleys can be epigenetic-antecedent or epigenetic-regressional. The evolution of epigenetic valleys is closely associated with covered karst formation.

On the mountain blocks no open karst dynamics (solution dolines, uvalas) and the related features can be observed. All the more typical is covered karst formation,

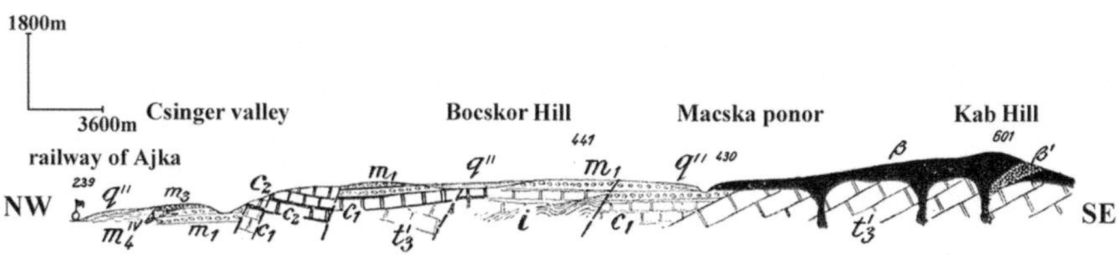

Fig. 2.4 Geological section across Kab Hill (Lóczy 1913). t'_3 Triassic Hauptdolomit, *i* Jurassic limestone, c_1 Middle Cretaceous limestone, c_2 Upper Cretaceous limestone, m_1 Middle Eocene limestone, m_3 Oligocene–Miocene gravels, m_4^{IV} Pannonian sediments, β basalt, β' basalt tuff; q'' loess

which, however, is not continuous in the mountains but forms major karstic patches, the largest of them are found on the Tés Plateau and in the Hárskút Basin and its environs (Middle Hajag). The covered karst features developed on loess mantle. Common features of the concealed karst are the suffosion dolines (or their fossilised varieties), buried dolines and depressions of the superficial deposit. On cryptokarst surfaces gorges developed. Typical karst features of Kab Hill are ponors, caprock and subsidence dolines.

Karstification primarily affected the horsts in medium elevation and those elevated to summit position and exhumed, particularly the terrains at 421–500 m elevation (Veress 1983). In the areas of horsts in summit position, concealed karst formation takes place on the cover sediment patches of paleokarst depressions.

The following areas of the mountains were studied in detail (Fig. 2.3): Tés Plateau (I), Hárskút Basin (II), Mester-Hajag (III), Kőris Hill (IV), Kab Hill and Bocskor Hill (V), near Devecser the environs of Pusztamiske (VI), gorges in the Sűrű hill group (Ördög-árok, Kő-árok, Csuha Valley; VII) and Égett Hill (VIII).

2.2.4 The Bükk Mountains

The Bükk Mountains represent a temperate middle-mountain karst in the North Hungarian Mountains, which is part of the Alcapa Macrostructural Unit (Csontos and Vörös 2004). Its formations point to Dinaric origin (Fülöp 1989): Carboniferous dolomites, limestones and shales, Permian sandstones and shales, Triassic (subordinately Jurassic) limestones, Lower Triassic sandstones and shales as well as Middle Triassic volcanic rocks. The Cainozoic formations include Eocene limestones, Oligocene clays and Miocene andesite (Balogh 1964). The Bükk is folded mountains with imbricated and nappe structures, where the nappes of southern dip and vergence have been truncated (Balogh 1964). The structure is of Cretaceous age and subdivided into the North-Bükk, Little and Great Plateaus and the South-Bükk. The North-Bükk is of anticline while the Little and the Great Plateaus are of synclinal character. The south-dipping folds of the South-Bükk are dismembered along faults.

The central part is the High-Bükk of 800–900 m elevation, divided by the Garadna Valley (developed along the sandstone zone of an anticline) into the Little Plateau in the N and the Great Plateau in the south, which are typical karst plateaus. To the N of the High-Bükk rises the North-Bükk, a former pediment dissected by valleys. To the S of the High-Bükk we find the Southwest-Bükk and the Southeast-Bükk. The structure of Southwest-Bükk is dominated by non-karstic rocks, and karstification is not typical. In contrast, the Southeast-Bükk is a karstic terrain of 600–700 m elevation dissected by valleys. The southern margin of the mountains is the Bükkalja, a pediment dissected by valleys. For the karst formation in the mountains, the Miocene (Tortonian) sediment cover was decisive. This sediment cover was removed by erosion during the Pliocene and Quaternary (Hevesi 1978), and this process has continued to the present. In the ruined sediment cover, inheriting epigenetic valleys are common, while on terrains without cover sediments, inherited epigenetic valleys occur. The following karst features are typical particularly in the areas of the Great and Little Plateaus.

- Doline rows, which are found on the floors of inherited epigenetic (dry) valleys not active today and can be isolated or connected to each other (uvalas). According to Hevesi (1980), doline rows originated from one-time ponors. Through the shifting of the rock boundary in the valley, new ponors developed and the former ponors were transformed into dolines. However, other scenarios for the formation of doline rows are also possible. According to Jakucs (1977), on the floor of the valley inherited over the limestone solution, dolines developed. An alternative way of development is that the inherited valley floor with its solution dolines was covered by sediments, and in the cover sediments subsidence dolines formed (Veress and Zentai 2009).
- On the elevations older and larger dolines without sidewalls (ruined dolines) are characteristic (Veress 1992). Both on the elevations and on the terrains between them, flat-floored dolines are common. The flat floors are due to lateral corrosion (Zámbó 1970). However, flat-floored dolines could also develop if uvalas and old blind valleys are buried under weathering residuals or loess and clays (Veress and Zentai 2009).
- Dolines merge into uvalas. Larger basins developed through the merging of several uvalas also occur in the mountains.
- Ponors are particularly common on the Little Plateau and in the Southeast-Bükk. In the area of the Great Plateau, ponor rows with blind valleys formed on the karst margin, where the karstic rock contacts with non-karstic rocks, while on the Little Plateau and in the Southeast-Bükk, ponors of the karst interior (valley floors) are characteristic.
- On the Little and Great Plateaus and in the Southeast-Bükk, karst gorges also frequently occur. They are mostly epigenetic valley sections, developed through ponor formation and the exposure of cavities (Hevesi 1978).
- Shafts of considerable size (100 m depth) are also found in the High-Bükk.

There is no contiguous covered karst in the mountains. Only patches of limited extension (some 100 m diameter) occur in the High-Bükk in the area of the former open karst partly or completely buried (Great Plateau) or on the subsequently covered floors of the epigenetic valleys (Little Plateau). In the patches both subsidence dolines and depressions of superficial deposit are found.

The research sites are the Nagymező (Great Plateau) and some valleys with doline rows (Little Plateau).

2.2.5 *Durmitor*

The mountains are a high-mountain karst (glaciokarst) in the N of Montenegro, in the Povrsi Brda region, in the central Dinaric Mountains between the gorges of the Piva and Tara rivers. The mountains of 360 km^2 area belong to the Central Dinarides and rise above a lower terrain (in the N, Jezerska Površ; in the S, Drobnjak; in the W and N, Pivska planina).

The basement is composed of Paleozoic schists and sandstones. The main building materials of the mountains are well-stratified Triassic and Cretaceous limestones with beds of rhythmically changing thickness. Thicker beds (of several metres) and thinner (10–20 cm) strata alternate. In the thinner sequences, the proportion of silica surpasses that for the thicker beds. SO_2 is concentrated in siliceous concretions on bedding planes. Cretaceous–Paleogene sediments constitute the Durmitor flysch. Among the Triassic sediments, sandstones, sandstone–limestone and andesite occur in limited extension (Zivaljevic et al. 1989).

The mountains belong to the Durmitor block of eastern Bosnia (Miljush 1978; Ustaszewski et al. 2009), where Triassic limestones are overthrust onto the Cretaceous flysch (Fig. 2.5). The block can be subdivided into at least two nappes: the upper is the Durmitor Nappe and the lower is the Bjelašica Nappe (Dimitrijevič 1983) or the Durmitor and the Kuči tectonic units (Zivaljevic et al. 1989). The nappes underwent further folding of various types (asymmetrical and recumbent folds) (e.g. Šareni Pasovi Hill). Lithological composition, stratification and secondary folding influenced glacial erosion and also karstification in the past and present.

The two levels of the Durmitor are the plateaus at 1400–1600 m and the higher surface of 1800–1850 m elevation, an uplifted segment of the former, which is heavily denuded by the present.

The ice of its glaciers accumulated in preglacial dolines and uvalas. An older (Fig. 2.6) and a younger glaciation can be identified in the mountains. The glaciers moved in three directions: towards the Jezerska Površ, the Tara and Piva gorges and the Komarnica Valley (Djurovič 2009). Moving primarily in northern direction, they left the mountains and approached the Tara and Piva valleys as piedmont glaciers, where they terminated at ca 1,200 m elevation (Marovič and Markovič 1972). They also accumulated till in the paleodolines of the lower surface. The southbound glaciers joined the glaciers of W to E alignment, which border the mountains in the S. The dolines and uvalas of the mountains are of preglacial origin and formed in interglacial time(s) (Cvijič 1899, 1913). Glacial action has transformed or deepened the paleodolines and uvalas or the carved rock basins to the degree that their floors are now incised into non-karstic rocks locally. The resulting depressions are filled by water and form permanent lakes (Črno and Škrčka lakes), but impoundment has also played a part in the formation of some of the lakes.

The karst type is holokarst with intensive fluvial erosion on the neighbouring lower surfaces (gorges) and with glacial erosion features on higher surfaces (Fig. 2.7 – Radulovic and Radulovic 1997). The recent karst features of the mountains include ponors, solution and subsidence dolines. According to the thickness of cover deposits, several varieties of paleodolines can be identified. Some of them are filled to their rims, while others have a thin cover, which only extends to some parts of the bottom. The cover can be contiguous (also occurring on the thresholds between dolines), but it can also be patchy (on doline floors) or, in the opposite way, burying the inter-doline thresholds. The floors of the paleodolines are also diverse: there are paleodolines and uvalas with flat bottoms or roches moutonnées, cuestas

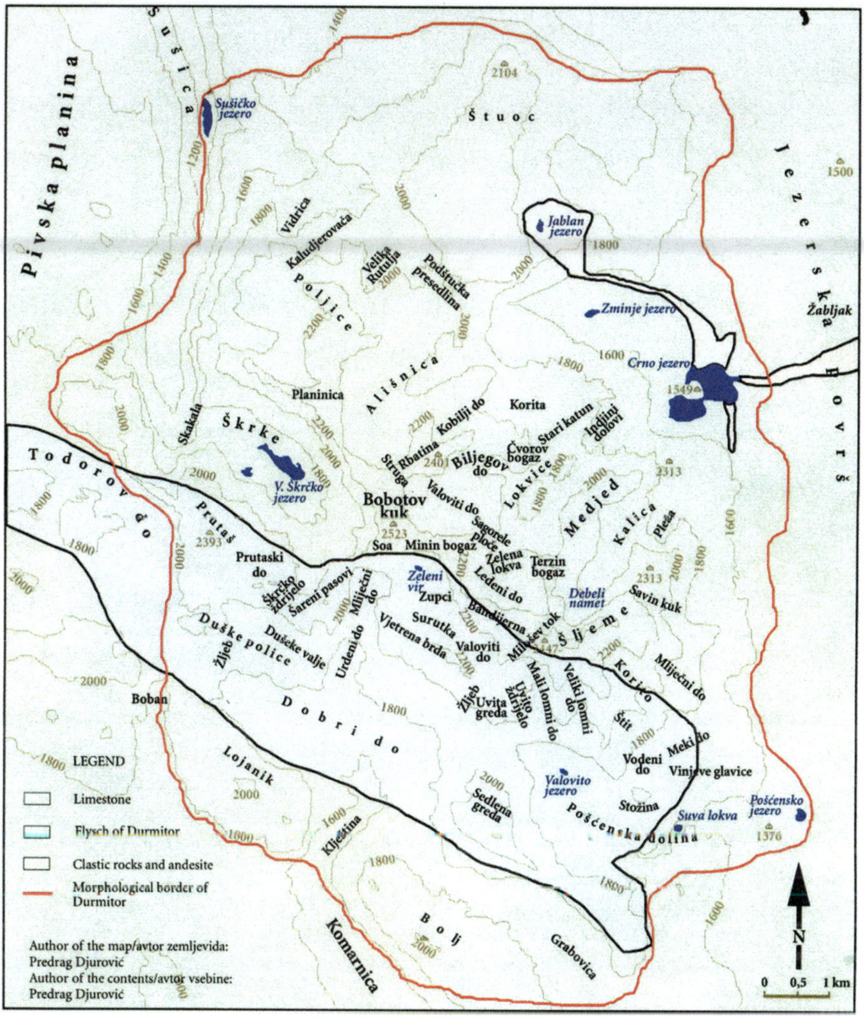

Fig. 2.5 The Durmitor and its geology (Djurovič 2009)

and bassets. The size of the dolines and the type of their floor can be highly variable. Basin-like or planated floors equally occur.

The covered karst features include suffosion dolines, shallow dropout dolines and depressions of superficial deposit, which developed in paleodolines filled up to various degrees. Suffosion dolines most commonly occur in partially filled-up paleodolines, but they can also be found on thresholds between paleodolines. In paleodolines where the cover developed only in patches, suffosion and solution doline types may occur side by side.

The research sites in the Durmitor were the Surutka (I), the Mlječni do (II, Fig. 2.7) and N of the mountains the plateau section between the Tara and Žabljak rivers (III).

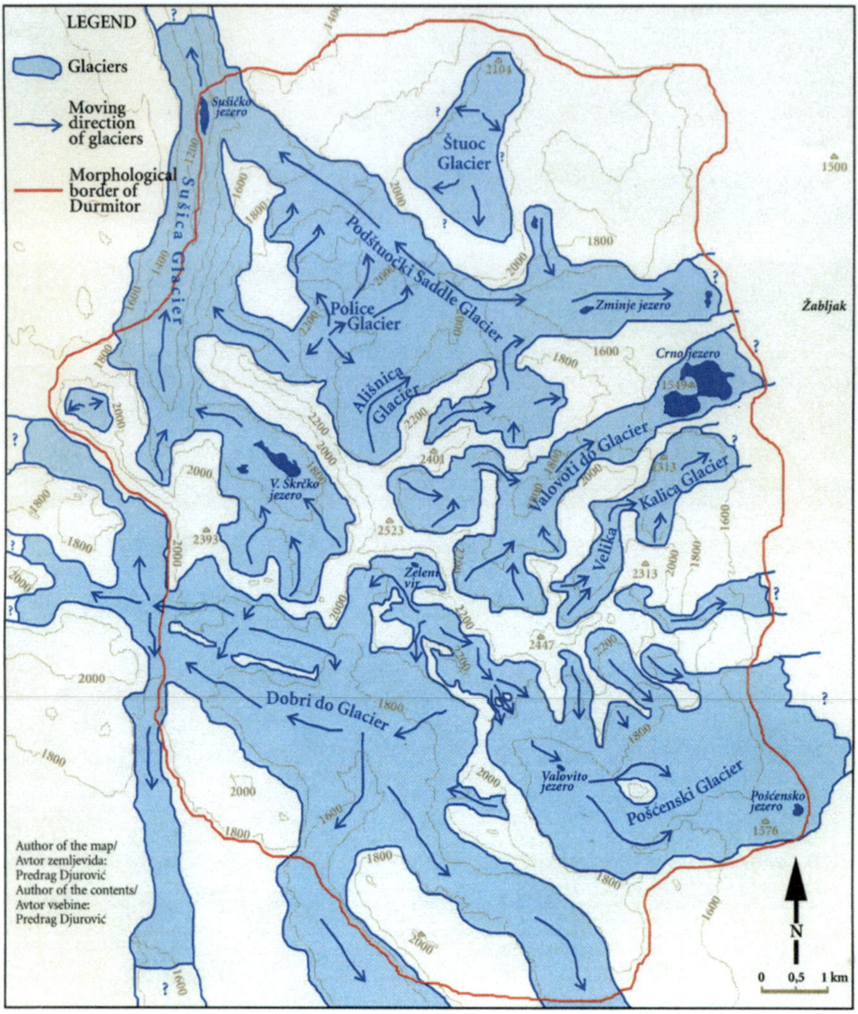

Fig. 2.6 The glaciers of the Durmitor in the older phase (Djurovič 2009)

2.2.6 *Northern Limestone Alps*

The Northern Limestone Alps are a high-mountain karst (glaciokarst) in the northern Eastern Alps in Austria. Their members are the Hochkönig, Steinernes Meer, Karwendelgebirge, Tennengebirge, Dachstein, Totes Gebirge, Hochschwab, Schneealpe, Raxalpe and Schneeberg. The mountains are the preserved frontal part of the Upper Austroalpine Nappe (Plan et al. 2009). Nappe formation began in early Tertiary times, and after the destruction of the nappes the Northern Limestone Alps

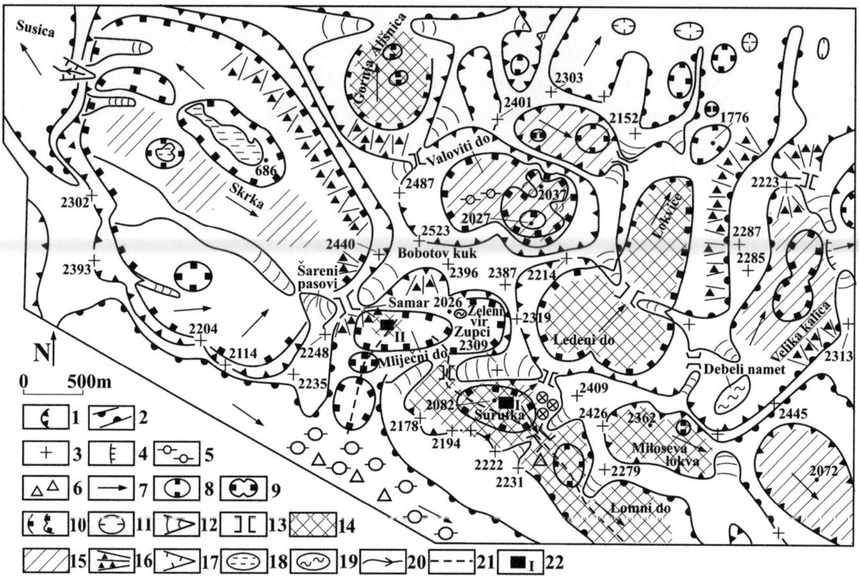

Fig. 2.7 Geomorphological map of part of the Durmitor (based on a tourist map) (Veress 2012b, modified) *1* cirque valley, *2* trough, *3* horn, *4* threshold, *5* terrain with roches moutonnées, *6* moraines, *7* direction of ice movement, *8* preglacial giant solution doline, *9* preglacial uvala, *10* opened-up giant doline or uvala, *11* postglacial solution doline, *12* karst threshold, *13* col, *14* covered karst (mostly moraine), *15* assumed covered karst, *16* debris fan, *17* river valley, *18* lake, *19* glacier, *20* watercourse, *21* section site, *22* research sites: *I* Surutka, *II* Mlječni do

were covered by a Lower Miocene gravel mantle ('Augensteine') (Bauer and Zötl 1972). The Late Miocene and Pliocene uplift was accompanied by peneplanation (Lichtencker 1926). The peneplain was dissected by river valleys and then recent tectonic movements dismembered it into blocks of various sizes (Bauer and Zötl 1972). The mountains (plateaus) of the Northern Limestone Alps are remnants of the Rax surface.

Where the gravel mantle was permeable or missing, karst processes were active during the Late Miocene – Pliocene uplift and ponors, dolines (probably also covered karst dolines) and caves were formed. The gravels were transported into cracks and caves (Bauer and Zötl 1972), and this, in our opinion, led to the formation of depressions of superficial deposit even at that date. According to Zötl (1963), a basin functioning as a polje also took shape in the Pliocene.

Before the Pleistocene glaciation, large-size dolines (preglacial giant dolines) with diameters of more than 100 m developed on the plateaus already deprived of their cover deposits (Bauer 1952). They are, however, less widespread than the interglacial giant dolines. It is probable that preglacial giant dolines influenced glaciation: the cirques of the Karwendel Mountains formed in such dolines (Fels 1929), which also occur in other mountains (e.g. in the Totes Gebirge) (Veress 2012b).

In the Pleistocene glaciations, the glacier networks typical of the Eastern Alps also developed in the area of the Northern Limestone Alps. The piedmont glaciers extended over the mountain foreland (Van Husen 2000).

Glacial erosion produced cuestas with mostly karren formation in the mountains of the Northern Limestone Alps (Schichttreppenkarst, Bögli 1964, 1978).

2.2.6.1 Totes Gebirge

The mountains rise in 590 km^2 area enclosed by the Ennstaler Alps and Dachstein and Traunstein mountains. They belong to the Tirolic facies of the Upper Austroalpine Nappe (Plan et al. 2009) and are built up of Triassic (Werfen, Dachstein) limestones and in the western part of Jurassic limestones. For covered karst formation, it can be concluded that in the building rock siliceous interbeddings (concretions and layers of some centimetres thickness) are common. Silica does not govern the motion of infiltrating waters only, but at outcrops, it also contributes to the formation of springs and surface watercourses in the interior of the mountains. It also promotes debris accumulation as limestone series with significant siliceous interbeddings are liable to mechanical weathering. Extensive debris fans could result at the base of cliffs. In the mountains along the exposed cliffs with siliceous interbeddings, springs issue in the sidewalls of giant dolines. Spring water infiltrates into the karst along the margins of surfaces with siliceous debris, where ponors and subsidence dolines which function as ponors develop. Ponors may also form in the exposed siliceous limestone series. In this case spring water (or rainwater through intermittent watercourses) reaches the karst along rock boundaries marked by siliceous beds.

The plateau of the mountains was dissected into arêtes by glaciers moving in S to N, NE to SW and E to W directions. The southeastern part of the mountains is less dissected by glacier troughs but abounds in cirques. The internal troughs are linked to the major glacial valleys of the mountain margins through steep scarps (Fig. 2.8).

Because of the former glacier network, the cirques of glacier valleys are less developed. The upper ends of troughs are often connected. In the internal glacial valleys, rock basins are less typical than in the main troughs of the margins.

Preglacial and interglacial giant dolines, shafts, shaft dolines and ponors are characteristic karst features, while on covered karst patches, mainly suffosion dolines appear, but covered karst ponors and fossilised karst features (dolines and ponors) are also present. The paleodolines of lower position are lined with till (e.g. Tauplitz alm), while in the higher-lying paleodolines, debris from collapses or frost weathering of siliceous material are found. The lakes in the mountains were formed in rock basins or paleodolines, and the Elm and Steyree lakes have an active outflow conduit (ponor). Their water appears again in the major karst spring of the mountain margin (Bauer and Zötl 1972).

The research sites were near the Pühringer Hütte (I), sections of the valleys S (II) of Appel-hause, area under Wildgössl peak (III) and a detail of the Tauplitz alm (IV, Fig. 2.8).

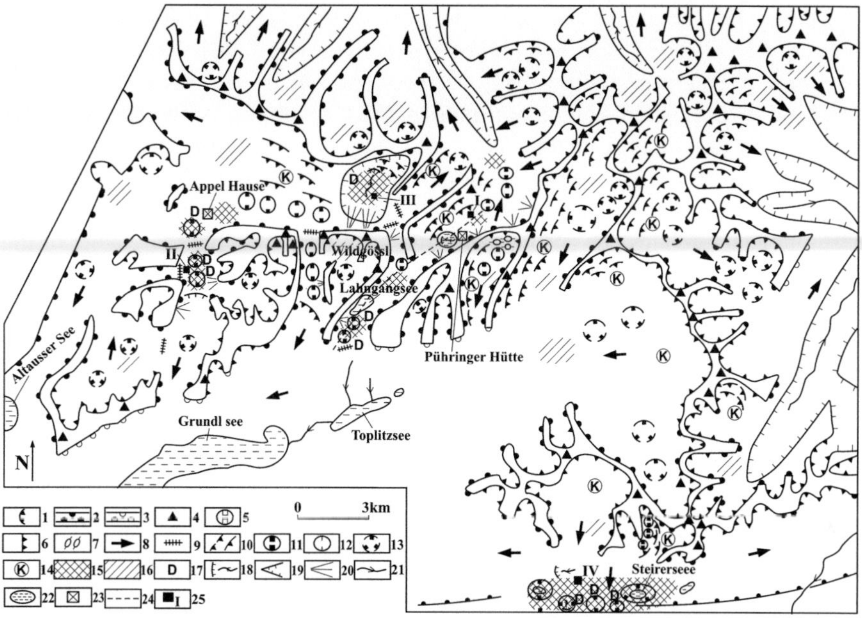

Fig. 2.8 Geomorphological map of part of the Totes Gebirge (based on tourist map) (Veress 2012b, modified) *1* cirque valley, *2* trough on plateau, *3* old trough between plateau sections, *4* horn, *5* rock basin, *6* step, *7* terrain with roches moutonnées, *8* direction of ice movement, *9* glacier bifurcation, *10* gentle slope of trough on bedding plane, *11* giant solution doline of interglacial age, *12* preglacial solution doline, *13* assumed interglacial or preglacial giant doline, *14* karren, *15* covered karst (mostly with subsidence dolines), *16* assumed covered karst, *17* depression of superficial deposit, *18* ponor, *19* erosion valley, *20* debris fan, mountain collapse, *21* watercourse, *22* lake, *23* tourist hostel, *24* site of section, *25* research site with identification symbol: *I* area near Pühringerhütte, *II* area south of Appel-hause, *III* area under Wildgössl peak, *IV* part of Tauplitz Alm

2.2.6.2 Dachstein

Dachstein is a high-mountain karst (glaciokarst) of ca 400 km^2 area between the Tennengebirge and the Totes Gebirge. It belongs to the Juvavic facies unit of the Upper Austroalpine Nappe (Schwarzacher 2005) and is built up of Upper Triassic Dachstein Limestone.

Dachstein is a characteristic plateau which is bordered by peaks of almost 3,000 m elevation on its southern margin and around 2100 m on the northern margin (Ochsen peak, Krippenstein, Speik B). The plateau surface tilts towards the N. It can be divided into three parts. In the S it is covered by a glacier group event at present. To the N a plateau section with giant dolines and glacial horns rises and glacier troughs are underdeveloped or missing. The third unit is the northern margin of the plateau, which is almost vertical, but at some places it is less steep and is dissected by glacier valleys of steep floor.

During the Riss glaciation, the alpine glacier network (Van Husen 2000) probably formed a plateau glacier here. Since glacial horns rise almost 200 m above their environs, if preglacial dolines had ever been present on the plateau, they were destroyed by glacial erosion. In our opinion, the giant dolines of the plateau developed in the Riss–Würm interglacial stage. They are, however, often arranged in rows or sometimes lie in deeper position than their environs. All these observations point to the origin of dolines on the floors of the remnants of former valleys. In the resulting giant dolines, ice accumulated and gave rise to the plateau glacier of the Würm glaciation, whose glacier tongues were connected to main glacier, which embraces the whole plateau. It is probable, however, that next to the plateau glacier, on the plateau or on its margin, independent valley glaciers, not recharged by the plateau glacier, also formed.

At present karst features form on arêtes, paleodolines and in the marginal glacier valleys. The recent open karst features of the second zone are the solution dolines, shaft dolines, ponors and karren, while the 'Schichttreppenkarst' is not common. In the first zone there is no covered karst, and in the second zone it primarily developed on the floor of giant dolines. In the third zone patches of covered karst also occur in the cirques of glacier valleys (formed in paleodolines). The site of research in Dachstein was such a cirque (along the tourist track number 601). In Dachstein suffosion dolines and depressions of superficial deposit are characteristic covered karst features.

2.2.6.3 Hochschwab

Hochschwab is a member of the Northern Limestone Alps with ridges rather than with plateaus. The plateau character is better shown in its eastern termination. Hochschwab is dissected by tributary glaciers running from N to S (bordering the plateau from northern direction) and regressing to the main ridge. It is part of the Juvavic Unit (Mandl et al. 2002), built up of Permian and Lower to Middle Triassic limestones. Its glaciers were part of the glacier network during the Riss glaciation, but in the Würm they were separated from the Alpine glacier network (Van Husen 2000). Karst features are represented by poljes (Plan and Decker 2006), probably depressions of superficial deposit, dolines (either solution or subsidence dolines), bogaz and karren (Plan and Decker 2006). Solution dolines can be paleodolines or recent features.

In addition to solution dolines, ponors and karren constitute the recent karst: for instance, kluftkarren, meanderkarren, rillenkarren, rundkarren and kamenitzas (Plan et al. 2012). The covered karst is first of all developed on the floors of paleodolines with cover deposit (one of them close to the eastern margin was a research site in Hochschwab) but also occurs elsewhere and takes the form of suffosion dolines and depressions of superficial deposit.

2.2.7 *The Julian Alps*

The Julian Alps (Slovenia) is a high-mountain karst (glaciokarst), member of the southern belt of the Eastern Alps in the headwater area of the Sava and Soča rivers. It is bordered in the N by the Fella Stream and in the S one of the headwater streams of the Sava River. Its area in Slovenia extends to 1500 km^2. In the E the Triglav mountain group rises above the Pokljuka Plateau between the two headwaters of the Sava River, and it is bordered by the Vrata in the N and by the Sava Bohinjka in the S. Its northwestern section is the Škrlatica–Razor (Kranjska Gora Dolomites) group. S of the Sava Bohinjka rises the Vogel range, W of the Soča the Mangart mountain group is found and to the S the Kanin peak and the plateau around it form the western section of the mountains (Fig. 2.9).

The basement of the mountains is built up of metamorphic rocks, overlain by Triassic limestones and dolomites. Structurally, the mountains are autochthonous or partly autochthonous (Miljush 1978; Ustaszewki et al. 2009).

The geomorphology and karst of the mountains were described by Kunaver (1973, 1976, 1984), Pirnat (2002) and – for the Canin Plateau – by Telbisz et al. (2011). Today, surfaces of different elevation can be identified. Glaciers are mostly developed in dolines and uvalas with the following morphological characteristics:

- The mountains are dissected by major glacial valleys (Fig. 2.9) with lakes which are formed in rock basins or behind terminal moraines.
- In glacial valleys of higher position, too, rock basins, roches moutonnées and cuestas are common. The Canin Plateau is also divided by cuestas (Szabó 2008), and rows of extensive debris fans are found in the valley sides. The valleys usually connect to the deeper and larger glacier valleys by transversal steps.
- Typical karst phenomena are the karren, shaft dolines, shafts and ponors on rock basins or at their outflow points. Paleodolines are less characteristic.

Out of the Pleistocene glaciers, only one had survived to the recent past and, according to data by Gams (2002), this filled the cirque below Triglav in the 1950s and 1960s and even reached beyond that. By the 1980s, however, it retreated to the slopes of Triglav and by the end of the century disappeared from there, too. It was replaced by a depression of superficial deposit as the till is transported into the karst through numerous shafts and giant grikes.

The research sites were located around Triglav (II) and in the Seven Lakes Valley (I, Fig. 2.9).

2.2.8 *Madagascar*

The basement of the island is of Precambrian age overlain by narrow strips of Mesozoic (Jurassic) and Tertiary limestone along the western margin (Schlüter 2008). The island is rich in karst phenomena: the most unique of the subtropical

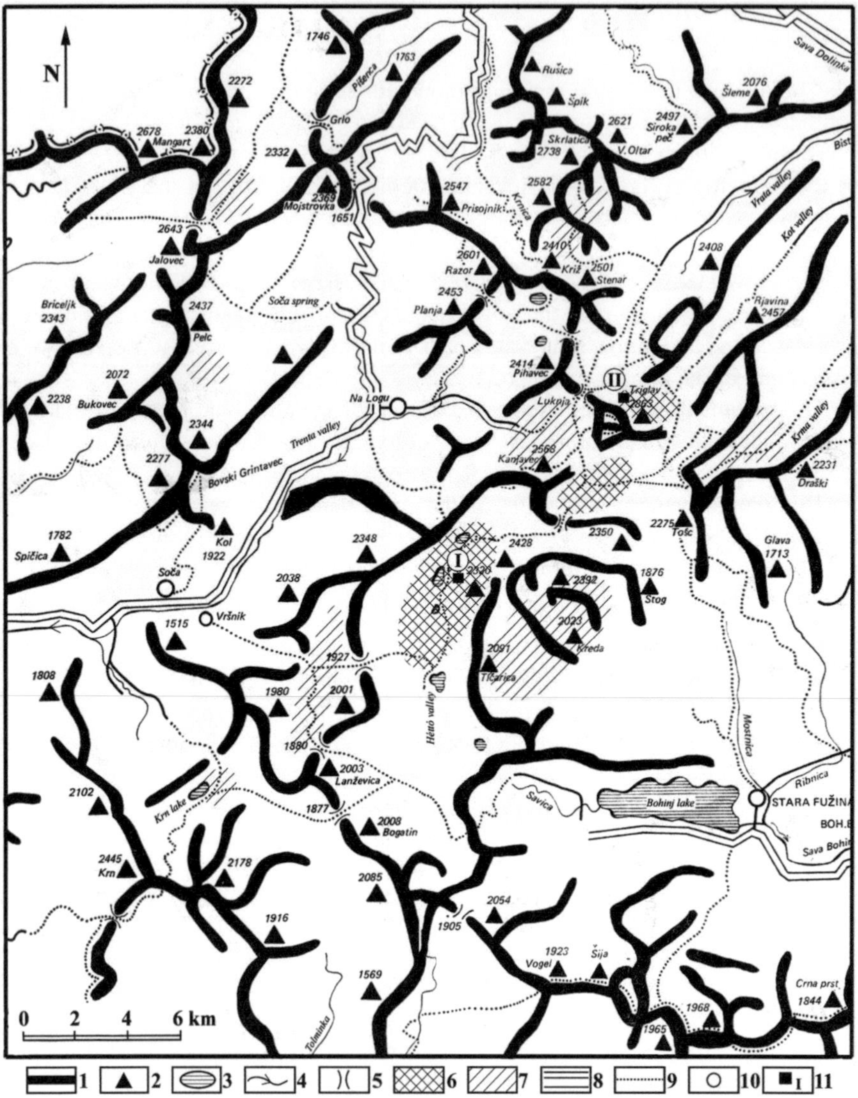

Fig. 2.9 The Julian Alps. *1* arête, *2* horn, *3* lake, *4* watercourse, *5* col, *6* covered karst, *7* assumed covered karst, *8* road, *9* tourist path, *10* built-up area, *11* investigation sites: *I* Seven Lakes Valley, *II* cirque valley below Mt. Triglav

karsts being the Tsingy studied in detail (Salomon 2009; Veress et al. 2008, 2009). Our research took place in the northern coastal zone of Jurassic limestones.

2.2.9 The Mecsek Mountains

The Mecsek of NNE to SSW strike is isolated mountains with a temperate middle-mountain karst in South Transdanubia (Hungary), a surface part of the Tisza Macrostructural Unit. This unit was detached from the European Plate to reach its present position during the Miocene (Stegena et al. 1975). The Mecsek–Villány zone of partly Late Paleozoic and mostly Mesozoic rocks was formed in the Tisza Migmatite zone. It shows a folded-imbricated structure (Fülöp 1989). The Mecsek Mountains are built up of Permian and Lower Triassic sandstones, Triassic and Jurassic limestones and Jurassic marls of great thickness (Fülöp 1989).

The mountains became a peneplain by the Miocene (Bulla 1968b) and were dismembered into a series of horsts by faults. Main landforms are the pediment in the margins reshaped by abrasion, plateaus, debris fans and erosional valleys. A lateral displacement across the mountains divides the area into Western and Eastern Mecsek.

The structure of the Western Mecsek is dominated by a W to E anticline with Permian and Triassic sandstones in the central and southern part and Middle Triassic limestones liable to intensive karstification in the N (Chikán et al. 1984). The Mecsek karst developed in this zone (Barta and Tarnai 1997), S of the settlements Abaliget and Orfű over 30 km^2 area, divided by a western and an eastern part by the epigenetic Szuadó Valley. There are two abrasional surfaces in this zone (Fig. 2.10):

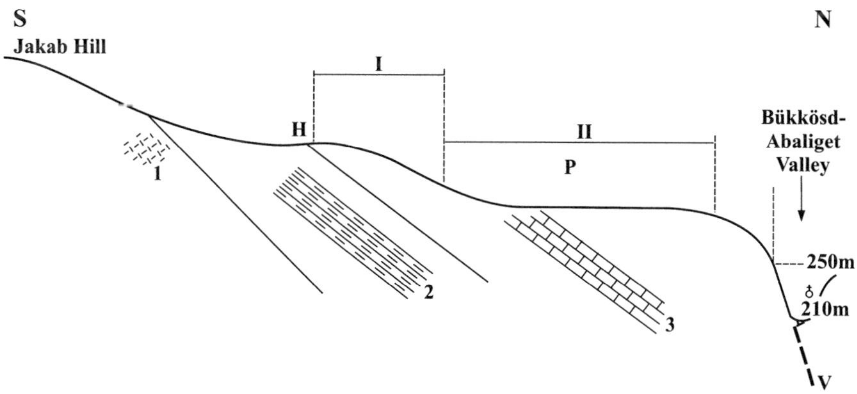

Fig. 2.10 Geological and geomorphological section across the Mecsek karst (Lovász 1971). *H* middle Miocene abrasion platform, *P* Pliocene abrasion platform, *I* zone of large dolines of low density, *II* zone of large dolines of high density, *1* Permian sandstone and Triassic dolomite, *2* lower Triassic limestone, *3* middle Triassic limestone, *V* fault

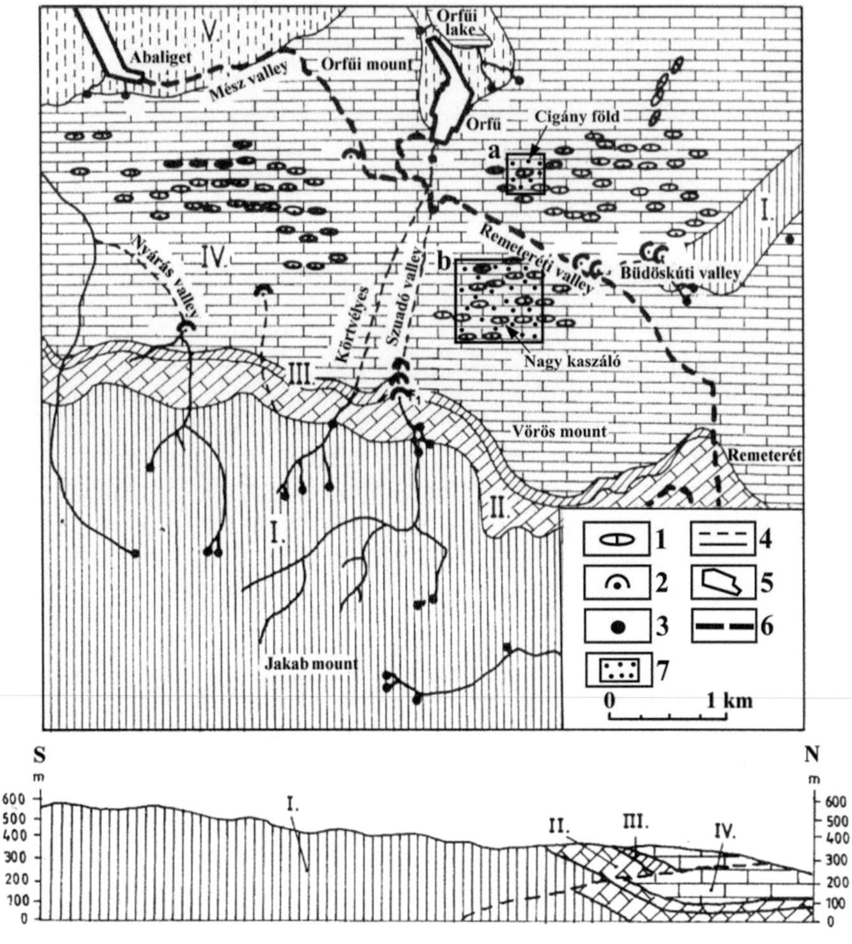

Fig. 2.11 Karst of the Mecsek Mountains (Barta and Tarnai 1997). *1* Doline field; *2* ponor; *3* spring; *4* intermittent and permanent watercourses; *5* built-up area; *6* surfaces road; *7* research sites, (**a**) Cigány-föld; (**b**) Nagy kaszáló; *I* non-karstic rocks (sandstones, conglomerates, siltstones); *II* poorly karstifying limestones; *III* dolomite; *IV* well-karstifying rocks below loess; *V* Tertiary and Quaternary deposits

the lower northern, Pliocene surface (of ca 325 m elevation) and the higher (375 m) southern, Middle Miocene abrasional platform (Lovász 1971).

Along the southern margin of the karstification zone, on the boundary between dolomite and limestone, N of the sandstone, ponors developed (Fig. 2.11, Barta and Tarnai 1997). The non-karstic landforms of the karst zone are represented by epigenetic valleys with flat ridges and plateau-like surfaces between them. The limestone is overlain by deep loess (several metres thickness) and reworked sands and clays. Decalcification has fully taken place in the loess to the present day (Hevesi 2001).

In the zone dolines occur in high density, including old, large (50–80 m diameter) solution dolines and subsidence dolines of smaller (some metres to several times 10 m) dimensions (Veress 2011), partly suffosion and partly dropout types. Depressions of superficial deposit are also present and called 'compartmented bundles' by Hevesi (2001).

On the older abrasional platform, solution dolines formed – according to Szabó (1968), in the Early Tertiary, and according to Hevesi (2001) since the Sarmatian, while on the younger surface, doline formation began in the Pliocene. Dolines are mostly arranged in rows and occur on both interfluvial ridges and valley floors.

Doline density is high: Hoyk (2002) mentions areas with 110 dolines per km^2 density, explained by the large number of subsidence dolines on interfluvial ridges, valley sides, head valleys and valley floors (Hevesi 2001), in solution dolines (lined or completely filled with loess) and on flat terrains between solution dolines (Veress 2011). Since dolines of this type developed in solution dolines and loess, they could not have formed before the Pleistocene and the Holocene.

The sites of investigation (Fig. 2.11) include the area called Cigány-föld, E of the Remeterét Valley (a), and the Nagy kaszáló between the Remeterét and the Szuadó Valleys (b).

2.2.10 Padis

The Padis is a temperate middle-mountain karst in the central Bihor (Bihar) Mountains (Apuseni Mountains, Romania). The Bihor Mountains are an exposed section of the Tisza Macrostructural Unit (Patrulius et al. 1971), similar to the Villány Mountains in Hungary (Villány–Bihor Unit), structurally autochthonous and a tectonic fenster in its environment (Fülöp 1989). The autochthonous basement is built up of metamorphic rocks, overlain by Mesozoic neritic sediments (Triassic and Jurassic limestones and dolomites and on the margins Permian sandstones and metamorphic rocks) (Bleahu 1976; Fülöp 1989). In the mountains of 34.5 km^2 area, karstic rocks are of nonuniform character: there are two zones of Jurassic and Permian non-karstic rocks (Móga 2004). During karstic processes it was divided into two surfaces: the upper with summits at 1200–1300 m is a group of isolated open karst terrains at variable altitudinal positions, while the lower is below 1,200–1,300 m, surrounded by higher terrains (Veress 1992). Both surfaces belong to De Martonne's 'Maguri-Marisel' peneplain (De Martonne 1907), which is, as pointed out by later research, a pediplain developed from a pediment (Berindei 1987). Remnants of an older (peneplain) surface ('Farcas-Cîrligatele') have been identified along the margins (Kék-Magura, Măgura Vânătă) (Berindei 1987). The relief between the two karstic surfaces is rather variable, ranging from some metres to 100–200 m. The lower surface derives from an older karstification phase, and, consequently, the elevations of the higher terrain are relict features of karstification.

Structurally it is monocline karst (Móga 2004). The landform assemblages of the plateau are the following (Fig. 2.12, Veress 1992):

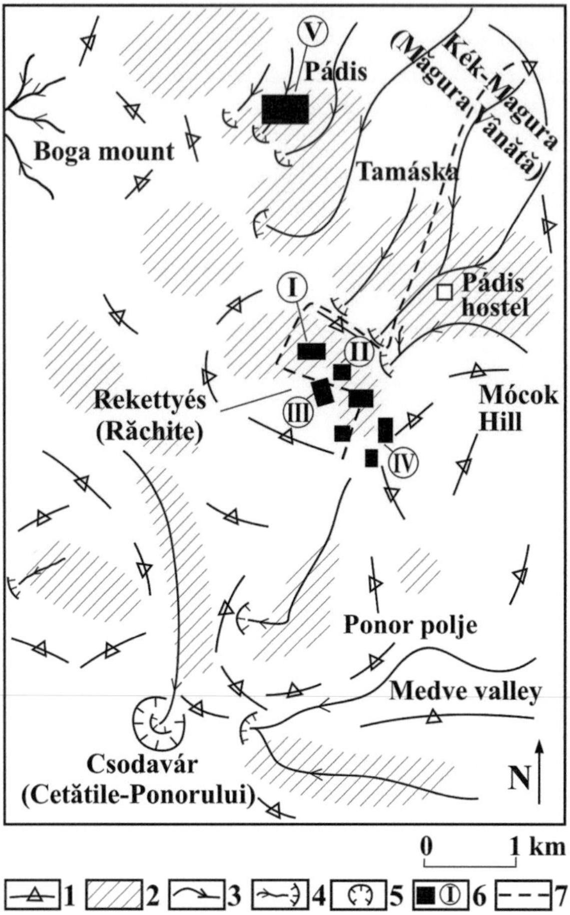

Fig. 2.12 Covered karst terrains on the Padis Plateau (Veress 1992). *1* Mountain ridge, elevation; *2* covered karst; *3* stream; *4* ponor; *5* collapse doline; *6* research sites, *I* Pádis1 (Răchite), *II* Pádis2, *III* Pádis3, *IV* Pádis4, *V* Pádis5; *7* sites of sections (see Fig. 4.37)

- The area of elevations with characteristic solution dolines, uvalas and shafts.
- ESE of the Padis hut the limestone is exposed to the surface, and on this lower surface large-size solution dolines are typical.
- W of the hut a terrain with solution dolines and uvalas presents a mosaic of open karst, cryptokarst and concealed karst features. Here, some of the dolines are only partly filled up.

The cover sediments of the covered karst derive from the Măgura Vânătă, where the intermittent and permanent watercourses spread out debris of sandstone origin in a fan (Veress 1992). The watercourses incised into the fan and transported its material further and, thus, the fans expanded in western direction. This kind of expansion is indicated by the fining of the grain size of the cover towards the W. The water must have percolated away from the watercourses along the alluvial fan margins or ponors could have developed there. When a section of an alluvial fan got into contact with a doline or uvala, doline infilling or burial started, in the karst depressions intermittent ponds developed even accompanied by lacustrine deposition.

On the Padis area there are uncovered, allogenic (cryptokarst) and concealed karst patches. An allogenic karst formed, where the reworked cover is thick, while a concealed karst developed where the cover is thin. There are ponors with blind valleys on its allogenic karst, while subsidence dolines and depressions of superficial deposit occur on its concealed karst.

On the plateau some epigenetic valleys, once also lined with sandstone debris, occur, too. To this day, however, these valleys have partly or wholly developed into depressions of superficial deposit.

As it has been mentioned, two non-karstic rock interbeddings are also present here. The watercourses issuing along their margins also give rise to ponor rows with blind valleys. During their erosion depressions of superficial deposit form even in solid bedrock (Groapa de la Barsa).

A single remarkable karst landform on the Padis plateau is a polje (Ponor-rét) and a system of collapse uvalas (Cetăţile Ponorului, Fig. 2.12).

Our sites of research (Fig. 2.12): Uvala Râchite, the areas marked as Padis1 (I) and its environs, Pádis2 (II), Pádis3 (III), Pádis4 (IV), the terrain between the Bogakő (Vf. Boghii) and Kék-Magúra (Măgura Vânătă), and Padis 5 (V) at Tăul Vărăsoaia Lake.

2.3 Observation Areas

2.3.1 Atacama Desert

In the area of the Atacama Desert Miocene and Quaternary eolian deposits are widespread with andesitic–dioritic volcanic, subvolcanic rocks in the basement (Scheuber and Andriessen 1990; Reutter et al. 1996; Riquelme et al. 2007). Since the region has been an extreme dry land since the Mesozoic, the issuing hydrothermal solutions evaporated and salts precipitated in the depressions of the surface. This way the rock salt deposits of the Atacama were produced.

The area studied was the environs of the Moon Valley (Valle de la Luna) near San Pedro, a salt karst in desert environment. The marginal zone is an area slightly dissected by watercourses rising towards a pediment. In the central part, where blown sand is deeper, depressions of superficial deposit formed, while where the blown sand cover is thinner or absent, karren formation affects the salt below the blown sand in the environs of a salt cave. The investigations took place in depressions of superficial deposit and on karren terrains.

2.3.2 *The Biokovo Mountains*

The Biokovo high-mountain karst (Croatia) belongs to the coastal zone of the Dinaric Mountains. The mountains of more than 300 km^2 area, the mountains terminate steeply to the Adriatic Sea, its direction corresponds to the Dinaric strike. The composing Jurassic and Cretaceous limestones and the Cretaceous–Oligocene flysch (Vlahovič et al. 2005; Korbar 2009) form part of the Apulian–Preadriatic Unit (Bognár 2001). The earthquakes (Herak et al. 1996) and the vertical and horizontal displacements of 2 cm/year rate point to recent tectonic activity (Cigrovski and Detelic 1998). In the Pleistocene and in recent times as well debris formation is at considerable rate, resulting in a series of debris fans along the Adriatic slopes of the mountains.

The narrow mountain range rises as a plateau, higher (ca 1,400 m) in the N and somewhat lower (ca 1300 m) in the S.

Doline size and density are high, and, therefore, now the original surface is restricted to a network of narrow crests between the dolines, with karst mounds at their nodal points (polygonal karst, Telbisz et al. 2005).

In the mountains traces of glaciation have been found (Telbisz et al. 2005), but it could not have been of major scale since glacier valleys are missing; at most nivation niches and minor cirques are recognisable (Telbisz et al. 2005). Ice probably did not fill all the dolines, but only those with favourable conditions for the accumulation and preservation of ice or those which were sufficiently deep. The glacial ice did not flow beyond the boundaries of the depression of ice accumulation (Veress 2012b).

As mentioned above, major paleodolines and, through the merging of neighbouring dolines, uvalas (paleouvalas) were also generated. Breaches of ridges between dolines and shafts are common – particularly in doline walls. Both depressions and elevations produced by mass movements are common, not only on hillslopes but also on the plateau (Telbisz et al. 2005). The floors of a number of dolines and uvalas have been filled up to form flat surfaces, which are part of the covered karst in the mountains appearing in isolated spots. The dolines of the plateau can be classified as:

- Dolines and uvalas with uncovered floors
- Dolines and uvalas with covered floors without covered karst features
- Dolines and uvalas with covered floor and some covered karst features (small subsidence doline)

2.3.3 *The Cerkniško polje*

The Cerkniško polje (Slovenia) is a mediterranean karst of ca 50 km^2 area, a member of the polje group of the Slovenian Karst and of the polje row of NW to SE alignment (Prezid, Loško, Planinsko and Logaško poljes).

The polje is aligned in the general (NW to SE) strike of the polje row. It was formed along the contact of Cretaceous and Jurassic limestones with Jurassic dolomite in its strike (Herák 1972). It has an allochthonous structure (Sebela 2012). The floor of the polje is mantled by Quaternary deposits (Herák 1972).

According to the classification by Gams (1977), it is an overflow polje and hydrologically a seasonally waterlogged polje. The water level of its lake shows major fluctuations. As a consequence of fluctuating karst water table, 12 katavothra developed in the polje (Roglič 1965). We investigated one of the groups of katavothra.

2.3.4 *The Dolomiti*

The Dolomiti is part of the Southern Alps, which is built up of the Upper Triassic dolomite, which gave its name to the mountains, which structurally belongs to the parautochthonous Apulian Plate (Mckenzie and Vasconcelos 2009).

The Dolomiti are dissected by cliffs of large vertical extension (with large debris fans) and by glacial valleys (with incision by fluvial erosion on their floors). The Dolomiti themselves are poor in karst features, but karst processes and landforms are of considerable scale on the plateaus of the margins of the Dolomiti (Bernadia, Cansoglio and Asiago Plateaus).

In some sections, even if at small scales, karstification takes place and karst features develop. The area studied by us (la Grava Lunga or Lange Alpe), bordered by the Tre Cime Lavaredo, Passo Alpe Mattina and Alpe Mattina, is among those sections. It is cirque valley divided into two parts by a regressing erosional valley and buried partly by mountain collapses, debris fans and accumulations of various mass movements (Fig. 2.13). On the buried terrains a number of subsidence dolines and one depression of superficial deposit developed.

2.3.5 *Iceland*

Iceland is a volcanic island of the Mid-Atlantic Ridge, built up of Miocene and Quaternary basalts and Pliocene–Quaternary glacial deposits (Thordarson and Hoskuldsson 2002; Guðmundsson 1996, 2012; Guðmundsson and Kjartansson 2007). The drifting of lithospheric plates resulted in the formation of faulted graben structures, and the island's volcanism is among the most active on Earth.

Investigations were carried out in two locations (Fig. 2.14). In the SW between Road No. 1 and Pulutjörn, ca 25 km SW of Mt. Hekla on the floor of an old valley (I), filled with a pre-Pleistocene lava flow of the Hekla, funnel-like depressions with gentle side slopes and of some metres diameter (similar to suffosion dolines) occurred in large numbers. Among them features with solitary and twinned passages and without passages can be found. Their sides are covered by soil and vegetation.

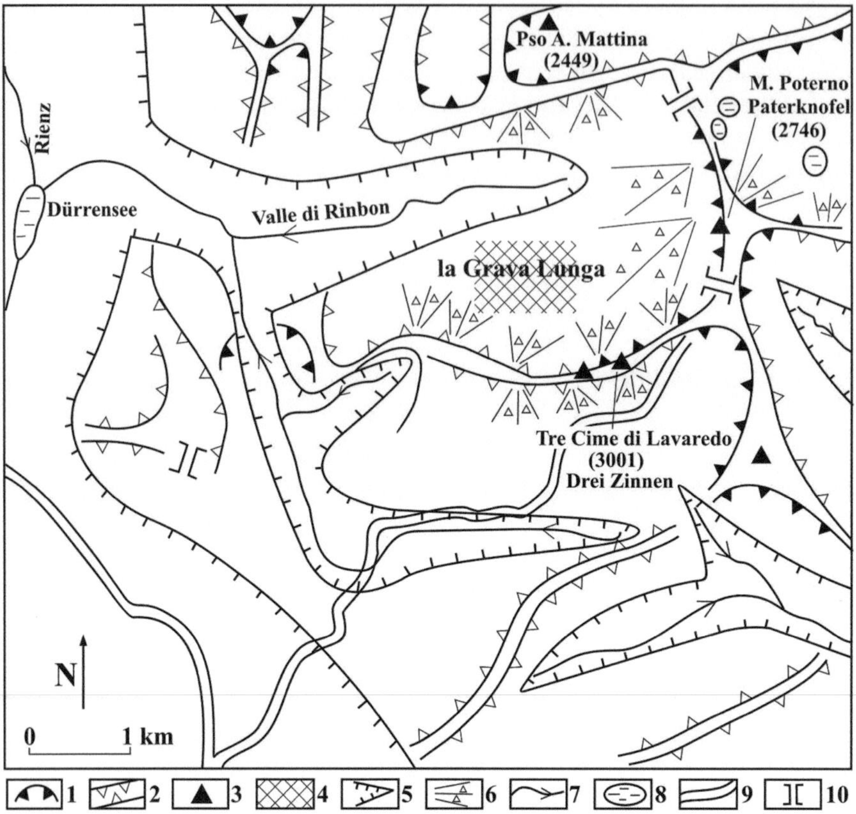

Fig. 2.13 Covered karst in the Dolomiti. *1* Cirque, *2* trough, *3* horn, *4* covered karst, *5* river valley, *6* debris fan, *7* watercourse, *8* lake, *9* road, *10* saddle

The other study area (II) falls to the Askja area, on the Icelandic plateau, NW of the Vatnajökull ice-field. The Askja volcanic system is part of the volcanic zone stretching across the island in NE to SW direction (Thordarson and Hoskuldsson 2002), which is 20 km wide and 200 km long and mostly included rift volcanoes. The Askja volcano (highest point: 1550 m) rises in the Dyngjufjöll Mountains and has double caldera, the outer being of 8–10 km diameter, and the enclosed inner caldera contains a permanent lake (Öskjuvatn). The outer caldera formed 4500–6000 years ago (Scarth and Tanguy 2001), and the inner in 1875. Both of them are of the subsidence type. The bottom of the outer caldera is an extensive flat terrain with a minor volcanic cone, which was produced by a maar eruption. The eruption on March 29, 1875, covered the outer caldera with volcanic ash and pumice, depositing over the snow cover. This was a phreatopluvial eruption (Francis and Oppenheimer 2004). The wet pyroclast could not be cast over longer distances and accumulated in great thickness in the caldera. The resulting thick welded tuff has a good insulating property. We investigated the depressions of vertical walls and of some metres depth and diameter developed on such surfaces.

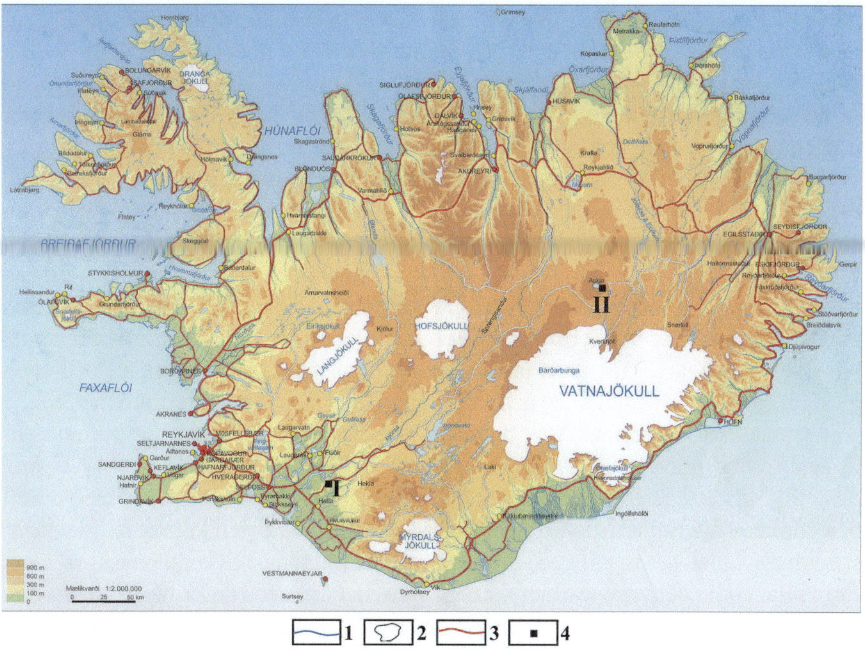

Fig. 2.14 Iceland. *1* Watercourse, *2* ice cap, *3* road, *4* research areas: *I* pseudokarstic dolines in the lava flow below Hekla, *II* caldera of Askja

2.3.6 *Middle Lena*

The Middle Lena area (in Yakutia Republic of Siberia) is a platform karst in tundra environment, bordered by the latitudes ρ_1 61° 03N and ρ_2 61° 12′N and the longitudes λ_1 126° 19E and λ_2 128° 17E. The Lena River follows the boundary between the Middle Siberian Plateau and the Baikal Range. The Middle Lena karst (situated between the Vitim and Buotama rivers), part of the Lena National Park, extends over 12,721.5 km^2; it is 500–700 m high on the right bank and 200–350 m high on the left bank. The karst along the river developed on Cambrian and Silurian limestones as well as dolomite (locally Jurassic limestone), which deposited on the Precambrian basement of the Angara Shield.

On the surface the locally 20 m deep sand and clay mantle has large extension. The Middle Lena karst is dissected by the Lena River and its tributaries. Among them, the Lena valley is a terraced valley, and the Sinyaya is incised meandering. V-shaped valleys (Diring Yuriakh) may occur as well as bowl-shaped valleys (Labiya, Bolshoy Taryn); in the case of the latter ones there are solution or subsidence dolines on the valley floor. Alases and subsidence dolines are common on interfluvial ridges. The slopes of some rivers like the Lena River are dissected by pillars.

The present climate of the Middle Lena region is subarctic with long winters (ca 7 months) and extreme temperature range: minimum temperatures in winter are at −50 to −60 °C and the maximum temperatures in summer at +30 to +35 °C. Annual precipitation amounts to 200–300 mm and mostly falls in the second half of the summer. Permafrost is 50–300 m deep and thaws to ca 1–1.5 m depth (Korzhuev 1961), although ice outcrops can be seen on the ground even in July. The frozen ground below the active layer melts in every 2 or 3 years (Ershov 2002). The permafrost layer is deeper under rivers (Ford and Williams 2007), deeper than 20 m under the Lena.

Structurally, it is an extensive platform karst, for climate a frost karst. Regarding the character of cover sediments, Korzhuev (1961) distinguishes between barren karst, covered karst and bush karst (on hillslopes with reworked cover deposits). The covered karst is cut up by the Lena and its tributaries, and the underlying rocks are exposed in sections or along the entire length in channel beds. The watercourses are either permanent or intermittent. Their water often seeps away from the channel but re-emerge again. This phenomenon can be explained by the presence of frozen ground: channel flow melts the permafrost and the water percolates into the rock.

Lasting frost conditions, low biogenic activity and, primarily, the presence of frozen ground prevent large-scale karstification, and the dimensions, density and diversity of karst features remain limited. Korzhuev (1961) identifies recent and paleodolines along the Middle Lena River. Among the paleodolines he mentions specimens which are manifested as depressions even today. Several phases of karst formation under climates warmer than the present have been described from Siberia (Korzhuev 1972; Pulina 2005; Vangengejm 1977; Belova 1985). Karst springs in the Middle Lena region point to paleokarst as their passage system must have developed before the accumulation of permafrost (Korzhuev 1961). The buried dolines on the islands Tas-Ary (in the channel of the Khatystakh River, a Lena tributary) also indicate paleokarst processes (Lungersgauzen 1966).

Korzhuev (1961) classified the karst features along the Middle Lena as dolines, karst lakes, caves, submergent streams, karst springs, ditches (with dolines), carbonate precipitations at the pillars of the Lena and clay karst (pseudokarst), the dolines of limited size and depth being predominant. There are dolines. Solution dolines exclusively occur in channel beds. Among subsidence dolines suffosion dolines are common but dropout dolines can also be found. Subsidence dolines occur on valley floors with cover deposit and on interfluvial ridges. Korzhuev (1961) also found dolines in ditches, which are the products of mass movements on the margins of the Lena valley. The mentioned author describes dolines from the bottoms of the ditches. On the interfluvial ridges, dolines occur as single features, in rows or groups. Grouped dolines appear on surfaces with closed drainage, and therefore, they are called 'kotlovinik' by Korzhuev (1961). Judging from their size and morphology, they are depressions of superficial deposit. The dolines which are not associated with the local thawing of frozen ground are paleodolines.

2.3.7 *Crimean Peninsula*

The karst of the Crimean Peninsula is a mediterranean karst, partly in high-mountain environment, on karstic Upper Jurassic, Paleogene and Neogene limestones. The Jurassic limestone overlies flysch rocks and forms an anticlinorium with monoclinal cuestas on their northern and southern wings. The northern Crimean Peninsula is a flat lowland steppe at 100–150 m elevation, while the southern part is mountainous (Gvozdetskiy 1981), lower in the W, and consists of karst plateaus in the middle and E. A characteristic karst plateau is the Chatir-dag. The wide, elongated depressions of the plateaus are remnants of the Neogene valley network (Gvozdetskiy 1981). The lower sections, as the Chatir-dag, show the following karst features: polje of 5–10 km diameter, numerous large solution dolines, uvalas, plateaus with polje-like basins in the W (with diameters up to some kilometres and of Late Neogene or Quaternary age), small, flat-floored depressions ('zapadinas', which are subsidence dolines) and ponors. On the Chatir-dag plateau, in addition to karren and dolines, Zabelin (2003) identifies uvalas of gentle side slopes of 150–200 m diameter and asymmetric cross-sections. The gentle slopes developed on bedding planes, and the steep walls are at the escarpments of cuestas. In their interior there is a ponor or shaft with channels leading to the ponors. (In our opinion they are depressions of superficial deposit and their 'ponor' is a subsidence doline.) Further filled-up 'ponors' may occur in the channels. The mediterranean karst continues in a terrain with cuestas.

The karst features at higher elevations in Crimea are extensive polje-like basins or smaller 'kalodetses' of some ten metres depth and 'sakhtas' (shafts) of several hundreds of metres depth. According to Dublyanskiy (1963, 1971, 1977), the 'kotly' (with slightly overhanging walls) of Pleistocene origin were formed by the meltwaters of snow fields.

Near the southern coast of the peninsula, asymmetrical dolines form rows along the bases of escarpments. Their floors are partially filled up and contains subsidence dolines. This was our investigation area.

2.3.8 The Lunan Region (South Chinese Mountains)

The South Chinese mountainous region is an extensive tropical–subtropical karst area of variable rock types and ages. Part of the area is a platform-type karst, while other parts have undergone major structural transformations and produced folded and nappe structures in the karst. By its structure the large karst region can be divided into the northern Yangtze Plateau and South China fold belt (Holland 1990). On the Yangtze Plateau, dolomites and limestones of Triassic to Cambrian (even Precambrian) age occur, while in the South China fold belt, rocks from the Triassic to the Devonian age are found (Barbary et al. 1991). Zhang (1980), Smart et al. (1986) and Daoxian et al. (1991) classified Chinese karsts, and this classification was borrowed by Sweeting (1995). They referred the tropical–subtropical karst of South China into inselberg and plateau canyon types (see Chap. 4).

Recent covered karsts are widespread in the intermountain plains of the fenglin and in the depressions of the fengcong. Karst processes in and under the cover sediment produced stone forests (Song and Liang 2009; Knez and Slabe 2009). In the present karstification continues under the weathering residues. Such karstification under superficial deposits was investigated near Lunan.

2.3.9 *The Salt Hill of Parajd*

The Salt Hill of Parajd is a temperate salt karst in the Transylvanian Basin (Romania) along the Korond Stream. The salt beds of the Transylvanian Basin are of Middle Miocene age (Ajtay 1989). Pressed out from the depth, the salt forms a diapir, which rises to the surface. Salt diapirs are exposed in several localities on the margin of the basin, and the Salt Hill of Parajd is one of these localities. If the diapir rises slowly or under humid climate, the diapir is truncated, while in the case of rapid upward motion and drier climate, a salt glacier forms instead of truncation (Talbot and Jackson 1987). If the salt plug rises very rapidly, truncation does not take place even under wet climate. The salt outcrops of the Transylvania Basin (including the Parajd outcrop) could remain exposed because of the rapid rise of the diapir. The rate of upward extension is 30 mm/year (Ajtay 1989). In spite of the missing truncation, a thin veneer covers the salt.

The Korond Stream has incised into the Salt Hill of 72 ha area and 60–80 m relative elevation. According to Ozoray (1963), the Korond Stream slipped from the uplifting vaulted surface and created an antecedent valley segment (gorge) in the western wing. The description by Orbán (1868), however, confirms that previously, the stream had been diverted around the hill from the W and in the eighteenth century carved through a cave in the salt. The opening of the cave ceiling probably also contributed to the formation of the antecedent valley.

In the diapir area the salt surface is either covered or barren. Barren sections occur in the valley sides, but not continuously, rather in patches. Mass movements (mudflows and slumps, Ozoray 1963) and runoff reworked the cover sediments from the valley margins to the valley sides. Salt outcrops are also observed on the floors and slopes of covered karst depressions. The surfaces on the Salt Hill between the valley and the depressions are covered by:

- The weathering residues of salt (components of alternating clay content)
- Reworked remnants of the local superficial deposit
- Alluvia transported here by the Korond Stream

According to the distribution of the alluvial deposits, cryptokarst and concealed karst terrains probably alternate in the area. Concealed karst may occur in the clay, too, if it thins out sufficiently or becomes interrupted (for instance, as a consequence of diapir uplift or human action). At the same time, concealed karsts or uncovered karsts may be transformed into cryptokarsts, for instance, in doline interiors or if part of the doline is buried under the cover reworked by mass movements.

On the barren salt patches, karren (locally with karren tables) and salt pinnacles (Benedek 1905; Veress et al. 2011) develop. On the covered surfaces, dropout dolines (Ajtay 1989), shafts, solution dolines and uvalas (Zentai 1994) occur. On the floor of dolines and uvalas, superficial deposit develops and on its margin ponors occur. In the dolines and on karren terrains, precipitations (Ozoray 1963) occur. Valley-like landforms are also observed on the Salt Hill. Creeks formed along abandoned dirt roads further incise by solution (Ozoray 1963). The development, transformation and destruction of features take place at a very rapid rate, within some years. In addition to high solubility, it is explained by surface and deep mining of salt and the sewage (from the households of Parajd) stored in the underlying sediments and reaching the rock salt mass.

References

Ajtay F (1989) Az Erdélyi-medence nagy kincse: A kősó (The treasure of the Transylvanian Basin: the halite). Természet Világa 5:203–206 (in Hungarian)

Arnaud-Vanneau A, Arnaud H (1990) Hauterivian to Lower Aptian carbonate shelf sedimentation and sequence stratigraphy in the Jura and northern Subalpine chains. In: Tucker ME, Wilson JL, Crevello PD, Sarg JR, Read JF (eds) Carbonate platforms, facies, sequences, and evolution. Blackwell Science Publications, pp 203–234

Balogh K (1964) A Bükk hegység földtani képződményei (Geological features of the Bükk Mountains). Magyar Állami Földtani Intézet Évkönyve 48(2):245–719 (in Hungarian)

Balogh K, Jámbor Á, Partényi Z, Ravaszné Baranyai L, Solti G (1982) A dunántúli bazaltok K/Ar radiometrikus kora (K/Ar Radigenic Age of Transdanubian Basalts)– Magyar Állami Földtani Intézet Évi Jelentése 1980-ról, Budapest, pp 243–260 (in Hungarian)

Barbary JP, Maire R, Zhang Shougue (1991) Gebine'89. Karstologia Mém 4, 232 p

Bárdossy G (1977) Karsztbauxitok (Karst bauxites). Akadémia Kiadó, Budapest, 413 p (in Hungarian)

Barta K, Tarnai T (1997) Karsztkutatás az orfűi Vízfő-forrás vízgyűjtő területén (Karst research on the catchment of the Vízfő spring, Orfű). Karszt és Barlang I–II:12–19 (in Hungarian)

Bauer F (1952) Zur Verkarstung des Sengsengebirges in Oberösterreich. Mitt. Höhlenkomm, pp 7–14

Bauer F, Zötl J (1972) Karst of Austria. In: Herak M, Stringfield VT (eds) Karst, important karst regions of the northern hemisphere. Elsevier Publishing Company, Amsterdam, pp 225–265

Belova VA (1985) Rasztitelnoszty i klimat pozdnego kajnozoja juga Vosztocsnoj Szibiri. Nauka, Novoszibirszk, 160 p

Benedek K (1905) Emlékirat a parajdi sóbányászat érdekében (Memoirs in the interest of the halite mining of praid). Bányászati Kohászati Lapok 41:240–242 (in Hungarian)

Berindei IO (1987) Muntii Bihor-Vladeasa. In: Odncea D et al (eds) Geografia Romaines. Edit Acad. R.S.R, Bucaresti, pp 434–453

Bleahu M (1976) Structural position of the Apesuni Mountains in the Alpine System – Revue Roumaine de Geologie. Geophys Geogr Ser Geol 20:7–19

Bögli A (1964) Le Schichttreppenkarst. Un exemple de complexe glaciokarstique. Rev Belg Geogr 1(2):64–82

Bögli A (1978) Die wichtigsten Karrenformen der Kalkalpen. Karst processes and relevant landforms. International Speleological Union, Commission on Karst denudation. Department of Geography, Philosophical Faculty, University of Ljubljana, Ljubljana, pp 141–149

Bognár A (2001) Geomorfološka regionalizacija Hrvatske. Acta Geogr Croatica 34:7–29

Bulla B (1968a) A magyar föld domborzata fejlődésének ritmusai, az újharmadkor óta a korszerű geomorfológiai szemlélet megvilágításában (The phases of the terrain development of hungary since the lower tertiary in perspective of the modern geomorphological view), Válogatott természeti földrajzi tanulmányok. Akadémia Kiadó, Budapest, pp 90–104 (in Hungarian)

Bulla B (1968b) Harmadkori elegyengetett felszínek maradványai Magyarországon (Remnants of tertiary planation surfaces in Hungary), Válogatott természetföldrajzi tanulmányok. Akadémia Kiadó, Budapest, pp 138–143

Chikán G, Chikán G-né, Kókai A (1984) A nyugati-Mecsek földtani térképe 1:25000 (A Geological Map of the Western Mecsek 1:25000). A Magyarhoni Földtani Társulat Dél-Dunántúli Területi

Szervezetének 25 éves fennállása alkalmából kiadta a Magyar Állami Földtani Hivatal, Budapest (in Hungarian)

Cigrovski-Detelic B (1998) Primjena GPS mjerenja i geotektonskih informacija u. obradi geodinamičke mrece Crodyn (Application of GPS measurements and geotectonic information in geodynamic network rooling Crodyn, 1994–96) – Unpublished PhD Thesis, University of Zagreb, 145 p

Csontos L, Vörös A (2004) Mesozoic plate tectionic reconstruction of the Carpathian region. Paleogeogr Palaeoclimatol Palaeoecol 210:1–56

Cvijič J (1899) Glacial and morphological studies of the mountains in Bosnia, Hercegovina and Monte Negro (in Serbian), vol LVII. Glas Srpske Kraljevske Akademije Nauka, Belgrade, 196 p

Cvijič J (1913) The ice age in the Prokletije and surrounding mountains (in Serbian), vol XCI. Glas Srpske Kraljevske Akademije Nauka, Belgrade, 149 p

Daoxian Y, Dehao Z, Xuewen Z (1991) Karst of China. Geological Publishing House, Beijing, 158 p

Dimitrijevič MD (1983) Geology of Eastern Yugoslavia: a short review. Guide-book for the field meeting of the int. working group of the IGCP Project 5:1–18

Djurovič P (2009) Reconstruction of the Pleistocene glaciers of mount durmitor in Montenegro. Acta Geogr Slov 49(2):263–289

Dubljanskiy VN (1963) O roli snega v zakarstovyvanii i pitanii karstovykh vod (The role of snow in karstification and in the recharge of karst waters.). Izv Akad Nauk SSSR Ser Geogr 2:69–75

Dubljanskiy VN (1971) O haraktere i intensivnosti himicheskoy denudacii v Gornom Krymu. Gidrogeologiya i karstovedenie (On the characteristic and intensity of chemical denudation in Crimean Mountains), vip. 4:25–33

Dubljanskiy VN (1977) Karstovyje peshshery i sakhty Gornogo Kryma, Leningrad, Karst Caves and Shafts of the Crimean Mountains. Nauka, Leningrad, 181 p (In Russian)

Ershov ED (ed) (2002) General geocryology: manual for high school. Moscow University Press, Moscow, 682 p

Fels E (1929) Das Problem der Karbildung in den Ostalpen. Petermanns Mitt. Ergänzungsh 202:1–84

Ford DC, Williams PW (2007) Karst hidrology and geomorphology. Wiley, Chichester, 562 p

Francis P, Oppenheimer C (2004) Volcanoes, 2nd edn. Oxford University Press, Oxford, 521 p

Fülöp J (1989) Bevezetés Magyarország geológiájába (Introduction to Hungary's geology). Akadémia Kiadó, Budapest, 246 p (in Hungarian)

Fülöp J (1990) Geology of Hungary II (Paleosoic). Hungarian Geological Institute, Budapest, 297 p

Gams I (1977) Towards the terminology of the polje. Proceedings of the Int. Speleol. Congress Sheffield, England, pp 201–203

Gams I (2002) Changes of the Triglav Glacier in 1945–94 period in the light of climatic indicators. http://ai.ijs.si/mezi/personal/triglav/

Guðmundsson AT (1996) Volcanoes in iceland – 10.000 years of volcanic history. Vaka-Helgafell, Reykjavík, 136 p

Guðmundsson A (2012) The geology of Iceland. Springer, New York, 300 p

Guðmundsson AT, Kjartansson H (2007) Living earth – outline of the geology of Iceland. Mál og Menning, Reykjavík, 308 p

Gvozdetskiy NA (1981) Karst. Izd-vo Mysl', Moscow, 214 p

Haas J (2001) Geology of Hungary. Eötvös University Press, Budapest, 318 p

Haas J (2004) Triassic. Eötvös University Press, Budapest, 384 p

Herak M (1972) Karst of Yugoslavia. In: Stringfield VT, Herak M (eds) Karst important karst regions of the northern hemisphere. Elsevier Publishing Company, Amsterdam, pp 25–83

Herak M, Herak D, Markušič S (1996) Revision of the earthquake catalogue and seismicity of Croatia, 1908–1992. Terra Nova 8:86–94

Hevesi A (1978) A Bükk szerkezet- és felszínfejlődésének vázlata (A sketch of the structure and landform development of the Bükk Mountains). Földrajz Értesítő XXVII(2):169–203 (in Hungarian)

Hevesi A (1980) Adatok a Bükk-hegység negyedidőszaki ősföldrajzi képéhez (Data to the quarternary paleogeographical features of the Bükk mountains). Földtani Közlemények 110(3–4):25–36 (in Hungarian)

Hevesi A (2001) A Nyugat-Mecsek felszíni karsztosodásának kérdései (Questions of the surface karstification of the West Mecsek). Karsztfejlődés VI:103–111 (in Hungarian)

Holland C (1990) The Yangtze platform: a gateway to Chinese geology. Proc Geol Assoc 101(1):1–17

Hoyk E (2002) A Nyugat-Mecsek karszt dolináinak morfometriai vizsgálata (A morphometric investigation of the karst dolines of Western Mecsek). Karsztfejlődés VII:161–171 (in Hungarian)

Irimus IA, Petrea D, Vescan I, Toma CB, Vieru I (2011) Vulnerability of touristic geomorphosites in Transylvanian saliferous areas (Romania). GeoJournal Tour Geosites 8(2):212–218

Jakucs L (1977) Morphogenetics of Karst regions. Adam Hilgar, Bristol, 284 p

Jámbor Á (1980) A Dunántúli-középhegységPannóniai képződményei (The Pannonian rocks of the Transdanubian Mountains). Magyar Állami Földtani Intézet Évkönyve LXII, Budapest, 259 p (in Hungarian)

Jugovics L (1954) A Déli-Bakony és a Balaton-felvidék bazaltterületei (Basalt terrains of Southern Bakony and of Balaton Upland). Földtani Intézet Évi Jelentése 1953-ról:65–88, Budapest (in Hungarian)

Kezsek CS, Bally AW (2006) The Transylvanian Basin (Romania) and its relation to the Carpathian fold and thrust belt: insights in gravitational salt tectonics. Mar Pet Geol 23(4):405–442

Knez M, Slabe T (2009) Lithological characteristics, shape and rock relief of the Lunan stone forests. In: Ginés A, Knez M, Slabe T, Dreybrodt W (eds) Karst rock features. Karren Sculpturing, vol Carsologica, 9. Zalozba ZRC. Institut za raziskovanje krasa ZRC SAZU, Postojna-Ljubljana, pp 439–452

Konda J (1970) A Bakony hegységi Júra időszaki képződmények üledékföldtani vizsgálata (Lithologishe und Fazies-Untersuchung der Jura – Ablagerungen des Bakony-Gebirges). Ann Inst Geol Public Hung 50(2):161–260 (in Hungarian)

Korbar T (2009) Orogenic evolution of the external dinarides in the NE Adriatic region: a model constrained by tectonostratigraphy of Upper Cretaceous to Paleogene carbonates. Earth Sci Rev 96(4):296–312

Korpás L (1981) A Dunántúli-középhegység oligocén-alsó-miocén képződményei (Oligocene-lower miocene formations of the Transdanubian Central Mountains in Hungary). MÁFI Évkönyve, Budapest, 140 p (in Hungarian)

Korzhuev SS (1961) Merzlotnyi karst Srednego Prilen'ya i nekotorye osobennosti yego proyavleniya (The Middle-Lena frozen karst and its characteristics). In: Sokolov NI, Gvozdetskiy NA, Balashov LS (eds) Regionalnoe karstovedenie. Izdatelstvo AN SSSR, Moscow, pp 207–220

Korzhuev SS (1972) Drevniy karst i tsikly karstoobrazovaniya Sibirskoy platformy (A paleokarst and karstification cycles on the Siberian platform). Tr Mosk Obshch Ispyt Prir XLVII:141–151

Kovács S (1984) Tiszia-probléma és lemeztektonika-kritikai elemzés a koramezozoós fácieszónák eloszlása alapján (The tisia problem, the plate tectonic concept. Contributions based on the distribution of early mesozoic facies zones). Földtani Kutatás 27(1):55–72 (in Hungarian)

Kövér Sz, Fodor L, Kovács S (2006) Structural position and sedimentary connections of Jurassic formations of the Rudabánya Hills – an overview of old conceptions and a new working hypothesis. Annual Report of the Hungarian Geological Institute, pp 97–117

Kunaver J (1973) The high mountainous karst of the Julian Alps in the system of alpine karst. IGU European regional conference, Symposium on karstmorphogenesis, Papers, Budapest, pp 209–225

Kunaver J (1976) Kotlič – a specific depression form of subnival alpine karst. Proceedings of the 6th International Congress of Speleology Academia, Prague, pp 223–230

Kunaver J (1984) Geomorphology of the Kanin Mountains with special regard to the glaciokarst. Geografski Zbornik 22:197–346

Less G (1998) Földtani felépítés (Geological characterization). In: Boross G (ed) Az Aggteleki Nemzeti Park. Mezőgazda, Budapest, pp 26–66 (in Hungarian)

Lichtenecker N (1926) Die Rax. Geogr Jahresber Österr 13:150–170

Lóczy L (1913) A Balaton környékének geológiai képződményei és ezeknek vidékek szerinti telepedése (The geological features of the surroundings of Lake Balaton and their regional distribution). A Balaton tudományos tanulmányozásának eredményei, I. K. Budapest, 617 p (in Hungarian)

Lovász G (1971) Adatok az Abaligeti-karszt geomorfológiai és hidrológiai jellemzéséhez (Data for the description of the Abaligetien Karst). Földrajzi Értesítő XX(3):283–296 (in Hungarian)

Lungersgauzen GF (1966) Inflyuviy-osobyi geneticheskiy tip materikovykh obrazovaniy (Genetical types of characteristic landform features) Dokl. Dokl AN SSSR 171(3):690–693

Mandl G, Bryda G, Kreuss O, Moser M, Pavlik W (2002) Erstellung moderner geologischer Karten als Grundlage für karsthydrogeologische Spezialuntersuchungen im Hochschwabgebiet. Unpubl. final report to the Viennese Waterworks, Geologische Bundesanstalt, Wien, 225 p

Marovič M, Markovič M (1972) Glacial morphology of the Durmitor Mount Wider area (in Serbian). Geolski anali Balkanskog poloustrva, Belgrade XXXVII:37–48

Martonne Emm.de (1907) Recherches surl'èvalution morphologique des Alpes de Transylvanie (Karpates mèridionales). Rev. de gèogr. annuelle, I (1906–1907) 286 p

Mauer F (2000) Growth mode of middle Triassic carbonate platforms in the Western Dolomites (Southern Alps, Italy). Sediment Geol 134(3–4):275–286

Mckenzie JA, Vasconcelos C (2009) Dolomite Mountains and the origin of the dolomite rock of which they mainly consist: historical developments and new perspectives. Sedimentology 56:205–219

Miljush P (1978) Tectonic framework and evolution of the Dinarides. Tectonophysics 44(1–4):321–344

Móga J (2001) A szerkezet és kőzetfelépítés szerepe a Szilicei-fennsík karsztos felszínformáinak kialakításában (The role of structure and lithology in the formation of karstic landforms of the silice plateau). Karsztfejlődés VI:143–159 (in Hungarian)

Móga J (2002) Felszínalaktani vizsgálatok a Galyaság területén (Geomorphological investigations in the area of Galyaság). Karsztfejlődés VII:173–186 (in Hungarian)

Móga J (2004) Erdélyi-középhegység karszttípusai (Karst types of the Apuseni Mountains). Karsztfejlődés IX:229–250 (in Hungarian)

Noszky J (1964) A Bakony-hegység északi részének földtani vizsgálata (A Geological Investigation of the northern part of the Bakony Mountains). Földtani Intézet Évi Jelentése 1961-ről, Budapest, (1):203–206 (in Hungarian)

Obenholzner JH (1991) Triassic volcanogenic sediments from the Southern Alps (Italy, Austria, Yugoslavia) – a contribution to the Pietra Verde Problem. Sediment Geol 74:157–171

Orbán B (1868) Székelyföld leírása (Description of Székely Land). Pest I. köt., Róth Mór Bizománya (Repring: 1982, Helikon Kiadó és a Magyar Könyvkiadó és Könyvterjesztő Egyesület) (Description of Transylvania), 1529 p (in Hungarian)

Ozoray G (1963) Újkősó-szakadék Parajdon (A new rock-salt precipice in Parajd). Földrajzi Értesítő 12(2):239–241 (in Hungarian)

Pataki A (1983) Karsztmorfológiai megfigyelések a nyirádi és iharkúti bauxit-előfordulás területén (Karstmorphological Observations in the Bauxite occurrence area of Nyirád and Iharkút). MÁFI Évi Jelentés az 1983 évről 121–133 (in Hungarian)

Patrulius D, Bleahu I, Bordea S (1971) The Triassic formations of the Apuseni Mountains and of the East Carpathian Bend – znd. Triassic Colloq. Carp Balk Assoc. Bucharest. Guidebook to excursions 8:5–54

Pécsi M (1980) A Pannóniai-medence morfogenetikája (The morphogenetics of the Pannonian Basin). Földrajzi Értesítő XXIX(1):105–127 (in Hungarian)

Pelikán P (2005) Geology of the Bükk Mountains. Hungarian Geological Institute, Budapest, 279 p
Pirnat L (2002) Speleology on Kanin-Soški razgovari 1. Proceedings for Regional Studies of the History Section of the Cultural Association Golobar, Bovec, pp 77–98.
Plan L, Decker K (2006) Quantitative karst morphology of the Hochschwab plateau, Eastern Alps, Austria. Z Geomorphol Suppl 147:29–56
Plan L, Filipponi M, Behm M, Seebacher M, Jeatter P (2009) Constraints on alpine speleogenesis from cave morphology – a case study from the eastern totes gebirge (Northern Calcareous Alps, Austria). Geomorphology 106(1–2):118–129
Plan L, Renetzeder C, Pavuza R, Körner W (2012) A new karren feature: hummocky karren. Int J Speleol 41(1):75–81
Pulina M (2005) Le karst et les phenomenes karstiques similaires des regions froides. In: Salomon JN, et Pulina M (eds) Les karsts des regions climatiques extremes. Karstologia Mémoires 14, Presses Universitaires de Bordeaux, pp 11–100
Radulovic V, Radulovic M (1997) Karst of Montenegro. In: Stevanovic Z (ed) 100 years of hydrogeology in Yugoslavia, Belgrade, pp 147–185
Raincsák E (1980) A Várpalota – Iszkaszentgyörgy közötti triász vonulat szerkezete és felépítése (The Structure of the Triassic Ridge between Várpalota and Iskaszentgyörgy). Földtani Intézet Évi Jelentése 1978-ról, Budapest, pp 187–196 (in Hungarian)
Reutter KJ, Scheuber E, Chong G (1996) The Precordillera faultsystem of Chuquicamata, Northern Chile: evidence for reversals along arc-parallel strike-slip faults. Tectonophysics 259:213–228
Riquelme R, Hérail G, Martinod J, Charrier R, Darrozes J (2007) Late Cenozoic geomorphologic signal of Andean forearc deformation and tilting associated with the uplift and climate changes of the Southern Atacama Desert (26°S–28°S). Geomorphology 86(3–4):283–306
Roglič J (1965) The delimitation and morphological types of the Dinaric Karst. Naše Jame 7(1–2):11–20
Salomon JN (2009) The tsingy karrenfields of Madagascar. In: Ginés A, Knez M, Slabe T, Dreybrodt W (eds) Karst rock features, vol Carsologica, 9, Karren Sculpturing Zalozba ZRC. Institut za raziskovanje krasa ZRC SAZU, Postojna-Ljubljana, pp 411–422
Sanders C, Huismans R, Van Wees JD, Andriessen P (2002) The Neogene history of the Transylvanian basin in relation to its surrounding mountains. EGU Stephan Mueller Special Publication Series 3, pp 121–133
Sásdi L (1990) Az Aggtelek-Rudabányai-hegység karsztjának földtani fejlődéstörténete (Evolution of the Karst Of the Aggtelek-Rudabánya mountains). Karszt és Barlang I:3–8 (in Hungarian)
Sauro U (1995) The study of morphokarstic unit of Sette Comuni Plateau (Venet Fore-Alps): State of the art. Study Tridentini geomorfologici e paesaggi dell' altopiano dei Sette Comuni fra il Pleistocene finale e l'Olocene. In Atti del Convegno: Comune di Gallio, pp 13–26
Scarth A, Tanguy JC (2001) Volcanoes of Europe. Oxford University Press, New York, 243 p
Scheuber E, Andriessen PAM (1990) The kinematic and geodynamic significance of the Atacama Fault Zone, northern Chile. J Struct Geol 12:243–257
Schlüter T (2008) Geological atlas of Africa. Springer Verlag, Berlin, 308 p
Schwarzacher W (2005) The stratification and cyclicity of the Dachstein Limestone in Lofer, Leogang and Steinernes Meer (Northern Calcareous Alps, Austria). Sediment Geol 181(1–2):93–106
Šebela S (2012) Postojna–Planina cave system. Slovenia encyclopedia of caves 2nd edn. ZRC SAZU Karst Research Institute, Ljubljana - Postojna, pp 618–624
Smart P, Waltham AC, Yang M, Zhang Y-j (1986) Karst geomorphology of Western Guizhou. Cave Sci (BCRA) 13(3):89–113
Solomonov N, Kolosov P, Kipriyanova L, Knapp HD, Zhuravlev A, Trofimova E, Maksakovskiy N, Butorin A, Petrovskaya E (2010) Nominaciya Prirodiny Park 'Lenskie Stolby' (Rossziyszkaya Federaciya). http://www.nhpfund.ru/files/lena-pillars-nature-park-nomination-ru.pdf

Song L, Liang F (2009) Two important evolution models of Lunanshilin karst. In: Ginés A, Knez M, Slabe T, Dreybrodt W (eds) Karst rock features. Karren Sculpturing Zalozba ZRC. Carsologica, 9. Institut za raziskovanje krasa ZRC SAZU, Postojna, Ljubljana, Eslovènia, pp 453–459

Stegena L, Geczy B, Horvath F (1975) Late Cenozoic evolution of the Pannonian Basin. Tectonophysics 26:71–90

Sweeting MM (1995) Karst in China. Springer, Berlin, 264 p

Szabó PZ (1956) Magyarországi karsztformák klímatörténeti vonatkozásai (Relations between climate and karst formation in Hungary). Dunántúli Tudományos Gyűjtemény, Pécs, pp 183–189 (in Hungarian)

Szabó PZ (1966) Újabb adatok és megfigyelések a magyarországi őskarsztjelenségek ismeretéhez (New data and observations to the knowledge of Hungarian paleokarstic phenomena). Dunántúli Tudományos Gyűjtemény, Pécs, pp 65–102

Szabó PZ (1968) A magyarországi karsztosodás fejlődéstörténeti vázlata (The History of the Development of Hungarian Karstification). Dunántúli Tudományos Gyűjtemény, pp 65–102 (in Hungarian)

Szabó L (2008) Barlangfejlődés a Canin-fennsík mélyén (Cave development in the depth of canin plateau). Karsztfejlődés XIII:5–22 (in Hungarian)

Talbot CJ, Jackson MPA (1987) Sótektonika (salt/halite techtonics). Természet Világa X:56–65 (in Hungarian)

Telbisz T, Dragašice H, Nagy B (2005) A horvátországi Biokovo-hegység karsztmorfológiai jellemzése terepi megfigyelések és digitális domborzatelemzés alapján (Doline morphometric analysis and karst morphology of Biokovo Mt (Croatia) based on field observations and digital terrain analysis). Karsztfejlődés X:229–243 (in Hungarian)

Telbisz T, Mari L, Szabó L (2011) Geomorphological characteristics of the Italian side of Canin Massif (Julian Alps) using digital terrain analysis and field observations (Geomorfološke značilnosti italijanske strani Kaninskega Masiva (Julijske Alpe), ugotovljene z uporabo digitalnega modela reliefa in terenskim opazovanjem). Acta Carsologica Postojna 40(2):255–266

Thordarson T, Hoskuldsson A (2002) Iceland (Classic geology in Europe). Dunedin Academic Press, London, 200 p

Ustaszewski K, Schmid SM, Lugović B, Schuster R, Schaltegger U, Bernoulli D, Hottinger L, Kounov A, Fügenschuh B, Schefer S (2009) Late Cretaceous intra-oceanic magmatism in the internal Dinarides (northern Bosnia and Herzegovina, Implications for the collision of the Adriatic and European plates. Lithos 108(1–4):106–125

Vakhrushev VA (2009) Regionalization of karst of the Crimean Peninsula. Speleol Karstol 3:39–46

Van Husen D (2000) Geological processes during the Quaternary. Mitt Österr Geol Ges 92:135–156

Vangengejm EA (1977) Paleontologicheskoye obosnovaniye stratigrafii antropogena Severnoy Azii. Nauka, Moscow, 170 pp

Veress M (1983) Eltérő magasságú tönkfelszínek karsztosodásának kérdései az Északi-Bakony keleti részén (Questions of karstification of the planation surfaces of different heights on the eastern part of the Northern Bakony Mountains). Folia Mus Historico Nat Bakonyiensis 2:29–44 (in Hungarian)

Veress M (1992) Karsztmorfológiai sajátosságok a Pádis fedett karsztjának példáján (Karstmorphological characteristics using an example of the covered karst of Pádis). Földrajzi Közlemények CXVI(3–4):125–141 (in Hungarian)

Veress M (2008) Adalékok az Aggteleki-fennsík völgyeinek fejlődéséhez (Some data to the valley evolution of Aggtelek plateau). Karszt és Barlang I–II:3–12 (in Hungarian)

Veress M (2010) A magyarországi eltemetett és rejtett karsztos térszínek felszínfejlődése (Development of cryptokarstic and latent karstic surfaces in Hungary). Földrajzi Közlemények 134(4):373–391 (in Hungarian)

Veress M (2011) Adatok a Mecsek-hegység fedett karsztosodásához a Cigány földi mintaterületről vett példák felhasználásával (Some data to the covered karstification of the Mecsek Mountains using examples taken from the Cigány földi research area). Karszt és Barlang I–II:9–30 (in Hungarian)

Veress M (2012a) Fedőüledékes depressziók típusai és kialakulásuk (Types and development of depression of superficial deposit). Földrajzi Közlemények 136(1):2–21 (in Hungarian)

Veress M (2012b) Glacial erosion and karst evolution (Karren formation on the surface formed by glaciers). In: Veress B, Szigethy J (eds) Horizons in earth science research, vol 7. Nova, New York, pp 1–94

Veress M (2012c) New data on the development of the baradla cave. Acta Carsologica 42(2–3):193–204

Veress M, Zentai Z (2009) Karsztjelenségek minősítése a Bükk-hegység néhány mintaterületén a mészkőfekü morfológiájának és a fedőüledékek szerkezetének értékelésével (Classification of karst phenomena on a few sample areas of the Bükk Mountains by using morphology of the limestone floor and the structure of the sedimentary rock). Karszt és Barlang 2007 I–II:37–54 (in Hungarian)

Veress M, Lóczy D, Zentai Z, Tóth G, Schläffer R (2008) The origin of the Bemaraha tsingy (Madagascar). Int J Speleol 37(2):131–142

Veress M, Tóth G, Zentai Z, Schläffer R (2009) The Ankarana tsingy and its development. Carpathian J Earth Environ Sci 4(1):95–108

Veress M, Puskás J, Zentai Z, Benkó Z (2011) Development of karren formation on the Saltic Hill of Praid (Transílvanian Basin, Romania). Carpathian J Earth Environ Sci 6(2):183–194

Veress M, Péntek K, Unger Z, Almási I (2012) Development of covered karstic dolines in ground ice environment (Eastern Alps, Austria). interests of experimental and mathematical modelling. Z Geomorphol 56(Suppl 2):79–104

Vlahović I, Tišljar J, Velić I, Matičec D (2005) Evolution of the adriatic carbonate platform: palaeogeography, main events and depositional dynamics. Palaeogeogr Palaeoclimatol Palaeoecol 220:333–360

Yuan D (1980) A brief introduction to China's research in karst. Institute of Karst, Guilin, 16 p

Yuan D, Zhu D, Zhu X (1991) Karst of China. Geological Publishing House, Beijing, 158 p

Zabelin M (2003) Fiziko-himicheskiye mekhanizmy karstoobrazovaniya. http://geo.web.ru/~teveleb/karst/karst.html

Zámbó L (1970) A vörösagyagok és a felszíni karsztosodás kapcsolata az Aggteleki-karszt délnyugati részén (The relation of red clays and karstification in the south-eastern part of Aggtelek Mountains). Földrajzi Közlemények 94(4):281–293 (in Hungarian)

Zámbó L (1998) Felszínalaktani jellemzés (Surface morphological characterisation). In: Baross G (ed) Az Aggteleki Nemzeti Park. Mezőgazda, Budapest, pp 70–96 (in Hungarian)

Zentai Z (1994) A Parajdi sókarszt geomorfológiája (Geomorphology of the halite karst of Praid). A BDTF Tud Közl IX. Természettud 4:233–248 (in Hungarian)

Zhang Z (1980) Karst types in China. GeoJournal 4(6):541–570

Zivaljevič M, Vujučič P, Stijovič V (1989) Tumačza list Żabljak K. 34–37. Osnovna geološka karta 1:100000, Beograd

Zötl J (1963) Morpogenese des Ennstales. Mitt Naturwiss Ver Steiermark 93:155–160

Chapter 3
Methods

Abstract This chapter presents the goals and descriptions of the investigations per-formed and the description of the sites of field studies. Methods included morphometric analyses, study of bedrock structure for karst features, mapping (topographic, karst morphological, denudational), aerial imaging, preparation of cross-sections, oblique views and block diagrams. For the investigation of the cover, engine-driven helical borer, geophysical techniques were applied and exploration pits were excavated in order to establish its thickness, structure and type. Changes in karst landforms were measured (mass movements, changes of depth). At some localities the particle size, clay and carbonate contents of the cover were analysed. Model experiments were run to reveal interactions between grike development, subsidence doline formation, capillary rise, settling velocity and water level sinking as well as to collect data on the sedimentation properties of subsidence dolines.

Keywords Morphometry • Function • Bedrock structure • Topographic map • Groundplan • Morphological map • Aerial photograph • Cross-section • Oblique view • Block diagram • Geoelectric–geological profile

3.1 Morphometry

The morphometric analyses are directed at the quantitative description and comparison of karst landforms (primarily dolines and karren).

Williams (1971, 1972a, b) studied the frequency of occurrence of dolines; their average depth, shape, volume, density, alignment and elongation; and the relationships between the groundplan, depth and shape of dolines. Clark and Evans (1954) investigated the spatial distribution of dolines. Day (1977) calculated surface roughness, while Brook (1981) used mathematical models to simulate the development of roughness in tropical karst regions.

Functions describe the relationships between parameters referring to the size of dolines. Analysing these functions, genetic conclusions are drawn. Jennings (1975) described the relationships between the horizontal extension, depth and shape of the doline with functions. The functions differed with karst areas.

M. Veress, *Covered Karsts*, Springer Geology,
DOI 10.1007/978-94-017-7518-2_3

Among the properties of dolines, distribution by elevation was studied (Telbisz 2004). The longitudinal axis of dolines and its azimuth were determined (Jeness 2003; Farsang and Tóth 1992; Péntek et al. 2000). Using GIS methods, for the individual study areas, the frequency distribution of doline depth, the lognormal distribution of doline area (Telbisz et al. 2009), the frequency distribution of doline longitudinal axes and the doline neighbour directions were determined (Telbisz and Móga 2005). In addition, the catchment geometry, the 'shift' of the doline centre point from its geometrical centre (Telbisz et al. 2005) and average groundplan area of dolines in function of elevation above sea level were determined (Telbisz 2010). The ratio of total area to the areas of dolines in function of slope inclination was also computed (Lippmann et al. 2008).

Morphometric techniques have been applied to karren too. Researchers investigated the relationships between meander karren (Zeller 1967; Hutchinson 1996; Veress and Tóth 2004; Veress 2010), between parameters of trittkarren (Vincent 1983), between slope angle and the frequency of occurrence of karren (Crowther 1996) and the frequency distribution of grikes by width (Rose and Vincent 1983). Furthermore, the roughness of various karren types (Crowther 1997), the relationship between the length and vertical distribution of rillenkarren (Ginés 1996) as well as the rate of karren formation were studied (Veress 2010).

We used morphometric methods in the study of dolines and karren with the purpose of ensuring the quantitative comparability of landforms in different karst areas. The following morphometric methods were used: construction and analysis of parameter space, depth to width function, investigation of the distribution of the number of dolines by elevation above sea level and calculation of various parameters of karren features.

3.1.1 The Parameter Space

Here, doline size parameters (width, depth, the ratio of concave to convex slope segments describing slope geometry) were represented in space in order to draw conclusions to the variations of these parameters for dolines of different sizes within the same karst area. The planar representation of parameter spaces is possible if the same characteristic appears twice in the coordinate system. During planar representation, the doline data shown on the third axis of the coordinate system is virtually rotated onto the opposite side of the other two axes.

At first, the dolines were surveyed in detail, their large-scale (1–50) map was prepared and the areas enclosed by contour lines were determined. Functions were identified to depict the relationship between doline depth and the area enclosed by contour lines (Péntek et al. 2007). The following area functions were written for the dolines (Fig. 3.1).

$$t(x) = \pi \cdot \left[\frac{1}{M} \cdot \frac{\ln x}{L} \right]^{\frac{2}{K}} \quad (0 < x \leq L)$$

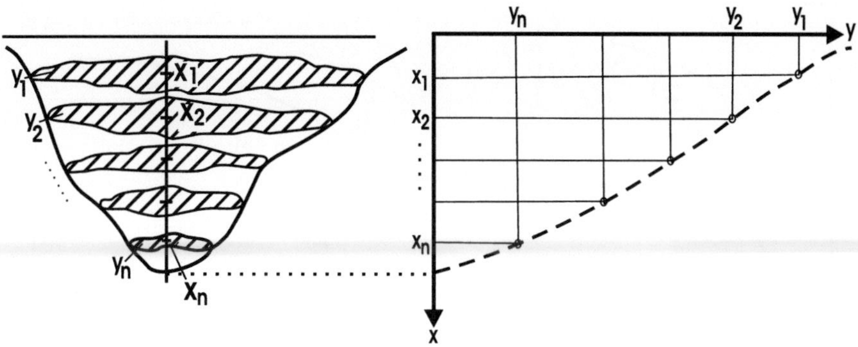

Fig. 3.1 The areal function (Péntek et al. 2007). *y* area enclosed by a contour line, *x* doline depth expressed by a contour line

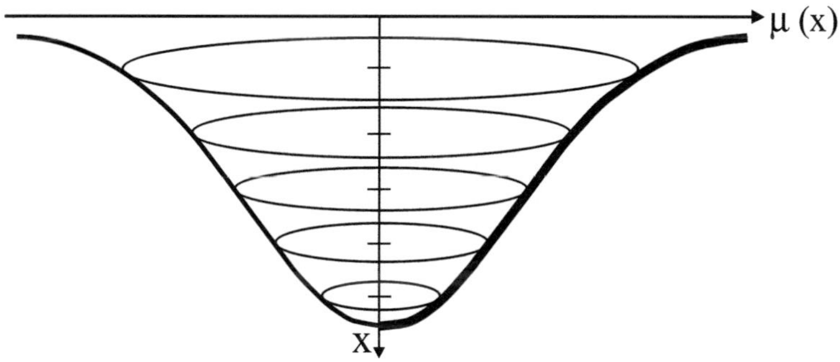

Fig. 3.2 The meridian function (Péntek et al. 2007)

where $L>0$, $M<0$ and $K\geq 1$ are characteristic parameters of the dolines. Next, an equal-volume transformation was performed to achieve rotational symmetry for the body. The resulting function for the side slope of the doline of rotational symmetry is called meridian function (Fig. 3.2) with the formula of

$$\mu(x)=\left[\frac{1}{M}\cdot\frac{\ln x}{L}\right]^{\frac{1}{K}} \quad (0<x\leq L)$$

The parameter $L>0$ refers to doline depth and $M<0$ to doline radius, which describes the horizontal extension of the doline (the lower it is, the more extensive is the form). $K\geq 1$ depends on the value of the inflexion point on the meridian function (Fig. 3.3). If K has a low value, the side slope of the doline consists of a long convex and a short concave segment; if *K* is high, the side slope is composed of a short convex and a long concave segment.

The parameters of the meridian functions of different dolines are represented in the parameter space of the coordinate system. The characteristics of subsidence

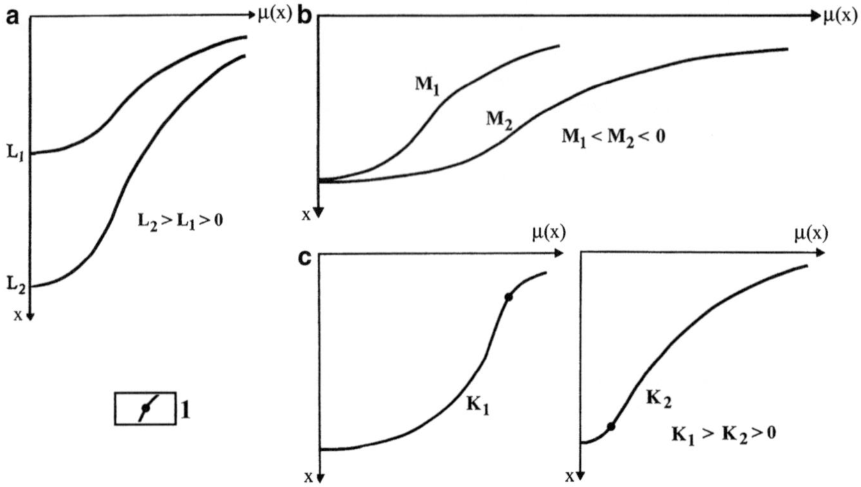

Fig. 3.3 The dependence of the meridian function on parameter *L* (**a**), on parameter *M* (**b**), on parameter *K* (**c**). *1*, inflexion point (Péntek et al. 2007), modified

dolines in the studied middle-mountain karsts (Bakony, Pádis) and glaciokarsts (Totes Gebirge, Julian Alps, Asiago Plateau) were represented separately in parameter spaces of the coordinate system.

3.1.2 Depth–Width Function

The function between doline depths and widths were also investigated. The steepness of the function indicates the relationship between vertical and horizontal growth of dolines. The functions received for various karst areas are comparable. Where the function is steeper, the same rate of deepening is accompanied with a higher rate of widening, where it is less steep, with lower rate of widening. If the functions for two areas differ in steepness, this indicates that the way of doline growth is different. Different manners of doline growth are due to variations in controlling factors (for instance, different denudation rates of doline slopes). Such data were collected and processed for the Pádis karst and the Bakony Mountains (Fig. 3.4).

3.1.3 Distribution of Dolines by Elevation Above Sea Level

We investigated the vertical distribution of dolines in the Bakony Mountains (Veress 1983). We identified elevation classes and referred the dolines into these classes. The frequency of dolines in the individual classes was expressed by the number of dolines.

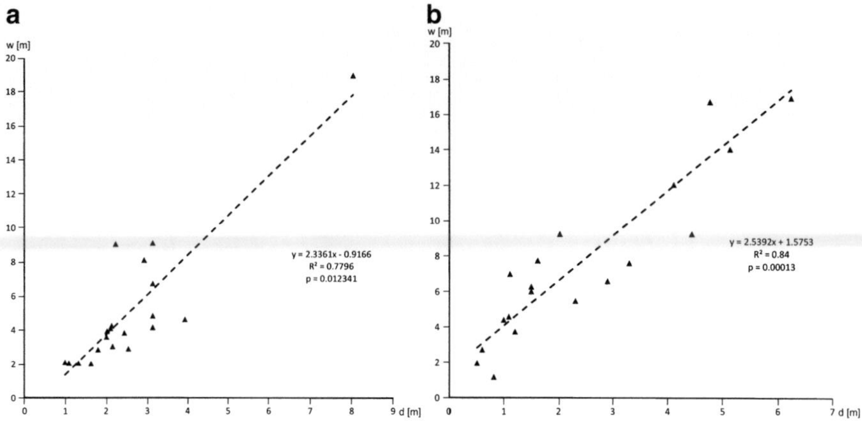

Fig. 3.4 Relationships between the depths (*d*) and diameters (*w*) of subsidence dolines in the Bakony (**a**) and on the Pádis (**b**)

The elevation of the highest frequency and the pattern and amplitude of distribution for individual areas are established. The highest frequency is represented by the class interval (surface elevation) where dolines occur in the highest number. Distribution pattern shows that moving away from the class of highest frequency what is the rate of decrease for doline frequency (doline number) in the ever lower and ever higher classes. If the rate of decrease is the same in both directions, the pattern is symmetrical; in other cases, it is asymmetrical. The amplitude of distribution is the elevation difference between the classes of highest and lowest elevations where dolines occur in a karst area.

3.1.4 Calculation of Karren Parameters

Karren formation on a terrain was characterised by various calculated parameters such as specific width, average composite character and dissection by karren (karren density) (Fig. 3.5). For the exposures, karren formation on the bedrock exposed from below the cover was calculated along profiles. The data necessary for calculations were determined from photographs. Therefore, dimensionless values were achieved. The averages of the above parameters were computed for study areas in various karst regions and compared them with each other.

Measuring or calculating specific width (Veress 2010), the degree to which a terrain is affected by karren formation is expressed. Along a profile, the widths of various karren features are measured. Specific width (*Sw*) is the average of the total width of karren features on 1 m distance of the profile. It is calculated by the following formula:

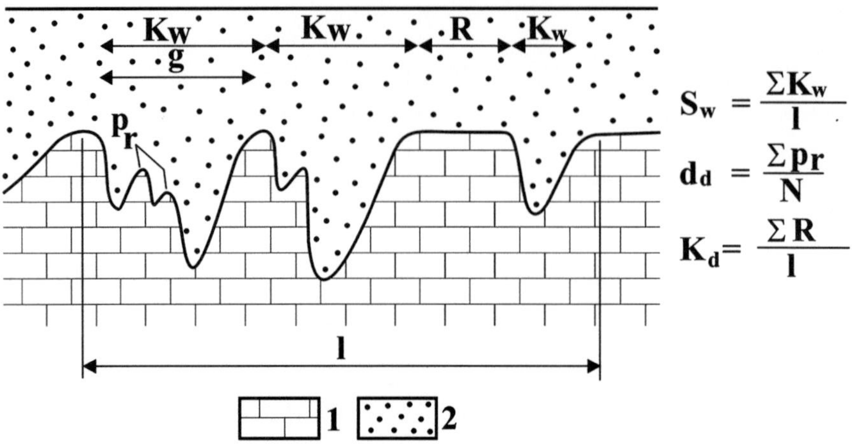

Fig. 3.5 Calculating karren parameters. *1*, limestone; *2*, cover; *g* grike; *pr*, crest between partial grikes; *Kw*, grike width; *R*, crest between grikes; *Sw*, specific width; d_d, average complexity; K_d, dissection with karren

$$Sw = \frac{\sum K_w}{l}$$

where

K_w is the width of a single karren feature.
ΣK_w is the total width of all karren features along the profile.
L is the profile length.

The higher the proportion of the surface is dissolved from the 1 m section of the profile, the higher is the specific width.

Average composite character (d_d) is obtained through dividing the total number of all mounds within the karren features (Σpr) by the number of grikes (N):

$$d_d = \frac{\sum pr}{N}$$

The higher is the d_d, the more composite in nature are the karren features of the profile.

Dissection by karren along the profile or karren density (K_d) is calculated through dividing the total number of ridges between the main karren features ($\sum R$) by the length of the profile (l):

$$K_d = \frac{\sum R}{l}$$

Values of the above-described parameters were calculated for a roadcut near Lunan (China), a stone quarry in Croatia, exposures along the road leading to the Pádis plateau and for karren exposed in the side of a doline of the area Pádis 2. For the calculation of the above parameters of uncovered karst terrain, data from karren on Totes Gebirge were used.

3.2 Investigation of the Structure of the Bearing Rock (Bedrock)

Our objective was to reveal the relationship between rock structure and covered karst formation. Rock structure (fractures and bedding planes) in relation to cavity formation in the phreatic or epiphreatic zone was studied by numerous authors (Glennie 1948; White 1960; Coase and Judson 1977; Palmer 1975; Quinlan 1981; Ford and Williams 2007). We were interested in the impact of rock structure on cavity formation in the epikarst (Veress 1982). The latter was also related to doline morphology. The investigations focused on the karst landforms of the Hárskút Basin and one of the dolines of the karst area around the Homód Valley.

Some typical dolines were selected, the shafts (cave) of which were accessible and the following parameters were measured:

- Dip slopes and directions of strata of the bearing rock.
- The number of fractures.
- The spatial position of shafts.
- Whether the shafts are continuous or composed of sections in different positions.
- The frequency of shafts.
- In addition, the groundplans (elongated or not) and cross-sections (symmetrical or asymmetrical) of the shape of the dolines belonging to the shafts were surveyed or mapped, and we distinguished dolines according to their occurrence as solitary forms or composed of partial dolines.

3.3 Mapping

3.3.1 Topographic and Groundplan Maps

Mapping serves the documentation of the form assemblages of karst regions, and the maps provide further data on the properties of covered karst formation. Topographic maps are often applied during the investigation of karst areas, including covered karsts (Roglič 1939; Crawford 1984; White 1988). The use of karst morphological maps is equally widespread (Warwick 1964; Hevesi 1980, 2001; Tan 1992; Racoviţă et al. 2002; Veress 2000; Ford and Williams 2007; Grimes 2012a).

We mapped various covered karst areas and landforms. In our research we surveyed the areas mostly to prepare large-scale (1–250 or 1–500) topographic maps. The available 1–10,000 scale maps were only seldom sufficient for our purposes. On the large-scale maps, karst features of small size (depth) could also be represented. Topographic maps were drawn for the following areas: the Bakony Mountains (Hárskút Basin, Tés Plateau, Kőris Mountain, Kab Mountain), parts of the Pádis plateau, the Bükk Mountains (Nagymező, Little Plateau), Durmitor (Mlječni do), the Totes Gebirge and the Dachstein Plateau. (Occasionally, individual dolines were surveyed. The contour maps of dolines were drawn at 1–50 scale.)

When mapping, karren groundplans were made at very detailed scale (1–20) using the network method (see below). Such maps were drawn on the pseudokarren features of the Asiago Plateau (Veress et al. 1999).

3.3.2 *Morphological Maps*

Supplemented with the mapping of landforms, the topographic maps could be developed into karst morphological maps. Such maps were drawn for the following mountains: the Bakony, Bükk, Aggtelek Karst, Durmitor, Pádis, Hochschwab, Totes Gebirge and Dachstein. Based on detailed surveys, morphological maps of dolines (e.g. from the Totes Gebirge) and doline groups (e.g. from Pádis) were prepared.

From the analyses of data from fieldwork, topographic maps (at 1:10,000 and 1:25,000 scales), the maps prepared by ourselves, karst morphological profiles (see below) and VES measurements, conceptual coveredness-karst morphological maps were drawn. On the latter, the site and type of coveredness, true rock boundaries (if relevant) as well as non-karstic and karstic landforms were represented through the generalisation of formations in individual characteristic karst areas. The enumerated elements of the map are not shown in the right position and size, but their fundamental configuration and interrelationships are emphasised.

3.3.3 *Denudation Maps*

For the reconstruction of the development of depressions of superficial deposit (DSDs), basement denudation maps had to be constructed. In order to analyse the process, the ways of denudation were typified on the floor of the depression selected as an example (Hochschwab), distinguished aerially and represented in extension (see Figs. 7.71 and 7.72). To the types, slope angles, obtained from the contour and slope inclination map of the floor, were rendered. In principle, the types of denudation are the following:

- Terrains denuded by pluvial erosion:

- With doline formation, terrains sloping towards the dolines emerge to the erosive effect of rainwater. This type comprises the terrains surrounding the dolines. Their extension is determined by delimiting the terrains sloping towards the dolines (Fig. 3.6a).
- The side slopes of gullies and valleys are also denuded by pluvial erosion (slope formation directly driven by linear erosion). The extension of such terrains is determined by delimiting the margins of ravines and valleys (Fig. 3.6b).
- As the gullies and valleys are incising, rainwater flows towards their margins, and the bordering terrains are also lowered by rainwater (slope formation indirectly driven by linear erosion). The extension of such terrains is delimited by the area of surfaces of 10–30° slope bordering gullies and valleys (Fig. 3.6c–e).

– Denudation surfaces of complex origin:

These surfaces emerge if gullies and valleys form on the terrains bordering the developing ponor (Fig. 3.6f). This terrain includes the side slopes of gullies and valleys and the terrains between them. Such a terrain is not found in the area of the studied DSD but existed before the ponor was infilled.

The floor of the depression is denuded by linear erosion along the watercourses. A terrain section without denudation was also identified where slope inclination is very low (0–10°). A more differentiated picture of the evolution of the depression is achieved by the fact that a lower accumulation surface and an upper accumulation surface are recognisable in the area. The lower surface is the area of the former ponor of less than 10° slope, while the upper is a 30–50° slope of the depression. From the upper slope section of the depression, sediments are redeposited here. Finally, the denuding slope of the depression is identified: an exhumed slope on limestone with more than 50° angle (Veress 2012).

3.3.4 Aerial Photographs

This technique is particularly useful for the investigation of the morphology of uncovered surfaces. The study of karst areas using aerial photographs was initiated by Williams (1971, 1972a, b). Recently, photographs were applied for the identification of subsidence dolines and measurement of their density (Currens et al. 2012). Radio-controlled model airplanes (UAVs) are also used for taking aerial photos because this technique is inexpensive and provides images from low elevation. Aerial photographs were taken with model airplanes, among others, by Jones et al. (2006) and Kalmár and Kozma (2012). We performed aerial photographing from a model airplane in the Totes Gebirge from several hundred metres of altitude.

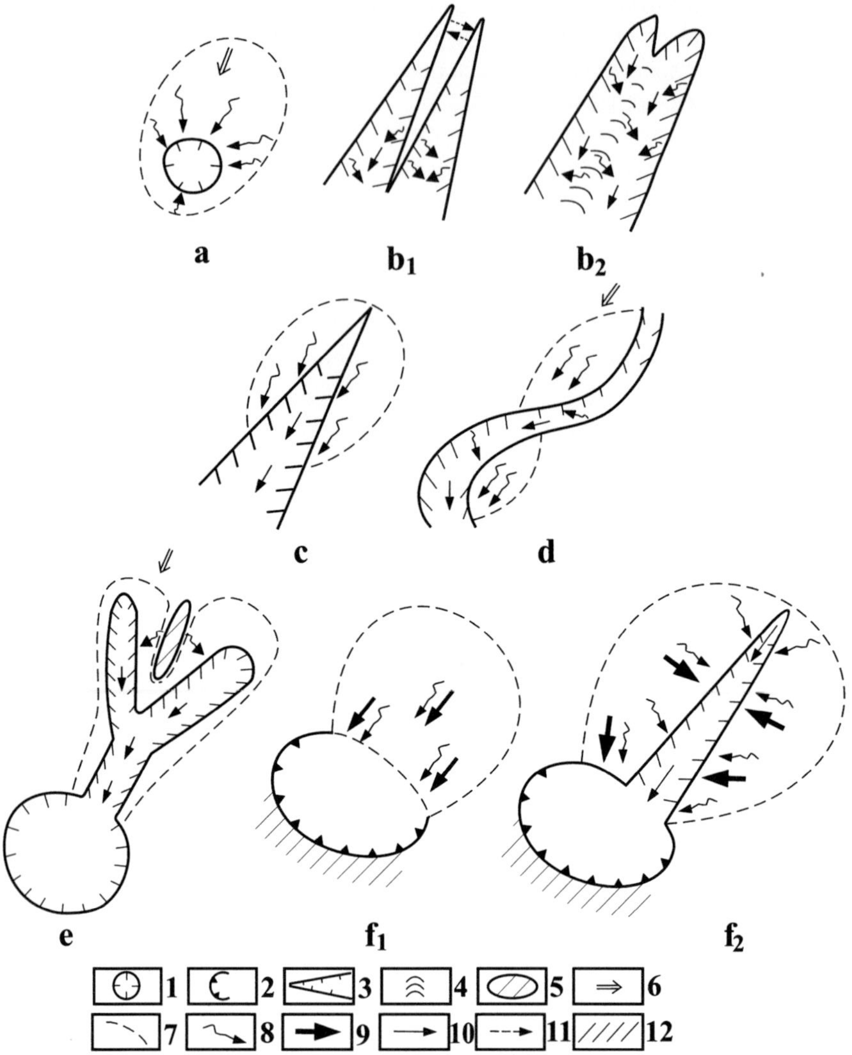

Fig. 3.6 Conceptual genetic types of pluvial erosion. (**a**) Slope formation by karst depressions. (**b**) slope formation directly caused by channels and valleys (**b_1**, erosion of valley sides; **b_2**, erosion of interfluvial ridges after valley margins are merged). (**c–d**) slope formation indirectly caused by channels and valleys: pluvial erosion active at the margin of the broadening valley (**c**), areal erosion at the margin of the meandering valley (**d**). (**e**) slope formation indirectly caused by channels and valleys (slope of residual surfaces affected by areal erosion situated between bordering forking channels). (**f**) complex slope evolution (**f_1**, areal erosion creating terrain sloping towards ponor; **f_2**, regressional valley evolution on this terrain, the directions of areal erosion are modified). *1*, subsidence doline; *2*, ponor; *3*, channel, valley; *4*, interfluvial ridge; *5*, residual terrain; *6*, initial surface slope; *7*, boundary of areally eroded terrain; *8*, areal erosion; *9*, slope formed due to areal erosion; *10*, linear erosion; *11*, valley broadening; *12*, limestone

3.3.5 *Karst Morphological Cross-Sections*

Karst morphological cross-sections representing the main properties were drawn on characteristic details of different karst regions. Conclusions can be drawn from the conceptual or proportional cross-sections on the type and evolution of karst. Karst morphological cross-sections are often used to characterise karsts. Conceptual karst morphological cross-sections were applied or presented – among others – by Wilford and Wall (1965), Williams (1966a), Sweeting (1973), Waltham et al. (1983), Jennings (1985), Ford and Williams (2007) and Grimes (2012a, b). Proportional karst morphological cross-sections were constructed by Sweeting (1958, 1966), Jennings and Sweeting (1963), (1980), Palmer (2004) and Veress (2009).

On the cross-sections we showed coveredness, the nature (permeable or impermeable) and extension of the cover, the bedrock, the various denudation surfaces, the landforms related to them and, if necessary, the hydrological conditions. Bedrock (limestone) outcrops were represented based on data from geological maps, field observations and in some cases from VES measurements. At sites with cover deposit, the morphology of the bedrock was also reconstructed from VES measurements or using geoelectric–geological profiles drawn from the data.

The here appearing sections are partly conceptual (for Kab Mountain) and partly proportional (Tés Plateau, southern Durmitor). In the case of this latter section type, the macroforms (e.g. paleodolines) are mostly proportionally depicted to a lower diameter limit of 200–300 m. Dolines smaller than this limit (e.g. subsidence dolines), however, cannot be shown true to scale. Exceptions are the large-scale cross-sections, where each and every karst features are represented in the true localities. Even on small-scale cross-sections, the solitary features (such as ponors) also appear in their actual localities. The small features occurring in groups (subsidence dolines) are represented by a symbol where they are typical. They are not shown proportionally and in the right spot; their number on the cross-sections is lower than in reality. Disregarding some sections at very large scale (Mester-Hajag, Tés Plateau), features in a strip are shown instead of those along a line – on both conceptual and proportional cross-sections. The sections only show those sections of karst regions which are relevant to the studied topic: the sections which constitute a single system with respect to their evolution. At the same time, several cross-sections were drawn on the karst region if they are representative of the different units of the karst region with regard to geomorphic evolution (e.g. for the Bakony Mountains). Some cross-sections are composed of segments of different alignment (Pádis). Such a representation of a detail is useful if it supplies the most information.

Such cross-sections were drawn for the Bakony Mountains (Tés Plateau, Mester-Hajag, Kab Mountain), Mecsek Mountains, the cliff coast of Madagascar, Durmitor, Totes Gebirge, Aggtelek Karst and Dachstein.

3.3.6 *Oblique Views and Block Diagrams*

Mostly to demonstrate single objects or groups of objects as well as their evolution, oblique views, series of views and block diagrams were constructed if it seemed to the purpose. Block diagram representations of karst are widespread in literature (for instance, Williams 1972a; Lehmann 1936; Jennings 1985 applied this option for representation), but oblique views can also be found (Wissmann 1954). Such representations were made of the Totes Gebirge, Dachstein and Pádis karsts. We also show numerous general landform types and processes in such figures.

3.4 Investigation of Cover Thickness, Composition and Structure

From the properties of the cover, conclusions can be drawn on the development of karst landforms, the process of karstification and on accompanying phenomena. On covered karst, measurements of cover thickness supply data for bedrock morphology (and thus on the size and nature of karst features in the cover); while establishing cover composition (and partly thickness), we are informed about karst evolution as well as the processes active on the karst. Cover thickness was studied using engine-driven spiral drill and metal rod and geophysical methods, while cover type and structure were investigated by geophysical methods and in exploration pits.

3.4.1 *Establishing Cover Thickness, Composition and Bedrock Map by Spiral Drilling*

Engine-driven spiral drill was applied in the Bakony Mountains (in the area of Mester-Hajag and on the catchment of the covered karst ponor G-6/b). At the latter site, it was not only used to establish cover thickness, but the presence of groundwater in the cover (collected in the drill hole) could be pointed out and, thus, a particular activity occurring at covered karst landforms, hidden activity, could also be identified (see Chap. 6). Taking advantage of the drilling data, bedrock maps were drawn, and geological profiles were constructed across the cover.

3.4.2 *Measuring Cover Thickness Using Metal Rod*

Metal rod was applied on a partially covered karren terrain on the Asiago Plateau. It could be used because the cover was not very thick (less than 2 m) and non-cohesive (redeposited soil and plant residue, needles of conifers). Thus, the rod could be pressed manually into the cover. The features recognised on the cover were surveyed in a grid (producing a groundplan). The mesh size of the grid was 20×20 cm. The grid was placed on the survey area (karst features). For mapping the marginal points of the feature were identified according to their positions in the grid cells or to their distance from one of the grid points. On the margins of landforms, cover thickness was also measured while pressing the metal rod down to the surface of the bedrock. Noting the length of the section of the rod below the ground surface, the thickness of the cover could be established.

3.4.3 *Constructing Geoelectric–Geological Profiles*

Applying this method, it is not only cover thickness and thus bedrock morphology but also the type of cover deposits that can be established.

There are numerous geophysical techniques known to measure cover and bedrock thickness: the seismic method (McCann et al. 1987; Hoover 2003), electric resistance (McDowell et al. 2002; Zhou et al. 2002) and electromagnetic (Jansen et al. 1993) and radar (McCann et al. 1987; Hoover 2003) techniques. A variety of vertical resistance measurement is VES measurement used by us (Veress 2009). The data obtained are used for making karst morphological profiles. Measurements were made in the Bakony (Mester-Hajag, Homód Valley, Kőris Mountain, at several sites of the Tés Plateau), Mecsek, Bükk (Nagymező, in a valley of the Little Plateau), Aggtelek Karst (Dász doline, Keserű-tó lápa, Zombor-lyuk, Bába Valley), Pádis (five sites), Julian Alps (Seven Lakes Valley), Totes Gebirge, Dachstein Plateau and Asiago Plateau.

Topographic maps were made of the study areas. VES measurements (vertical electric sounding) served to measure bedrock depth and cover thickness at different sites. In VES measurements through two earthed electrodes, electric current is conducted underground, and two other electrodes are used to measure the potential difference caused by the resulting electricity distribution. The distribution of electricity and the potential difference and the calculated apparent specific resistance are a function of the specific resistance of the individual layers and their thicknesses. From the measured values of potential difference, in function of the distance of current electrodes, curves can be constructed. Using an inversion programme, in ideal case, from the curves bedrock depth, the type of cover deposits and their thicknesses can be identified (Veress 2009).

Fitting the sediment series calculated for the individual sites together, profiles can be constructed along the alignment of measurements (geoelectric–geological profile), which represent the ground surface (with covered karst depressions), the limestone bedrock, the nature of sediment series, the position of series boundaries (including those of cover deposits), the structure of cover deposits as well as the calculated resistance values of the different rocks. The position of the limestone bedrock can be supplemented or rectified on spots where the limestone outcrops. Using elevation data for the limestone, topographic maps of the limestone bedrock can be drawn.

From the geoelectric–geological profiles, the following data on dolines are established:

- The position of the depressions relative to the bedrock topography.
- The thickness of cover deposit was given for the environs of the depression. The thickness of cover deposit was identified in two ways. Minimum cover deposit thickness (or inner thickness) was measured between the deepest point of the depression and the bedrock surface. Maximum cover deposit thickness (or outer thickness) was determined: two marginal points of the depression (along the profile) was connected with a straight line and then the distance between this line and the bedrock was measured across the deepest point of the depression. From the data, the average thickness of the inner and outer cover deposit for various covered karst dolines was calculated (above a mound of the bedrock surface, its slope or depression) in the Bakony Mountains (Veress 2009).

The structure of the cover was also investigated with multi-electrode measurements in the study area of the Homód Valley in order to establish the change in the pore volume of the cover.

With the multi-electrode geoelectric technique, two of the electrodes among the many (in our case 60) stuck into the cover deposit are for current input, while two others function as potential electrode. The current input and the potential-measuring electrodes are continuously exchanged. Eventually, measurements along profile (showing the horizontal variability of specific resistance) and soundings (depicting vertical variability of specific resistance) can be performed without moving the electrodes during measurement. The outcome of the measurement is a so-called pseudo-profile under the measurement line. Applying a suitable inverting program me, a profile of the distribution of specific resistance can be constructed from this profile. Down to some metres depth, the detailed and continuous resistance 'structure' of the cover deposits can be determined. Recording deviations in resistance, the moisture content of cover deposits and, indirectly, the changes in their pore volume values are measured. The changes in pore volume are represented along a profile.

3.4.4 Analysis of the Composition of the Cover

3.4.4.1 Particle Size

Particle size distribution was identified for the cover in a subsidence doline of a paleodoline of the Hochschwab and in the environs of dolines in the Bakony Mountains (Tés Plateau) in order to conclude on the water conductivity of the cover.

3.4.4.2 Calcium Content

Because of the calcareous cover infiltrating waters may be completely or partly saturated. Therefore, the karstification of the bedrock is influenced or occasionally even prevented by the carbonate content of the bedrock (Williams 1966b; Trudgill

1985). From the carbonate contents of the cover in different karst regions, conclusions can be drawn to the reasons behind the differences in karstification. If the carbonate content of the cover is high, there are hardly any subsidence dolines, while if it is low, there are more of them. For comparison the carbonate content of loess was determined in two areas of the Tés Plateau (Bakony Mountains), from samples collected in the environs of the dolines I-28 (eight samples) and I-31 (four samples). The data were published by Füzesi (2007). The carbonate content of the cover of the mentioned subsidence doline in the paleouvala of the Hochschwab was determined The carbonate content of the loess on the Mecsek karst was borrowed from literature (Hevesi 2001). On the Tés Plateau, the depth of sampling ranged from 0 to 200 cm. Carbonate contents were determined by the Scheibler method: the sample was treated with hydrochloric acid, and the amount of CO_2 released was measured by calcimeter. This allowed the determination of total carbonate expressed in percentage of mass.

3.4.4.3 Exploration Pit

The exposure of sediments in karst landforms is important to reveal their development and evolution. Exposures are mostly made in solution dolines (Aubert 1966; Hall 1976). We established exploration pits in the fills of subsidence dolines and exposed the cover series for direct observation. Exploration pits were dug in the study area of the Homód Valley in the Bakony Mountains, in doline Ho-8 and fossil doline D-14 as well as in the dolines of the Hárskút Basin study area (also Bakony Mountains) (e.g. in dolines Hu-1 and Gy-9).

From the cover thickness above the buried objects in the doline fill, the rate of doline infilling is also established if the time of occupying their present positions can be estimated. Thus, in doline Hu-1 of the Hárskút Basin, under 70 cm of fill, a tin manufactured at the earliest in the 1950s and that got into the doline in that time (with some years of error) was found in 1980. The rate of infilling of this doline can be established in knowledge of the burial of the tin and the thickness of cover. The time of burial is finding date minus burial date. Burial date was assumed to be the earliest date of manufacturing.

3.5 Changes of Karst Features

Pointing out changes, the processes and characteristics of covered karst can be revealed. On covered karst changes take place at relatively high rates (see Chap. 6). The investigations which are related to intense rainfalls have been presented in detail by Veress (2013).

Changes are demonstrated by measurements, observations and photographing. Qualitative changes are well indicated by vegetation, the alterations of which are detected by observations of uprooted, crooked trees, deformations of roots or plant

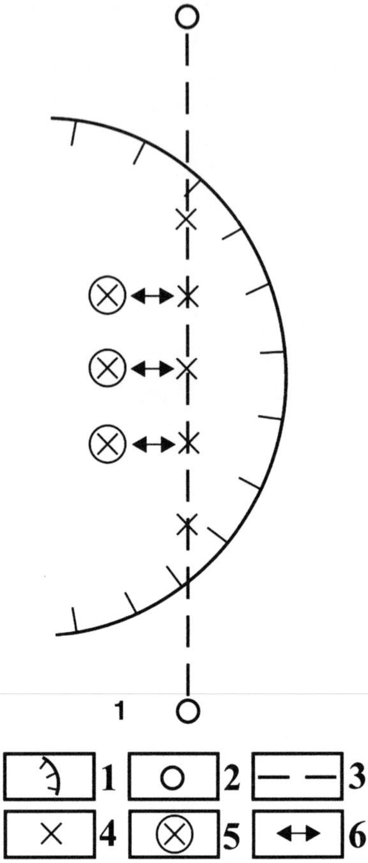

Fig. 3.7 Concept of measuring mass movements. *1*, doline margin; *2*, outer fixed point; *3*, direction between two fixed points determined by instrument; *4*, spikes placed in the doline side; *5*, sites with spikes displaced due to mass movement; *6*, extent of mass movement represented by the spikes if the movement is perpendicular to the marked direction

ruins due to displacement. The former refer to mass movements on the side slopes of depressions, while the latter represent doline formation (see Chap. 6). With the help of the latter, the duration of change is also estimated, for instance, if arable crops or grassland is found on the doline floor, the collapse which caused the above change must have taken place in the growing season of the year of observation. Human activities can also be helpful in dating: for instance, if the doline was formed in arable soil, its development is dated to times after the last ploughing. Buried objects or trees buried to their foliage could point to doline infilling.

Through observations, the intermittent lakes and overflow phenomena deriving from water recharge to the dolines can be observed. At stagnant water level, plant waste coatings or zones are created on doline slopes, and on objects in the doline (e.g. on trees), coatings or coating rings develop. Photos were taken repeatedly from the same sites. The differences between the photos show the changes which happened between the two dates.

3.5.1 *Measurements of Mass Movements in Dolines*

The measurement of mass movements (Veress 2000) was partly aimed at monitoring changes in the dolines and partly their material turnover.

Two fixed points were established on the terrain bordering the dolines by setting iron spikes into the ground. At one of the points, we set up a theodolite and targeted the other fixed point. Then we placed iron spikes in the doline side along the designated direction (start direction) (Fig. 3.7). The measurements were repeated once or twice a year: the start direction was reconstructed using the outer and the other fixed points. Then we measured the distances between spikes in the side of the doline related to the start direction. If the value between the spike and the line of the start direction was more than 0 cm, it means that the soil with the spike was displaced. Thus, these values are indicative of the displacement of cover deposit from the side of the doline to its bottom. The displacement of the spike at a certain date equals total displacement since the beginning of the measurement. From the difference between data obtained at two measurements of different date, the displacement between the dates of the two measurements can be calculated. The rate of displacement can also be calculated. The measured value will be equal to the actual mass movement if the start direction is perpendicular to the direction of material displacement. If the angle is lower, the measured value will be lower.

Mass movement measurements have been performed for 10 years for three dolines in the Hárskút Basin as well as in the Homód Valley and on Kab Mountain for two dolines each.

3.5.2 *Measuring Size Changes of Dolines*

The primary goal was the establishment of the rate of depth change.

3.5.2.1 Point-Like Depth Change

Here, the depth change of a single (the deepest) site of the doline is determined.

Between 2003 and 2006, the covered karst subsidence dolines of the Bakony and Bükk mountains as well as some solution dolines of the Aggtelek Karst were mapped. However, mapping was executed in various karst regions and at various dates. Further, these mapping dates are identified as the starting dates of measurement. In 2010 we measured the maximum depths of the mapped dolines using GPS. From the difference between the maximum depth obtained from the map and the depth data of the 2010 GPS measurements, the depth changes of the studied dolines were determined for the period between the start of measurements and 2010. Since in 2010 depth was only measured at a single site, information is only obtained on a single point and the depth change for the deepest point. This way from the data for the start of measurements and for 2010, we determined the maximum value of doline deepening (if the 2010 value is the higher) or infilling (if the 2010 value is the lower). From deepening/infilling between the dates of measurements, the average annual rate of deepening/infilling was also calculated. The rates of depth change were compared for dolines with various environs (grassland, arable land, forest) (Veress 2013).

The changes of some dolines on the Tés Plateau with known morphology due to the 2010 rains (mainly in September) were recorded on the occasion of the field work of October 8, 2010 (the changes in the forms of the dolines were recorded annually.) According to our GPS, measurements show that there were both deepening and infilling dolines among them. According to observations of October 8, 2010, in some of both the deepening and infilling dolines, denudational and accumulational features occurred. The denudational features indicate material transport from the doline floor into the karst, while the accumulational features point to deposition on the doline floor. For deepening dolines, material input into the doline is of lower extent than transport from there into the karst, but for infilling dolines input is of higher extent.

Regarding jointly the features and depth changes of the studied dolines, the dolines were classified by their material turnover. The bases of classification were the following. With deepening dolines, material transport into the karst is higher than transport into the doline, while it is the opposite for infilling dolines. With deepening dolines, the features indicating outward transport show that material is transported from the doline into the karst, while the lack of accumulational features points to no sediment input into the doline, or if there is, it cannot accumulate on the floor. If there are accumulation features in deepening dolines, they must have formed before outward transport. With infilling dolines, if features due to material removal are present, there is material transport into the karst, but its rate is lower than transport into the doline. If there are no features of material removal, there happened to be no transport from the doline into the karst, or it was interrupted for the time of infilling.

3.5.2.2 Areal Depth Change

Here depth change is determined for the entire area of the depression. At the depression I-31 (Tés Plateau), the detailed survey performed in 2004 was repeated in 2010. From the survey, data contour maps of the depression were drawn (The contour interval was 30 cm). The difference between the 2010 and 2004 contours on maps was digitally derived. (The 2004 contours were subtracted from the 2010 contours.) The map of depth change was produced for the depression. In those sections of the doline where the contour lines have negative values (i.e. the depth values in 2010 were lower), deepening in the area enclosed by the contours occurred; where the values are positive (the depth values in 2010 were higher), the floor was elevated, hence infilling occurred.

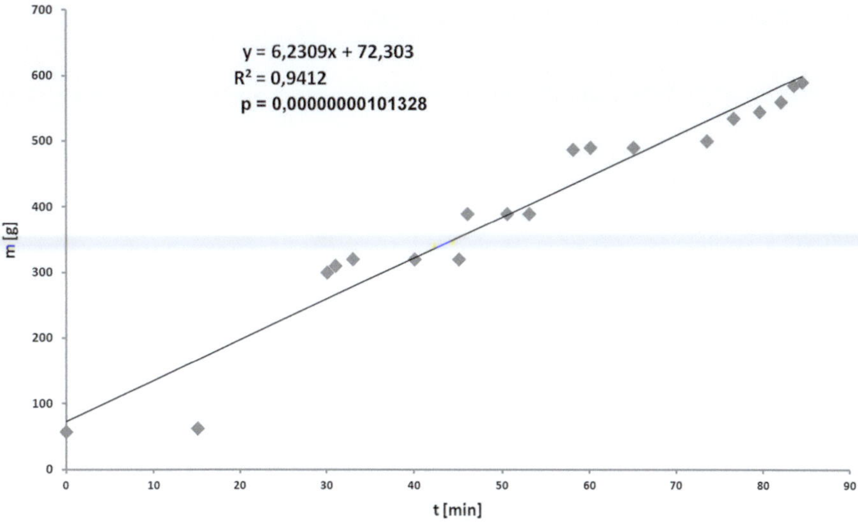

Fig. 3.8 Function showing the quantity of fallen particles versus time due to heating (Veress et al. 2012); *m* mass of the fallen pieces

Fig. 3.9 Solution experiment of grikes filled with superficial deposit in the laboratory

3.6 Laboratory Model Experiments

Laboratory experiments are important in karst research since the simulated phenomena take place rapidly and the conditions can be defined and changed. Rudnicki (1960), Curl (1966), Allen (1972), Quinif (1973), Fabre and Nicod (1982), Veress et al. (2012) and Deák et al. (2013, 2014) investigated various karst phenomena using model experiments. Model experiments on plaster were carried out by Glew and Ford (1980), Dzulynski et al. (1988), Veress et al. (1998, 2014) and Slabe (2005, 2009), because plaster is easily soluble and this creates favourable conditions for the generation of various phenomena.

We executed model experiments in four topics: the development of grikes filled with cover deposit, doline formation on cover with ground ice, formation of subsidence dolines on plaster plate covered with deposit without ground ice and sedimentation in flood lakes created in laboratory. On cover with plaster, the role of particle size of the cover in water current and in the nature of water flow onto the bedrock was investigated. In relation to the flood lake sedimentation, the influence of the rate of water level sinking on the nature of deposition and the relations between settling particle size and the rate of water level sinking were investigated.

3.6.1 Modelling the Formation of Subsidence Dolines on Cover with Ground Ice

The changes induced in the cover by the air ascending and melting ice in the cover were investigated.

Frozen samples of different particle size were placed into the experimental equipment. Air of varying temperature and current velocity was blown onto the samples from below. The sample melted, and fragments were detached from it and fell upon a balance allowing to measure the mass of fragments in function of time (Fig. 3.8). It can be seen that with increasing duration of air flow, more and more material drops as a result of partial melting (Veress et al. 2012).

3.6.2 Modelling Grike Evolution

We investigated the evolution of grikes filled with cover deposit which have their floors above the karst water table and where the karst water table is at the level of the grike floor.

To study grike evolution under cover deposits, grikes with vertical walls (their lower end was closed) were created on a plaster plate of 5° slope (Fig. 3.9). Such imitated grikes were made on two plaster plates, which were dried by air and framed by polystyrene. The grike floors on one of the plaster plates were not impermeable.

(Here, plaster on the grike floor was 1 cm thick. On this plaster plate three grikes were formed and coded as XIII.1, XIII.2 and XIII.3.) Onto the grike floors of the other plaster plate, a clay coating was applied to make the plaster impermeable. (Also on this plaster plate, three grikes were formed and coded as XIV.1, XIV.2 and XIV.3.) The length of grikes was 45 cm, while their widths ranged from 21 to 29.3 mm. The depths of grikes with impermeable floor ranged from 14 to 25 mm, while the depths of those where plaster was placed on clay to prevent the presence of impermeable material on the grike floor ranged from 16 to 22 m. Water feeding took place at the upper part of the grikes in the plaster plates onto the cover. The rate of dripping was 50 cm^3/min. Into each grike 100 dm^3 of water was fed over 10 h daily (only on working days). The three grikes on each plate were filled with cover deposit of the following particle sizes: 5–2.5 mm, 1–0.5 mm and 0.125–0.063 mm. The solution experiment was not repeated (Veress et al. 2014).

3.6.3 *Modelling the Development of Subsidence Dolines in Cover Deposit Without Ground Ice*

It was investigated how the material deficit generated on the bedrock is inherited over to the cover and what factors influence the process.

For the experiments gypsum plates of 45 × 30 × 3 cm size were made. The gypsum plates were dried on air and framed with polystyrene. The solution experiment was not repeated since it requires much time, and the conditions of this experiment could not be precisely reconstructed. The experiments were executed on covers of 1 cm (on nine gypsum plates) and 5 cm thickness (on two gypsum plates). (The experiment and the water supply were applied at two sites for each of the plates.) The particle size classes of the cover placed on the gypsum plates were the following: 2.5–5 mm, 2.0–2.5 mm, 0.5–1.0 mm, 0.25–0.5 mm, 0.125–0.25 mm, 0.063–0.125 mm and smaller than 0.063 mm. The slope of the gypsum plate was set to 5° or 0° in the experiments.

During the experiments 100 dm^3 of distilled water was dripped at each site. Water amounts of 5 dm^3 per day were fed to each site. In the morning 2 dm^3 and in the evening 3 dm^3 of water was filled into the vessel used for dripping. Water feeding only happened on working days. There were 2–3 h breaks during the day and 3–4 h breaks in the night. For water feeding burettes were used. Water was dripped at 80 cm^3/min rate. Drop sizes ranged from 0.1 to 0.075 cm^3. The drop size was calculated by measuring how many drops amounted to 1 cm^3 volume, and the drop size was the ratio of 1 cm^3 to the number of drops. Water feeding had various ways: onto the cover or directly onto the bedrock. In both cases it could take place both vertically and laterally. In the case of vertical water feeding onto the bedrock, the burette crossed the cover. If water was applied laterally, the water was conducted from a funnel through a glass pipe onto the cover or bedrock.

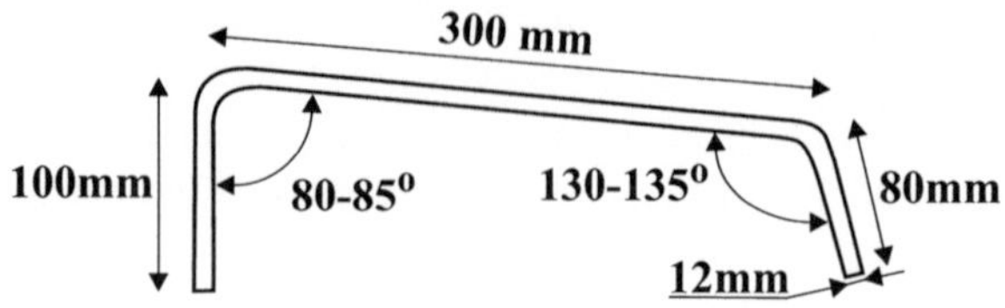

Fig. 3.10 Experimental equipment for water overlifting

3.6.4 Influence of Particle Size on Water Lifting and Water Overlifting

The changes observed on the cover of the gypsum plate during water feeding pointed to the decisive influence of the particle size of the cover deposit on water motion in the cover, the size of the emerging water body, its changes and water transfer from the cover onto the bedrock. Therefore, we investigated the influence of the particle size of the cover on water overlifting and water lifting and their rates.

3.6.4.1 Water Overlifting

A curved glass pipe was used. Its length was 480 mm, its inner diameter was 12 mm. The length of its ascending limb was 100 mm, its lateral conduction length was 300 mm and the length of its descending limb was 80 mm (Fig. 3.10). The angle between the lateral conductor and the ascending limb was 80–85°, and its angle with the descending limb was 130–135°. The glass pipe was filled with sediment of 0.125–0.350 mm and then 0.5–1.00 mm particle size dried at 105 °C. The glass pipe was set so that the lateral conductor had an angle of 5° with the horizontal. The ends of the pipes were closed by filter paper. The ascending limb was put to 2 cm depth into a water-filled vessel in vertical position. The vessel stored 30 cm^3 water. At the other end of the pipe, a receiving vessel was placed where filter paper indicated the appearance of the first drop of water. The time of appearance of the first drop and the time of transfer of 10 cm^3 of water were measured. The water overlifting experiment was meant to answer the question whether water is capable of rising in the ascending limb and at what particle size, i.e. whether the process depends on particle size.

3.6.4.2 Water Lifting

A glass pipe of 1,050 mm length and 12 mm inner diameter was filled with sediments of different particle size classes: 0.001–0.063 mm, 0.063–0.125 mm, 0.125–0.250 mm and 2.5–5.000 mm. The sediment-filled pipe was placed into the water-storing vessel in vertical position. With different sediment fills, the rate of water rise was measured during various time intervals (1 h, 5 h, 24 h and infinite time). The question was: what is the rate of water rise in sediments of different particle size?

Table 3.1 Measurement data in case of suspension A_1 (ek)

Sample number	Time (min)	Logarithm of time	Measured mass in 5 cm³ (mg)	Mass of L for 1 dm³ (mg)	Mass of lk for 1 dm³ (mg)
1	1	0	9.80	1.96	0.04
2	5	1.60	8.00	1.60	0.40
3	15	2.70	6.30	1.26	0.74
4	30	3.40	5.50	1.10	0.90
5	60	4.09	4.20	0.84	1.16
6	120	4.78	3.70	0.74	1.26
7	180	5.19	3.00	0.60	1.40
8	360	5.88	2.50	0.50	1.50
9	600	6.39	1.70	0.34	1.66

Remark:
L, suspended sediment of suspension taken by pipette in the given moment
ek, constant (2 g clay/1 dm³ water), initial concentration of suspension
lk: concentration of suspended matter in the glass cylinder at the give depth, which is the difference between the initial concentration of the suspension and actual concentration

3.6.5 *Sedimentation in the Flood Lakes of Subsidence Dolines*

At intensive snowmelt or rainfall, intermittent lakes emerged in the karst landforms. In the flood lakes the amount of suspended sediment deposited from the lakes increased. Deposition was studied in a basin in the laboratory. Using experience from artificial settling and field observations, conclusions can be drawn for the environments of lakes in the depressions, including the denudation properties of the environment (hinterland) or the surface of the catchment and the depletion characteristics of the lake.

3.6.5.1 Investigation of the Relationship Between the Settling Rate of Suspended Material and Water Level Sinking of Flood Lakes

The applied technique was meant to reveal whether the settling of suspended grains depends on the rate of water level sinking of lakes formed in the dolines.

The settling rate of grains was established in measuring cylinder. In a settling basin, how the relations between the rate of water level and particle sinking influence deposition was investigated. In the settling basin at the settling sites (where the amount of deposited matter was measured), settling rate equals the rate measured in the cylinder at similar depth. This parameter (the settling rate) was constant in the settling basin during the investigations. If in the basin sediment of such particle size settles which has a higher settling rate than the rate of water level sinking, it can be claimed that settling depends on the rate of water level sinking. This can be evidenced if at a given rate of water level sinking, it is checked whether there is deposition of matter of a given particle size (with settling rate known from the measurement cylinder) onto the object slide (Deák et al. 2014). The experiment was made in the following way.

The sample from a non-cohesive rock was dried and sieved into fractions by particle size. The suspensions (A_1 and A_2) were prepared from the fractions of the smallest particle size (below 0.063). From this material 2 g each was brought into suspension, and the suspension was filled to 1 dm^3, receiving two 0.2 m/m % solutions (A_1 and A_2). The suspension system A_2 was further loaded with fibrillary plant chips (1–0.5 mm size fragments of wheat straw) until the solution had a concentration of 0.05 m/m % for plant waste (Table 6.2).

In the homogenised solutions, the pipette method (Stefanovits 1981) was used to measure suspended matter contents at 5.0, 7.5 and 10.0 cm fluid layer depth at exponentially growing time intervals (1, 5, 15, 30, 60, 120 min, etc.). On every occasion, samples of 5 cm^3 volume were taken and dry matter contents were determined. The dry matter content of the 5 cm^3 samples measured at depths 5.0, 7.5 and 10.0 cm and at different times (1, 5, 15, 30, 60, 120 min, etc.) was referred to 1 dm^3 volume (Table 3.1).

A function was fit to the data series which is graphically shown on curve 2 in Fig. 6.15. The other function in Fig. 6.15, which is represented on curve 3 of Fig. 6.15, was produced through subtracting the suspended matter concentrations at each observations times from the initial concentration. The measurements were made for both suspensions (A_1 and A_2). From the data settling velocities were determined in the following manner.

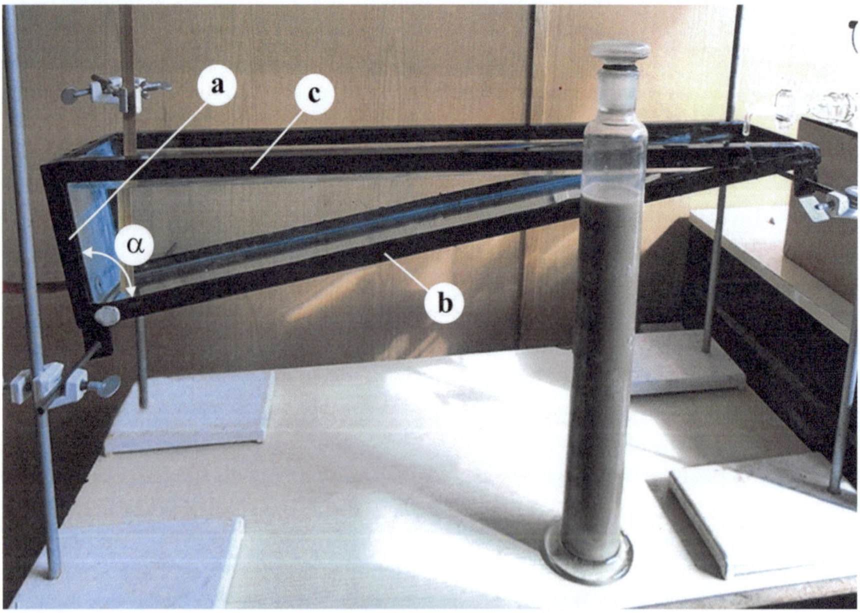

Fig. 3.11 Settling basin. (**a**, **b**), catheti of the right-angled triangle formed by the sides of the side walls of the settling basin; (**c**), hypotenuse of the right-angled triangle; α, the right angle formed by the two catheti

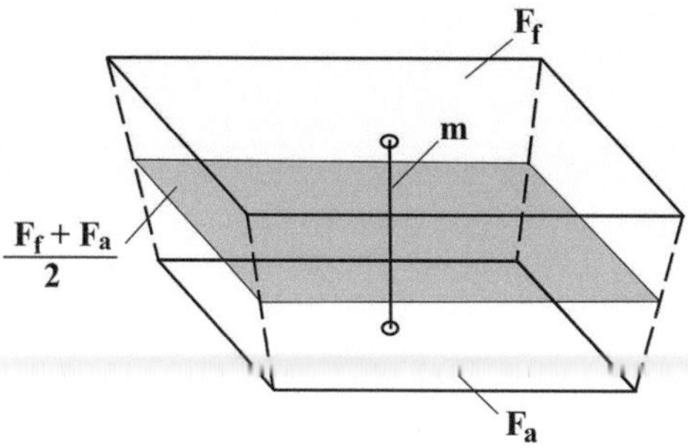

$$V\,[cm^3] = \frac{F_{f_{1\text{-}6}}[cm^2] + F_{a_{1\text{-}6}}[cm^2]}{2}\, m_{1\text{-}6}\,[cm]$$

$$m_{1\text{-}6}\,[cm] = \frac{2\,V\,[cm^3]}{F_{f_{1\text{-}6}}[cm^2] + F_{a_{1\text{-}6}}[cm^2]}$$

Fig. 3.12 Determining the rate of water level sinking. *V*, amount of water released (1 dm^3 of water). $m_{1\text{–}6}$: water level sinking in case of 1 dm^3 water released because of the shape of the basin. $F_{f_{1-6}}$ and $F_{a_{1-6}}$: size of water surfaces in case of various water level sinkings

Table 3.2 Rates of the water level sinking (cm/min)

$\Delta V[dm^3]$	0.25 min	0.5 min	0.75 min	1 min	2 min	5 min	10 min
1. –dm^3	2.9	1.48	0.98	0.74	0.37	0.14	0.07
2. –dm^3	3.2	1.56	1.03	0.78	0.39	0.15	0.07
3. –dm^3	3.32	1.66	1.10	0.83	0.41	0.16	0.08
4. –dm^3	3.56	1.78	1.18	0.89	0.44	0.18	0.08
5. –dm^3	3.92	1.96	1.30	0.98	0.49	0.19	0.09
6. –dm^3	4.36	2.18	1.45	1.09	0.54	0.21	0.10
7. –dm^3	4.92	2.46	1.63	1.23	0.61	0.25	0.12
8. –dm^3	5.96	2.98	1.98	1.49	0.74	0.29	0.14

Remark: The rates of water level sinking marked in grey fields are applied in the study of suspension systems of different composition

The suspended matter content of suspension A_1 (based on data from 10 cm layer thickness) is exponentially reduced with time, while the value calculated for the same measurement point (the difference between the initial concentration and the actual concentration) takes the shape of the saturation curve with time. This technique was also applied to water depths of 5.0 and 7.5 cm.

Half of the initial concentration (2 g) is received if the intersection of the curves is read on the y-axis. The x value of intersection indicates the time necessary for the reduction of concentration to its half at the studied point. The half-time of concentration can be determined more precisely if concentration changes are represented in the natural logarithm of measurement times.

If the distances measured from the surface of the suspension to the sampling sites (5.0, 7.5, 10.0 cm) are divided by the half-times of the concentration typical of them, the settling velocities at the points of observation are achieved.

Table 3.3 Composition of the suspension systems

The name of the suspension	Qualitative composition of the suspension	Quantitative composition of the suspension in %
Suspension A	Distilled water (10 dm^3) + clay (20 g)	Onto clay: 0.2 m/m % 0 N°
Suspension B	Tap water (10 dm^3) + clay (20 g)	Onto clay: 0.2 m/m % 20 N° permanent hardness
Suspension C	Tap water (10 dm^3) + clay (20 g) + $CaCl_2$ (3.96 g)	Onto clay: 0.2 m/m % 40 N° permanent hardness
Suspension D	Tap water (10 dm^3) + clay (20 g) + $CaCl_2$ (3.96 g) + $KHCO_4$ (3.57 g)	Onto clay: 0.2 m/m % 60 N° permanent and carbonate hardness
Suspension E	Tap water (10 dm^3) + clay (20 g) + organic matter (2 g)	Onto clay: 0.2 m/m % 20 N° organic matter 0.02 m/m %
Suspension F	Tap water (10 dm^3) + clay (20 g) + $CaCl_2$ (3.96 g) + organic matter (2 g)	Onto clay: 0.2 m/m % 40 N° total hardness organic matter 0.02 m/m %
Suspension G	Stream water (from a river) (10 dm^3) + suspended matter (clay + dissolved salt) (ca 1.5 g)	Dry matter content: 0.15 m/m % 14 N°

In the settling basin, deposition from suspended matter was measured at given rate of water level sinking (water level sinking was ensured by water conduction at the bottom of the basin). The rate of sinking can be determined and regulated in the settling basin (Deák et al. 2013). The water level related to 10 dm^3 volume was marked. Relative to this level, microscope object slides capable of capturing the settling matter were placed at 5.0, 7.5 and 10.0 cm depths. Then the basin was filled up with 10 dm^3 of suspension A_1 and first slow then rapid water level sinking was generated. The depositions on the object slide were determined at the given depths. Settling velocity was known from measurements in the measuring cylinder. Its determination was described above. The same measurement series was also executed with suspension A_2 also containing plant waste.

From the joint application of both methods, it is claimed whether there is deposition from matter with known particle size or composition with a given rate of water level sinking.

3.6.5.2 Investigation of Sedimentation in Flood Lakes

The settling basin, where the experiments were implemented, is a wedge-shaped vessel with a glass plate (Fig. 3.11).

The settling basin was installed so that the hypotenuse of the right-angled triangle of vessel walls should be horizontal. This way the rectangular glass plate (deposition surface) meets the horizontal plane at 9° angle and reflects a gently sloping doline floor. Before starting to lower water level, the initial state was documented in a photograph, and upon releasing 1 dm^3 fluid (at different velocities), after 20 min of quietude, the change was documented in another photo. This was repeated on six occasions in every experiment series.

Following its installation, the above-described equipment was filled with 10 dm^3 of water. In the first step, water level sinking was measured: the actual water level was recorded (F_f), then 1 dm^3 of water was released from the settling vessel, and the new surface (F_a) was determined. In knowledge of the width of the settling vessel (16 cm) and the lengths of water surfaces, the rate of water level sinking (m) was calculated in centimetres from the equation in Fig. 3.12. The values received were the following: m_1, 0.74; m_2, 0.78; m_3, 0.83; m_4, 0.89; m_5, 0.98; and m_6, 1.09 m.

The extent of water level sinking was related to the time necessary to release 1 dm^3 of water (10, 5, 2 and 1 min and 45, 30 and 15 s, resp.). Thus, the velocity of water level sinking [cm/min] was determined (Table 3.2). It is visible that with advancing water level sinking and repeatedly conducted fluid of the same volume, the rate of lake water level dropping has an increasing trend. It is explained by the shape of the basin. It is to be emphasised, however, that there were interruptions of stagnant (unchanged) water levels between water level sinking.
The investigated systems were water–clay suspension systems with waters with different hardness, and they occasionally also contained organic matter. (The various suspension systems were marked with capital letters 'A', 'B', 'C', etc.).

For each system clay content was 0.2 m/m %. The composition of the different suspension systems is shown in Table 3.3.

References

Allen JRL (1972) The origin of cave flutes and scallops by enlargement of inhomogeneities. Rassegna Speleol Ital 24(1):3–20

Aubert D (1966) Structure, activité et évolution d'une doline. Bull Soc Neuchateloise Sci Nat 89:113–120

Brook GA (1981) An approach to modeling karst landscapes. South Afr Geogr J 63:60–76

Clark PJ, Evans PC (1954) Distance to nearest neighbour as a measure of spatial relationships in populations. Ecology 35:445–453

Coase A, Judson D (1977) Dan yr Ogof and its associated caves. Trans Br Cave Res Assoc 4:244–344

Crawford N (1984) Karst landform development along the Cumberland Plateau escarpment of tenessee. In: LaFleur RS (ed) Groundwater as a geomorphic agent. Allen and Unwin, Boston, pp 294–339, A Geological Guide to Mammoth Cave National Park, Zephyrus Teaneck, NJ

Crowther J (1996) Roughness (mm-scale) of limestone surfaces: examples from coastal and mountain karren features in Mallorca. In: Fornós JJ, Ginés Á (eds) Karren landforms. Universitat de les Balears, Palma de Mallorca, pp 149–159

Crowther J (1997) Surface roughness and the evolution of karren forms at Lluc, Sierra de Tramuntana, Mallorca. Zeitshrift Geomorphol 41(3):396–407

Curl RL (1966) Scallops and flutes. Trans Cave Res Group Great Brit 7:121–160

Currens JC, Paylor RL, Beck FG, Davidson B (2012) A method to determine cover – collapse frequency in the Western Pennyroyal karst of Kentucky. J Cave Karst Stud 74(3):292–299

Day MJ (1977) Surface roughness as a discriminator of tropical karst styles. Z Geomorphol Suppl Band 32:1–8 de Hall R (1976) Investigations of sinkhole stratigraphy and hydrogeology, south-central Indiana. Bull Nat Spel Soc 38:88–92

Deák GY, Samu SZ, Veress M (2013) Bevonatképződés vizsgálata szuszpenziós rendszerekből modellkísérletekkel (Investigation of veneer development of suspension systems with model experiments). Karsztfejlődés XVIII:49–64 (in Hungarian)

Deák GY, Veress M, Mitre Z, Szemes M (2014) Ülepedés- és vízszintcsökkenés sebességének viszonya a szuszpenziókban (Relations between rates of deposition and water level sinking in suspensions). Karsztfejlődés XIX:147–158 (in Hungarian)

Dzulynski S, Gil E, Rudnicki J (1988) Experiments on kluftkarren and related lapis forms. Z Geomorphol 32(1):1–16

Fabre G, Nicod J (1982) Lapiés, modalités et rôle de la corrosion, crypto-karstique. Phénomèn Karstique III Mémoires Doc Géog 3:115–131

Farsang AM, Tóth T (1992) Bükki dolinák morfometriai vizsgálata (A morphometric investigation of dolines in the Bükk Mountains). In: A Bükk karsztja, vizei, barlangjai I. (Karst waters and caves in the Bükk) Conference volume, University of Miskolc, Miskolc, pp 39–50 (in Hungarian)

Ford DC, Williams PW (2007) Karst hydrogeology and geomorphology. Wiley, Chichester, 561 p

Füzesi I (2007) Loess tests carried out in the surroundings of some covered karstic depression (Tés Plateau). Carpathian J Earth Environ Sci 2(2):39–44

Ginés A (1996) Environmental approach to the typology of karren landform assemblages in a Mediterranean mid–mountain karst: the Serra de Tramuntana, Mallorca, Spain. In: Fornós JJ, Ginés A (eds) Karren landforms. Universitat de les Balears, Palma de Mallorca, pp 163–176

Glennie EA (1948) Some points relating to Ogof Flynnon Ddu. Trans C R G 1(1):13–25

Glew JR, Ford DC (1980) Simulation study of the development of rillenkarren. Earth Surf Proc 5:25–36

Grimes KG (2012a) Surface karst features of the Judbarra/Gregory National Park, Northern Territory, Australia. Helectite 41:15–36

Grimes KG (2012b) Epikarstic Maze cave development: Bullita cave system, Judbarra (Gregory Karst, Tropical Australia). Helectite 41:37–66

Hevesi A (1980) Adatok a Bükk-hegység negyedidőszaki és földrajzi képéhez (Data on the quaternary and geography of the Bükk mountains). Földtani Közlöny 110(3–4):540–550 (in Hungarian)

Hevesi A (2001) A Nyugati-Mecsek felszíni karsztosodásának kérdései. Karsztfejlődés VI:103–111 (in Hungarian)

Hoover RA (2003) Geophysical choices for karst investigations. www.saic.com/geophysics/downloads/karstChoices.pdf

Hutchinson DW (1996) Runnels, Rinnenkarren and mäanderkarren: form, classification and relationships. In: Fornós JJ, Ginés A (eds) Karren landforms. Universitat, de les Balears, Palma de Mallorca, pp 209–223

Jansen J, Anklam J, Goodwin C, Roof A (1993) Electromagnetic induction and seismic refraction surveys to detect bedrock pinnacles. In: Beck BF (ed) Applied karst geology. Balkena, Rotterdam, pp 115–122

Jeness J (2003) Longest straight line V.1.3. Jeness Enterprises. http://www.jenessent.com/arcview

Jennings JN (1975) Doline morphometry as a morphogenetic took: New Zealand examples. N Z Geogr 31:6–28

Jennings JN (1985) Karst geomorphology. Basil Blackwell, New York, 293 p

Jennings JN, Sweeting MM (1963) The limestone ranges of the Fitzroy Basin, Western Australia. Bonner Geographische Abhandlungen 32:60 p

Jones GP, Pearlstine LG, Pearcival HE (2006) An assessment of small unmanned aerial vehicles for wildlife research. Wildl Soc Bull 34(3):750–758

Kalmár S, Kozma K (2012) A demonstration of the geomorphological value of radio-controlled aerial vehicle imaging techniques in the study of the Hernad River. Z Geomorphol 56(Suppl 2):121–132

Lehmann H (1936) Morphologische studien auf Java. Geog. Abhandl. III. Stuttgart, 114 p

Lippmann L, Kiss K, Móga J (2008) Az Abaligeti-Orfűi karszt karsztos felszínformáinak vizsgálata térinformatikai módszerekkel (GIS investigation of the karst landforms in the Abaliget-Orfű Karst). Karsztfejlődés XIII:151–166 (in Hungarian)

McCann DM, Jackson PD, Culshaw MG (1987) The use of geophysical surveying methods in the detection of natural cavities and mineshafts. Q J Eng Geol 20:59–73

McDowell PW, Barker RD, Butcher AP, Culshaw MG, Jackson PD, McCann DM, Skipp BO, Matthews SL, Arthur JCR (2002) Geophysics in engineering investigations. Construction industry research and information association report C592 (and Geological society engineering geology special publication 19). CIRIA, London, 252 p

Palmer AN (1975) The origin of maze cavers. Natl Speleol Soc Bull 37:56–76

Palmer NA (2004) Mammoth cave region, United States. In: Gunn J (ed) Encyclopedia of caves and karst science. Fitzroy Dearborn, New York, pp 495–499

Péntek K, Veress M, Szunyogh G, Dazamits R, Tengelits A (2000) A karsztos mélyedések morfometriájának függvénytani eszközökkel történő leírása (Description of the morphometry of karst depressions by functions). BDF Tudományos Közlemények Szombathely 7:73–96 (in Hungarian)

Péntek K, Veress M, Lóczy D (2007) A morphometric classification of solution dolines. Z Geomorphol 51(1):19–30

Quinif Y (1973) Contribution a l'étudc morpholoqiquc dcs coupoles. Ann Spéléologie 28(4):565–573

Quinlan JF (1981) Hydrologic research techniques and instrumentation used in the Mammoth Cave Region, Kentucky. In: Roberts TG (ed) GSA Cincinnati'81 field trip guidebooks III, pp 502–504

Racoviţă G, Moldovan O, Onac B (2002) Monografia carstului din Munţtii Pădurea Craiului. Cluj University Press, Cluj-Napoca, 264 p

Roglič J (1939) Geomorphologische Studien von Duvanjsko (Polje von Duvno) in Bosnien. Mitt Geog Ges 83:152–176

Rose L, Vincent P (1983) Some aspects of the morphometry of grikes a micture model approach. In: Paterson K, Sweeting MM (eds) New directions in karst, proceedings of the Anglo–French Karst Symposium. Geo Books, Norwich, pp 497–513

Rudnicki J (1960) Experimental work on flutes development. Speleologia 2(1):17–30

Slabe T (2005) Two experimental modelings of karst rock relief in plaster: subcutaneous 'rock teeth' and 'rock peaks' exposed to rain. Z Geomorphol 49(1):107–119

Slabe T (2009) Karren simulation with plaster of Paris models. In: Ginés Á, Knez M, Slabe T, Dreybrodt W (eds) Karst Rock Features. Karren Sculpturing Zalozba ZRC. Carsologica 9. Institut za raziskovanje krasa ZRC SAZU, Postojna-Ljubljana, pp 47–54

Stefanovits P (1981) Talajtan (soil science). Mezőgazda Kiadó, Budapest, 382 p (in Hungarian)

Sweeting MM (1958) The karstlands of Jamaica. Geogr J 124:184–199

Sweeting MM (1966) The weathering of limestones, with particular reference to the carboniferous limestones of northern England. In: Dury GH (ed) Essays in geomorphology. Heinemann, London, pp 177–210

Sweeting MM (1973) Karst landforms. Columbia University Press, New York, 362 p

Tan M (1992) Mathematical modelling of catchment morphology in the karst Guizhou, China. Z Geomorphol N F 36(1):37–51

Telbisz T (2004) Digitális domborzatmodellek használata a karsztkutatásban (Digital elevation models in karst research). Karsztfejlődés IX:21–33 (in Hungarian)

Telbisz T (2010) A montenegrói Sinjajevina-karsztfennsík felszínalaktani vizsgálata terepi és térinformatikai módszerekkel. Geomorphological investigation of the Sinjajevina Plateau, Crna Gora, using feldwork and GIS). Karsztfejlődés XV:85–101 (in Hungarian)

Telbisz T, Móga J (2005) Töbör-morfometriai elemzések a Szilicei-fennsík középső részén (Doline morphometric analyses in the central Szilice Plateau). Karsztfejlődés X:221–228 (in Hungarian)

Telbisz T, Dragasiče H, Nagy B (2005) A horvátországi Biokovo-hegység karsztmorfológiai elemzése terepi megfigyelések és digitális domborzatelemzés alapján (Karst morphological investigation of the Biokovo mountains, Croatia, using fieldwork and DEM). Karsztfejlődés X:229–243 (in Hungarian)
Telbisz T, Móga J, Kósik SZ (2009) A Pelsőci-fennsík digitális domborzat elemzése és töbörmorfometiriai jellemzése (DEM and doline morphometry of the Pelsőc Plateau). Karsztfejlődés XIV:121–137 (in Hungarian)
Trudgill ST (1985) Limestone geomorphology. Longman, London, 196 p
Veress M (1982) Adatok a Hárskúti-fennsík morfogenetikájához (Data on the morphogenesis of the Hárskút Plateau). Karszt Barlang I:71–82 (in Hungarian)
Veress M (1983) Eltérő magasságú tönkfelszínek karsztosodásának kérdései az Északi-Bakony keleti részén (Problems of karstification on peneplains of different elevation in the eastern part of the North-Bakony mountains). A Bakonyi Természettudományi Múzeum Közleményei 2:29–44 (in Hungarian)
Veress M (2000) Covered karst evoluton in the Northern Bakony Mountains, W-Hungary. A Bakony Természettud. Kut. Eredményei Bakonyi Természettudományi Múzeum, Zirc 23, 167 p
Veress M (2009) Investigation of covered karst form development using geophysical measurements. Z Geomorphol 53(4):469–486
Veress M (2010) Karst environments – karren formation in high mountains. Springer, Dordrecht, 230 p
Veress M (2012) Glacial erosion and karst evolution (Karren on the surfaces formed by glaciers). In: Veress B, Szigethy J (eds) Horizons in earth science research. Nova Science, New York, pp 1–94
Veress M (2013) Intense rainfall on subsidence karst doline. In: Lóczy D (ed) Geomorphological impacts of extreme weather. Case studies from central and eastern Europe. Springer, Dordrecht, pp 327–345
Veress M, Tóth G (2004) Types of meandering karren. Z Geomorphol 48(1):53–77
Veress M, Pidl K, Mantler M (1998) A gipsz karsztosodásának modellezése laboratóriumi körülmények között (Modelling the karstification of gypsum under laboratory conditions). Berzsenyi Dániel Tanárképző Főiskola Tudományos Közleményei XI. Természettudományok 6:147–166 (in Hungarian)
Veress M, Zentai Z, Kovács GY (1999) Álkarros formák a Bosco Seccói forrás (Asiagói-fennsík) környékén (Pseudokarren near the Bosco Secco spring, Asiago Plateau, Italy). Karsztfejlődés III:19–29 (in Hungarian)
Veress M, Péntek K, Unger Z, Almási I (2012) Development of covered karstic dolines in ground ice environment (Eastern Alps, Austria). Interests of experimental and mathematical modelling. Z Geomorphol 56(2):79–104
Veress M, Gárdonyi I, Deák GY (2014) Fedett karsztosodás vizsgálata fedővel borított gipsztáblán (Investigation of covered karst formation on gypsum plate with cover). Karsztfejlődés XIX:159–171 (in Hungarian)
Vincent PJ (1983) The morphology and morphometry of some arctic Trittkarren. Z Geomorphol 27:205–222
von Wissmann H (1954) Der Karst der humiden heissen und sommerheissen Gebiete Ostasiens. Erdkunde 8:122–129
Waltham AC, Smart PL, Friederich H, Eavis AJ, Atkinson TC (1983) The caves of Gunung Sewu, Java. Cave Sci 10(2):55–96
Warwick GT (1964) Dry valleys in the southern Pennines, England. Erdkunde 18:116–123
White WB (1960) Termination of passages in Appalachian caves as evidence for a shallow phreatic origin. Bull Nat Speleol Soc 22:43–53
White WB (1988) Geomorphology and hydrology of karst terrains. Oxford University Press, New York, 464 p
Wilford GE, Wall JRD (1965) Karst topography in Sarawak. J Trop Geogr 21:44–70

Williams PW (1966a) Morphometric analysis of temperate karst landforms. Irish Speleol 1:23–31

Williams PW (1966b) Limestone pavements: with special reference to western Ireland. Trans Inst Br Geogr 40:155–172

Williams PW (1971) Illustrating morphometric analysis of karst with examples from New Guinea. Z Geomorphol 15:40–61

Williams PW (1972a) Morphometric analysis of polygonal karst in New Guinea. Geol Soc Am Bull 83:761–796

Williams PW (1972b) The analysis of spatial characteristics of karst terrains. In: Chorley RJ (ed) Spatial analysis in geomorphology. Methuen, London, pp 135–163

Zeller J (1967) Meandering channels in Switzerland. IUGG/IASH Symposium on river morphology, Bern, pp 174–184

Zhang ZH (1980) Karst types in China. Geo J 4(6):541–570

Zhou W, Beck BF, Adams AL (2002) Effective electrode array in mapping karst hazards in electrical resistivity tomography. Environ Geol 42:922–928

Chapter 4
Classification of Covered Karsts

Abstract Covered karsts are classified according to the character of the cover (cryptokarst if the cover is impermeable and concealed karst if it is permeable), to the origin of cover deposits (locally deposited or transported there) and to the age of karstification (syngenetic if the depression in the cover and the form in the bedrock are of the same age and postgenetic if not). Covered karsts may develop in structural landforms (synclines, tectonic graben, horsts of various elevations) and in depressions formed by the powers shaping the surface, which could be of karstic origin (doline, ponor, polje, fengcong depressions and intermountain plains of fenglin karst) or of non-karstic origin (valley, abrasional platform and river terrace). The appearance and pattern of covered karst depend on the type of karst. When characterising pattern geosyncline or glaciokarst (folded-nappe structure), block mountains, platform, salt diapir, tundra, taiga, temperate, mediterranean and tropical covered karsts are distinguished. The geosyncline (glaciokarst) and block mountains karst types mainly occur under temperate climate, while the platform and salt diapir may appear on any climatic karst type. According to pattern, covered karst developed from allogenic karst (which can be recent allogenic, rejuvenating allogenic or semi-allogenic), nappe allogenic, horst, cirque, glacial trough, polje, karst hill, cueasta, tropical depression, polygonal, intermountain plain, petrified forest, platform and salt diapir covered karsts are identified. On the recent allogenic karst, in slope direction from the bordering non-karstic surface, cryptokarst, mixed and open karst zones are found. On the rejuvenating allogenic karst, in slope direction from the former terrain of sediment supply, open, mixed and buried karst zones follow each other. On the semi-allogenic karst, in lack of the accumulation of fluvial deposits, the cover sediment does not derive from the non-karstic terrain bordering the karst and is of continuous distribution. On the nappe allogenic karst, buried karst, cryptokarst (with karst windows), concealed karst and open karst zones occur. On horst covered karst, on the neighbouring blocks, concealed, crypto and buried karst terrains alternate (in patches or in continuous distribution). On the glacial trough covered karst, striped or striped-patchy patterns are typical. On the polje karst, continuous or zonal (so-called internal zonal) covered karst develops. On the karst hill covered karst, the patches of covered karst may coalesce, while on cueasta covered karst, the patches of covered karst are arranged in stripes parallel to the cuestas. The tropical depression covered karst and the polygonal covered karst show patchy distribution, while the intermountain plain covered karst is continuous,

M. Veress, *Covered Karsts*, Springer Geology,
DOI 10.1007/978-94-017-7518-2_4

but can show internal zonation with crypto and concealed karst zones. No or poor zonation can be identified on the platform karst. The covered karst of salt diapir shows patches and develops on the cover uplifted by the rising salt diapir.

Keywords Cryptokarst • Concealed karst • Patterns of covered karst • Patchy • Continuous • Zonal and striped patterns • Geosyncline • Block mountain • Platform • Salt diapir • Tundra • Taiga • Recent allogenic • Rejuvenating allogenic • Semi-allogenic • Nappe allogenic • Polje • Karst hill • Cuesta • Tropical depression • Polygonal • Intermountain plain • Petrified forest covered karst

4.1 Introduction

Covered karsts can be classified according to the bedrock and cover material (or its origin), the age of covered karst formation, the mode of development and pattern of the cover.

The covered karst is a type of karst where phenomena, landforms and processes emerge because of the material deficit on the karstifying bedrock. The karstic nature of a covered karst may be occasionally very marked. It is common that on the covered karst doline density is higher than on open karst (Thomas 1974). Dolines developed in cover were first described by Cvijič (1893), but a covered karst feature is also mentioned by Carr (Sweeting 1973) and Lucas (1872). Katzer (1909) identified, in addition to naked and soil-covered karsts, the karst covered by non-karstic rock and called it ‘unterirdische Karst’.

When classifying karsts according to cover, Gvozdetskiy (1965) distinguished between bare karst, covered karst, soil-covered or ‘boddy’ karst (with terra rossa) and buried karst. Covered karsts can have any cover not deriving from karst (either unconsolidated or consolidated, but developed independent from the limestone, the bedrock underlying the karst), while on soil-covered or ‘boddy’ karst, not only soil but also solution residual (terra rossa) can occur. The class of Gvozdetskiy’s covered karst cannot include karsts covered with morainic deposit, where the cover derives from the bedrock, but it is not a solution residue. On the buried karst, the underlying karstic rock is buried to the extent that on the cover surface, no karst landforms develop; the karstifying rock is only known from boreholes. In the literature, buried karsts are not counted among karst types.

The non-karstic rock covering the karstic bedrock cannot be qualified unambiguously as covered karst in all cases. In principle, the following cases are possible:

- On the cover, karst features are observable, and in this case, the terrain can be classified with covered karsts.
- In the cover, processes are active which result from the karstification of the bedrock, on which indirect information is available (e.g. from borehole data).
- In the cover, no karst features are found, and there is no information on karstification in the cover. In this case, karst formation in the bedrock is in its initial stage, and, thus, no visible (or measurable) signs are observed either on the

surface or in the cover, but this process is expected to be active in the near future. It is also possible, however, that no karstification will take place in the future (no surface karst features will occur). Such areas can be regarded buried karsts.

All those terrains where no covered karst formation is detected now but will occur in the future are potential covered karsts. Such terrains, however, cannot always be distinguished from those terrains where karstification is not active under the present climatic, morphological and geological conditions and will not be active in the future either. Particularly in tundra and taiga areas, extensive terrains may occur where these processes are not only unobservable but does not and will not occur under present conditions.

Covered karsts are defined by Sweeting (1973) as karsts where the limestone is covered by soil, vegetation or debris, i.e. it has a cover of unconsolidated rock. Jennings (1985) does not clearly distinguish between karsts covered by soil and those covered by unconsolidated deposits. The bare karst is also interpreted in different ways. This karst type may mean a karst where the ice removed the cover (glaciokarst) and also a karst where the removal of the (primarily soil) cover was due to human action, as in the case of mediterranean karsts (Sawicki 1909). Both on glaciokarsts and mediterranean karsts, however, patches with soil and those with unconsolidated cover deposits alternate with each other. The pattern and extension of such patches may be highly variable, and, under certain conditions, such patches may form a covered karst.

Quinlan (1967, 1978) identified the following varieties of covered karst:

- Subsoil karst, where the cover is weathering residual carried over the karst or produced in situ.
- Mantled karst, where the cover is some allochthonous rock or sediment. The karst is older than its cover. The mantled karst is part of the karst landscape.
- Buried karst, where the cover is also some allochthonous rock or sediment. The karst is older than the cover, but it is not part of the karst landscape.
- The interstratal karst is covered by autochtonous rock or sediment, and it may be part of the karst landscape. The karst is younger than its cover. On karst of this type, no cave is found.
- The subaqueous karst is a subsided karst. Thus, it is either located under the sea level or under a river.

Jennings (1985) used the name covered karst for karren fields and distinguished bare karst (no cover and no soil on the surface with karren) and partly covered karst (with partial covering of soil, plant detritus, humus, sediment and moss). Patchy covering may extend to individual karren features or over larger terrain segments, and it is also possible that the karren feature is uncovered and its environs are covered. Finally, Jennings identifies covered karst with a uniform cover of the terrain, which can be only soil or also sediment.

Other authors (Salomon et al. 1995; Knez et al. 2003; Ford and Williams 2007) use the term cryptokarst as a synonym for covered karst to describe terrains where the bedrock with karren is covered by soil, vegetation and occasionally by non-karstic cover. The stone forest karst covered by non-karstic rock (Lunan Petrified Forest) is called cryptokarst (Maire et al. 1991), subjacent (Chen et al. 1986) or subcutaneous karst (Slabe 1999). The cryptokarst denomination was first used by Nicod (1976).

A classification of karsts based on the relationship with non-karstic rock was made by Jakucs (1977), who identified allogenic and authogenic karsts (Jakucs 1977). An allogenic karst receives water from the bordering non-karstic terrain, while the authogenic karst (which rises above its environs) does not. In his typology, he identified, although not yet named, that type of allogenic karst where consolidated non-karstic rock borders the karst and another type where, for tectonic reasons, such a non-karstic rock is located inside the karst (described by Jakucs as 'of authogenic character in globality'). He also mentions that the authogenic karst shows a gradual transition towards the allogenic karst since in its area weathering (solution) produces cover sediment.

Williams (1987) and then Ford and Williams (1989, 2007) further developed the above typology: in addition to the authogenic karst, they also identified allogenic and mixed allogenic–authogenic karsts. The karst is of allogenic type if it is covered by non-karstic rock, and it is of mixed allogenic–authogenic type if the cover occurs in stripes and patches. The two latter types essentially differ in the extent the cover is removed. The two typologies have the additional difference that while Jakucs (1977) emphasises the variations in the assemblage of landforms (dolines on authogenic karst and ponors on allogenic karst), Ford and Williams (2007) focus on differences in the nature of drainage. On authogenic karsts, drainage is diffuse (through seepage), and on allogenic karsts, it is pointlike (through ponors surface waters reach the karst where the karstic rock outcrops from below the cover), while on mixed karsts both types of water transfer are typical.

Gams (1994) classified the karsts according to the position of the karst in relation to the surface and according to the elevation of the karst.

We use the denomination covered karst for all kinds of karst where the karstic rock is covered by non-karstic rock under the soil – irrespective of the character of this rock (consolidated or unconsolidated, impervious or permeable cover) – or where the non-karstic rock governs or entirely controls the processes in the karst area. We also regard a karst covered karst if the non-karstic cover has developed in small thickness and small extension or if it is widespread and has a great thickness.

In our opinion, various types of covered karst can be identified according to the impervious or permeable nature of the cover (Veress et al. 2013). Consolidated rocks (like, for instance, volcanic rocks or sandstones) are only permeable along faults, while unconsolidated rocks are usually permeable – although to various extents. The latter may be cohesive or non-cohesive, but show significant variations in respect to water conductivity (for instance, the clay is impervious). For impermeability cohesive rocks are akin (transitional) to consolidated rocks. At the same time, geomorphic evolution on cohesive rocks is double-faced. If the material deficit is rapidly inherited from the cover, the evolution of forms is similar to consolidated rocks or similar forms develop on unconsolidated but cohesive rock. If the process is not rapid, form evolution will be rather similar to that on non-cohesive rocks – even in the case of cohesive rocks.

Covered karsts are of great practical importance as on such surfaces depressions (occasionally in large numbers) can rapidly develop. The resulting landforms, particularly if they are of large size, may damage human structures (roads, railways, buildings, etc.). In order to prevent their development or to mitigate damage, the engineering practice applied numerous solutions. Prediction, protection and mitigation are described in the works by Sowers (1986, 1996), Kannan and Nettles (1999), Waltham and Fookes (2003), Waltham (2004), Xeidakis et al. (2004), Waltham et al. (2005), Ford and Williams (2007) and Veress et al. (2013).

4.2 Crypto and Concealed Karsts

The karst covered by impervious rock (the cover consolidated or not) is called cryptokarst, while if the cover is permeable, it is concealed karst (Hevesi 1986; Veress et al. 2013, Table 4.1). The cryptokarst can develop on both consolidated and

Table 4.1 Theoretical relationship between the type of covered karst and cover rock

Covered karst type	Character and properties of cover	Bedrock type	Water transfer into karst	Side of landform
Cryptokarst	Consolidated rock	Any rock	Local (at faults) or pointlike (at ponors)	Stable collapsed surface
	Unconsolidated, cohesive rock (thick clay or rock with very high clay content)	Any rock	Local (at faults) or pointlike (at ponors)	Unstable collapsed surface
	Consolidated and unconsolidated, cohesive	Gypsum, rock salt	None	Stable collapsed surface (consolidated rocks), unstable collapsed surface (unconsolidated rocks)
	Consolidated and unconsolidated cohesive rocks	Any rock	From groundwater or karst water	Stable collapsed surface (consolidated rocks), unstable collapsed surface (unconsolidated rocks), cohesive
Concealed karst	Unconsolidated, cohesive rock (low clay content or thin clay)	Any rock	Large extension, slow and hindered	Unstable collapsed surface
	Unconsolidated, non-cohesive	Any rock	Large extension, rapid	No collapsed surface

unconsolidated cohesive rocks. In the latter case, it is possible if the cover is thick clay or very thick rock with high clay content. In both cases, water reaches the surface of the cover along fractures and ponors from either groundwater or karst water. It can also develop if no water comes to the surface of the bedrock, but the material deficit in the cover is due to cavity formation through the dissolution of gypsum or rock salt. The sides of landforms developed on consolidated rocks are shaped by collapse and stable. On unconsolidated but cohesive rocks, the slopes of landforms are also of collapse origin but unstable. On the non-cohesive rocks of a concealed karst, the slopes are not shaped by collapses. Cohesive rocks can also constitute concealed karst if they are not perfectly impervious. This is possible if the clay content of the cover is not high or the clay cover is thin.

The cryptokarsts are transitional towards fluviokarsts. Fluviokarsts are shallow karst where the thickness of the limestone is smaller than the maximum relief of the terrain. On this karst type, both fluvial geomorphic action and karstification are present or it shows a mixed character with bare and covered patches, a chaotic drainage network, allogenic valleys, blind valleys, dry valleys, ponors, marginal poljes, caves of horizontal extension and dolines (Roglič 1964, 1965; Sweeting 1973; White 1988; Gunn 2004). Such landform assemblages occur in numerous karst regions in Europe (Dongus 1962; Hevesi 1980) and North America, on temperate (Malott 1939, 1945; Racoviţă et al. 2002), mediterranean, high-mountain (Roglič 1965) or tropical karst (Tan 1992).

From the aspect of covering, it is difficult to delimit the terrains of the same type (cryptokarst or concealed karst). The larger is the extension of the karst area considered, the greater is the chance for the occurrence of covered patches on bare karst and, in the opposite case, bare patches on covered karst. A reliable typology according to covering could be made if the cover conditions of individual landforms (such as valleys, dolines and poljes) are investigated. At the same time, in karst regions, bare and covered zones, stripes and patches which are of larger extension than that of the individual landforms mentioned above can be identified. When the karst is investigated according to the covering pattern (see Sect. 4.6), the presence, extension and distribution of such zones are taken into consideration. When describing the cover conditions, covered (and bare) karst patches and zones are distinguished. The former are more limited and rather homogeneous according to their cover, while the latter are of larger extension and with heterogeneous cover. In their area of the zones, both impervious (cryptokarst) and permeable (concealed karst) patches may occur, covered and bare equally.

Cover sediments (for instance, solution residuals) can be present on bare karsts (or on their landforms). In spite of its covered character, it is predominantly a bare karst where the cover only extends at most to some percent of the area. On partially covered karst, the extension of the cover may reach 50 %. On completely covered karst, covering may reach 100 %. In this case, the cover is of uniform appearance. The completely covered karst can be homogeneous if it is concealed karst or only cryptokarst. It is composite karst if both crypto and concealed karsts are present and mixed karst if in the zone uncovered parts also occur. In this case, cover is patchy. Thus, the zone of the karst may be mixed composite karst (concealed karst and cryptokarst details are alternating with uncovered terrain), mixed cryptokarst (cryptokarst interrupted by uncovered patches) and mixed concealed karst (concealed karst is interrupted by uncovered spots). The extension of the patches can be variable, ranging in diameter from some metres to several kilometres. There may be cryptokarst and concealed karst details on the area of the patches as well.

In general, the more elevated parts of the karst (in case closed forms are lacking) if they are eroding are mostly bare karsts and only converted into covered karst if there is aeolian deposition on their surface. Then the covered karst will be concealed karst. On lower terrains and in depressions, cryptokarsts develop.

Covered karst can form through denudation and accumulation – through denudation if the process exposes the non-karstic stratum/strata of the fold or – in the case of non-folded structures – if unconstricted glacial erosion affects the surface. From denudation, crypto covered karsts result. The burial of karst can result from (marine or lacustrine) sedimentation, volcanism, transport and accumulation of sediment. In the former cases (denudation) cryptokarst, while in the latter (accumulation), mostly concealed karst develop.

With its development, covered karst is becoming more and more complex. The concealed karst mostly forms on the entire cryptokarst or on its portions of various sizes during the denudation of the consolidated rock. In the latter case, the cryptokarst is partially exhumed. On the karst, open karstic terrains of variable size emerge. Concealed karst terrains develop on the latter ones on the surfaces buried under alluvium derived from cryptokarst terrains. They may develop in the way that uncovered terrains are buried under non-fluvial deposits (solution weathering, morainic and subaerial dust deposits). In this case, the cover does not derive from the cryptokarst terrain.

The cover may regulate the karst water system as it retains or conducts water with delay (Ford and Williams 2007). Because of the character of the cover deposit but also of other karst properties, water budget on the covered karst can be very diverse. The water content of the cover may be of rainwater origin and reach the bedrock seeping through the cover. Water influx into the karst is possible over the entire extension of the cover, along fractures or along the impervious intercalations of the cover. In the last case, water percolates into the karst along the impervious bed of the cover (after it infiltrates at its surface outcrop). The cover may also receive water from springs or groundwater. Unconfined groundwater may be fed by seepage from river channels (Sweeting 1973) or by karst water. On permafrost, underground ice conditions horizontal water flow can intensify. The water here derives from seepage from intermittent or permanent watercourses, which further increase permafrost thawing. Horizontal flow (seepage) may take place in the cover but also in the bedrock – depending on the upper surface of permafrost. It may happen, however, that the material deficit in the cover (through solution) is not caused by water from the cover but, if the karst is of allogenic character, by the watercourses arriving from the non-karstic terrain.

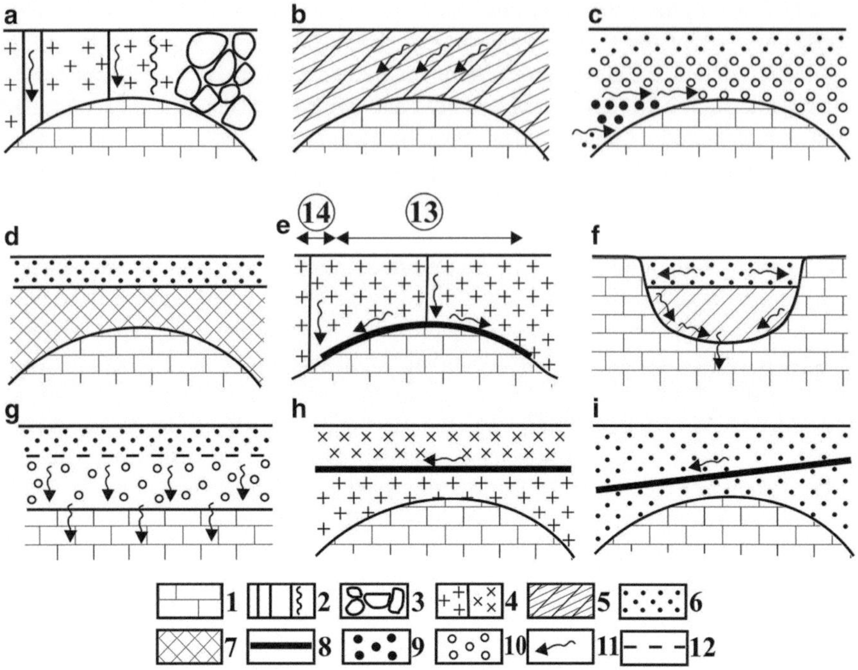

Fig. 4.1 Ways of water transfer into the karstic rock on covered karst (**a**) through the impermeable cover water percolates locally at several sites along cooling cracks, faults, atectonic fractures among fragments; (**b**) water percolates into the karst along bedding planes; (**c**) water percolates into the karst over more clayey deposits; (**d**) water temporarily reaches the karst (when ground ice melts); (**e**) the seeping water locally, at few sites, reaches the karst; (**f**) the water percolating in the less impermeable terra rossa reaches the doline side above the more impermeable terra rossa and then reaches the karst at the deepest point between the terra rossa and the doline floor; (**g**) groundwater from the cover percolates into the karst or the karst water of the cover sinks there; (**h, i**), no water influx into the karst, *1* limestone, *2* cooling crack, fault, atectonic fracture, *3* frost-shattered, collapsed debris, *4* various impermeable rocks (e.g. basalt), *5* well-stratified rock (e.g. sandstone), *6* permeable rock, *7* ground ice, *8* impermeable intercalation (e.g. clay), *9* gravel with low clay content, *10* gravel with higher clay content, *11* water percolation, *12* groundwater and karst water table, *13* no water influx into karst, *14* water influx into karst

The water budget of the concealed karst is different from that of the cryptokarst in the following aspects (Fig. 4.1):

- On the concealed karst, the entire amount of the infiltrating water reaches the karst, while on the cryptokarst, only some portion reaches it – and only under certain conditions and in some sites.
- On the concealed karst, water transfer from the cover is rapid and of short duration, while on the cryptokarst, it is slow and long-lasting. For this reason, although the solution capacity of the water percolating through the cover of the cryptokarst can be reduced because of the carbonate content of the cover, the chance is still greater that it reaches the bedrock unsaturated. On the cryptokarst,

slow water transfer significantly increases the chance for the infiltration of already saturated water into the karstic bedrock.

– On concealed karsts, water transfer takes place over a large surface, while, in contrast, on cryptokarsts, it is only possible in certain spots. It is to be noted that even a permeable rock (cover) can divert water flow. According to our laboratory experiments (see Chap. 7), because of the capillary rise of water, soil moisture develops in the cover mainly if the cover is fine-grained. Here the cover will be water diverter and thus locally impermeable.

The above primarily refers to non-cohesive cover. In the case of cohesive rocks, the water budget is transitional (or increasingly similar to) to the budget of the cryptokarst.

The extent and nature of the material transport from the cover also varies with different water budgets and also with the different properties of cover sediments. In the case of a consolidated cover, the removal of cover material by seeping waters is of local nature and small degree or totally absent since water transfer into the karst can only happen along the fractures of the cover and independent grains can only be detached from the rock to a small extent. On a non-cohesive cover, material transport is less local, and its degree is controlled by the amount of water stored in the cover, its seepage rate and the extent of cavity formation in the cover. In the case of cohesive rocks, the removal of the cover material can reach a relatively large scale. At the same time, because of its cohesion, cavities and passages can emerge in the cover. Although the latter can also form on non-cohesive cover, there is lower probability for that, and their size and duration are more limited.

The impervious non-karstic cover deposit of the cryptokarst falls into three categories according to its spatial position. One of them is when it is next to the karstic rock due to tectonic reasons. The non-karstic rock borders the karst terrain and dips towards that. In the second case, the non-karstic rock is intercalated in the karstic rock. In the latter case, covered karst most often develops if the anticline of the folded structure is truncated. The non-karstic strata intercalated between the limestone strata are exposed in various widths in the area of the truncated anticline. Similar to a syncline, their strike is identical with that of the fold. The non-karstic rocks of the syncline, however, are not necessarily of the same age as the karstic rock but could develop syngenetically with folding (during the denudation of the anticline) or much later. In the first case, the non-karstic terrain is built up of consolidated impervious rock (flysch) and in the latter – not necessarily – the non-karstic terrain can be composed of non-impervious non-cohesive rock. In the third case, the non-karstic rock overlies the karstic rock due to volcanic processes or marine transgression (sedimentation).

Four cryptokarst varieties are distinguished (Table 4.2): allogenic cryptokarst, authogenic cryptokarst, transitional cryptokarst and cryptokarst developed from buried karst.

– On the allogenic cryptokarst (which corresponds to the allogenic karst as described in literature), karstification along rock boundary takes place along the margin of the impervious rock, where blind valleys and ponors, depressions with

Table 4.2 Types of covered karst

	Covered karst				
		Cryptokarst			
Property	Concealed karst	Allogenic	Autogenic	Transitional	Cryptokarst developed from buried karst
Cover sediment	Unconsolidated	Consolidated	Consolidated, the bedrock is gypsum or rock salt	Consolidated	Consolidated, if not: impermeable
Thickness	Thin	Thick	Thick	Thin	Thick
Site of karstification	Interior of covered terrain	Margin of sediment cover	Interior of covered terrain	Interior and margin of covered terrain	Epigenetic valley floors
Landform	Subsidence doline, depression of superficial deposit	Ponor, depression of superficial deposit	Caprock doline	Caprock doline, depression of superficial deposit, ponor	Ponor

superficial deposit (further: DSD) develop, and river valleys dissect the non-karstic terrain. In the interior of the non-karstic terrain, no karstification is observed and it can be regarded a buried karst.

- On the authogenic cryptokarst, karstification takes place within the impervious rock, where there is locally water transfer into the karst from the cover. Covered karst formation can also happen because of the fluctuation of karst water table or of the effect of cavities of well-soluble rocks (like gypsum and rock salt) or of the paleokarstic cavities of less soluble rocks (limestone and dolomite) on the cover. In easily soluble gypsum and rock salt, even with very thick (more than 100 m) cover, karst formation in the cover is possible. In North America, breccia pipes extend upwards from gypsum or rock salt beds at 1200 m depth (Quinlan et al. 1986). In Shanxi and Hebei provinces of China, breccia pipes penetrate through a cover of 500 m thickness (Lu and Cooper 1997).
- On the transitional cryptokarst, karstification can take place within the impervious cover but also along its edges. In the interior of the cryptokarst, karstification can be similar to that on autogenic karst, where it is inherited over the cover, but it can be of allogenic character, too, when it happens where the non-karstic rock is thinning out or the bedrock outcrops. In the former case, caprock dolines result, while in the latter, ponors (or caprock dolines which develop into ponors) come about.
- On cryptokarst developed from buried karst, there is karstification within the impermeable rock where epigenetic valleys reach the limestone area. On the valley floors, ponor formation takes place along the rock boundary in the valley (Jakucs 1977). Recent and past ponor formation is typical in the Bükk Mountains of Hungary (Hevesi 1980) and occurs in other karst regions, including the Appalache Mountains (Worthington 1984).

Karstification may take various ways. Karst forms in the interior of the impermeable cover, but the water from springs issuing along the cover margins also contributes to concealed karst evolution, such as it is observed on the Central Kentucky Karst, where the Big Clifty sandstone represents the cryptokarst, on which (inherited) dolines developed (Quinlan and Pohl 1967; Balázs 1970). The water from the sandstone springs reaches the Mammoth Cave through the subsidence dolines (concealed karst) of the Sinkhole Plain (White et al. 1970).

Corbel (1947) compared landform evolution on bare and covered karsts. In his opinion on bare karsts karren, vertical landforms are typical and the filling of caves is negligible. On covered karst, karren is less widespread and horizontal landforms are predominant, collapses are rare and cave sedimentation is rapid. According to Sweeting (1973), because of the washed-in cover deposit, the landforms of covered karst are of moderate depth and of wider floors.

Landform development also differs on both types of covered karst. On cryptokarst (particularly if the cover is consolidated rock), changes follow the changes in the bedrock with a delay, but occur abruptly and have a short duration. Therefore, collapses are observable (Table 4.1). In the case of a consolidated cover, structural changes (fractures, faults) can also be present and influence both water seepage and geomorphic evolution. On concealed karst, changes in the unconsolidated cover immediately but slowly follow the karstification of the bedrock. The changes, however, are of small-scale subsidences instead of collapses. Karst phenomena and features on the unconsolidated cover may substantially vary according to the cohesive or non-cohesive nature of the cover. The processes on a cohesive cover resemble to those on consolidated cover or are transitional between them. Geomorphic evolution often takes place through collapses.

On concealed karsts, subsidence dolines and DSDs are characteristic. Smaller-scale erosion features (gullies, ravines), however, also occur here. If they exist, they are often generated by the dolines under formation. At the same time, it is also possible that the subsidence dolines queue up on the floors of valleys, gullies or ravines. The density of landforms may largely differ with the various types of covered karst: it is mostly higher on concealed karst and less on cryptokarst.

On concealed karst, the destruction and inactivation of landforms are more typical than on cryptokarst. On cryptokarst (of autogenic type), erosional cave development becomes decisive during karst development. Caves are of larger size and more complex in planform than on concealed karst. On concealed karst catchments of surface karst, features cannot be precisely delimited by divides, but on cryptokarst (with the exception of autogenic cryptokarsts), they are more marked. Rainwater may flow from the same site in several directions and into several dolines.

For the systemisation of covered karst and the understanding of their processes, knowledge on the mutual position of karstic and non-karstic rocks is indispensable. From the aspect of karstification, the rock boundary between karstic and non-karstic rocks is regarded active if the non-karstic strata dip towards the karstic. This is mostly the case if non-karstic rocks overlie karstic rocks (Fig. 4.2I.a). The evolution of the crypto covered karst and the character of its processes are controlled by the thickness of non-karstic rock and the change of this thickness moving away from the rock boundary (Fig. 4.2I.b, c). The types of rock boundaries are the following.

The non-karstic rock can be of marginal position or intercalated into the karst. Both consolidated and unconsolidated impermeable rocks can be of mantle nature when this subsequently covered the karst. In the case of the consolidated non-karstic rocks, outer true rock boundary forms on the karst margin (Fig. 4.3a), while in the case of internal non-karstic rock, inner true rock boundary (Fig. 4.3b) is formed. In the case of overlying cover too, a true rock boundary is formed (Fig. 4.3c). If the karst is covered by permeable rocks, where it is thinning out, concealed rock boundary develops (Fig. 4.3d). The cover can thin out (and form a concealed rock boundary) in the interior anywhere (on valley floor, above the elevation of the bedrock), but also along the margins. If there are clay intercalations in the permeable cover, interrupted or half rock boundary may emerge at the terminus of the intercalation (where waters infiltrating from the surface reach the bedrock) (Fig. 4.3e). Through impervious rocks, for instance, along faults, water can infiltrate into the karst and

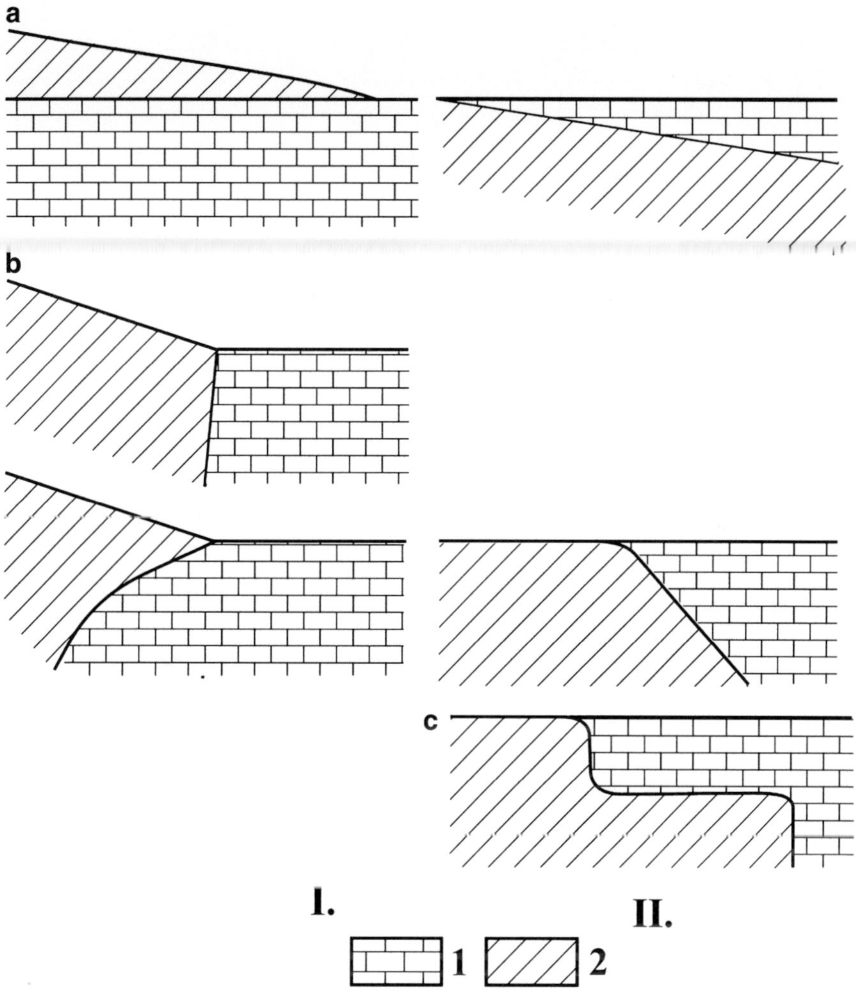

Fig. 4.2 Positions and thicknesses of karstic and non-karstic rocks at their contact or superposition: *I* non-karstic rock overlies karstic rock: (**a**) the thickness of non-karstic rock slightly changes with distance from the rock boundary, (**b**) the thickness of non-karstic rock remarkably changes with distance from the rock boundary. *II* Karstic rock overlies non-karstic rock: (**a**) the thickness of limestone slightly changes with distance from the rock boundary, (**b**) the thickness of limestone remarkably changes with distance from the rock boundary, (**c**) the thickness of limestone abruptly changes with distance from the rock boundary, *1*, limestone, *2* non-karstic rock

generates a structural rock boundary (Fig. 4.3f). Finally, a buried rock boundary is generated (Fig. 4.3g), if the impervious rock (which can also be ground ice) is overlain by a less impervious deposit and the infiltrated waters are conducted just by this impervious rock to the limestone.

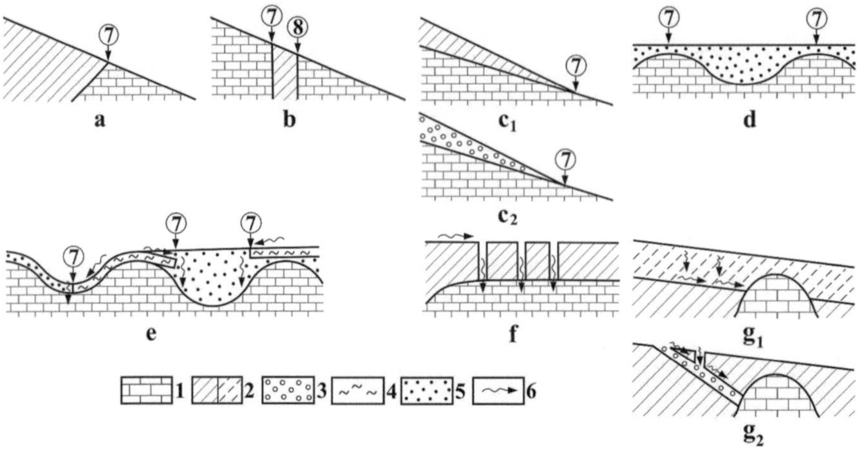

Fig. 4.3 Rock boundaries *1* limestone, *2* consolidated, impermeable and less impermeable non-karstic rock, *3* unconsolidated, impermeable non-karstic rock, *4* partially impermeable cover sediment (unconsolidated), *5* permeable cover sediment, *6* water seepage, *7* rock boundaries, *8* inactive rock boundary (**a**) true outer rock boundary of non-karstic rock in marginal position, (**b**) true inner rock boundary of intercalated non-karstic rock, (**c**) inner true rock boundary (in case of overlying non-karstic rock), (**c_1** of consolidated cover rock, **c_2** of unconsolidated cover rock), (**d**) hidden rock boundary, (**e**) (interrupted) half rock boundary, (**f**) structural rock boundary, (**g**) buried rock boundary, (**h**) exhumed rock boundary

4.3 Age of the Covered Karst

The covered karst can develop through either syngenetic covered karst formation or postgenetic covered karst formation (Veress 2000a, 2009). In the former case, the geomorphic evolution of the bedrock and the cover takes place simultaneously. Parallel to the formation of the shaft (grike) on the bedrock, the cover is transported into it. Parallel to the transportation of the cover, a doline develops. In the case of postgenetic karstification, the landform developed on the bedrock is older than the one on the cover. In the case of postgenetic karstification, the shaft can originate in two ways. One of the possibilities is that the shaft develops on bare karst. Later the karst is buried, and a doline comes about when the substance removed from the cover fills the shaft. In the other way, the shaft is formed under the cover. Here, too, doline develops on the bedrock as the deposit transported from the doline fills the shaft and then the cover with its doline is destroyed and a repeated burial follows. A new doline is generated on the surface of the new cover if the previous fill of the shaft of the bedrock is reworked by percolating waters into the cavities deeper in the karst.

Covered karsts can be recent or covered paleokarsts. The recent karst is developing or modified at the present time, while the covered paleokarsts do not develop further. This karst type is either buried or eroded. Both the covered paleokarst and the bare karst can play a role in or influence recent covered karst formation. Still they exert a joint control on the formation of present-day karsts or in the development of covered karst phenomena. Two types are common: one is a covered karst terrain constituted by the cover created or preserved in the area of the surface paleoform and the other type are chimneys, shafts, grikes which are filled up and then the material deficit is inherited over the cover through the loss of the fill (see above). Covered paleokarsts are not easily distinguished from bare paleokarst since it is difficult to establish the date of origin of bedrock form under the cover. Thus, the bare karst forms could have been subsequently buried or the cover could have been eroded after covered karst formation.

Stone forest karst is a covered paleokarst (cryptokarst). Stone forest karst has developed under the basalt cover very early, and in the present, it develops further under laterite (Song and Liang 2009). Consequently, the previous covered karst became differentiated genetically. In the exhumed parts of the Lunan stone forest, open karst formation also takes place (Fig. 4.4), while under sediment covered karren formation is active (Song and Liang 2009), which is now different from the previous process since it happens under laterite instead of basalt. The development of the small-size features formed under laterite is modified or they lost their active character as a consequence of further exhumation (Fig. 4.5).

Fig. 4.4 Karren of the Lunan stone forest formed under barren, non-karstic cover (*1*)

Fig. 4.5 Exhumed karren features of the Lunan stone forest, formed under cover sediment *1* crack along bedding plane, *2* major chimney ruin, *3* minor chimney ruin

4.4 Classification of Covered Karst According to Their Rocks

4.4.1 Bedrock Material

Bedrocks can be limestone, chalk, dolomite, gypsum, anhydrite, halite or occasionally marls and highly calcareous conglomerate. The character of the bedrock determines the extension and landforms of the covered karst. Covered karst on gypsum and halite occurs more rarely and of more limited areal extension than covered karst on limestone. At the same time, the density and size of landforms on gypsum and halite covered karst types mostly exceeds those of covered karst on carbonate rocks. The more soluble is the bedrock (gypsum, halite), the deeper solution will cause karstification on the surface. Breccia pipes lead from well-soluble rocks in greater depths to the surface. The explanation is that the debris of well-soluble rocks is also dissolved and further, and further debris accumulation ensues, enhancing the upward growth of the pipe.

On dolomite covered karst seldom develops, probably due to the poor intensity of solution of this rock type. The exceptions are the tropical karsts and the covered karst of high mountains (glaciokarst) in limited extension, e.g. in the Dolomites (Figs. 4.6 and 4.7).

Since the features formed in bedrock are short-lived and rapidly inherited, on gypsum and particularly on rock salt, syngenetic karstification is predominant and postgenetic karstification is less typical. Deep-seated gypsum and rock salt, however, are exceptions. The hollows formed in them are more enduring and also allow postgenetic karstification. After inheritance, new features are created in the bedrock and may cause renewed form generation in the cover. With more intensive solution process, in the case of gypsum and rock salt covered karsts, forms develop more rapidly even on the cover. During the solution of such rocks, the salinity of water increases, and under tundra or taiga conditions, salinity contributes to the survival of the flow system since saltwater does not freeze under 0 °C temperature either.

Some rocks (like dolomite) can be bedrock, and, in the case of rocks particularly susceptible to karstification (gypsum, rock salt), they can also constitute the cover. The karstifiable cover rocks are permeable and conduct water to the rock salt or gypsum beds.

Fig. 4.6 Subsidence dolines on dolomite, in moraine in the Grava Lunga cirque (Dolomites)

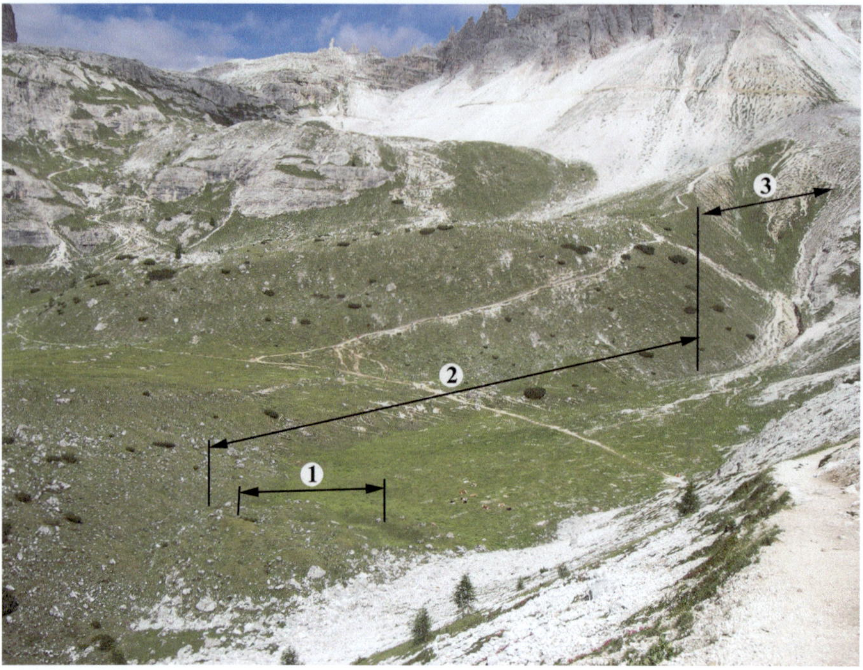

Fig. 4.7 Depression of superficial deposit in the Grava Lunga cirque (Dolomites) *1* doline zone of depression where material transport takes place into the karst, *2* depression, *3* ravine in morainic deposit

4.4.2 The Cover Material

The cover of the karst can be homogeneous (no lateral change in material properties) or heterogeneous (horizontally variable material properties). A vertical heterogeneity of the non-karstic cover is also observable: the rocks overlying each other differ in porosity, permeability and stability. With laterally changing porosity and permeability, the site of water transfer from the cover to the bedrock may also vary and influence the site and intensity of karstification as well as the density and size of landforms generated on the cover. The variable stability of the cover leads to variations in the way, extent and rate of inheritance.

Zuffordi (1976) classified the cover sediments of karst by their place of origin. He identified autochthonous deposits (if the cover derives from the karstic rock), allochthonous deposits (the cover derives from a terrain outside the karst) and para-autochthonous deposits (the cover forms in situ from non-karstic rock). According to the way of transport (and origin), Fillippov (2004) distinguished covers of alluvial, deluvial, proluvial and eluvial origin.

4.4.2.1 Autochthonous Deposits

The autochthonous deposit can be solutional residue of the cover or can form through the mechanical weathering of the bedrock material. Terra rossa is a typical solutional residue, formed in humid regions (Lang's rainfall factor of 30–60) with 14–19 °C mean temperature (Zámbó 1998) from the insoluble, mainly clay contaminations of limestone. According to some authors, temperate terra rossa varieties are alien to the karst, weathering residues of non-karstic rocks (Andrusov et al. 1958) or mixed karst-alien deposits and solutional residues. On mediterranean karst, it can be regarded a typical in situ product (Zámbó 1998). Under tropical climate, the covering laterites and laterite bauxites have iron oxide contents of 80–90 % and aluminium oxide of 5–10 %; in the case of bauxites, the latter can be even over 60 % (Fekete 1988). Laterites are of several times 10 m depth and consist of different zones under the tropical soils, from top to bottom: duricrust, mottled zone and pallid zone (Grimes and Spate 2008; Grimes 2009). In temperate karsts, laterites are a relict cover (Zámbó 1970) since, according to Stefanovits (1976), the properties of terra rossa and red clay soils in Hungary indicate mediterranean environments of origin and have been largely redeposited. This is confirmed by the investigations of Zámbó (1970), and redeposition is also observed elsewhere, e.g. on the Pádis plateau. While the higher terrains lack terra rossa, it is exposed in roadcuts in several metres of thickness on lower terrains (Fig. 4.8). The same is observed in

Fig. 4.8 Terrain with karren below terra rossa (Pádis)

mediterranean karst. At higher elevations, terra rossa is missing and grikes are filled by clayey soils of darker colour (for instance, in the Livno polje and its environs in the roadcuts of road number 76). The formation of terra rossa or laterite, however, does not necessarily happen even under suitable climate. Thus, if the limestone is very pure, no or very little residue is produced. The terra rossa can be washed into the joint network of the limestone (Zámbó 1970) – particularly if it is of little thickness. In terra rossa, the ratio of clay minerals is high (Zámbó 1998); decalcification, the leaching of silica and the accumulation of iron and aluminium oxides are of remarkable extent. Terra rossa of the Aggtelek Karst (Hungary) is silty clays. Since limestone disintegrates during dissolution, considerable amounts of limestone debris can be present even in terra rossa of small thickness. Such terra rossa with limestone debris is often observed in mediterranean karst areas in Croatia and Bosnia.

Terra rossa can be partially but not totally impermeable. According to Zámbó (1970), in the Aggtelek Karst, karren formation took place under the terra rossa. This is favoured by the low carbonate content of the (clayey variety of) terra rossa, its CO_2 production and the fact that impermeability can only develop gradually in this rock in wet weather. Zámbó (1986) claimed that the upper part of the terra rossa is partially saturated with water and percolation between the grains is possible. At the same time, in the lower part of the terra rossa cover, the saturation of pores is complete, clay minerals swell and prevent percolation. As it is known, solution is not only possible under terra rossa, but even under laterites (Song and Liang 2009; Slabe and Liu 2009; Grimes and Spate 2008) or bauxites (Papastamatiou 1964; Businszkij 1964; Sweeting 1973). The karstification under bauxite is called secondary karstification. Bauxite formation and karstification were often parallel processes (Bárdossy 1977). When the groundwater got into contact with the bauxite body, secondary karstification stopped, and, following uplift and the sinking of groundwater table, it resumed (Peljtek 1971). Karstification under bauxite is indicated by the dislocation of bauxite blocks towards the bedrock and reaching a position between the bauxitic clay and kaolinite (Bárdossy 1977) as well as the oblique position of bauxite beds, its cover and bedrock, while the dolines of the bedrock are vertical (Combes 1969). This means that they formed after the sequence acquired an oblique position. Karstification under bauxite does not only involve the generation of dolines but also that of cavities (Knechtel 1963). According to Bárdossy (1977), secondary karstification takes place with cover thickness less than 10 m.

Terra rossa thicknesses can be highly variable because of erosion and redeposition. Terra rossa primarily fills up paleokarst features or acquires greater thicknesses there. In the paleodolines of the Aggtelek Karst, the thickness of terra rossa exceeds 10–15 m (Zámbó 1970), but in the mediterranean karst, thicknesses vary greatly in exposures not distant from each other. Thickness depends on the duration of solution, the contamination of the limestone, the rate of redeposition and, in the case of partially alien origin, on the rate it was transported onto the karst.

On terra rossa soils (red brown earth, rendzina) may form and their CO_2 production (combined with that of terra rossa) contributes to the karstification of the underlying rock (Zámbó 1970, 1993, 1998). At the same time, the terra rossa debris inhibits solution and if it is washed into the underlying joint system reduces the rate of water flow and eventually also inhibits solution. According to the data by Zámbó (1970), the terra rossa of the Aggtelek Karst fills in the joint and solution cavity network to 5–7 m depth below the dolines.

The role of terra rossa in covered karst formation can be summarised as follows:

- Percolating waters cause solution on the contact of terra rossa with the karstic rock, and thus karstification below the terra rossa can begin or resume. The side slopes of the doline are affected by solution on the surface of contact with the terra rossa fill resulting in lateral corrosion (Zámbó 1970). At the terra rossa/rock contact, the downward percolating water reaches the karst at the deepest point of the doline and produces a system of passages and chimneys in the bedrock. This process can induce the formation of covered karst features (ponors, subsidence dolines functioning as ponors, depressions of superficial deposit) on the area of dolines where there is terra rossa fill.
- The rainwater percolating through the terra rossa transports the passage fills deeper into the karst and induces material deficit and postgenetic covered karst formation in the cover.
  - The rainwater percolating through the terra rossa generates solution on the bedrock and syngenetic covered karst formation on the cover.

The generation of the cover of karstic origin from the bedrock can happen through frost shattering, mountain failures, rock avalanches and glacial erosion. The above sediments primarily occur in high-mountain karsts, but frost-shattered debris is also typical in tundra and temperate karsts. Debris also generates during solution processes (Veress and Péntek 1996). The morainic deposit can be affected by considerable transport and redeposition, while frost-shattered or solutional debris mostly remains in situ.

The reworking of debris and particularly of the morainic deposit mostly takes place by fluvial water. During reworking or the ensuing solution, the grain size of morainic deposit reduces, clay content increases and carbonate content decreases, affecting solution under the cover. According to Williams (1966), if the cover has large amounts of carbonate debris, the percolating waters dissolve the debris and reach the bedrock saturated. Thus, no solution can occur there, and no syngenetic karstification takes place on the cover – or only under certain circumstances. (However, since the seeping water is able to flush the cavities of the bedrock and removes their fills, postgenetic karstification is observed.)

The solution residue of rock salt in rock salt karsts is clay.

4.4.2.2 Allochthonous Deposits

The debris alien to the karst can be of fluvial, glacial, marine, aeolian or volcanic origin. Consequently, the cover can be clay (weathering residue, of fluvial or marine origin), loess, silt (fluvial, marine or aeolian) sand (fluvial, marine) gravel, sandstone, conglomerate, flysch, molasse as well as volcanic rocks, volcanic tuffs, volcanic debris and blocks. Because of the changes in the conditions of rock formation, the cover can be constituted – both vertically and horizontally – of not a single but several rocks. Reworked or mixed varieties of the above-mentioned rocks also occur. Any rock (one or more) can be mixed with any other, and the proportions of constituents can vary on a wide range, and, therefore, the cover of the covered karsts can be variable. Loess, clay and their mixed varieties and rocks of various clay contents are common.

4.4.2.3 Paraautochthonous Deposits

In the frost-shattered debris, fragments embedded in the limestone and alien to the karst (e.g. siliceous debris, sandstone and clay), which increase in amount during the partial or complete dissolution of limestone fragments. It is common in high mountains that on the steep slopes and floors of large-scale karst landforms (paleodolines) – particularly if those are located on the floors of glacial valleys – the debris of mountain failures and rock avalanches is accumulated. In these accumulations silica or siliceous debris may occur. On karst of cool and wet climate (British Isles, Scandinavian Peninsula, but in the Alps too), peat may form which may create a cover too.

Consolidated rocks are exposed in anticline structures or through glacial erosion. The removed material contributes to the accumulation of paraautochthonous deposits.

4.5 Covered Karst of Landforms

It is common that in the same karst region, karst landforms of similar age, originated from open karst, occur next to each other but at different elevations. Those at lower elevations could be lined or partially filled by the non-karstic cover sediment if there are non-karstic sediments in their environs, while in those of higher elevation, there is no such fill. The karstic depressions (e.g. dolines) of higher position with solution residue and intensive solution could have also been buried. The development of dolines lined or filled with non-karstic sediments ceases or modifies. In the area of lined or filled dolines, covered karst formation occurs and transforms them into relict features, paleodolines. The eventual paleoform creation is explained by the presence of cover sediment in it. Within the same karst region, in the neighbourhood of such dolines, recently active solution dolines of similar age and higher position and without cover sediment occur, indicating autogenic karstification.

Covered karst can also develop in the area of different landforms through the accumulation or erosion of the non-karstic cover. The removal of the cover deposits may change the conditions and character of the covered karst. The following changes are observable: the non-karstic rock cover thins out (consequently, the rate of karstification increases), it is dissected into separate parts (a mosaic of covered karst and bare karst terrains result) or rocks originally in lower position are exposed which differ in cohesion and conductivity from the rocks eroded from above. Thus, to the effect of glacial erosion, a consolidated non-karstic rock is exposed (crypto-karst formation). Finally, the local erosion of the cover, with transport directed into the karst leads to the formation of depressions of superficial deposit.

The covered karst associated with landforms can originate from the burial of landforms of non-karstic origin and of paleokarst landforms. The resulting patchy covered karst, the extension and shape of the patch, is largely controlled by the size and shape of the enclosed depression. The burial of the landform can be only partial. Then the extension of the covered karst is smaller than that of the bearing landform. The size, shape and location of the covered karst patch depend on a range of factors: the rate of sediment transport, its agent (glacier, mountain failure and others) and the dimensions and morphology of the bearing landform. If the bearing landforms are arranged in rows, the covered karst terrains also acquire a similar pattern. In polygonal karst, the pattern of the covered karst details is less oriented; it is irregular.

On the karst, accumulation attains different degrees. It may happen that during burial no infilling occurs in the case of the burial of karren terrains or plateaus as well as in the case of large-scale landforms (poljes, intermountain plains). If the sediment is transported into the minor depressions (e.g. dolines) of the karst, the process is infilling. The cover deposit can develop in one (or even more) part(s) of the depression or continuously. In the case of continuous occurrence, it could line (cover of small thickness), fill (in great thickness) or bury (in thickness greater than the depth of the filled landform) the depressions. The cover can extend over the depression in uniform or variable thickness. In the case of fill (e.g. at mass movements), the fill of the depression is often of variable thickness. On the margins of the depression, the cover is thick and is thinning out with increasing distance from there. Uniform cover thickness mostly results from fluvial accumulation.

The sediment can cover plain surfaces (Fig. 4.9a) or – as it has been mentioned – non-karstic depressions of tectonic or erosional origin and karstic depressions. Covered tectonic depressions include tectonic grabens (Fig. 4.9c) and synclines (Fig. 4.10). As it has been mentioned, the resulting covered karst terrains are elongated in strike direction. Covered karst develops on horsts of different elevations. The cover, along with the covered karst, can also be uplifted to different altitudes after its formation (Fig. 4.9b–d). In an uplifted position, the cover of the covered karst can be dissected (Fig. 4.9b).

If the covered karst develops in the area of a salt diapir, its shape and extension coincides with the shape and extension of the exposed salt diapir. Such covered karsts occur in the Transylvanian Basin (at Parajd) (Fig. 4.9e) and at other places on Earth (for instance, near the Caspian coast, in Texas and Iran).

The landforms of erosional origin on covered karst include river valleys (Fig. 4.11a–g, k), glacial valleys (Fig. 4.11h–j) and marine terraces. Covered karst most often occur on the floors of river valleys but sometimes occurs in valley sides. Such covered karst can be cited from the taiga karst of the Middle Lena valley, the Neretva valley and the valleys of midlatitude medium mountains (e.g. the Bakony

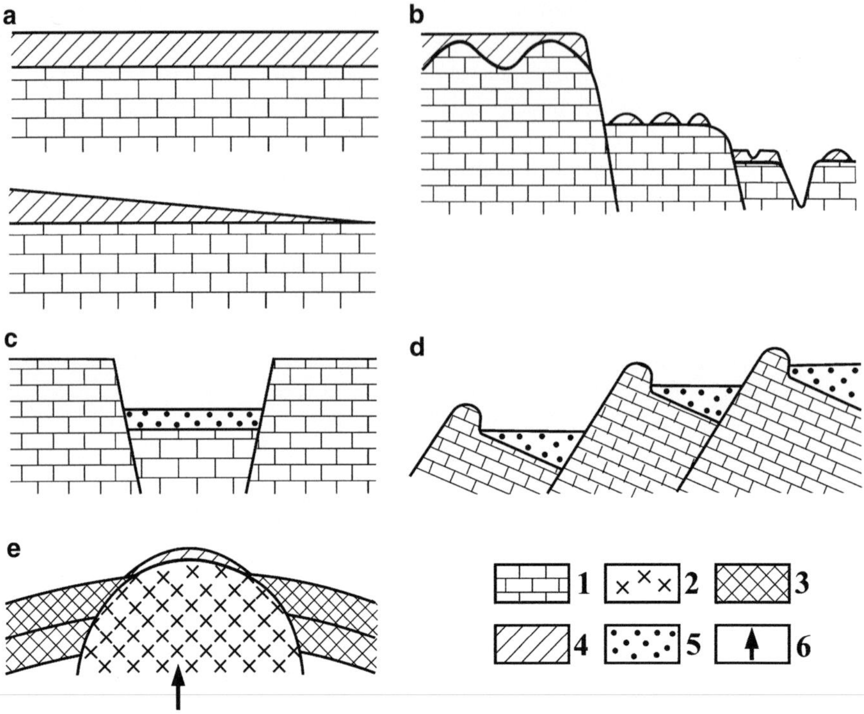

Fig. 4.9 Theoretical types of covered karst governed by structure (**a**) flat limestone surface where the cover is either continuous or wedges out; (**b**) the blocks acquired various elevations (with uneven depressions and elevations of bedrock on some blocks, while flat on other blocks, where the cover can be continuous, detached into patches and dissected by valleys to various extents; (**c**) the block forms a graben or basin, where the limestone is under cover sediment; (**d**) the blocks subsided in step-like manner or occasionally tilted, and the cover is preserved on the blocks since the marginal karst elevations hinder denudation; (**e**) rock salt diapir is exposed and covered by the contamination of salt and the sediment transported there, *1* limestone, *2* rock salt, *3* non-karstic rocks cut through by rock salt, *4* impermeable cover sediment, *5* permeable cover sediment, *6* rising salt

Mountains). In these cases, the covered karst terrain is elongated and similar in shape to the valley floor. The cover is fluvial deposits. Covered karsts develop on fluvial terraces (Gvozdetskiy 1965, 1981; Soriano and Simón 2001) and also on accumulational landforms, for instance, on alluvial fans (Fig. 4.12) and debris fans.

Covered karsts often occur in glacial valleys (Fig. 4.11h–j). In these valleys, cover sediments are glacial deposit, frost-shattered debris, lobe of mountain failure or a mixture of these (Fig. 4.13). On glaciokarsts morainic material is the most important, which can be exclusively of glacial origin, unworked, of variable grain size with the predominance of major blocks (of several tens of centimetres diameter). In the glacial deposit reworked by fluvial processes, fine grain sizes (maximum several centimetres diameter) are predominant and the grains are worked. The glacial deposit can be reworked once or several times. The material of the transformed

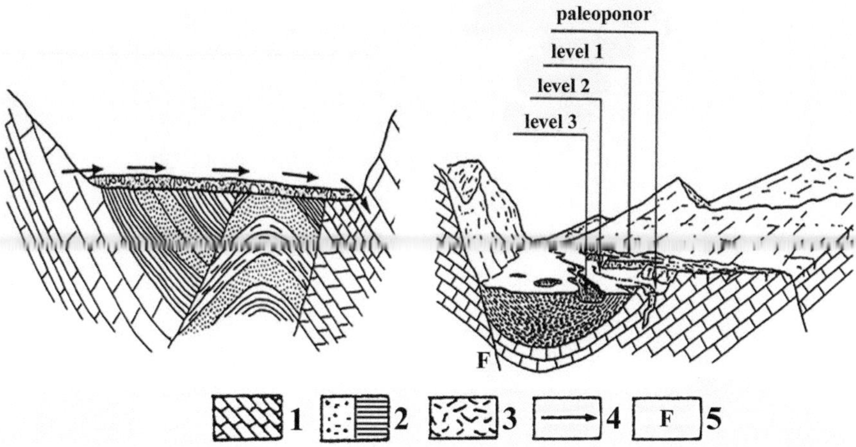

Fig. 4.10 Structures controlling the distribution of non-karstic rocks in tectonic poljes (Delannoy 1997; Ford and Williams 2007, modified) *1* limestone, *2* flysch or dolomite, *3* alluvium, *4* permanent watercourse, *5* fault

moraine is further disintegrated in situ following weathering processes. Such moraines are enriched in silt and clay. In mixed glacial deposits, the original rock material is mingled with frost-shattering products and the material of mountain failures. If it is only mixed with frost-shattered debris, grain size composition is shifted towards finer grains; if it is mixed with the material of mountain failures, it is shifted towards the coarse grains. Moraines can be of different ages. On the floor of the same glacial valley, two or three moraines of various ages and clay contents can occur (Fig. 4.14). Older moraines are usually of higher clay content and transformed to a greater degree than younger moraines. On the floor of the glacial valley, moraines of various ages often occur next to each other. If the moraine is built of siliceous limestone, it includes silica blocks. All these can cause the transformation of the primary (not reworked) moraine in both horizontal and vertical direction (Fig. 4.13).

The extension of the covered karst does not only depend on the accumulation of the moraine (influenced by ice motion, amount of moraine material and bedrock morphology) but also on the dimensions of the valley and its relative positions in the mountains. The dimensions and shape of the covered karst terrain are largely controlled by the position of cirques relative to the glacial trough, the size of cirques and the extent of connectivity and character of neighbouring cirques (Fig. 4.15). The cover is seldom continuous and mostly patchy (Fig. 4.16). In case of continuous cover, it is rather of circular extension in the cirques.

In the glacial troughs, the shape of the covered karst is elongated in the direction of the valley axis. The cover is of variable thickness as on the glacially denuded surface depressions (rock basins, paleodolines), and elevations (roches moutonnées, cuestas and hogbacks) alternate. On such terrains, the moraine often thickens out in depressions, but the opposite is also possible: the moraine is thicker on inter-

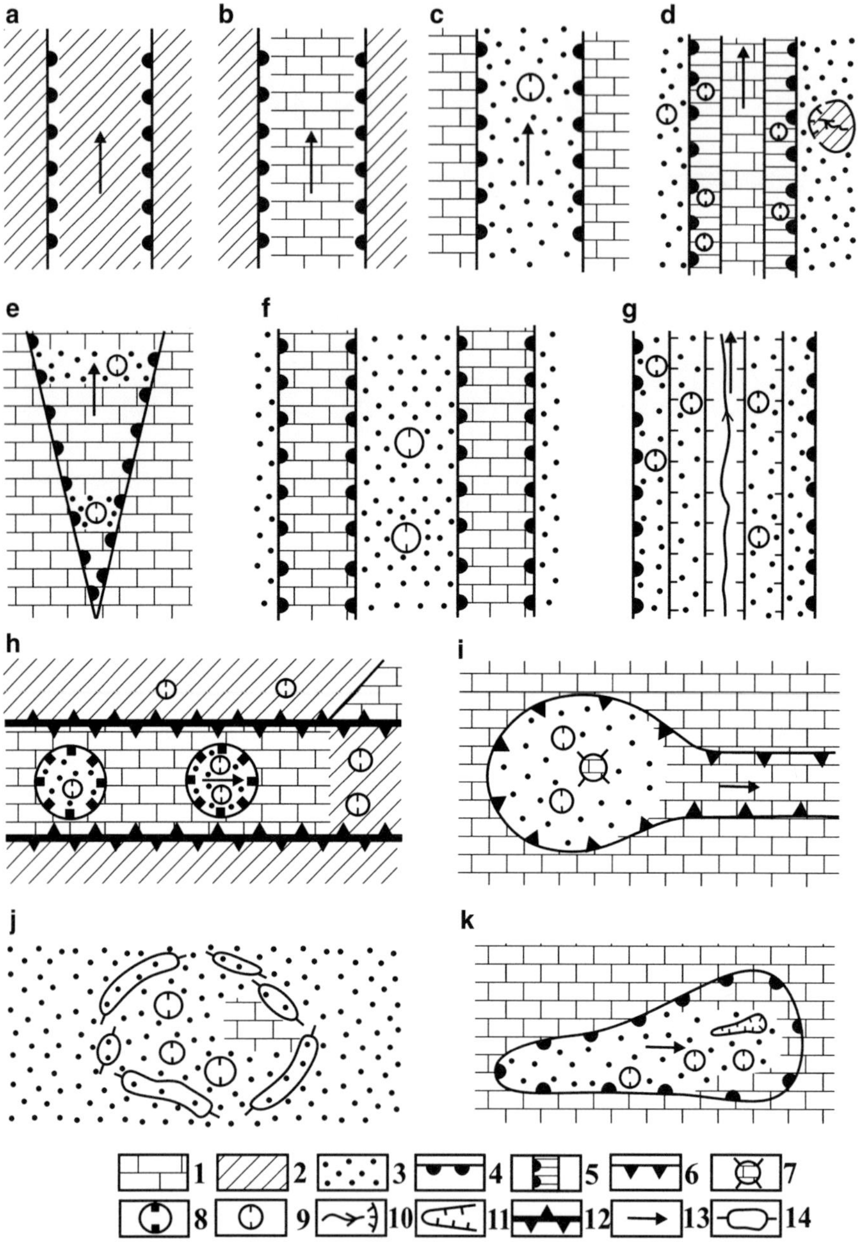

Fig. 4.11 Covered karst in valleys (covered karst patterns determined by valleys): (**a**) the valley has not yet cut through the impermeable cover, uniform cryptokarst or buried karst; (**b**) the valley floor incises into the limestone, non-uniform cryptokarst; (**c**) the valley floors are under permeable cover, striped-concealed karst results; (**d**) the cover develops in the valley side and its environs (the latter occasionally with impermeable patches), concealed karst in the valley side, both concealed karst and cryptokarst on interfluves; (**e**) patches of permeable cover on the floor of epigenetic valleys,

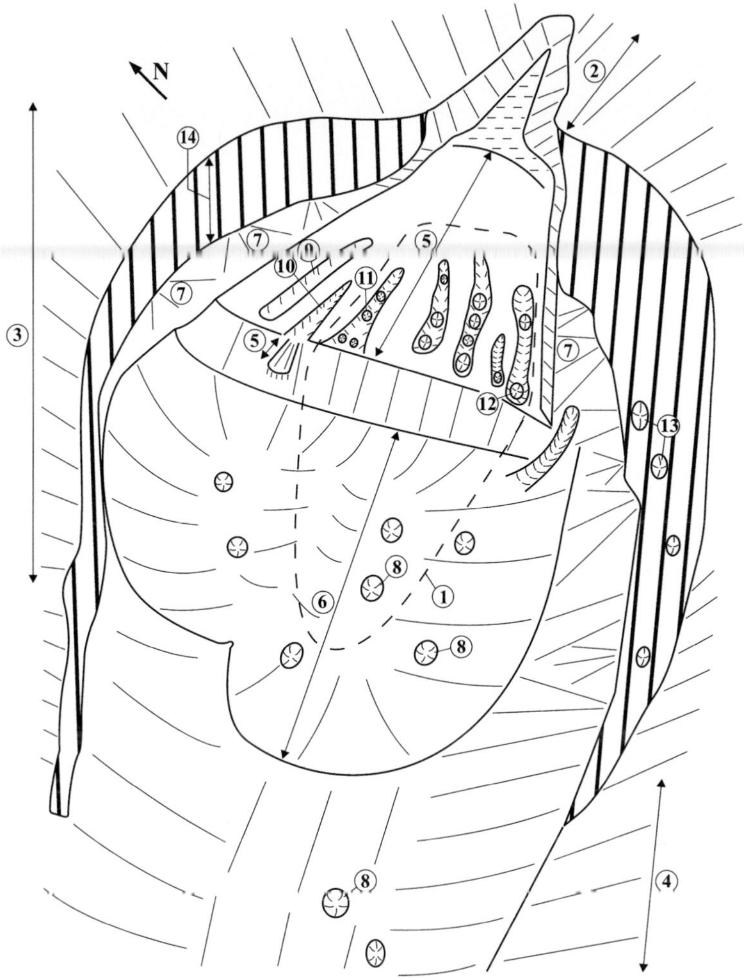

Fig. 4.12 Karstification on the alluvial fan of a cirque in plateau margin position (Dachstein, Austria) *1* boundary of mapped area, *2* valley, *3* cirque, *4* glacial trough, *5* dry alluvial fan, *6* half filled uvala, *7* debris fan, *8* subsidence doline, *9* blind-ended gully, *10* gully, *11* doline on gully floor, *12* doline in gully end, *13* cavity, cave *14* side slope of cirque valley

Fig. 4.11 (continued) patches of concealed karst on the valley floor; (**f**) the cover is preserved on interfluves between epigenetic valleys, concealed karst stripes on interfluvial ridges; (**g**) concealed karst on river terraces; (**h**) alternating open (with lined and buried paleodolines) and covered valley floor sections in the glacial troughs, striped or patchy covered karst; (**i, j**) cirque covered karst buried under glacial deposits; (**k**) local covered karst on the floor of blind valley, *1* limestone, *2* impermeable cover, *3* permeable cover, *4* river valley, *5* side of river valley, *6* glacial valley, *7* roche moutonnée, *8* solution doline (paleodoline), *9* subsidence doline, *10* ponor, *11* ravine, *12* aréte, *13* surface slope, *14* morainic hill

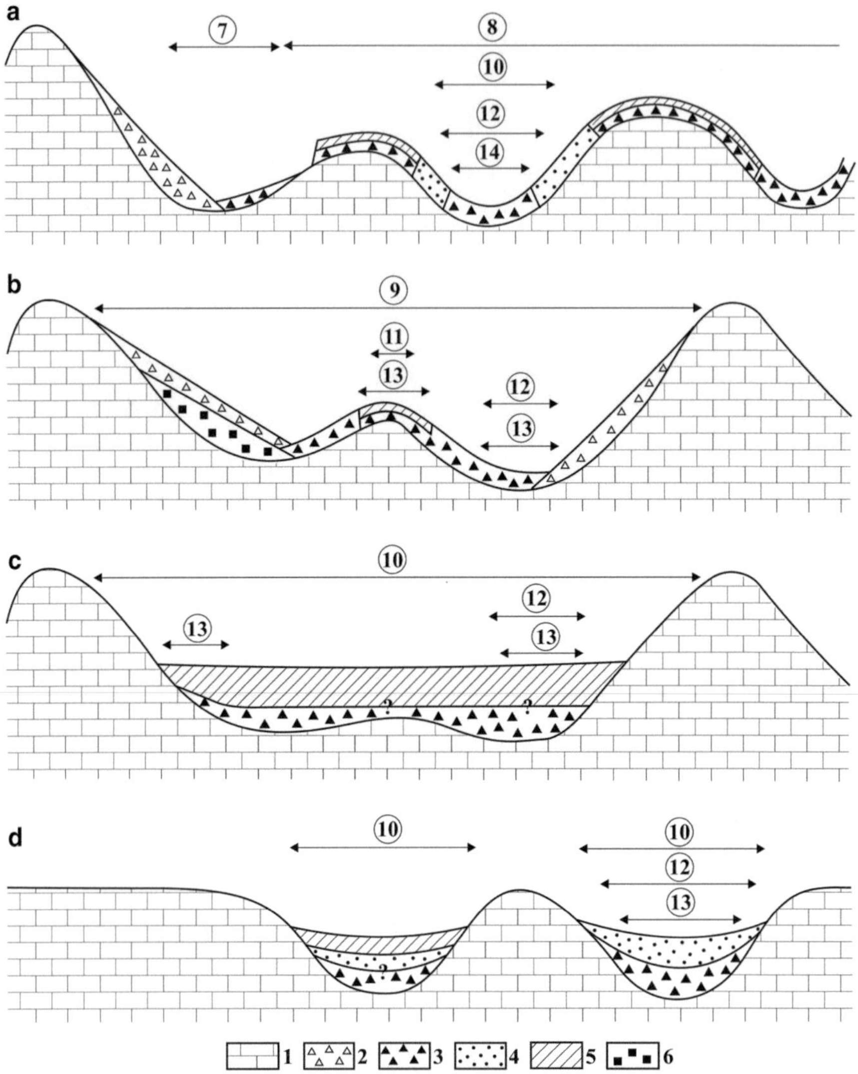

Fig. 4.13 Cover sediments of high-mountain karst: (**a**) cover sediments along glacial valley, (**b**) cover sediments in the cross-section of the glacial valley, (**c**) in a single paleouvala on plateau, (**d**) in doline row on plateau, *1* limestone, *2* frost-shattered debris, *3* coarse-grained morainic deposit, *4* fine-grained morainic deposit, *5* morainic deposit with high clay content, *6* rock blocks of mountain failure, *7* cirque, *8* longitudinal profile of glacial trough, *9* cross-section of glacial trough, *10* paleodoline and paleouvala, *11* roche moutonnée, *12* bedrock shafts and giant grikes, *13* deep subsidence dolines with gentle slopes, *14* shallow subsidence dolines

Fig. 4.14 Tongue of young moraine (in the cirque of a glacial valley, Hochschwab, Austria): *1* young moraine, *2* subsidence doline on older morainic deposit, *3* fossil subsidence doline (with lakes) of older morainic deposit

depression terrain. Variations in the thickness of the moraine are also caused by rows of lobes of mountain failures, regularly repeated rockfalls and rock avalanches (Fig. 4.17).

The covered karst on marine terraces and abrasional peneplains has a greater extension parallel with the coast, smaller in a direction perpendicular to the coast and usually developed on mountain margins (Fig. 2.10).

Particularly in the area of medium mountain karsts, blocks and horsts (plateaus) are often buried. Mountain covered karst develop where the blocks are covered. The resulting covered karsts are controlled by the dimensions and shapes of the bearing blocks. The cover sediment is often fallen dust (loess). Plateaus are often partially covered in a patchy pattern. Patchy cover may result from denudation too, which leads to the alternation of covered and uncovered terrains (mixed composite karst). Patchy cover occurs in the case of two types of cover sediment, where impermeable patches (cryptokarst) alternate with permeable patches (concealed karst) and uncovered patches (bare karst) (mixed karst). In the area of an uplifted block, the older, probably impermeable, cover is dissected into patches during denudation, and the covered karst is interrupted by valleys. In this case, mixed cryptokarsts adjusted to the extension of blocks form. Subsequently the terrain of patchy cover can be repeatedly covered, for instance, by permeable material (loess). This results in the

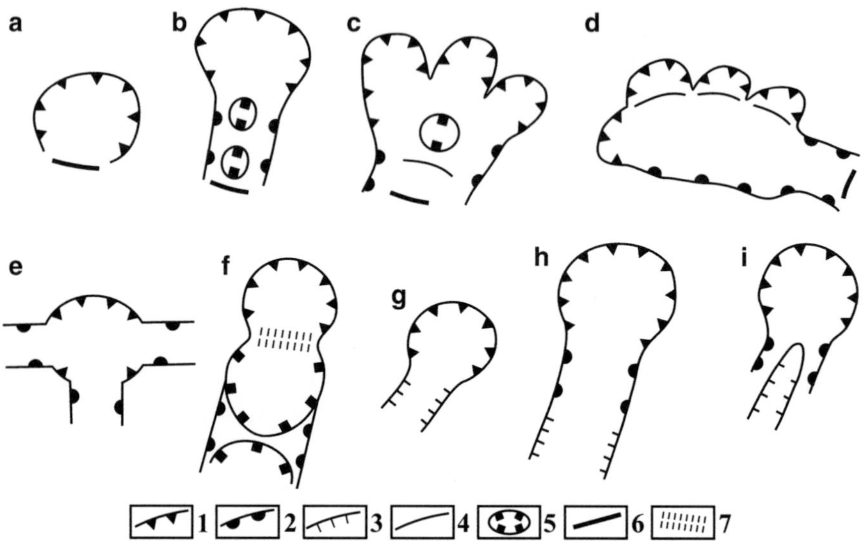

Fig. 4.15 Glacial trough and cirque types (Modified after Veress 2012b): (**a**) no glacial trough, (**b**) cirque and glacial trough, (**c**) glacial trough from merged cirques, (**d**) cirques aligned on the margin of a glacial trough, (**e**) several glacial troughs branching out from a cirque, (**f**) glacial valley formed from preglacial giant dolines, (**g**) river valley linked to cirque, (**h**) river valley linked to glacial trough, (**i**) river valley partially formed in glacial trough, *1* cirque, *2* glacial trough, *3* river valley, *4* transverse scarp, *5* interglacial giant doline, rock basin, *6* moraine, *7* threshold between karst depressions

generation of covered karst of mixed karst, mixed cryptokarst or mixed concealed karst types, e.g. in the Bakony Mountains, where the Oligocene–Miocene gravel mantle of high clay content (Korpás 1981) and low permeability has been dissected into isolated patches during denudation (mixed cryptokarst). Where the terrains between the patches of gravel mantle are covered by loess, it is called covered karst with loess and gravel patches with open karst patches (mixed composite karst). Both mixed karsts and mixed composite karsts are widely spread in midlatitude medium mountains and on glaciokarsts.

Covered karst can also develop from karst terrains dissected by karst landforms (Fig. 4.18) and on the area with karst features. Since the locality of karstification depends on cover thickness (see Chap. 7), topography is a major control on the distribution of covered karsts. With the partial removal of the cover sediment, the cover is dissected as isolated patches and preserved in karst depressions. In the area of these patches, covered karst develops. According to the character of the cover, it can be cryptokarst or concealed karst or, where the cover is very thick, buried karst. Between covered karst patches, open karst patches are found, and their extension increases during the denudation of the cover.

In the area of karst landforms, covered karst primarily develops if the karstic rock is partially or entirely buried, and thus sediment is transported into the karst depression. The intensity of covered karst formation is enhanced by the karstic origin of

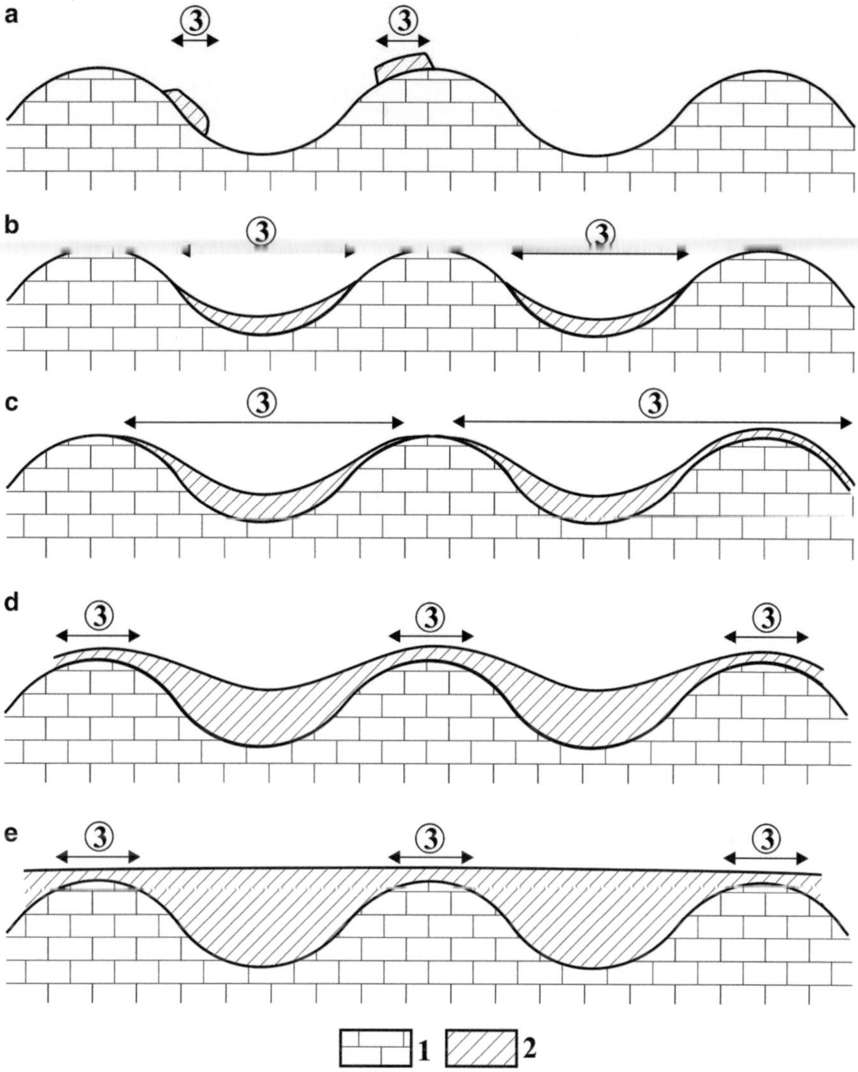

Fig. 4.16 Morphological types of moraine on floors of glacial troughs with paleodolines: *1* limestone, *2* moraine, *3* covered karst formation, (**a**) moraine patches, (**b**) moraine on doline floor, (**c**) ridges between dolines free of morainic cover, (**d**) uniform moraine cover, (**e**) morainic deposits filling and burying dolines

the bearing landform of covered karst. The reason for this is that there are previously created passages, chimneys on the bottom of paleoforms which convey the cover deposit into the karst. The rate of karstification is also increased by the fact that rainwater fallen onto the area of the bearing karst landform can only infiltrate or flow into the karst at this locality. The mode of karstification is also specific in the

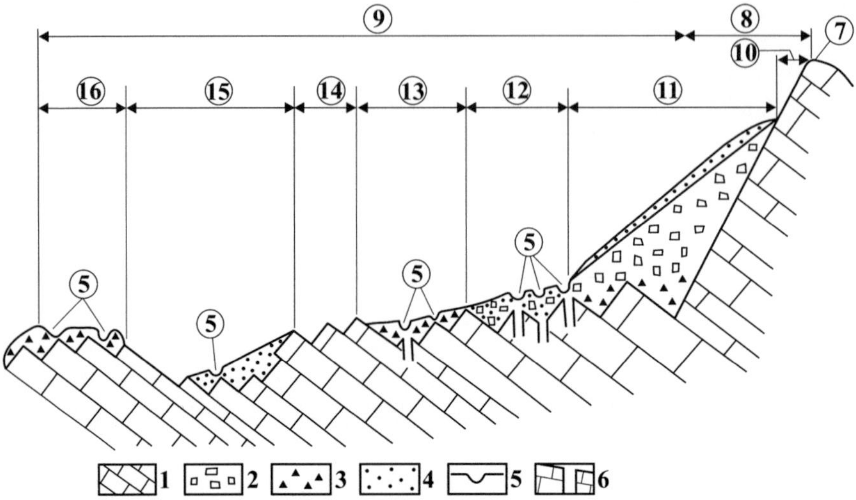

Fig. 4.17 Sites of covered karst formation in the cross-section of a glacial trough: *1* limestone; *2* blocks of mountain failure; *3* moraine; *4* frost-shattered debris; *5* subsidence doline; *6* chimney, giant grike, shaft; *7* aréte; *8* side slope of glacial valley; *9* floor of glacial valley; *10* slope segment with karren; *11* large and thick heap of mountain failure (no karstification); *12* thin front of mountain failure with karstification; *13* moraine ruin under karstification; *14* karren zone without debris (schichttrippenkarst); *15* rock basin or paleodoline with debris accumulation on its side slope; *16* zone of reworked morainic deposit

areas of close depressions of glacial origin (rock basins). The special character of the process is manifested in the higher intensity of karstification or in the exposure of non-karstic bedrock which favours cryptokarst formation.

Covered karsts occur in high-mountain glaciokarsts, dolines of medium mountains and tropical karsts (Figs. 4.19, 4.20, 4.21, 4.22 and 4.23), uvalas (Fig. 4.24a–c), doline and uvala groups (Fig. 4.25), ponors and blind valleys (Figs. 4.11k, 4.24d, 4.26 and 4.27). Giant grikes, shafts and shaft dolines can also be buried and form covered karst in the high mountains of Europe (Alps, Pyrenees, Dinaric Mountains). Covered karst features also develop in the fills of caves which lost their ceiling (Knez and Slabe 1999), in the poljes of the mediterranean karst (Figs. 4.24e and 4.28), in the depressions of the polygonal karst (Fig. 4.24f), in both tropical and subtropical varieties, in the intermountain plain areas of fenglin (Fig. 4.24h), in the depressions of fengcong (Fig. 4.24g) and in cockpit karst (Fig. 4.23).

Covered karst extends to the entire area of such landforms or to its sections of variable size. The covered karst patch may extend beyond the margin of the paleoform over to parts of its environs. As a consequence, extremely variable covered karst details are generated. As it has been mentioned above, the karstification of the bearing karst depression is modified just because of the cover formation or glacial erosion, and this modification is not restricted to the appearance of covered karst. Thus, on glaciokarsts solution, dolines generated after glaciation occur in preglacial or interglacial dolines. The depressions of the tropical karst remain active even after the formation of covered karst and the dolines keep growing.

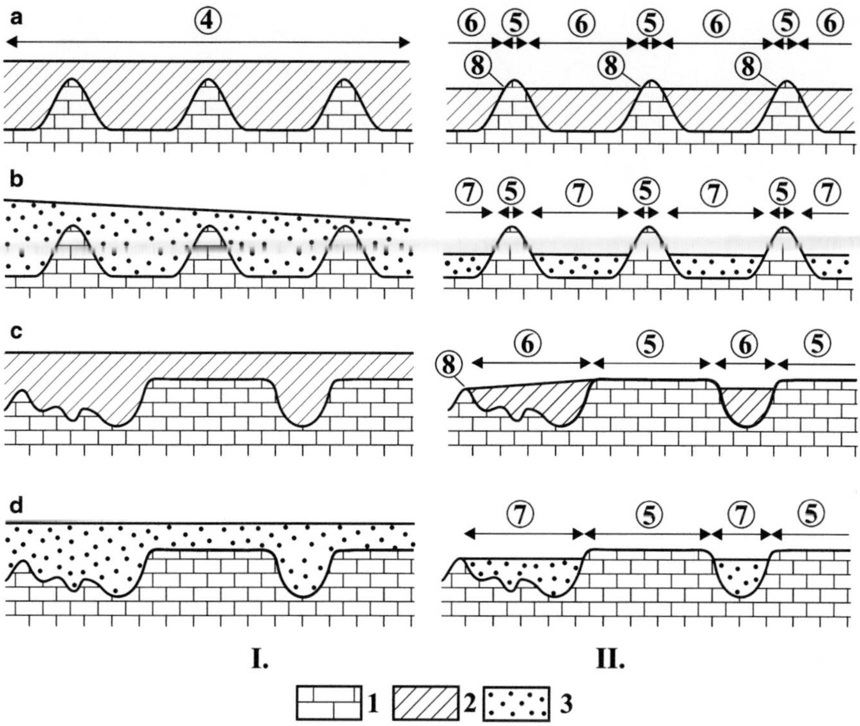

Fig. 4.18 Covered karst patterns controlled by paleokarst landforms (**a**, **b**) the bedrock is dissected by mounds, (**c**, **d**) the bedrock is dissected by dolines: *I* initial stage, *II* advanced stage, *1* limestone, *2* impermeable cover, *3* permeable cover, *4* buried karst, *5* open karst, *6* cryptokarst, *7* concealed karst, *8* ponor

Covered karst is not continuous in the area of poljes but interrupted by patches of limestone outcrops. This makes the pattern of the covered karst irregular, and cryptokarst and concealed karst patches often alternate. Therefore, in the cryptokarst sections of poljes, valleys, ponors, depressions of superficial deposit occur, while in the concealed karst sections, ravines, gullies, subsidence dolines, dropout dolines and depressions of superficial deposit are found. Whole poljes or polje sections are transformed into depressions of superficial deposit (Fig. 4.28).

In the area of intermountain plains, covered karst can develop if the cover is underlain by karstic rock. The covered karst, however, is uniform neither in this case but constituted of the mosaic of covered karst terrains and open karst patches of various sizes. Along or between rivers, there are concealed karsts and cryptokarst patches (Fig. 4.29). On the area of concealed karsts, suffosion dolines and dropout dolines occur to which ravines and gullies are connected.

In the inherited valley sections of the depressions of the fengcong karst and polygonal karst, isolated patches of covered karst can occur with subsidence dolines and associated networks of ravines (Fig. 4.30, Williams 1971). Major bedrock depressions are dissected by depressions of superficial deposit (Figs. 4.23 and 4.24f–g).

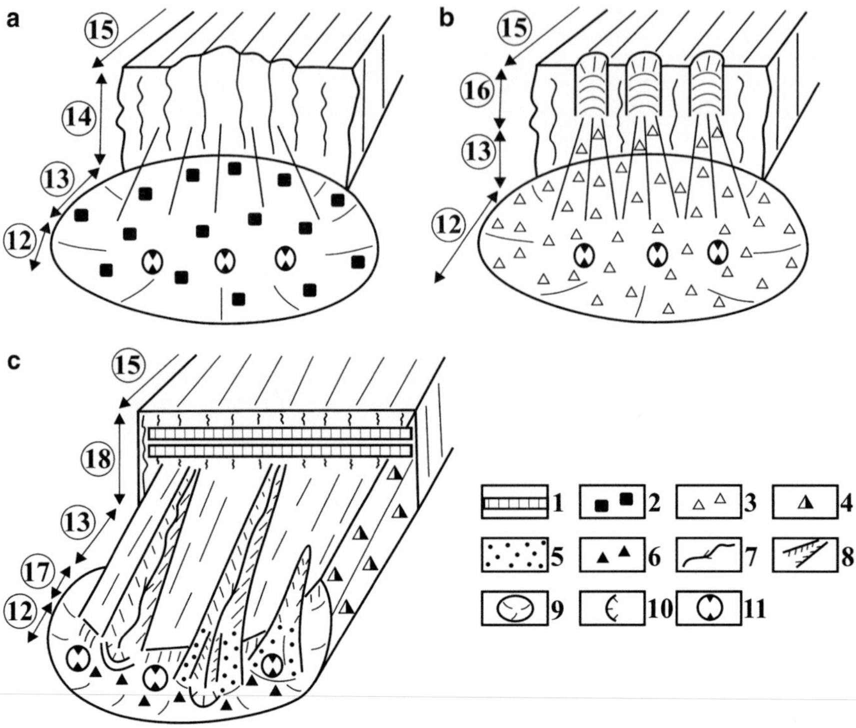

Fig. 4.19 Karstification of paleodolines partially covered on the material of mountain failures (**a**), on debris fans of rock avalanches (**b**) and on fluvial accumulation (**c**) *1* siliceous layer; *2* collapsed blocks; *3* frost-shattered debris with limestone; *4* frost-shattered debris partly of limestone, partly of siliceous material; *5* reworked debris; *6* morainic deposit; *7* watercourse; *8* ravine, valley; *9* paleodoline; *10* ponor; *11* subsidence doline; *12* outer part of cover, where the cover is thin and therefore karstification takes place in the doline; *13* heap of mountain failure, zone of debris fans, buried section of paleodoline (no karstification); *14* side of glacial valley with mountain failure; *15* plateau, aréte; *16* slope of glacial trough dissected with cliff niches and rock avalanches; *17* debris reworked by watercourses (alluvial fan); *18* slope of glacial trough with siliceous intercalations

4.6 Patterns of the Cover Sediment of Covered Karst

4.6.1 General Description

Covered karst developed on various karst types. The type of karst is primarily controlled by karst structure and climate. Therefore, the character and pattern of the covered karst depends on karst structure and climate. According to structure, geosyncline or glaciokarst (folded and nappe structure), block mountain (faulted structure), platform (no structural change of rocks) and salt diapir covered karst are distinguished. According to the climate, covered karsts formed on tundra, taiga,

Fig. 4.20 Subsidence dolines in paleodolines lined with mountain failure heaps and morainic deposits (Durmitor, Surutka, Montenegro) *1* suffosion doline

temperate, mediterranean and tropical karsts are identified. (Tundra karst is distinguished from the temperate karsts.)

The karst regions of folded and nappe structure were substantially uplifted and transformed to glaciokarsts. Covered karst formation was primarily controlled by glacial geomorphic processes and the accumulation of moraines. On the moderately elevated surfaces of platform karsts, there were favourable opportunities for the generation and survival of extensive covered karsts. Block mountain karsts developed on blocks of various elevations. The diapir karst formed as a result of rock salt pushing upwards, and with the salt domes the cover of the salt mass (e.g. its solution residue or cover sediments transported there) was also uplifted. The covered karsts dependent on structure which are also bound to climatic types are treated in the chapter on karsts associated with climate, such as the karsts of geosynclines and block mountains. The former type is widespread on temperate karsts and mediterranean karsts, while the latter is on temperate and tropical karsts. The karsts which can be generated on any climate (platform karst, diapir karst) are described separately.

The present pattern of covered karst is the result of either accumulation or denudation. In both cases, erosional discontinuity surfaces interrupt the cover. Erosional discontinuities of the cover can also develop during covered karst formation. In this case, the lower (older) part of the cover sequence is dissected by depressions which are filled with younger cover sediments.

Fig. 4.21 Subsidence dolines in paleodolines (Durmitor, Surutka) *1* subsidence doline, *2* rainwater rill

As mentioned above, in a karst region, the cover sediment can be continuous (uniform) or interrupted. In the latter case, the extension, shape, arrangement, density and location of cover sediment patches generate a pattern of cover.

Continuous sediment cover is possible on platform karst or induced by loess mantle, marine inundation and lava flow. Noncontinuous sediment cover can be striped (in river valley, on interfluvial ridge) or striped with patches if the sediment accumulated on the valley floor is partly eroded or reworked or if there is a row of paleodolines on the valley floor. In the latter case, the patches of cover sediment are represented by doline fills. The cover sediment patches (and thus the covered karst) sometimes show a chaotic pattern, typically on inselberg karst and mediterranean karst (see below). The pattern of the cover sediment can be banded or zonal. Banded covered karsts occur on river terraces or marine platforms. Among the patterns, we regard the banded to be of the narrowest, the zonal the widest, while the width of the striped pattern is between that of the banded and the zonal pattern.

On temperate medium mountains karsts, the zonal pattern is widespread. In case of such a pattern of cover sediment, moving away from the interior of the karst or from the non-karstic terrain covered and non-covered zones alternate. The shape and extension of zones is controlled by the size of the karst. The zone width is measured along the slope of the karstic surface, while its length is perpendicular to that direction.

Fig. 4.22 Subsidence dolines in a paleouvala (Dachstein)

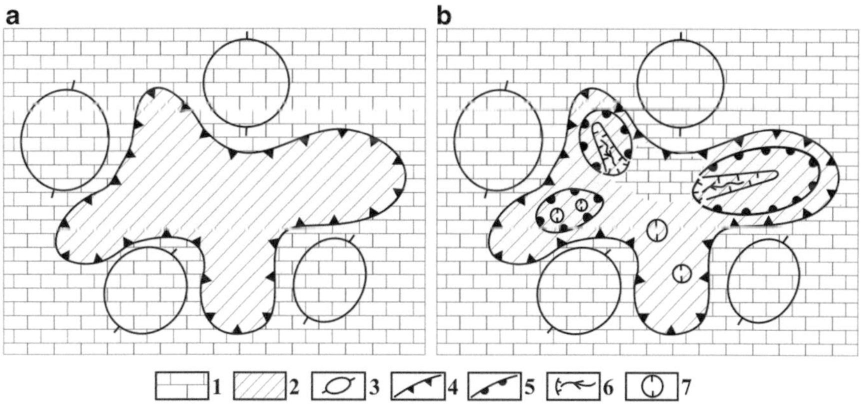

Fig. 4.23 Covered karst of star doline *1* limestone, *2* cover sediment, *3* elevation, *4* star doline, *5* depression of superficial deposit, *6* ponor, *7* subsidence doline, (**a**) initial stage, (**b**) advanced stage

The cover sediment of the zones can also be continuous and patchy. Between the zones of cover sediment, uncovered zones can also be present. The material of the covered karst zones is constituted of non-karstic rocks underlying the limestone but also of cover transported there or originated in situ. The location and width of

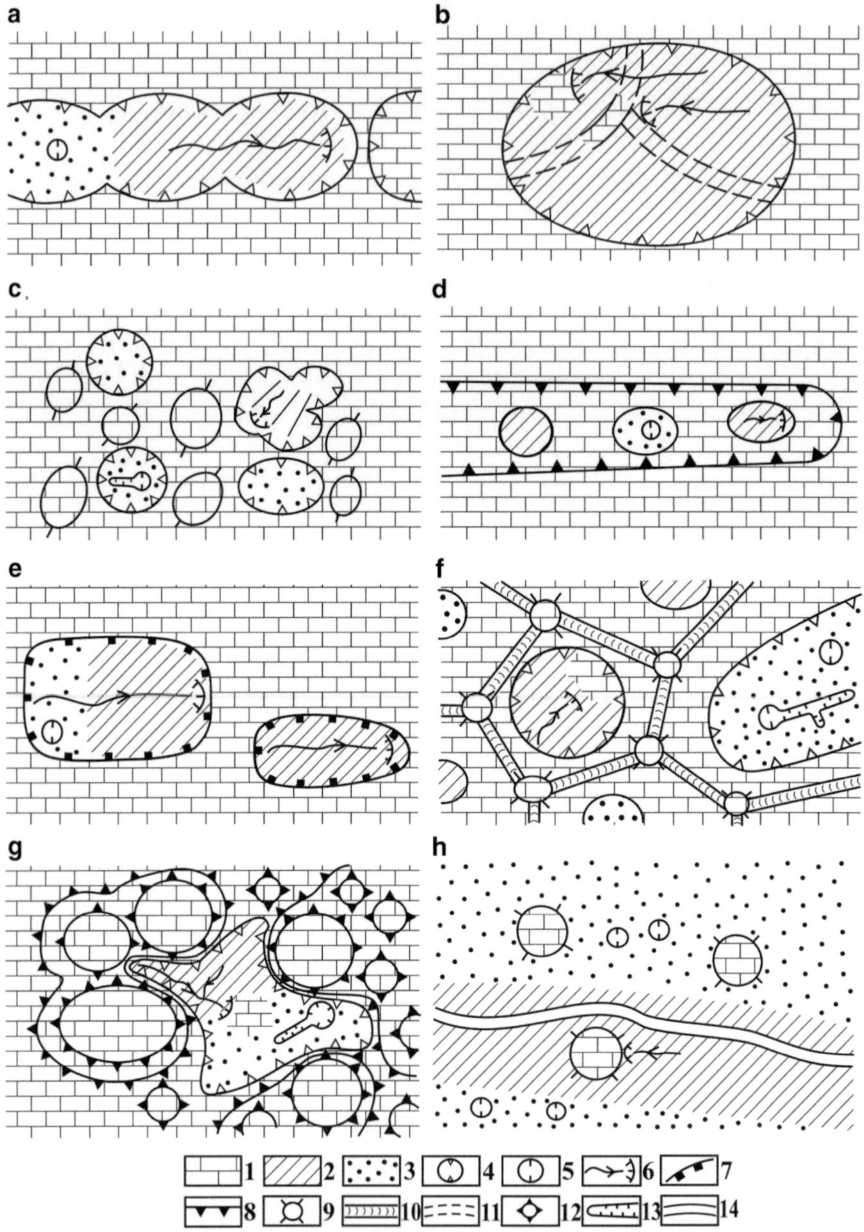

Fig. 4.24 Covered karst within karst landforms (covered karst patterns controlled by karst landforms) (**a**–**c**) covered karst on floor of uvala, (**d**) covered karst formed in solution dolines and ponors of an epigenetic blind valley, (**e**) covered karst in polje, (**f**) covered karst formed in dolines of polygonal karst, (**g**) covered karst of dolines of fengcong, (**h**) covered karst of fenglin intermountain plain, *1* limestone, *2* impermeable cover, *3* permeable cover, *4* solution doline, *5* subsidence doline, *6* ponor, *7* polje, *8* epigenetic blind valley, *9* karst hill, *10* karst dividing wall, *11* buried karst dividing wall, *12* karst inselberg, *13* ravine, *14* river

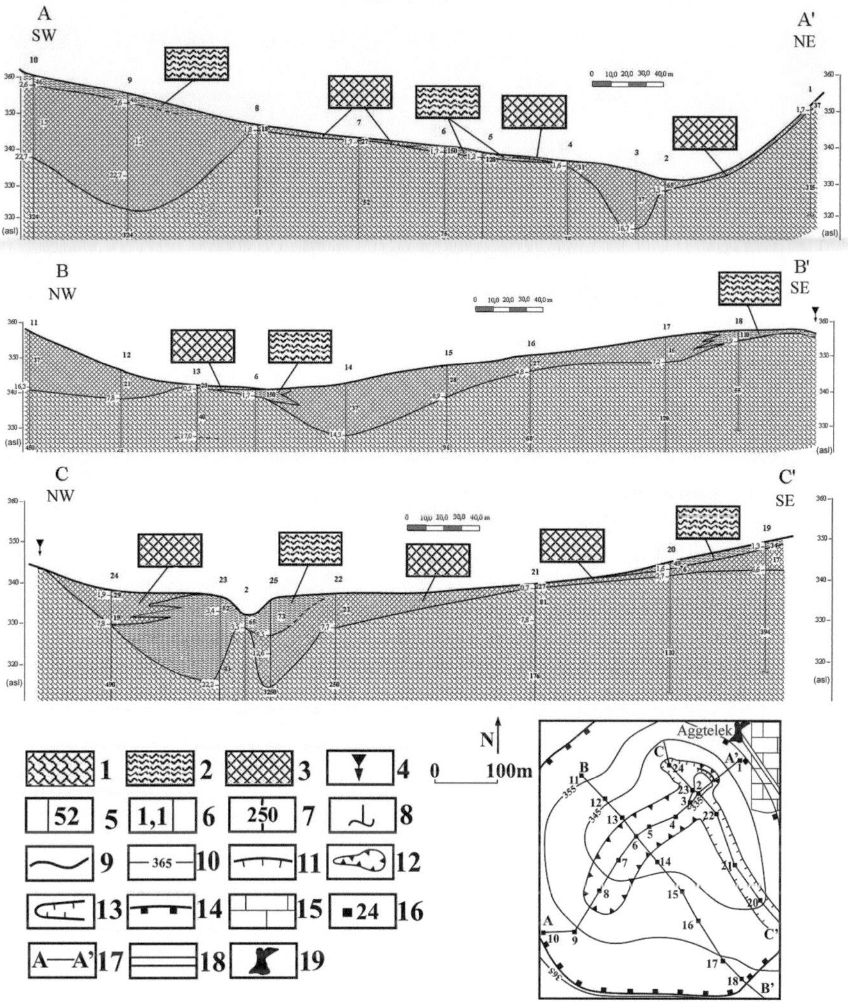

Fig. 4.25 Depression of superficial deposit bearing the Zombor-lyuk ponor (Aggtelek Karst, Hungary): *1* limestone debris, fragmented limestone, *2* (clayey) limestone debris, *3* clay (with limestone debris and sand), *4* limestone outcrop, *5* geoelectric resistivity (Ohmm), *6* depth to base of geoelectric series (m), *7* geoelectric resistance of bedrock (Ohmm), *8* approximate depth of VES measurement, *9* geoelectric series boundary, *10* contour line, *11* ponor, *12* blind valley, *13* ravine, *14* depression of superficial deposit, *15* limestone, *16* site and identification code of VES measurement, *17* profile location, *18* Aggtelek-Jósvafő road, *19* built-up area

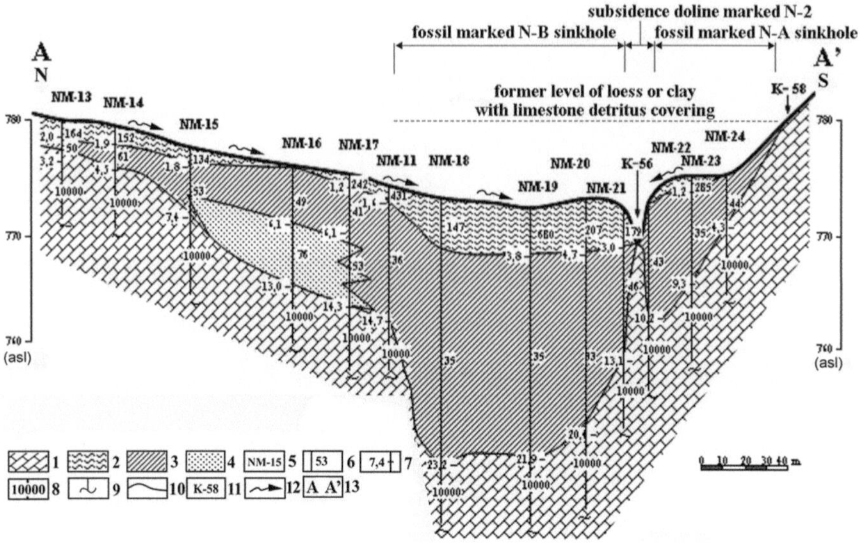

Fig. 4.26 Geoelectric–geological profile A-A′ of the infilled blind valley of the Nagymező (site of profile is in Fig. 4.27, Bükk Mountains, Hungary, Veress and Zentai 2009): *1* limestone, *2* (clayey) limestone debris, *3* (clayey-silty) loess or clay with limestone debris, *4* loess (with sand or limestone debris), *5* identification code of VES measurement, *6* geoelectric resistivity of the series (Ohmm), *7* base depth of the geoelectric series (m), *8* geoelectric resistivity of bedrock (Ohmm), *9* approximate depth of penetration of VES measurements, *10* boundary of geoelectric series, *11* number of limestone outcrop, *12* water flow, *13* symbol of profile

covered karst zones is gradually shifting because of the denudation of the cover sediment or the sediment input to the area of the zone. The pattern of the cover sediment within the zone also changes from the continuous towards the patchy or in the opposite direction.

A covered karst zone is built up of permeable rocks (cryptokarst zone) or of impermeable rock (concealed karst zone). The zone can be composite if impermeable and permeable patches alternate and mixed composed of covered and uncovered patches and mixed composite if in the area of the zone impermeable and permeable cover sediment patches and uncovered patches are found.

The zonal pattern depends on the position of the karst, its elevation, slope direction and angle, the type and extent of denudation, the history of karst development, the karst environment (whether there is a non-karstic terrain in the environs of the karst), the extension of the karst and the morphology of the bedrock at the given site, the structure of karst, the mode, extent and rate of transport of cover sediment onto the karst. The covered patches of the covered karst zones are mostly generated because in the bedrock depression the cover is preserved.

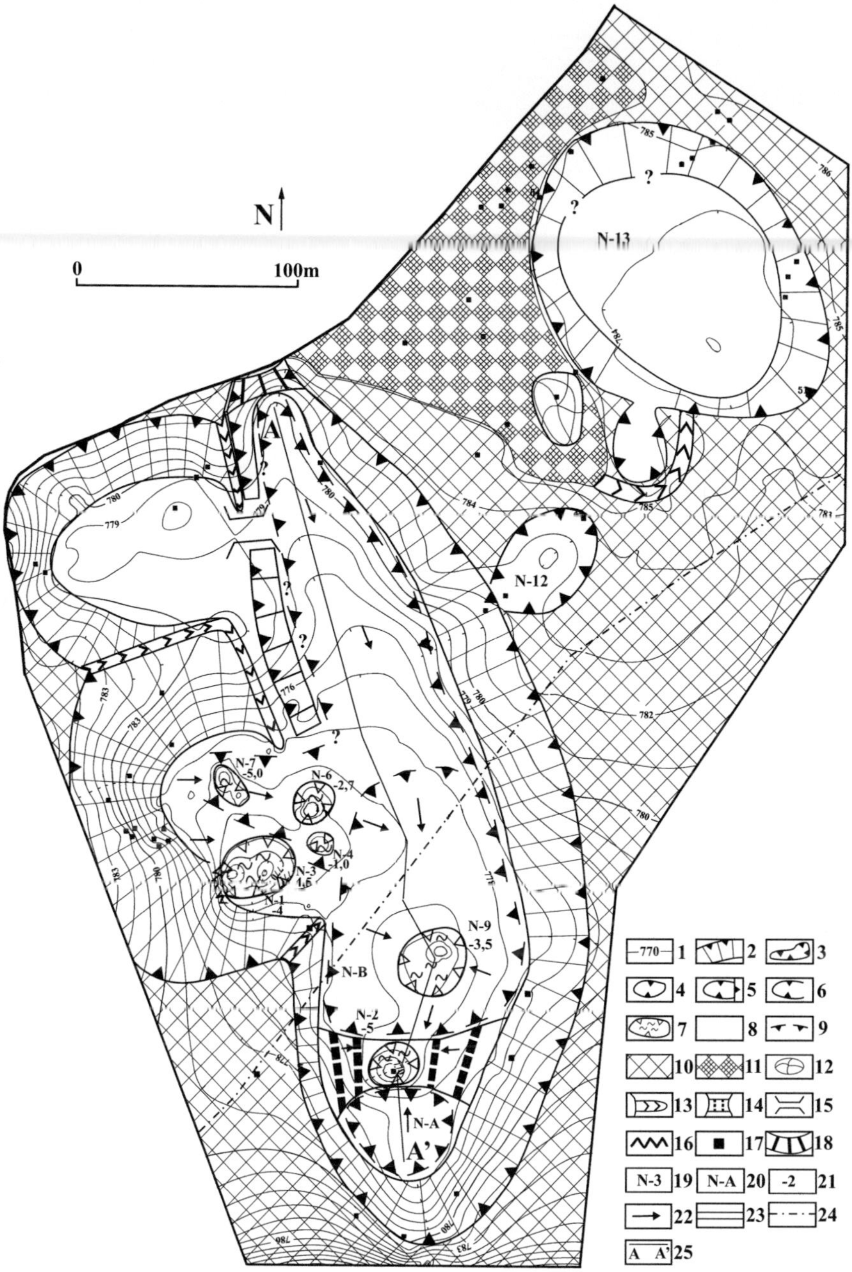

Fig. 4.27 Geomorphological map of the infilled blind valley and environs of the Nagymező (Veress and Zentai 2009): *1* contour line; *2* margin and slope of doline, blind valley; *3* blind valley; *4* doline with flat floor; *5* half open hanging doline; *6* half open doline; *7* subsidence doline; *8* floor of karst landforms and blind valley; *9* margin of infilled and buried gully or ponor; *10* relict karstic terrain; *11* karst slope; *12* karst hill; *13* remnant of dividing wall between dolines; *14* dividing wall between dolines; *15* col; *16* cliff; *17* rock outcrop; *18* road embankment; *19* symbol of karst depression; *20* symbol of buried karst depression; *21* depth of subsidence doline; *22* assumed water flow on terrain with cover sediment; *23* road; *24* fence; *25* geoelectric–geological profile

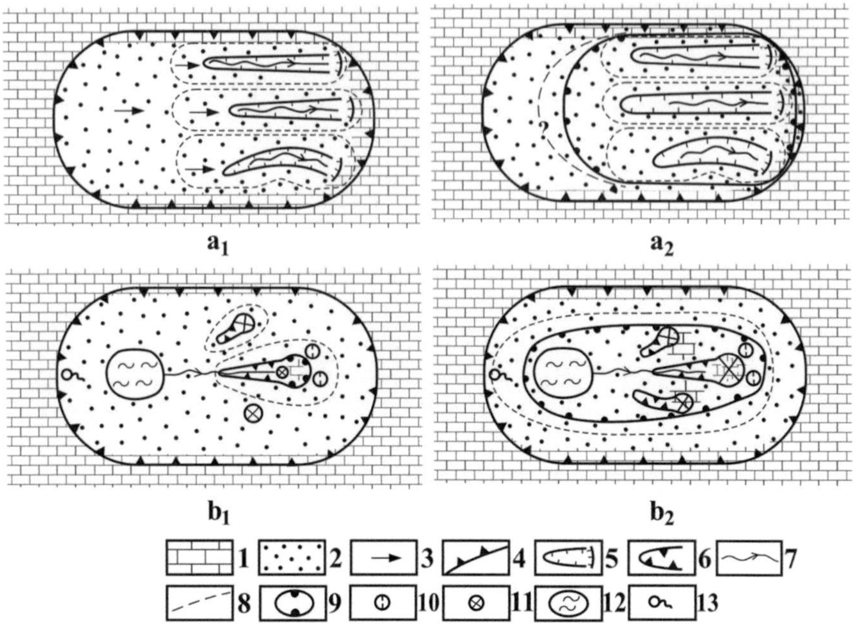

Fig. 4.28 Evolution of depressions of superficial deposit in poljes: *1* limestone, *2* cover sediment, *3* surface slope, *4* polje, *5* blind valley with ponor (embryonic depression of superficial deposit), *6* ravine, *7* watercourse, *8* topographic divide, *9* depression of superficial deposit, *10* subsidence doline, *11* katavothron, *12* lake, *13* spring, (**a**) in the peripheral permanently dry (or temporarily waterlogged) polje blind valleys develop (**a_1**), through the widening of blind valleys and lowering of interfluvial ridges a depression of superficial deposit emerges with blind valleys (**a_2**), (**b**) in the overflow or temporarily (or permanently) waterlogged polje depressions of superficial deposit and dolines form at the katavothra (**b_1**), they merge and, as a consequence, a single major depression of superficial deposit develop (**b_2**)

4.6.2 Cover Pattern of Climatic Covered Karst

4.6.2.1 Tundra-Covered Karst

On tundra karst surface, denudation is of moderate rate and therefore the cover is less denuded. Consequently, on the tundra karst, the covered karst is potentially mostly of continuous, but in reality of patchy appearance. The reason for this is that during the Pleistocene glaciations, the present-day tundra areas were covered by ice sheet, and the thick ground moraine buried the karst (Smart 2004) and prevented karstification (buried karst). Karstic patches, mostly cryptokarst (autogenic cryptokarst), primarily formed under gypsum and rock salt. In tundra karst, the paleodepressions were only preserved locally and mostly destroyed by the ice sheet. The cover sediment of karst is of morainic, marine or possibly fluvial origin.

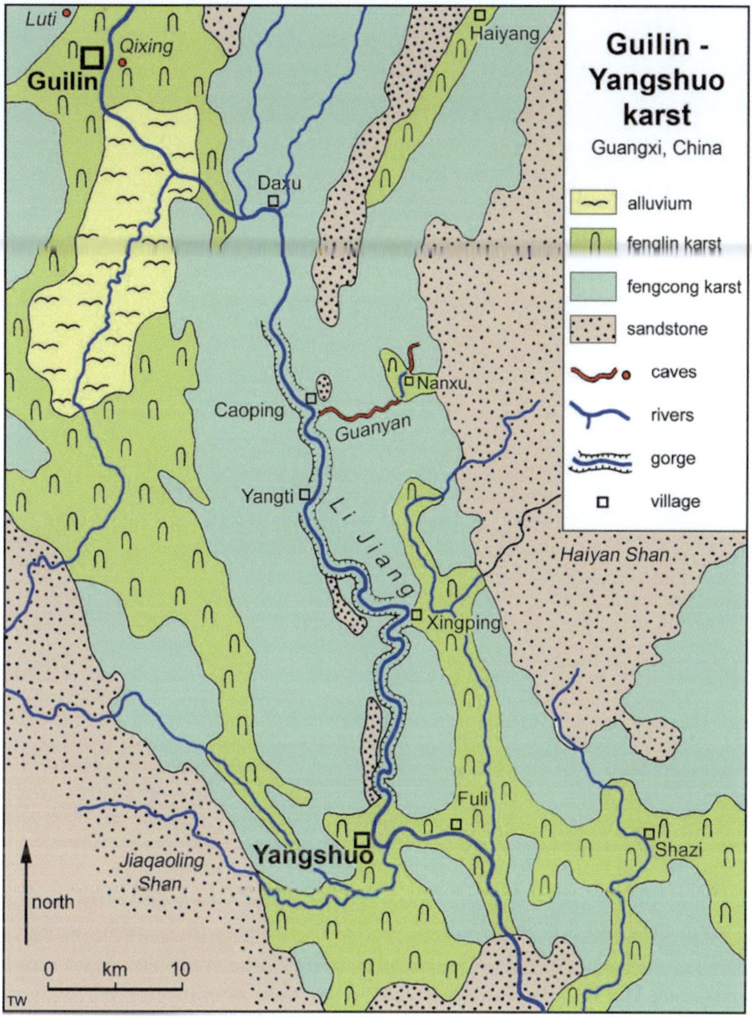

Fig. 4.29 A very generalised map of the distribution of fengcong and fenglin in the karst between Guilin and Yangshuo (Waltham 2008)

The rate of karstification decreases for both open and covered karsts towards the poles. This is explained by the late date of the cease of ice cover (although no ice covered most of the Siberian areas) and primarily the presence of permafrost and low CO_2 production. Solution can be expected where latent heat transfer occurs from below (Van Everdingen 1981), pointlike water recharge takes place, taliks interrupt the permafrost, surface watercourses emerge (Ford and Williams 2007) or the groundwater is saline (Pollard et al. 1999). Therefore, the density and dimensions of recent covered and open karst landforms reduce towards the poles. On carbonate rocks, karstification is mostly observed at steep cliff walls (Lauriol and

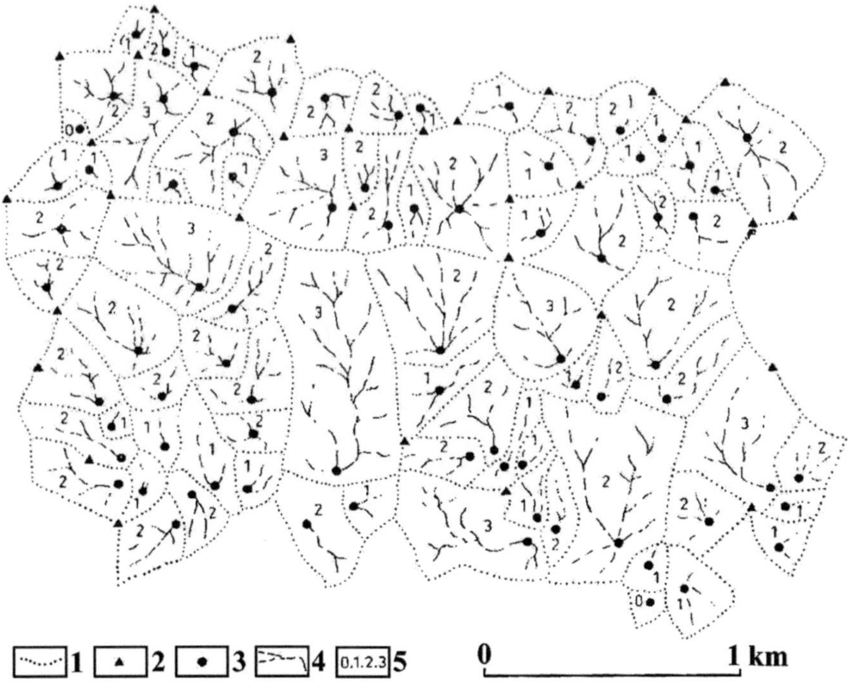

Fig. 4.30 Polygonal karst (Williams 1971) *1* topographic divide, *2* summit, *3* swallet, *4* stream channel, *5* basin order

Gray 1990) and on breccia terraines (Ford 2004). The karstification of carbonate rocks is possible where heating generates flow cells (Ford and Williams 2007), which reach down to 500 m depth (Ford 2004). Taliks, which are important in the water budget of the karst, may result from karstification (Ford and Williams 1989). The older karst landforms (paleodolines) are partially or completely filled with glacial sediments or debris. On these fills, ponor-like features (probably subsidence dolines) develop (Ford and Stanton 1968). Considerable karstification is observed on gypsum and rock salt karsts (Van Everdingen 1981), partly because of the higher solubility of rocks, partly because of the freezing point of saline water below 0 °C. As a consequence, the covered karsts of the tundra are primarily bound to the distribution of gypsum and rock salt. On gypsum, covered karst formation is generated by temporary marine inundation. Covered karst with gravel cover is described from Svalbard, where permafrost melting is caused by seawater action (Salvigsen and Elgersma 1985).

In the sense of the above, covered karst only occurs sporadically on tundra karst. Its sporadic patches are primarily found on gypsum and rock salt or on carbonates where local influences (marine inundation, thermal effect of meltwater, activation of paleokarst, presence of talik) make the permafrost melt.

4.6.2.2 Taiga-Covered Karst

In this karst type, although climatically it belongs to the temperate belt, the presence of permafrost causes quite different karstification. The erosion of the surface can be considerable, while permafrost is widespread. The taiga karst of Siberia was not affected by glaciation. The surface is dissected by river valleys. Gvozdetskiy (1981) described karst and covered karst from the region of the Nizhnyaya Tunguska, Vilyuy and Olenek Rivers, the interfluves of the Letnyaya Tunguska and Suhaia Tunguska valleys and from the catchments of the Amga, Aldan, Olekma, Vitim, Yenisey, Angara and Lena Rivers.

The valleys of the karst are either inherited from the cover sediment or developed due to the presence of the permafrost. The karst shows the following pattern type in these areas. In the main valleys, permanent streams are found, while their tributaries are intermittent along the whole length of the valley (e.g. the Labiya, a Lena tributary) or partially intermittent (e.g. the Bolsoj Taryng, another Lena tributary). In the latter case, water seeps away from the watercourse and re-emerges at some distance on the valley floor. Between the sites of seepage and re-emergence, there is water flow on the surface only during snowmelt, which causes thawing in the upper layers of the permafrost. The water percolates below the channel, in the sediments overlying the permafrost or, where there is no cover deposit on the valley floor, in the limestone. Patches of non-karstic rocks appear on the wide and almost flat valley floors in the tributary valleys, mostly along the upper sections, in headwater areas. Among the patches, the limestone bedrock outcrops on the valley floor. The extension of the cover sediment on the valley floor is rather limited, but more remarkable in the valley sides and along the margins of the valley floor, where debris fans accumulate due to frost weathering. On the interfluves, uniform unconsolidated sediment cover, mostly sand, is typical. Cover thickness is several metres on the interfluves, and it is only locally interrupted by limestone outcrops. As a consequence, sediment patches occur on valley floors, while on interfluves, the superficial deposit is continuous. Covered karst appears in river valleys in striped and patchy pattern, while on the interfluves in striped pattern (Figs. 4.31 and 4.32). Covered karsts occur both on valley sides (Korzhuev 1961; Gvozdetskiy 1981) but also occur on river terraces, for instance, on the terraces of the Angara River, where covered karst developed on dolomite, limestone and gypsum (Gvozdetskiy 1981).

The landforms of the covered karst are either recent or paleokarst features (Korzhuev 1961, 1972). The latter – similarly to the tundra karst – could form before the glaciations or in one of the interglacials (Korzhuev 1972; PulinaPulina 2005). In Siberia, because of the moderate surface denudation, young paleokarst depression could be preserved. Major karst landforms are mostly subsidence dolines (suffosion and dropout dolines), uvala-like, compartmental dolines, kotlovinas and suchodols (Korzhuev 1961; Gvozdetskiy 1981). Kotlovinas are major depressions with doline groups in their interior (Korzhuev 1961), i.e. depressions of superficial deposit. Valley dolines occur in channels on valley floors (Fig. 4.33) or on infilled valley floor (Fig. 4.34). Suchodols are valleys with doline rows containing ponor-like depressions and with intermittent lakes in the dolines (Pulina 2005).

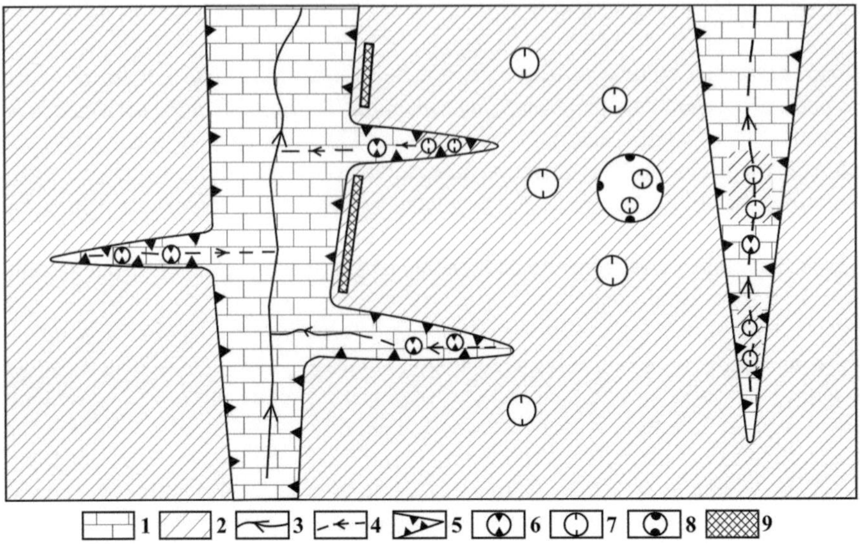

Fig. 4.31 Karst zones of tundra karst (not to scale) example of the Middle Lena area *1* limestone, *2* cover sediment, *3* permanent watercourse, *4* intermittent watercourse, *5* valley, *6* solution doline, *7* subsidence doline, *8* depression of superficial deposit, *9* pillars

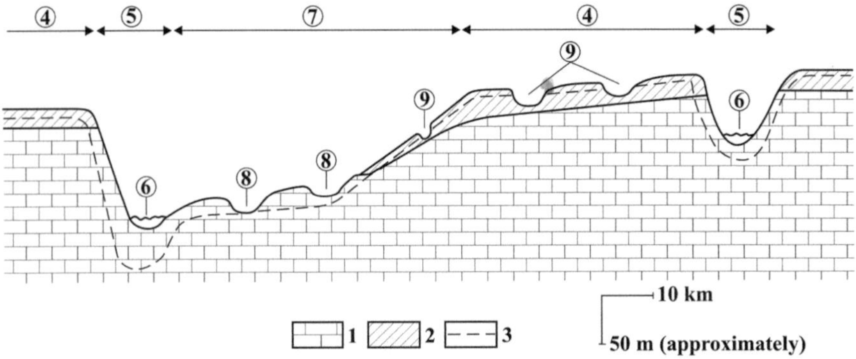

Fig. 4.32 Theoretical cross-section of the tundra karst (examples of the Lena area, not to scale, karst landforms are not proportional to their environments): *1* limestone, *2* cover sediment, *3* permafrost, *4* plateau, *5* valley, *6* channel, *7* tributary valley, *8* solution doline, *9* subsidence doline

Valley floor dolines develop where the water of the intermittent watercourse on the valley floor thaws ground ice in the cover sediment. In the covered karst of interfluvial ridges, subsidence dolines of various origins occur where air from the paleokarst passages melts the ice in the passage and the cover sediment, but they also emerge at sites of pointlike water conduction (Ford and Williams 2007). It is probable that water conduction sites under lakes (formed in association with permafrost melting) are created due to the pressure and thermal effect of lake water. The seeping

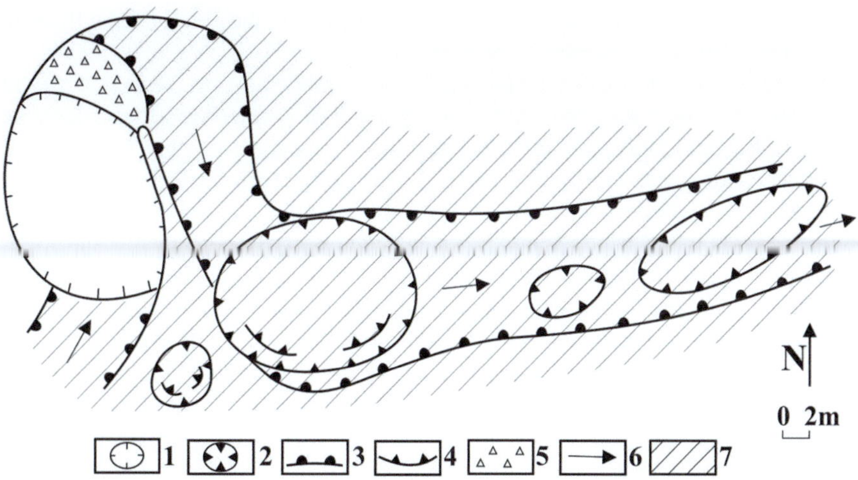

Fig. 4.33 Suffosion dolines from the upper part of Bolsoj Taryng Valley, Sakha Republic: *1* solution doline, *2* suffosion doline, *3* margin of channel, *4* mass movement, *5* debris accumulated by a stream, *6* slope of channel floor, *7* superficial deposit

Fig. 4.34 Dropout doline in the Bolsoj Taryng valley

water could be in connection with talik water, such as in the case of Lake Borulak on the Lena-Buotoma interfluvial ridge in Yakutia.

The subsidence dolines of the taiga karst are of moderate size (several metres in diameter), and their density is only estimated at 1–2 dolines/km^2. Particularly on valley floors, the dolines are arranged in rows of moderate length. The groups of depressions are separated by also linear doline groups of solutional origin on the uncovered valley floors (or channels on valley floors).

In Canada, the glacially transformed Nahanni karst is climatically and topographically similar to the Middle Lena karst (Ford and Williams 2007). The Nahanni is a doline and mixed karst. The major depressions are 1000 m long and their bottoms are under cover sediment (Brook and Ford 1980). The hydrostatic pressure of water flowing into these depressions induces collapses in the ice of the passages and the formation of dolines, also called "cenote-form point-recharge dolines" (Ford and Williams 2007).

4.6.2.3 Temperate Belt Karsts

The medium mountain areas of the true temperate belt belong to this, where the fluvial erosion is dominant at the beginning of karst formation. In medium mountain karst areas, epigenetic valley evolution takes place if the karst water table is close to the surface (or the developing valley floor). Valley formation is replaced by karstification if karst water table sinks relative to the valley floor. Subsequently, the covered karst gradually loses its cover sediment and open karst develops resulting in a surface dissected with valleys and karst landforms. This surface can be buried again under various rocks partially or completely, to various extents and thickness, and covered karst is produced. Burial and the accompanying karst formation take the following forms.

Covered karst developed from allogenic karst results if the karst is bordered by impermeable rocks, mantled allogenic covered karst if the impermeable rock covers the karst as a mantle and horst covered karst if the karst is dismembered into covered blocks of various elevations and evolution histories.

Three varieties of karst developed from allogenic karst are distinguished: recent allogenic covered karst (see Sect. 4.6.2.3.1.1), renewed allogenic covered karst (see Sect. 4.6.2.3.1.2) and semi-allogenic covered karst (see Sect. 4.6.2.3.1.3). The types of temperate medium mountain karst are represented in Fig. 4.35.

4.6.2.3.1 Covered Karst Developed from Allogenic Karst

The karst is bordered by terrains of non-karstic rocks, from where fluvial transport of sediments is or was possible.

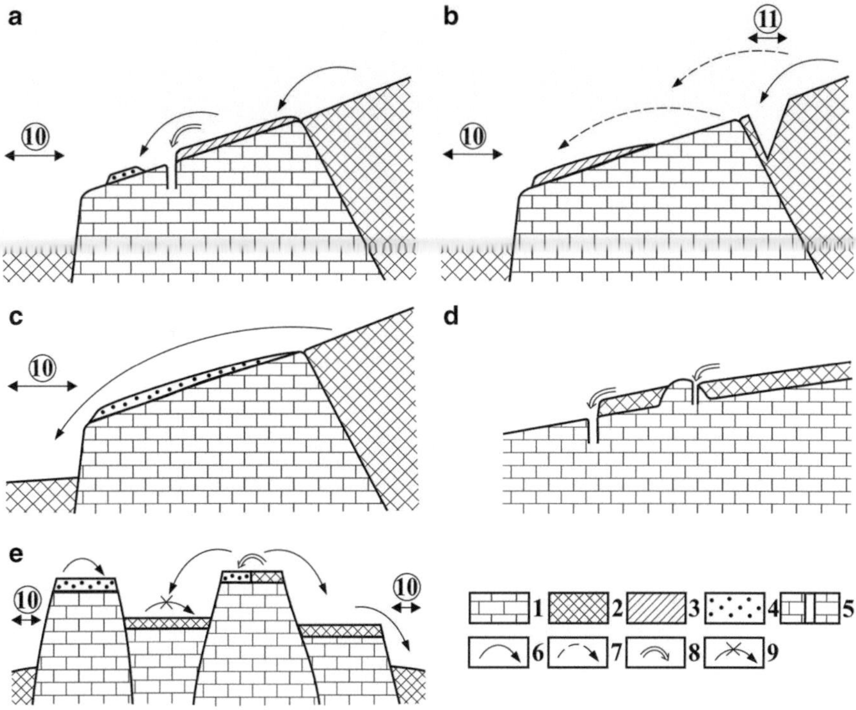

Fig. 4.35 Theoretical profiles of medium-mountain covered karst types: *1* limestone; *2* consolidated non-karstic rock; *3* unconsolidated, impermeable cover sediment; *4* permeable cover; *5* chimney, shaft; *6* sediment transport; *7* former sediment transport; *8* transport of cover sediments into the karst hollows; *9* no transport of cover sediment from the block; *10* mountain foreland, *11* valley, (**a**) recent allogenic covered karst, (**b**) renewed allogenic covered karst, (**c**) semi-allogenic covered karst, (**d**) mantled allogenic karst, (**e**) horst covered karst

4.6.2.3.1.1 Recent Allogenic Covered Karst

In recent allogenic covered karst, watercourses actually transport sediment from the bordering non-karstic terrain. The condition for that is not only that the karst should be bordered by non-karstic surface but also that the latter should be in lower position than the former and the surface of the karst should not slope towards the non-karstic terrain (Fig. 4.35a). A further condition to the development of covered karst and its zones is that there should be no valleys on the open karst or if such are present they should not lead beyond its limits. If such valleys were found, the watercourses would transport the sediment input from the non-karstic terrain out from the karst. In addition, no ponors should form, or if they do, their capacity for sediment capture and transport into the caves of the karst should be minimal. The 1.2 and 1.4 type allogenic karsts that were classified by Gams (1994) do not transform into recent allogenic karst as the cover is transported into the karst through the well-developed ponors. If all these requirements are met, the watercourses of the karst spread the sediment over the karst surface, as exemplified by the Pádis plateau (Figs. 4.36 and 4.37).

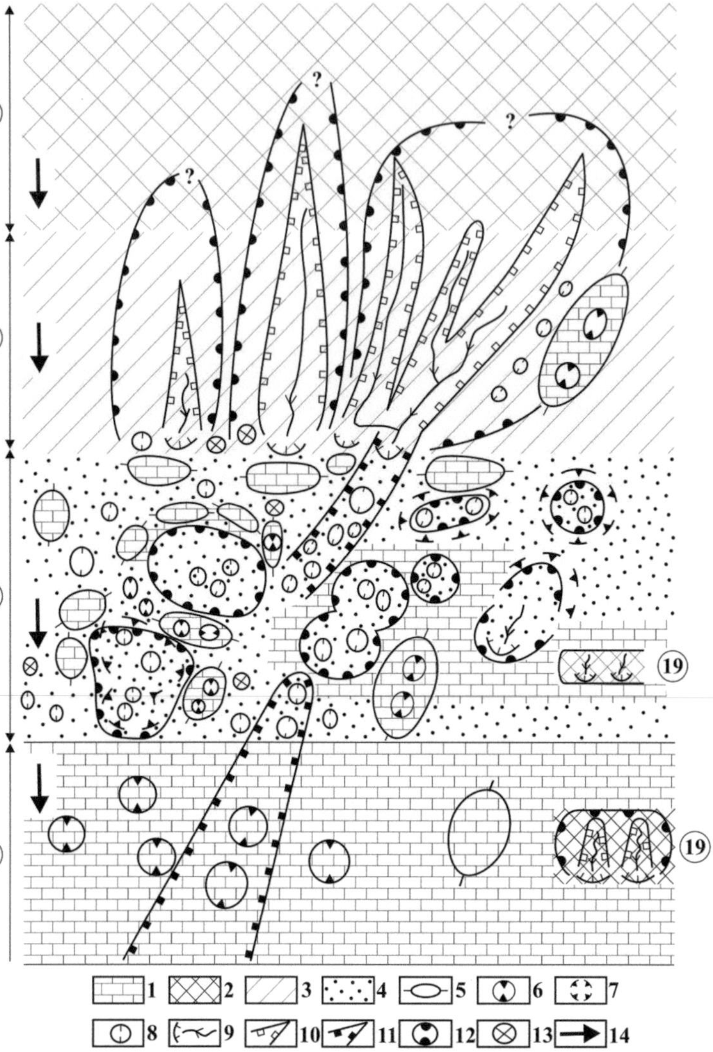

Fig. 4.36 Karst zones of recent allogenic covered karst (example of the Pádis, Romania, not to scale): *1* limestone; *2* consolidated non-karstic rock; *3* thick, virtually impermeable unconsolidated cover; *4* thin, virtually permeable cover; *5* elevation; *6* solution doline; *7* buried paleodoline, paleouvala; *8* subsidence doline; *9* ponor; *10* valley in the cover; *11* epigenetic valley; *12* depression of superficial deposit; *13* fossil subsidence doline; *14* surface slope; *15* zone of non-karstic rock (buried karst); *16* allogenic cryptokarst zone; *17* mixed karst zone; *18* open karst zone; *19* karstification with intercalated non-karstic rock zone

The zones on the Pádis follow in order from the Blue Magura (Măgura Vânătă), in NE to SW direction: buried karst zone, cryptokarst zone, mixed composite karst zone and finally open karst zone. (In the area of the last, where non-karstic rocks wedge in, further cryptokarst zones of lesser extension occur.) The karst of the Pádis has a higher and a lower level (Fig. 4.38), which are the products of karstification

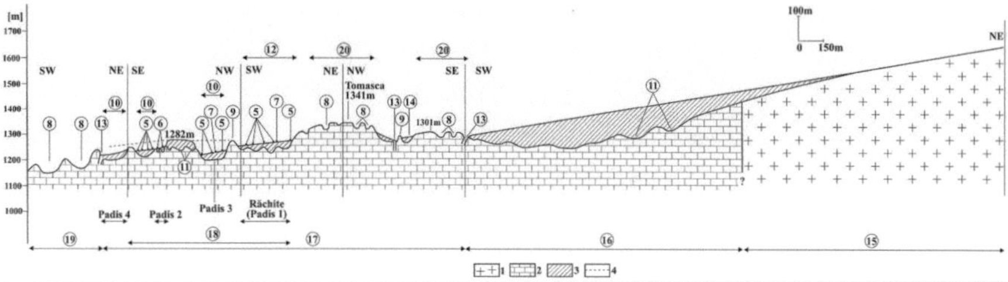

Fig. 4.37 Cross-section of recent allogenic covered karst (Pádis): *1* sandstone, *2* limestone, *3* rock debris reworked by watercourses (with sand and clay), *4* former surface of sediment cover, *5* suffosion doline, *6* dropout doline, *7* fossil doline (lake), *8* active solution doline, *9* threshold between uncovered karst features, *10* inactive infilled solution doline (depression of superficial deposit), *11* (assumed) buried solution doline, *12* inactive covered uvala (depression of superficial deposit), *13* ponor, *14* epigenetic valley perpendicular to profiles, *15* buried karst, *16* allogenic cryptokarst of the lower level, *17* mixed karst, *18* concealed karst of the lower level, *19* open karst of the lower level, *20* elevations of the upper level (site of section can be found in Fig. 2.12)

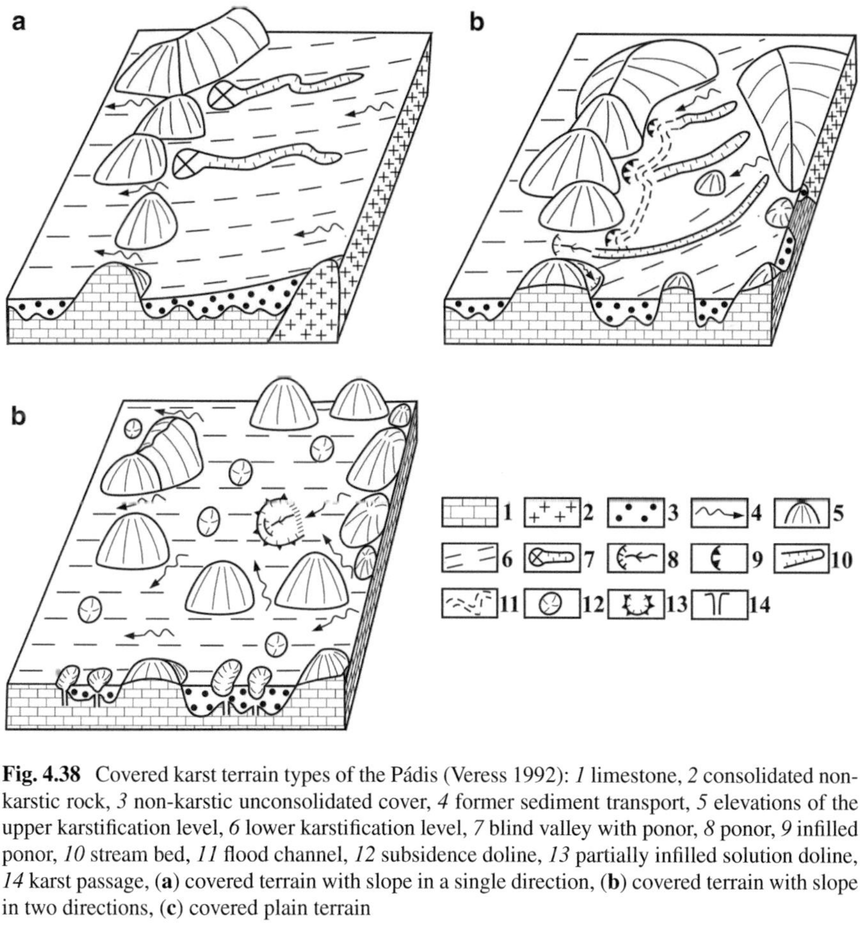

Fig. 4.38 Covered karst terrain types of the Pádis (Veress 1992): *1* limestone, *2* consolidated non-karstic rock, *3* non-karstic unconsolidated cover, *4* former sediment transport, *5* elevations of the upper karstification level, *6* lower karstification level, *7* blind valley with ponor, *8* ponor, *9* infilled ponor, *10* stream bed, *11* flood channel, *12* subsidence doline, *13* partially infilled solution doline, *14* karst passage, (**a**) covered terrain with slope in a single direction, (**b**) covered terrain with slope in two directions, (**c**) covered plain terrain

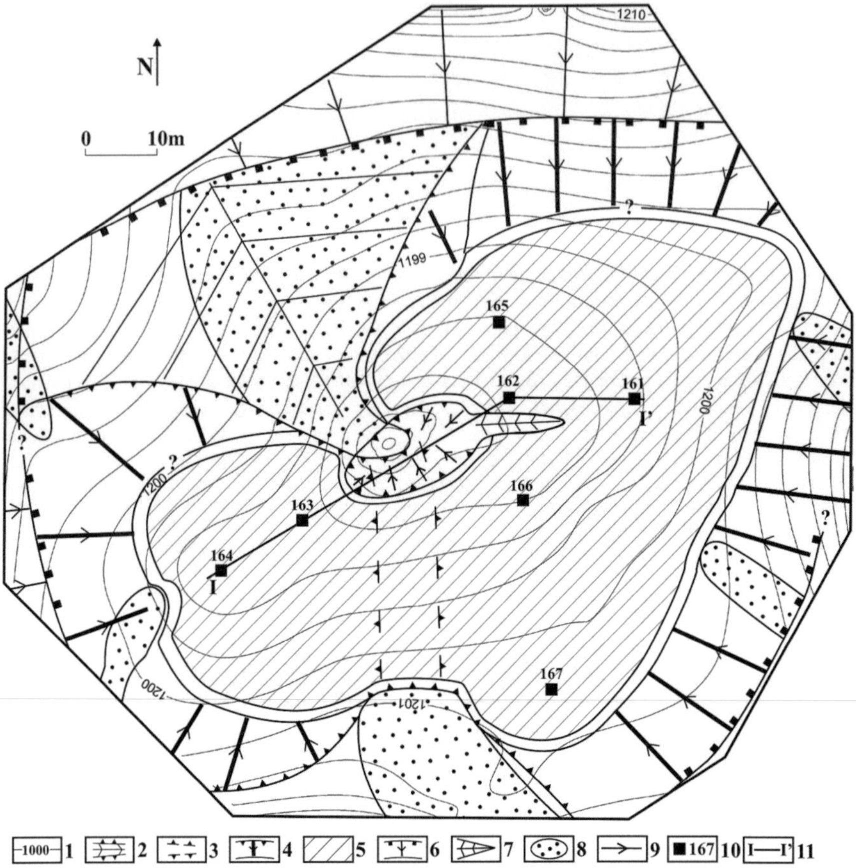

Fig. 4.39 Karst morphological map of the Pádis-4 area: *1* contour, *2* dividing wall between dolines, *3* buried dividing wall between dolines, *4* margin and slope of ponor, *5* floor of depression of superficial deposit, *6* margin and side slope of depression of superficial deposit (paleouvala), *7* ravine, *8* limestone, *9* karst terrain beyond the margin of the depression, *10* site and symbol of VES measurement, *11* geoelectric–geological profile

(Veress 1992). The upper level is uncovered and neither the lower level is of continuous cover, but it occurs in patches of various sizes. The cryptokarst sections of allogenic character developed in the sections of greater cover thickness. Cryptokarst, however, can also appear in patches where the fills of paleodolines and paleouvalas are thicker (Figs. 4.39 and 4.40). In the mixed composite zone open, covered (cryptokarst and concealed karst) patches equally occur (Fig. 2.12). Patches of minor extension could develop or be preserved in individual dolines, uvalas (Figs. 4.39 and 4.40) or on an inherited valley floor section (Figs. 4.43 and 4.44). In the dissection of the cover into patches, several factors could be influential. The transported and accumulated sediment did not bury the elevated parts of the uneven surface (Fig. 4.38), and the large patches of the cover (developed in the above

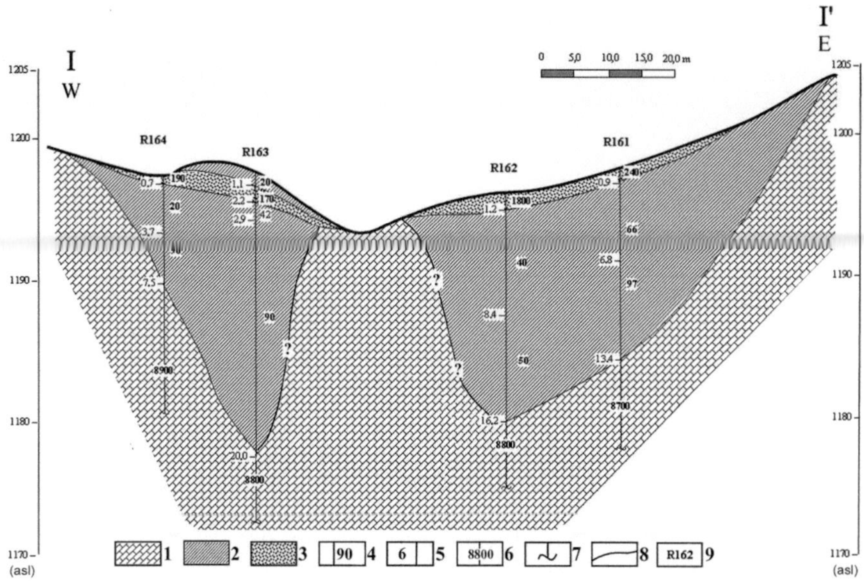

Fig. 4.40 Geoelectric–geological profile I-I′ of the Padis-4 area: *1* limestone, *2* clayey silt, *3* rock debris (sandstone or limestone, locally siliceous, clayey), *4* geoelectric resistivity of the series (Ohmm), *5* base depth of the geoelectric series, *6* geoelectric resistivity of bedrock (Ohmm), *7* approximate penetration of VES measurements, *8* boundary of geoelectric series, *9* identification code of VES measurement

described way) were cut up by denudation into smaller patches. In addition to denudation, it was due to the transport of part of the cover sediment into the karst through karst landforms. The covered karst is renewed since further amounts of sediment continually arrive from the bordering non-karstic terrain (Măgura Vânătă). This is evidenced by at least one older stage of covered karstification in the superficial deposit identified in some covered karst sections (see Chap. 7).

Along the margin of the allogenic cryptokarst of the Pádis, ponors (Figs. 4.41 and 4.42) occur, on the concealed karst patches of its mixed karst, subsidence dolines (Figs. 4.43, 4.44, and 4.45) can be found, while on the cryptokarst patches of its mixed karst, ponors (Figs. 4.39 and 4.40) occur. The depressions of superficial deposit are of the true type since they formed in dolines and uvalas (Figs. 4.39 and 4.45). The latter are either marginal or of the karst interior.

4.6.2.3.1.2 Renewed Allogenic Covered Karst

The renewed allogenic covered karst has been isolated from all non-karstic sediment sources (Fig. 4.35b). Thus, this type has developed from recent allogenic karst. During the denudation and reworking of cover sediments, karstification starts anew or modifies in this type of the karst. The reason for isolation can be geomorphic evolution (for instance, valley incision) or tectonic (uplift of the karst or

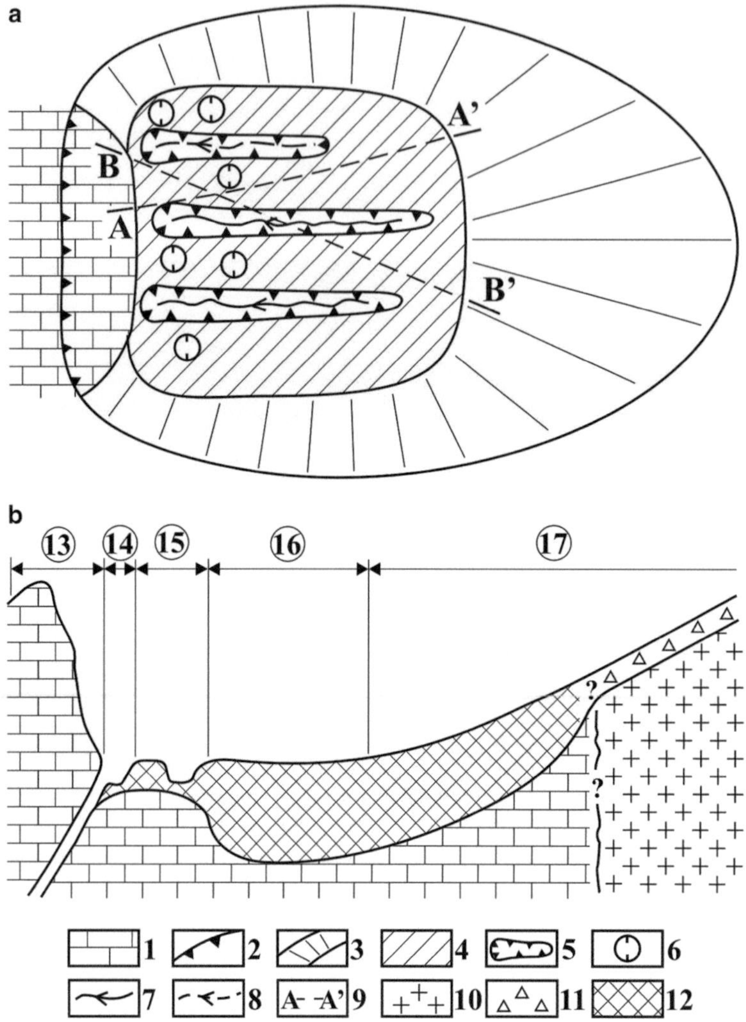

Fig. 4.41 Theoretical groundplan (**a**) and cross-section (**b**) of the depression of superficial deposit near Lake Tăul Vărăsoaia (Padis 5 area): *1* limestone, *2* margin of depression on limestone, *3* side slope of depression on non-karstic rock, *4* floor of depression, *5* blind valley, *6* subsidence doline, *7* permanent watercourse, *8* temporary watercourse, *9* approximate sites of VES profiles (see Fig. 4.42), *10* sandstone, *11* sandstone debris, *12* unconsolidated cover sediment, *13* slope of depression on limestone, *14* ponor, *15* zone of dolines, *16* floor of depression, *17* side slope of depression

subsidence of the non-karstic terrain). This type evolves like an island and is less dependent on the environment, as exemplified by the Aggtelek Karst (Fig. 2.2).

The northern and more elevated part of the Aggtelek Karst (N of the Jósva Valley) is the zone of open karst, where the cover sediment is only reserved in patches and it is partly autochthonous deposit (e.g. weathering residue or its reworked variety).

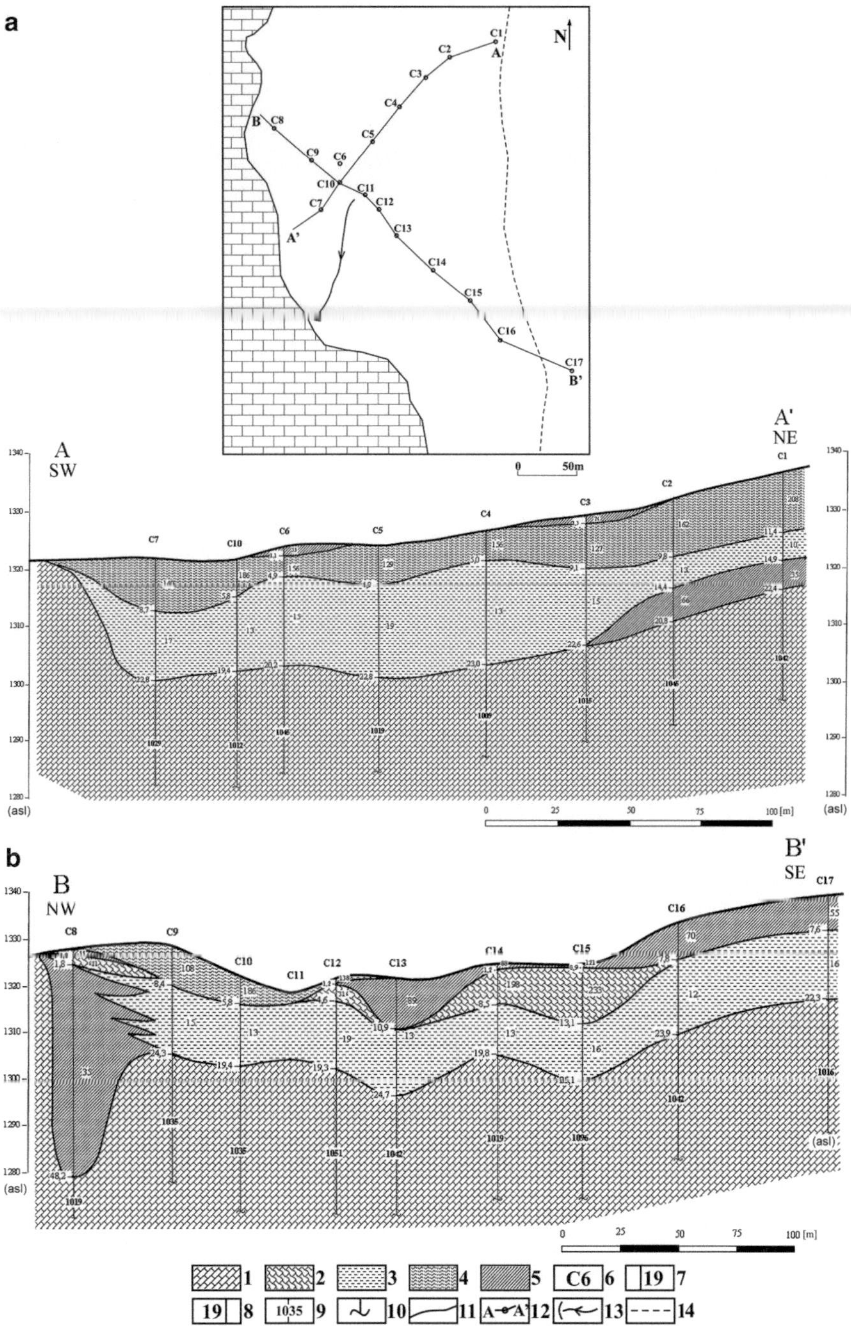

Fig. 4.42 Alignments (**a**) and geoelectric–geological profiles (**b**) of VES measurements in the depression of superficial deposit near Lake Tăul Vărăsoaia (Padis 5 area): *1* limestone, *2* fragmented limestone, *3* clay, *4* (clayey) limestone debris, *5* clay with limestone fragments, *6* symbol of VES measurement, *7* geoelectric resistivity of the series (Ohmm), *8* base depth of geoelectric series, *9* geoelectric resistivity of bedrock (Ohmm), *10* approximate penetration of measurement, *11* geoelectric layer boundary, *12* site of VES measurement and geoelectric-geological profile, *13* ponor, *14* road

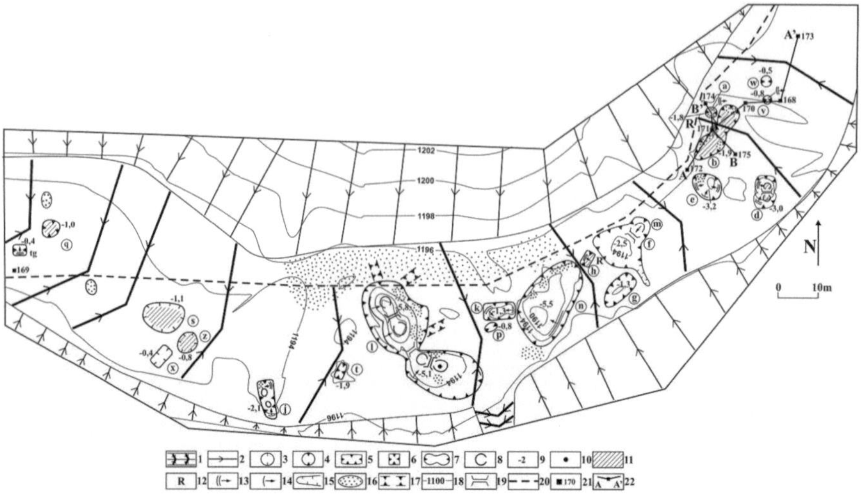

Fig. 4.43 Geomorphological map of the Pádis-2 area (dolines formed on the cover with sandy, clayey and sandstone debris on the floor of the epigenetic valley): *1* epigenetic valley, *2* slope of karstic elevation, *3* circular suffosion doline, *4* circular dropout doline, *5* elongated dropout doline, *6* angular dropout doline, *7* uvala-like dropout doline, *8* partial depression, *9* doline depth, *10* non-karstic pipe, *11* plain floor, *12* doline formed in 2012, *13* collapse, *14* landslide, *15* gully, *16* limestone, *17* tectonic crack widened by solution, *18* contour line, *19* col, *20* road, *21* site and identification code of VES measurement, *22* alignment of VES measurement, (**a**–**z**) identification mark of the doline

To the S, the zone of mixed karst is found (Galyaság, the area between the Jósva Valley and the Rét Stream), followed by the zone of buried karst (S of the Rét Stream valley) (Figs. 4.46 and 4.47).

As it was mentioned in Chap. 2, the karst became tilted on the Plio-Pleistocene boundary, and the watercourses coming from the N spread gravels over its S, SE sloping surface (Sásdi 1990). Probably the gravels covered the northernmost part of the karst in stripes, and the northern open karst was not covered completely at that time. Later, watercourses developed in the gravel mantle stripes, the watercourses formed valleys and where the watercourses reached the limestone, rock boundaries of valleys developed and at these places ponors developed. In their continuation, the predecessor of Baradla Cave formed. Even at present, the whole karst shows an S, SE slope.

In the northern open karst, patches of covered karst emerged in older doline fills (terra rossa) or on non-karstic rock stripes (the sandstone stripe of the Bába Valley) (Figs. 4.48 and 4.49). Surface slope is opposite only in the area of Galyaság, both for the limestone (elevation of limestone summits decreases in northern direction) and for the surface with cover sediment. The first is explained by the tilt of the Klippe (secondary nappe) of Galyaság, while the latter by the denudation of the surface with cover sediment. The sediment cover was transported towards the ponors located on the northern margin of Galyaság and through them into the karst.

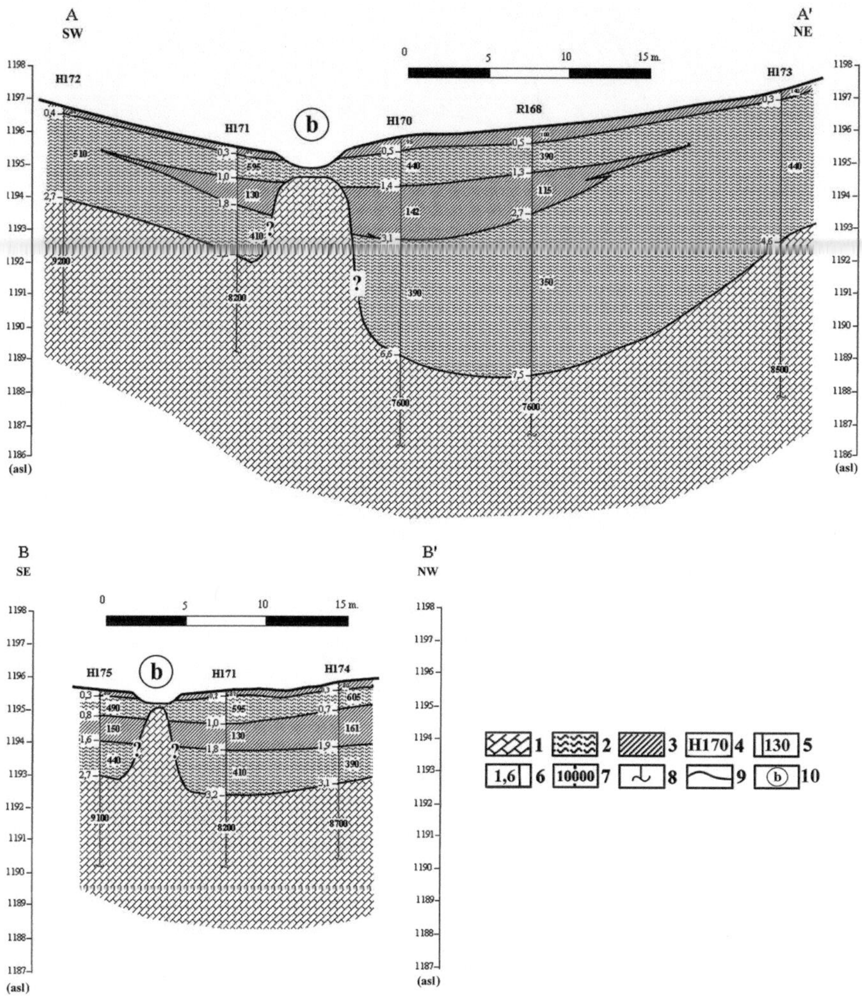

Fig. 4.44 Geoelectric–geological profiles from the area Pádis-2: *1* limestone, *2* mixed rock debris (sandstone or limestone, siliceous, clayey), *3* clayey silt, *4* identification code of VES measurement, *5* geoelectric resistivity of series (Ohmm), *6* base depth of geoelectric series, *7* geoelectric resistivity of bedrock (Ohmm), *8* approx. penetration of VES measurement, *9* boundary of geoelectric series, *10* code of dropout doline (**b**) that can be seen in Fig. 4.43

With denudation extending in southern direction, the surface of the sediment cover changed its slope from southerly to northerly (Veress 2010, 2012a). Galyaság is a mixed composite karst of the Aggtelek Karst, where impermeable and permeable (loess) cover sediments mantle the karst in patches of variable size and shape (Veress 2010, 2012a), including limestone patches. The spatial variations of coveredness are caused by several factors, among them paleokarst formation (limestone elevations and paleodepressions dissect the surface) and denudation. The reworking

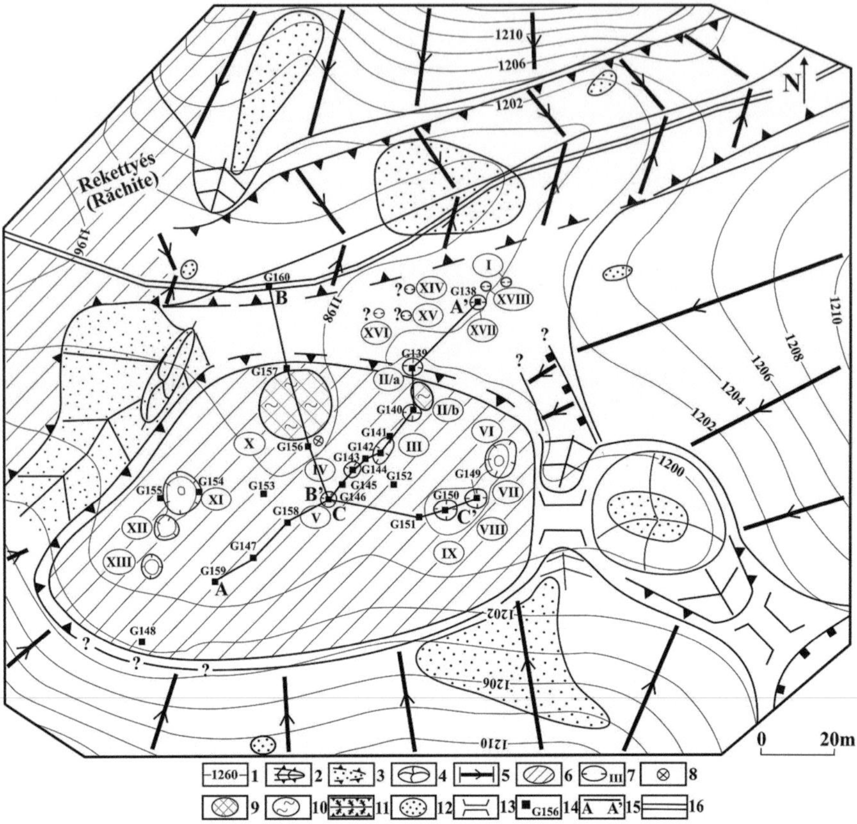

Fig. 4.45 Karst morphological map of the area Pádis-3: *1* contour line, *2* dividing wall between dolines, *3* buried dividing wall between dolines, *4* elevation on dividing wall, *5* solution doline and its slope, *6* infilled floor of solution doline (and depression of superficial deposit), *7* suffosion doline with code, *8* opening formed between August 2011 and July 2012, *9* infilled suffosion doline, *10* lake, *11* epigenetic valley, *12* limestone outcrop, *13* col, *14* site and code of VES measurement, *15* alignment of geoelectric–geological profile, *16* road

of the cover deposit does not take place through surface transport. The ponors developed along the margin of paleodepressions conveyed part of the cover sediment into the cavities and caves of the karst. The once uniform cover was dissected during denudation; the previous depressions of superficial deposit were transformed or destroyed, and new depressions, which are true depressions, formed. They emerged in areas of paleodoline groups (Figs. 4.25 and 4.50) or a single paleouvala (Figs. 4.51 and 4.52). On these covered karst patches, both cryptokarst formation on gravelly, clayey cover (with ponors) and concealed karstification in the area of loess patches (with subsidence dolines) occur.

The S and SE part of the karst (S of the Rét Stream valley) is buried karst, from where the watercourses do not flow to the N, towards the interior of the karst, but in

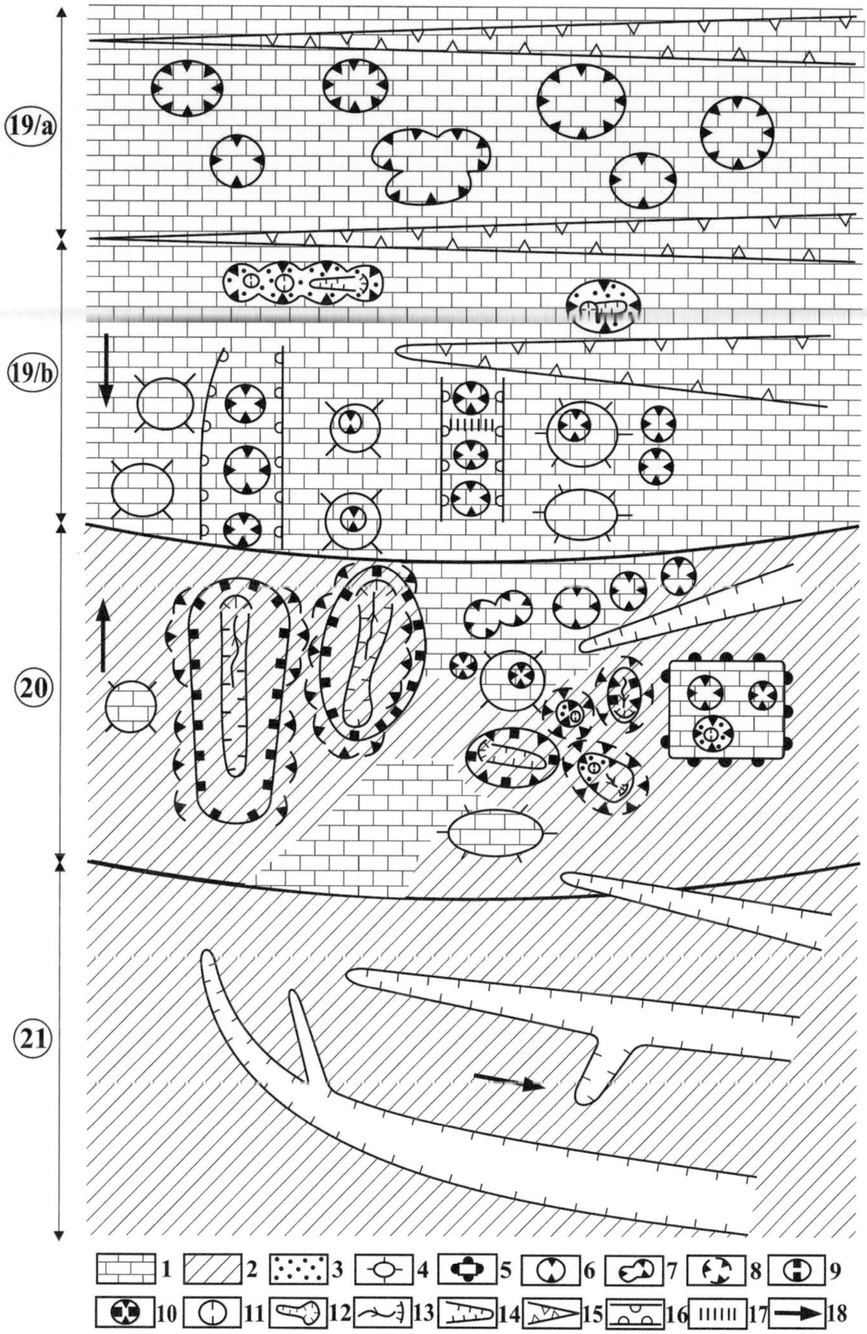

Fig. 4.46 Karst zones of renewed allogenic covered karst (not to scale, example of the Aggtelek Karst): *1* limestone, *2* impermeable cover sediment of marine or fluvial origin, *3* cover sediment of weathering residue origin, *4* karstic elevation, *5* block, *6* solution doline, *7* uvala, *8* buried doline or uvala, *9* depression of superficial deposit, *10* partially exhumed paleodoline, *11* suffosion doline, *12* suffosion doline with gully, *13* ponor, *14* valley in cover deposit, *15* valley formed along anticline, *16* epigenetic valley, *17* valley divide, *18* surface slope, *19a* uncovered karst zone, *19b* patches formed in the uvalas and dolines of the uncovered karst zone through the accumulation of weathering residue (concealed karst and cryptokarst), *20* mixed composite covered karst zone, *21* buried karst zone

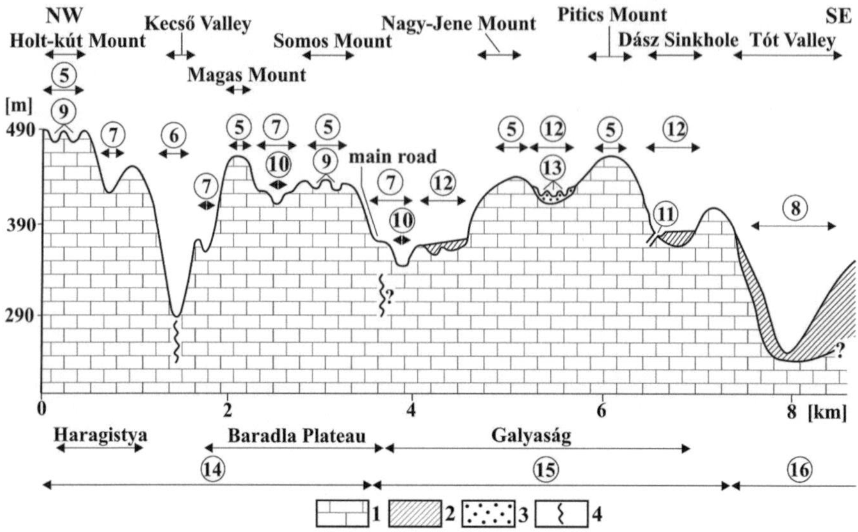

Fig. 4.47 Cross-section of renewed allogenic covered karst (example of the Aggtelek Karst): *1* limestone, *2* impermeable cover, *3* permeable cover, *4* imbrication boundary, *5* summit, elevation, *6* valley formed on structural boundary (imbrication, anticline), *7* epigenetic valley (moulded by corrosion), *8* erosional valley, *9* solution doline, *10* solution doline on valley floor, *11* ponor, *12* depression of superficial deposit, *13* subsidence doline, *14* uncovered karst, *15* partially covered (mixed composite) karst, *16* buried karst

SE direction. This part of the karst was tilted into a lower position, and its surface acquired SE slope. For the former reason, the cover sediment in the S and SE part of the karst was not denuded or to a lesser degree, while for the latter reason, the watercourses do not flow towards the interior of the karst, such as on the Pádis, but towards the valley bordering the karst (Bódva Valley).

A special variety of renewed karst, partly with properties typical of platform karst, is the Central Kentucky Karst. There the lower downslope section of the karst forms composite karst zones with the remnants of the one-time sandstone cover with caprock dolines and inherited hanging valleys on its cryptokarst (White et al. 1970; Székely and Szentes 1981). On the counterslope upper section, with mixed composite karst zone further away from the local base level of erosion, cryptokarst, concealed karst and open karst patches also occur.

4.6.2.3.1.3 Semi-allogenic Covered Karst

Semi-allogenic covered karst develops if the epigenetic valleys of non-karstic terrain which borders the karst cross the karst. Thus, the watercourses move their sediment in the older epigenetic valleys of the karst and transport it out of the mountains. Deposits of fluvial origin do not accumulate on the karst even if the ponors of the karst convey the sediment of the watercourses into the karst. The karst will be semi-allogenic because the rocks of the non-karstic terrain do not

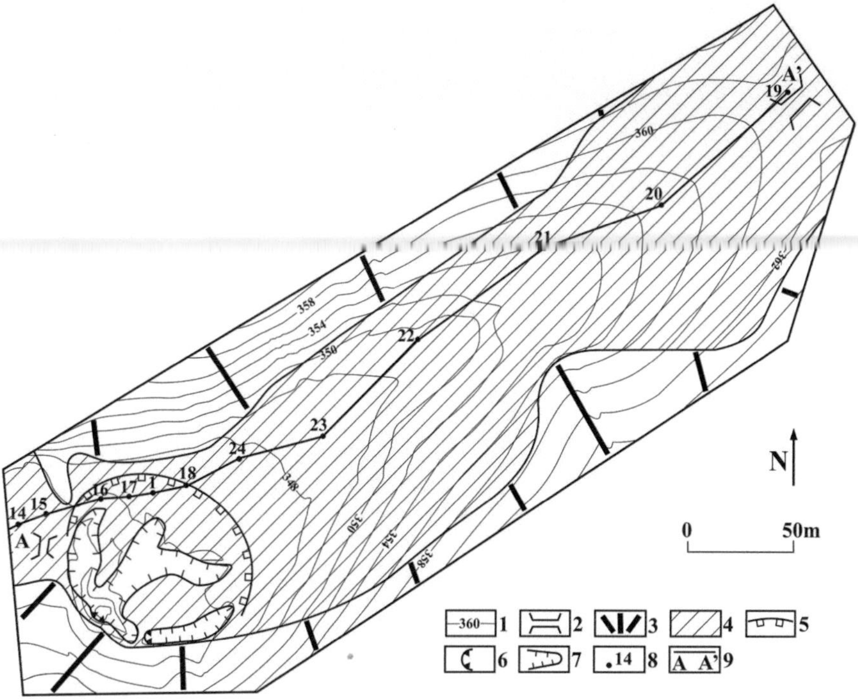

Fig. 4.48 A map of a part Bába Valley (environs of the ponor Nr. 2) (Aggtelek Karst): *1* contour line, *2* col, *3* uncovered side slope of depression of superficial deposit, *4* floor of depression of superficial deposit, *5* internal depression of depression of superficial deposit, *6* ponor, *7* ravine of ponor, *8* site and identification code of VES measurement, *9* alignment of geoelectric–geological profile

influence the coveredness of the karst and its karstification. This type develops if after the formation of the epigenetic valley, the cover sediment of the karstic terrain is removed areally (e.g. through abrasion) or the cover sediment had been eroded previously, but epigenetic valley formation could take place in its absence. This is possible if the karst water table lay on the karstic terrain or in its vicinity. The semi-allogenic covered karst is exemplified by the karst of the Mecsek Mountains (Figs. 4.53 and 4.54).

The coveredness conditions of the karst is controlled by the in situ produced solution residue, the accumulation and distribution of deposits of fallen dust and its chances of its survival. (The chance of survival for the cover sediment depends on morphology and the angle of surface slope.) Solution residue and particularly loess favour the development of concealed karst, which is homogeneous, not isolated into zones but at most interrupted by uncovered patches. If the karst had been created before the accumulation of the cover sediment, the cover fills or lines the paleodolines. (If no karstification had taken place, the cover overlies a flat or nearly flat surface.) In the case of the variety with paleodoline, true depressions of superficial

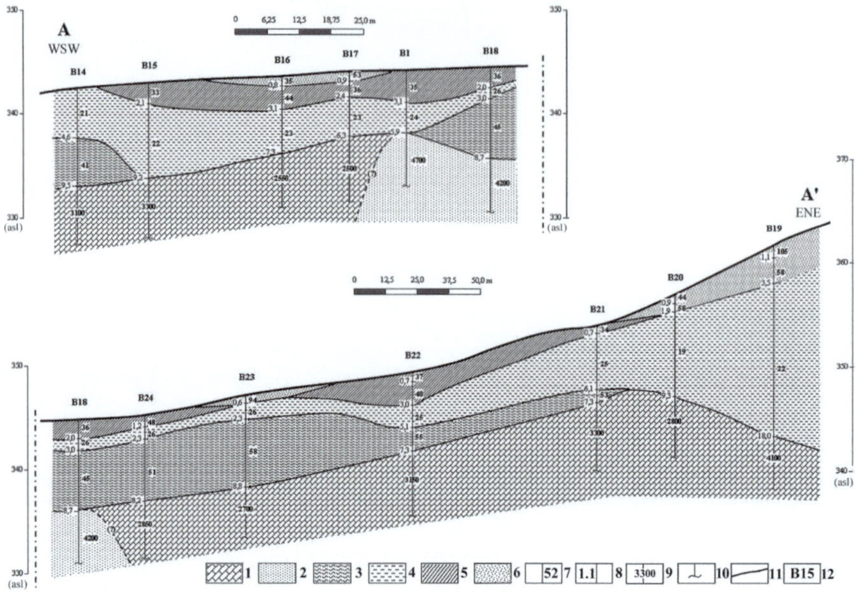

Fig. 4.49 The geoelectric–geological profile A-A′ of the Bába Valley: *1* limestone, *2* sandstone, *3* limestone debris (clayey), *4* clay, *5* clayey sand and gravel (loess), *6* sand and gravel or loess (with limestone debris), *7* geoelectric resistivity of the series (Ohmm), *8* base depth of geoelectric series, *9* geoelectric resistivity of bedrock (Ohmm), *10* approx. penetration of measurement, *11* geoelectric layer boundary, *12* site and identification code of VES measurement

Fig. 4.50 Depression bearing the Zombor-lyuk ponor (Aggtelek Karst)

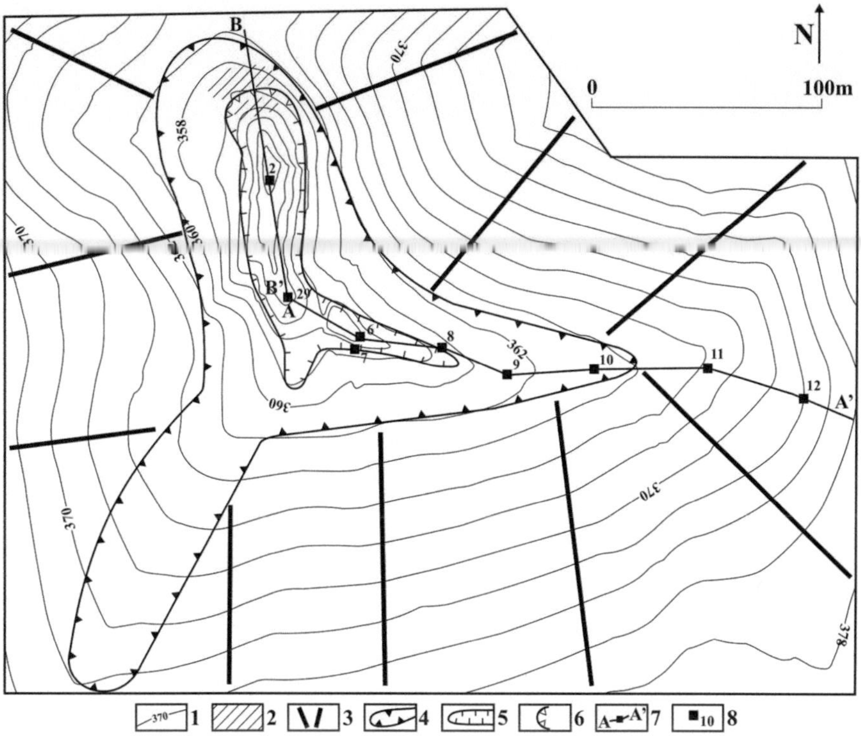

Fig. 4.51 Morphological map of the Dász doline (Aggtelek Karst): *1* contour line, *2* limestone outcrop, *3* slope of paleodoline, *4* blind valley, *5* ravine (internal valley), *6* ponor, *7* alignment of geoelectric–geological profile, *8* site and identification code of VES measurement

deposit have come about in the paleodolines. In the lined or filled paleodolines of the karst and among them, subsidence dolines occur – often in great density (W. 4.55).

4.6.2.3.2 Mantled Allogenic Covered Karst

The non-karstic impermeable consolidated rock on the karst can form a mantle, which results from marine inundation or lava flow. A typical mantled allogenic covered karst is Kab Mountain (Veress and Unger 2015), where the mantle is represented by basalt (Figs. 2.4 and 4.56). On Kab Mountain, buried karst, cryptokarst, concealed karst (with loess cover) and uncovered zones (where surface karst landforms are absent) are distinguished (Figs. 4.57 and 4.58). On the true rock boundary of the basalt sheet margin, a row of ponors came about (e.g. one feature is the Macska-lik). The locations of ponors can shift away from the margin of the basalt if the basalt filled up the upper section of a valley formed on limestone. The stream of the valley lined the section of the uncovered valley floor with basalt debris, in the

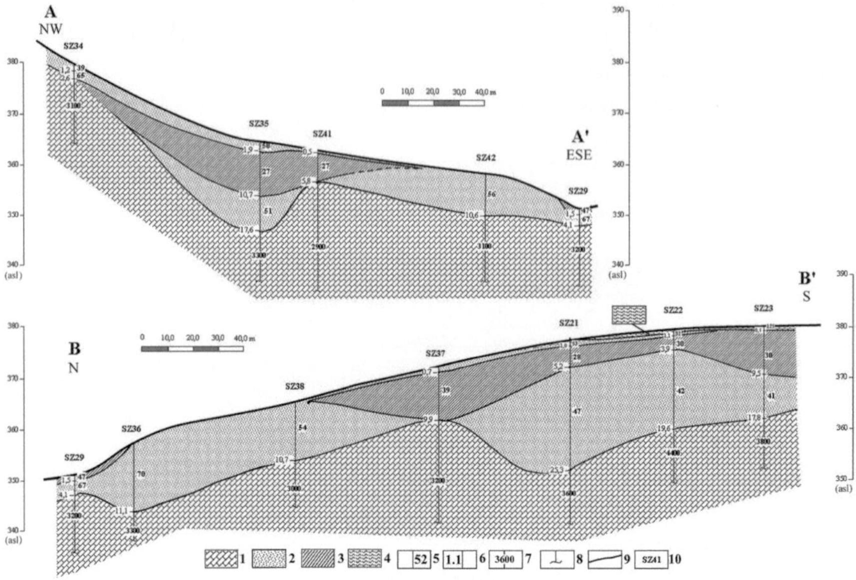

Fig. 4.52 Geoelectric–geological profiles A-A′ and B-B′ of the Dász doline: *1* limestone, *2* sand and gravel (with limestone debris), *3* clayey sand and gravel, *4* limestone debris (clayey), *5* geoelectric resistivity of the series (Ohmm), *6* base depth of geoelectric series, *7* geoelectric resistivity of bedrock (Ohmm), *8* penetration depth of VES measurement, *9* geoelectric series boundary, *10* identification code of VES measurement

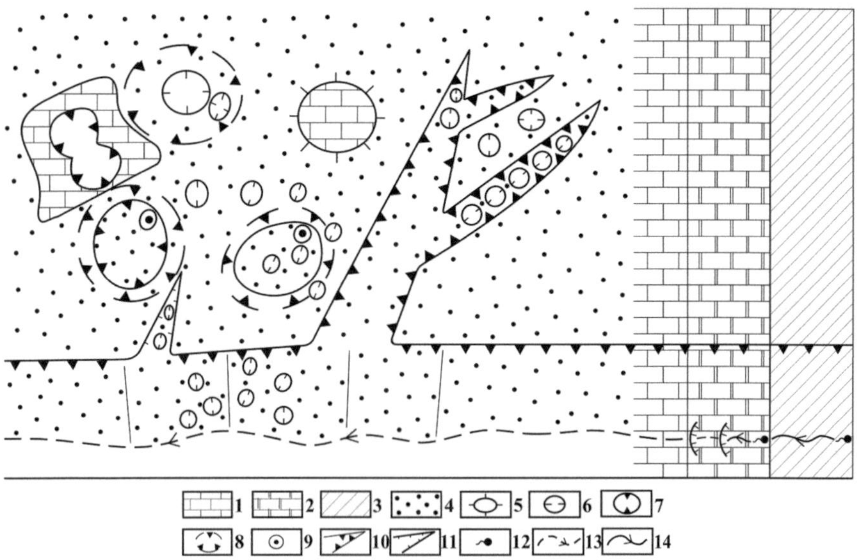

Fig. 4.53 Karst zones of semi-allogenic covered karst (example of the Mecsek Mountains, not to scale): *1* limestone, *2* poorly karstified limestone and dolomite, *3* sandstone, *4* permeable cover (loess), *5* elevation, *6* subsidence doline, *7* solution paleodoline, *8* buried paleodoline, *9* shaft, *10* epigenetic valley, *11* ravine, *12* spring, *13* intermittent watercourse, *14* permanent watercourse

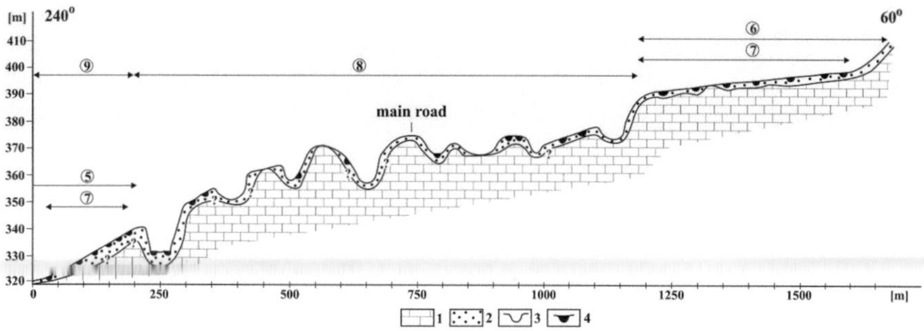

Fig. 4.54 Profile across a detail of the Mecsek karst (Cigány-földek, E of Szuadó Valley): *1* limestone; *2* cover (loess); *3* solution doline; *4* subsidence doline; *5* slope of epigenetic valley (Szuadó Valley); *6* slope of surfac elevation; *7* zone of subsidence dolines; *8* in the zone of solution dolines, where subsidence dolines occur in dolines or on remnant terrains between dolines; *9* Szuadó Valley

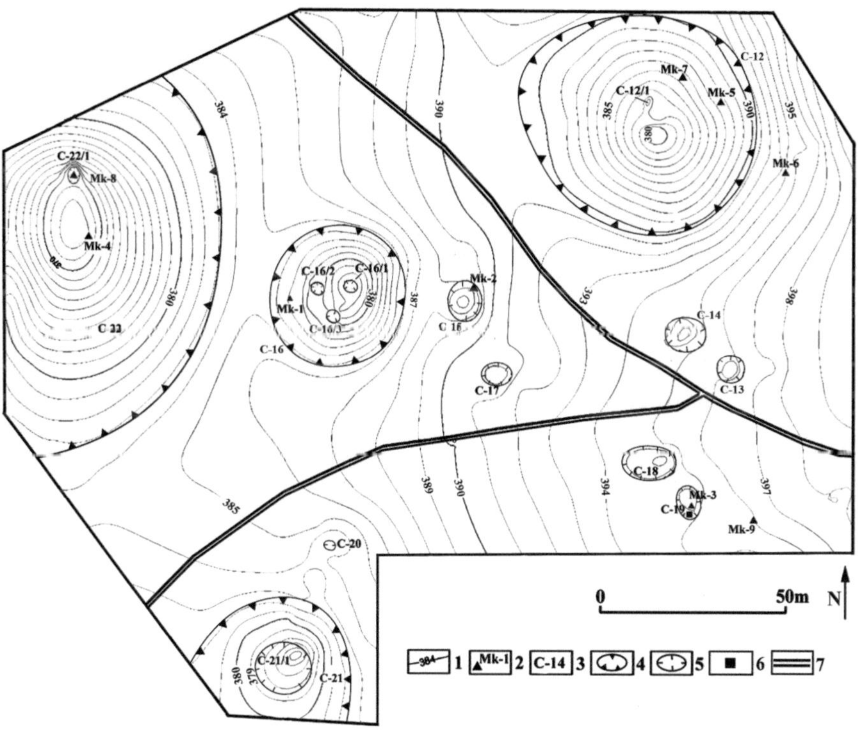

Fig. 4.55 Detail of map of Cigány-földek (Mecsek Mountains, Veress 2011): *1* contour line; *2* rock outcrop and its identification code; *3* code of karst depression; *4* solution doline; *5* subsidence doline; *6* non-karstic pipe, shaft; *7* road

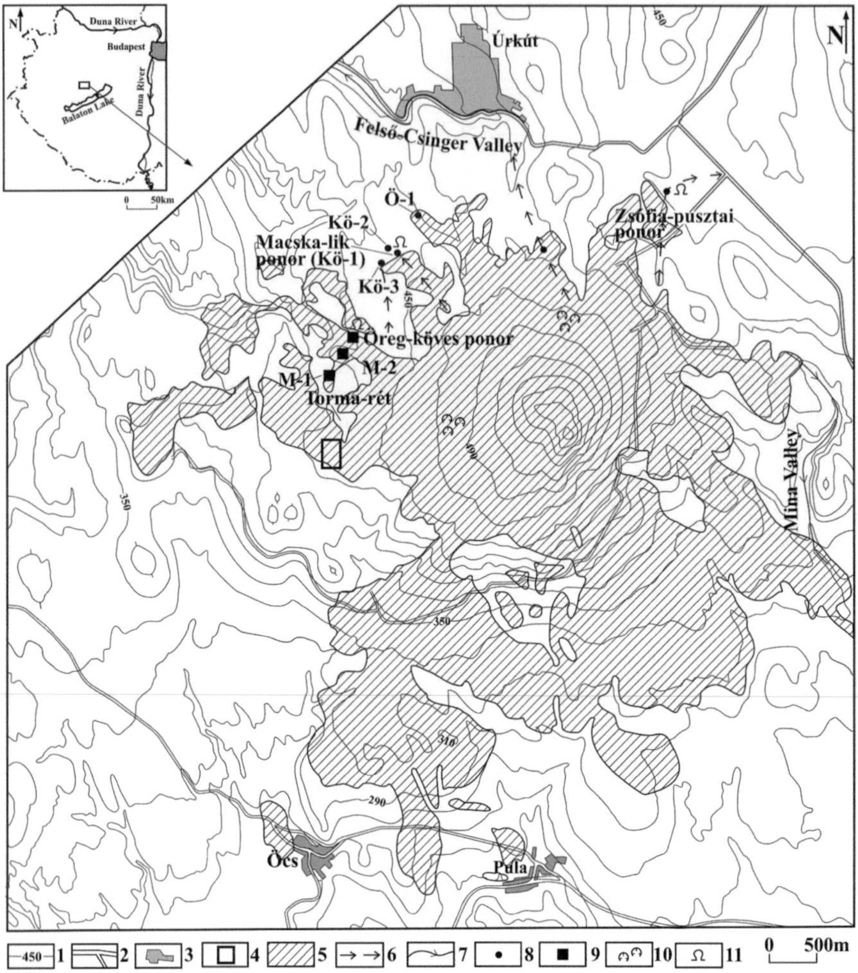

Fig. 4.56 Kab Mountain (South-Bakony Mountains, Hungary): *1* contour line; *2* road; *3* built-up area; *4* extract (see Fig. 4.60); *5* basalt; *6* intermittent watercourse; *7* permanent watercourse; *8* karst landform; *9* composite karst landform, depression of superficial deposit (composed of subsidence doline, caprock doline, ponor); *10* doline in debris; *11* ponor cave, letters (Ö, M, Kö) are the identification marks of karst features

vicinity of the margin of the basalt sheet. Where the basalt debris lining ends, ponor develops (e.g. Zsófia-puszta covered karst ponor).

Karstification may happen in the interior of the basalt sheet (in the cryptokarst zone) if the bedrock is dissected by paleokarstic depressions and elevations. (Since karstification takes place both in the interior and on the margins of the basalt sheet, this zone of Kab Mountain is transitional cryptokarst.) Karstified patches form above the elevations of the limestone bedrock (Németh 2005; Móga and Németh 2005, Veress and Unger 2015). Such karstic terrains of local extension bordered by

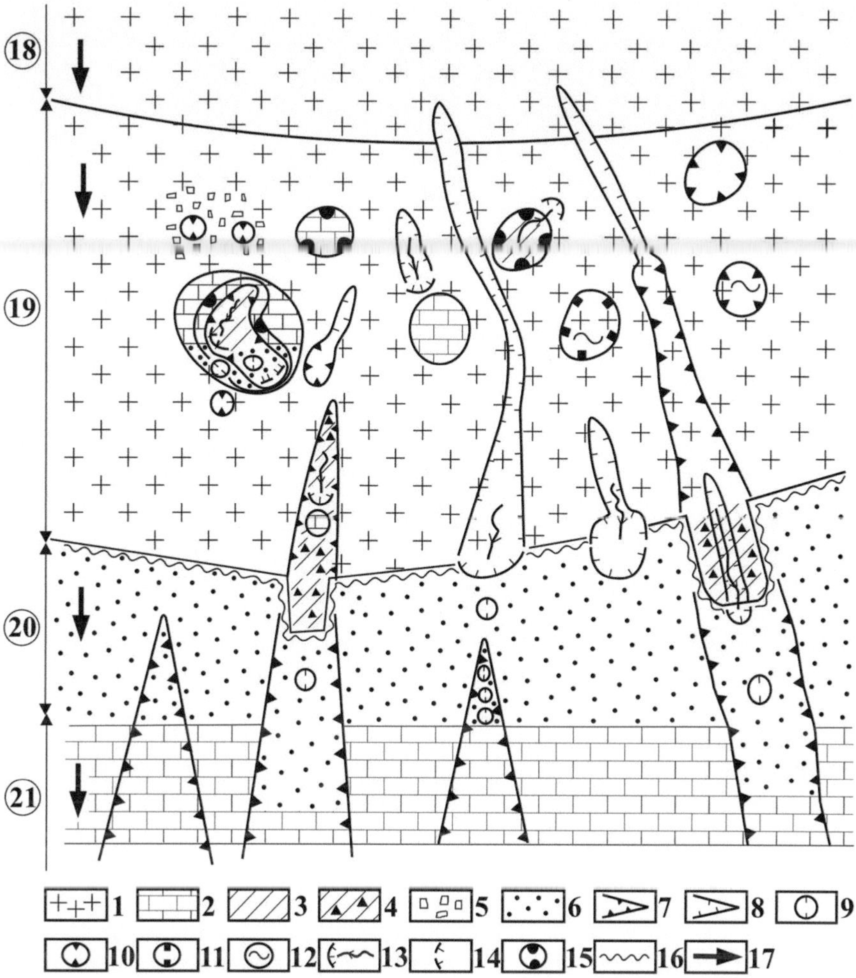

Fig. 4.57 Karst zones of mantled allogenic covered karst (example of Kab Mountain, not to scale): *1* impermeable rock (basalt); *2* limestone; *3* weathering residue; *4* fluvially reworked rock debris; *5* in situ accumulated basalt debris; *6* permeable cover (loess); *7* epigenetic valley; *8* not yet inherited valley, ravine; *9* subsidence doline; *10* caprock doline; *11* former karst depression filled with basalt; *12* lake, waterlogged area; *13* ponor; *14* inactive ponor; *15* depression of superficial deposit; *16* true rock boundary; *17* direction of surface slope; *18* buried karst; *19* transitional cryptokarst interrupted by karst windows; *20* concealed karst; *21* uncovered karst

basalt or covered by a thin basalt layer are called karst windows. In karst windows, concealed karst can also develop.

The karst windows of Kab Mountain could develop in the following (Fig. 4.59):

- The limestone elevations were not covered by the basalt sheet. At a limestone outcrop, a ponor emerges (Figs. 4.57, 4.59a and 4.60 – ponor Bk-1).
- The watercourse eroding the basalt exposes the limestone. Where the limestone is exposed, a ponor comes about on the valley floor (e.g. Tönkölyös ponor in

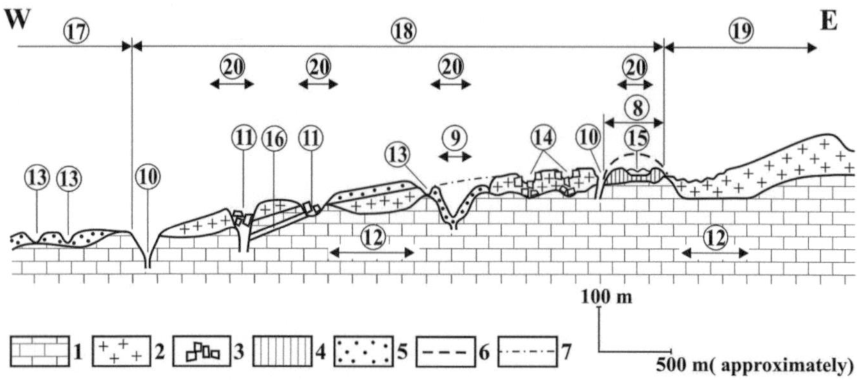

Fig. 4.58 Theoretical cross-section of a paleokarst terrain covered with basalt (example of Kab Mountain): *1* limestone, *2* basalt, *3* basalt blocks, *4* weathering residue, *5* loess, *6* former limestone surface, *7* former basalt surface, *8* depression of superficial deposit, *9* epigenetic valley inherited from basalt (depression of superficial deposit), *10* ponor, *11* caprock doline, *12* paleodoline, *13* subsidence doline, *14* doline formed in collapsed basalt, *15* lake, *16* karst passage, *17* concealed karst, *18* transitional cryptokarst, *19* buried karst, *20* karst window. Note: different sections of the profile are of different direction

Figs. 4.56, 4.57, and 4.59d). On the valley floors, caprock dolines (Fig. 4.59) or alternating caprock dolines and ponors (Fig. 4.59f) can develop at the thinning basalt cover.

- If the basalt sheet is thin, above the chimney of the limestone elevation, caprock doline develops (Fig. 4.59b) if the limestone gets water through the fissures of the basalt. The caprock dolines develop in several varieties according to the stage of development depending on the size of the feeding area. (A doline is regarded to be in a more advanced stage of development if its morphology presents properties typical of ponors: it has a water conduit and an associated ravine or valley.) Caprock dolines with ponor properties particularly develop if their environs are without outflow. That mainly happens if they are surrounded by ramparts of basalt debris of several metres height in a wreath-like fashion (Fig. 4.59c). The terrain surrounded by ramparts mostly tilts from all directions towards the caprock dolines. A ponor can form in the interior of a terrain surrounded by ramparts if the limestone is exposed from below the basalt (Fig. 4.60).
- On the limestone elevations rising above the basalt sheet, dolines developed, and thus the elevations were lowered. Subsequently, solution residue accumulated on the surface. When the elevation of the surface of the karst window falls below the level of the neighbouring basalt, the burial of the window was enhanced through the transport of sediment from the neighbouring basalt terrain. The previous elevation, now buried, became a cryptokarst patch. Water runoff from the patch generates a ponor in the vicinity of the basalt margin (e.g. the Torma-rét area).
- There is an impermeable intercalation in the basalt (red clay or the underlying basalt bank is impermeable), which diverts percolating water towards the elevation of the limestone bedrock, where karstification ensues (Németh 2005).

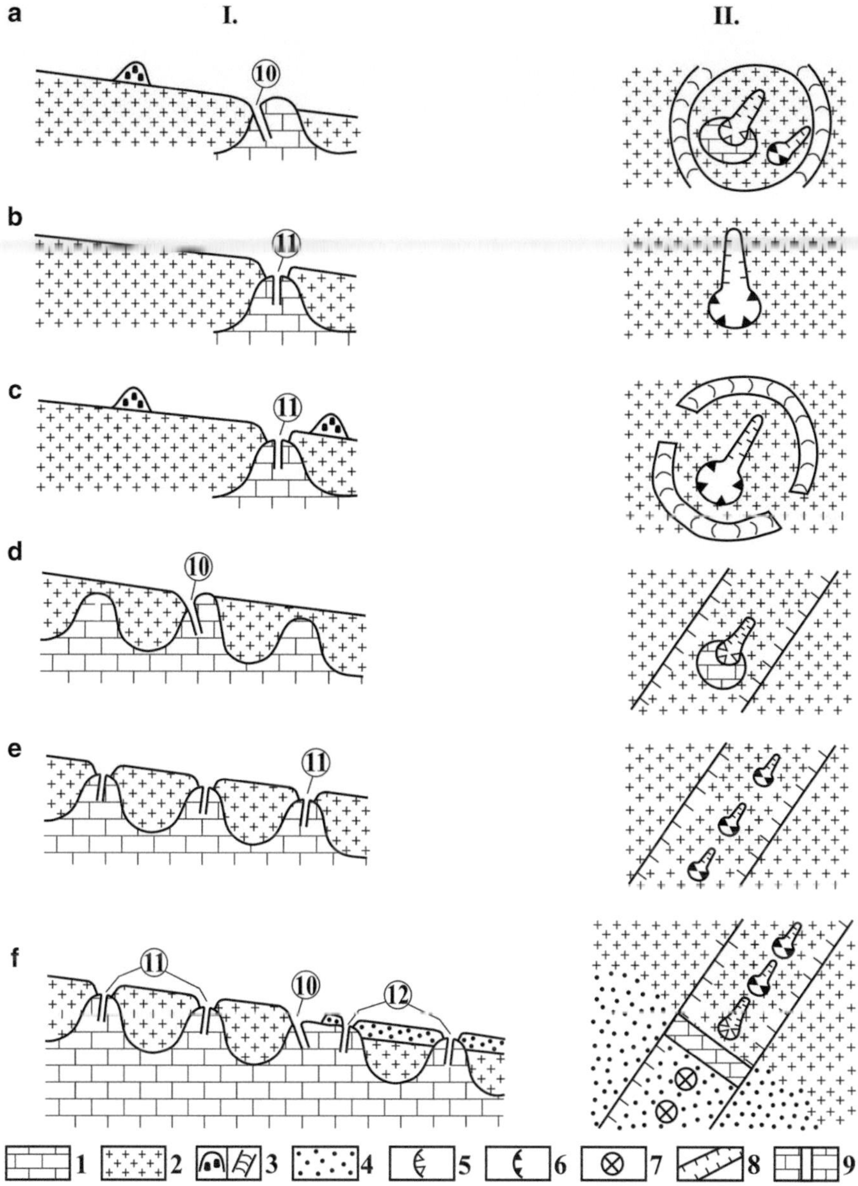

Fig. 4.59 Karstification at karst windows (based on examples from Kab Mountain): *1* limestone; *2* basalt; *3* rampart or scarp built from basalt blocks; *4* loess; *5* ponor; *6* caprock doline; *7* subsidence doline; *8* valley; *9* cross-section of chimney, shaft; *10* ponor; *11* caprock doline; *12* subsidence doline, *I* cross-section, *II* top view, (**a**) ponor formation in the limestone outcropping in karst window, (**b**, **c**), caprock doline formation on basalt thinning out at karst window, (**d–f**) ponor formation at karst windows created by fluvial action: ponor forms at limestone outcrop (**d**), caprock dolines form on thinned-out basalt (**e**), ponor and caprock dolines form at limestone outcrops or thinned-out basalt (**f**)

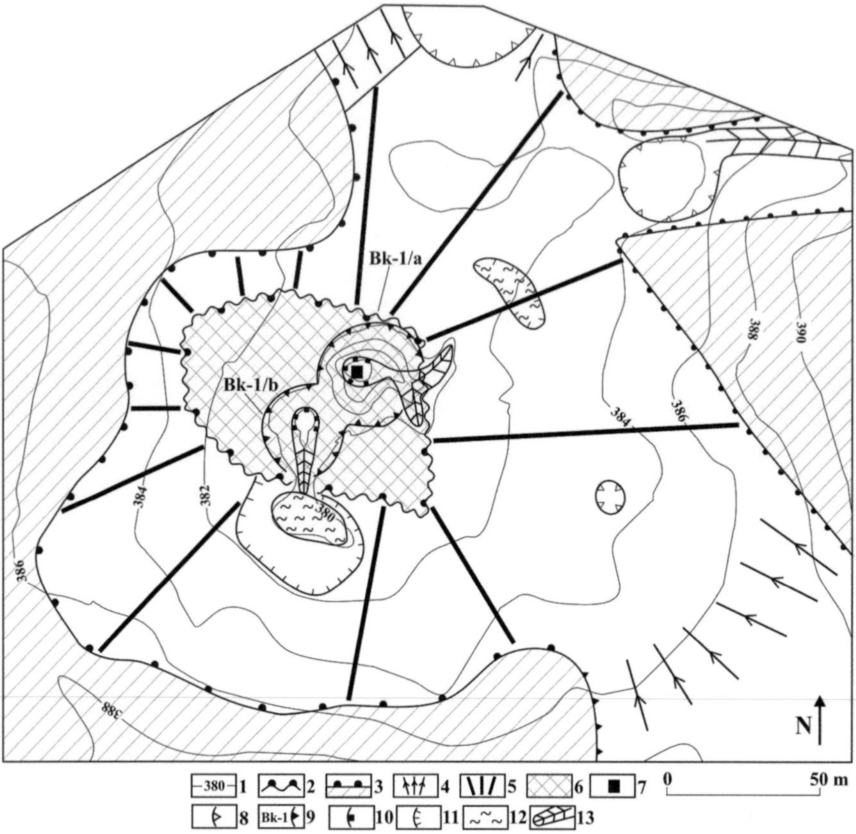

Fig. 4.60 Morphological map of a karst window and its environs (Kab Mountain): *1* contour line; *2* boundary of distribution of basalt blocks; *3* scarp, rampart created from basalt blocks; *4* slightly sloping surface without scarps; *5* gently sloping terrain between scarp and karst window; 6 karst window; *7* limestone outcrop; *8* caprock doline; *9* ponor and its code; *10* dropout doline; *11* fossil doline; *12* lake; *13* gully, ravine

There are at least two varieties of mantled allogenic karst. If a karstic terrain is dissected by mounds and depressions and it becomes covered, the above-presented karst types with karst windows develop (first variety). If the bedrock is less dissected and no karst window appears, the second variety, typical of platform karsts, is generated. If there are gypsum and rock salt intercalations on platform karst, however, the higher solution capacity of such rocks leads to the formation of local karstified patches above them, which can even be regarded large-scale karst windows.

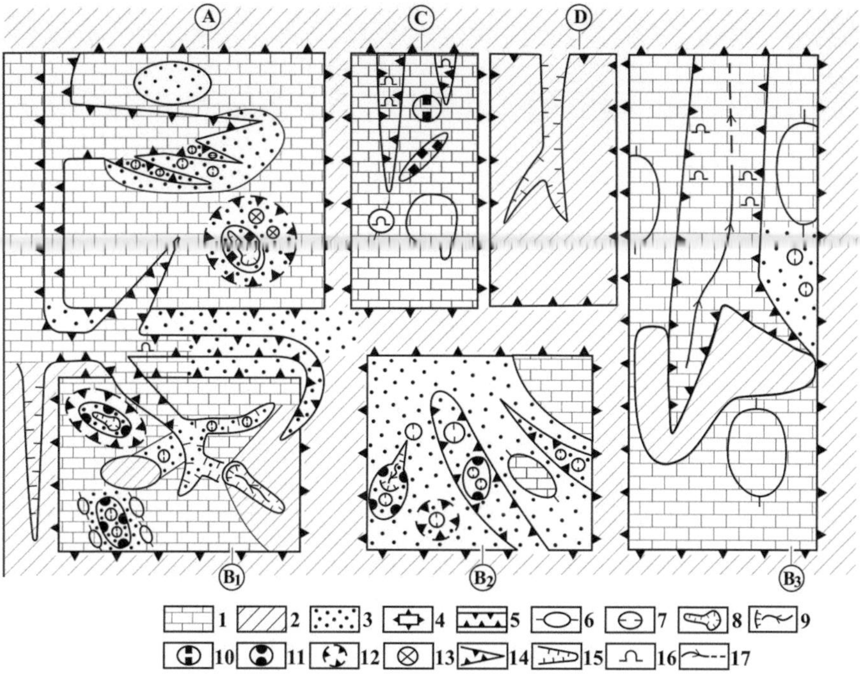

Fig. 4.61 Horst covered karst (based on the example of the North-Bakony Mountains, not to scale): *1* limestone; *2* impermeable cover (clayey gravel); *3* permeable cover (loess); *4* horst; *5* graben, basin; *6* elevation; *7* subsidence doline; *8* subsidence doline with gully; *9* ponor; *10* collapse doline-like landform; *11* depression of superficial deposit; *12* buried paleodoline or ponor; *13* fossil subsidence doline; *14* epigenetic valley; *15* ravine, not yet inherited valley; *16* opened-up cavity (remnant cave); *17* watercourse of permanent flow and then intermittent because of seepage, (**a**) high elevated horst with patchy cover sediment of loess, its subsidence dolines are formed mostly in loess on the valley floor, on interfluve ridge (with concealed karst patches); (**b**) horst of medium elevation, its subsidence dolines and depressions of superficial deposit are created on valley floors, between exhumed elevations, above the elevations of buried limestone or in paleodolines (with patches of concealed and cryptokarst), for variety (**b**$_1$) former gravel mantle and loess constitutes the covered karst, for variety (**b**$_2$) loess cover is dominant, for variety (**b**$_3$) epigenetic valley formation, in the valley sides opened-up cavities; (**c**) block of medium elevation with cavity formation next to the carbonate surface, superficial depressions created from the cavities through collapses; (**d**) block in low position with valleys (buried karst)

4.6.2.3.3 Horst Covered Karst

Horst covered karsts are dismembered into blocks of different geological history and elevation. A typical example is the Northern Bakony Mountains. In the medium mountains of horst structure, the previously accumulated cover sediment is removed; the younger cover – disregarding local fluvial reworking when the blocks rise above their environs – can only form from fallen dust. Karstification varies with blocks or block types, and sometimes it is only present on certain block types (Fig. 4.61).

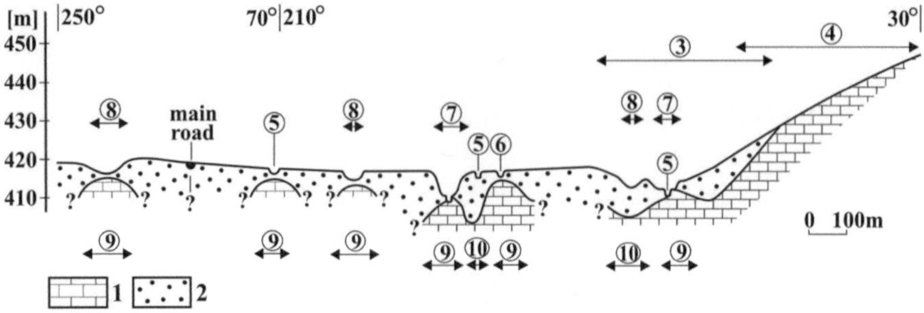

Fig. 4.62 Cross-section of the Tés Plateau (Bakony Mountains, for data on limestone bedrock morphology see Veress 2005, 2006): *1* limestone, *2* loess, *3* old epigenetic valley inherited from the former gravel sheet, *4* elevation, *5* suffosion doline, *6* covered karst ponor, *7* depression of superficial deposit in valley, *8* depression of superficial deposit on plain surface, *9* elevation on bedrock, *10* depression in bedrock (paleoponor or paleodoline)

On loess-mantled surfaces, concealed karst formation takes place. It relatively seldom happens in the area of paleodolines in the Bakony Mountains since geological research (Szabó 1964; Bárdossy 1977; Pataki and Nyírő 1983) indicates that the older dolines are mostly infilled (e.g. with bauxite) and buried under Eocene limestone (Bárdossy 1961), Oligocene clays, sands, sandstones (Jaskó 1957a) or Pliocene travertine (Jaskó 1957b). Syngenetic dolines only emerge where the cover is thinner (concealed rock boundary) at the elevations of the buried bedrock (Fig. 4.62). At the patches of the gravel mantle, ponors and covered karst ponors occur. Before the accumulation of loess, ponor and in their continuation cave, formation happened on valley floors of the gravel mantle (e.g. on the Tés Plateau). On the Tés Plateau, the subsidence dolines and depressions of superficial deposit created by recent karstification often form in the area of previous ponors through postgenetic karstification as the passages of the previous ponors lost their sediment fill.

Karstification on the blocks of different types takes the following courses (Fig. 4.61, Veress 2000a, b).

- On blocks of ca 550 m elevation, only patches of the loess mantle are preserved in paleodolines or interfluvial ridges. Occasionally cover sediment patches transported into the valleys may also occur. Concealed karst formation takes place in the area of such patches (Fig. 4.61a).
- Karstification on blocks of medium elevation (300–400 and 550 m) happens in different ways. If there is only a loess or patchy loess mantle on the block or block group, concealed karst develops in a uniform manner all over the block on the loess patch (Fig. 4.61b$_2$). If, in addition to loess, there are also impermeable patches on the block, both concealed karst and cryptokarst may be present (Fig. 4.61b$_1$). On impermeable cover sediment patches, ponors or covered karst ponors occur. On blocks of medium elevation where epigenetic valleys developed, valley incision opens up karst cavities and creates remnant caves. For such karst landforms, water outflow from the karst never happened. This kind of valley evolution happens in intensively incising valleys of blocks covered with

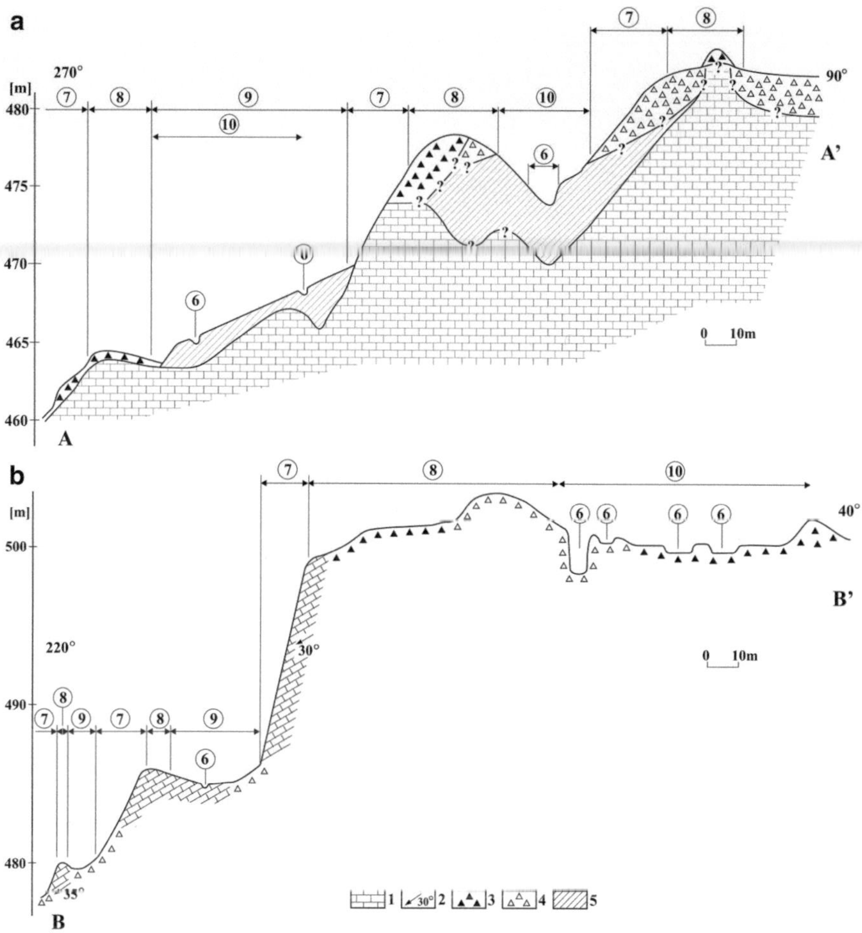

Fig. 4.63 Cross-section of the northern (**a**) and southern (**b**) Mester-Hajag (Bakony Mountains): *1* limestone, *2* dip of strata, *3* high density of limestone debris in the cover, *4* low density of limestone debris in the cover, *5* superficial deposit with silt, *6* subsidence doline, *7* steep SW slope, *8* ridge, elevation, *9* gently sloping terrain with cover sediment between ridges, *10* depression of superficial deposit

impermeable rocks over extended areas. The watercourses in the valleys feed karst water by seepage and contribute to cavity formation under channels. The cavities created under valley floors promote, accelerate valley formation (Fig. 4.61b$_3$).

- Within blocks and groups of blocks of medium elevation, there is also great variation in karstification styles according to lithology and tectonism. Because of the uplift and tilting of Middle Cretaceous limestone blocks, cover sediments are denuded and remnant paleokarst landforms (elevations, ridges) are exhumed (Figs. 4.61b$_1$ and 4.63). The cover is transported in NE to SW (as on Égett

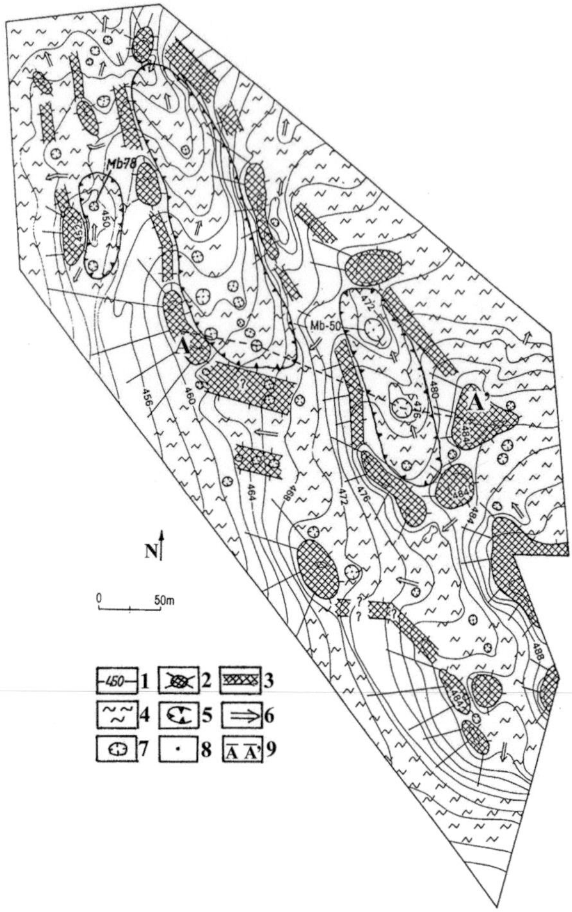

Fig. 4.64 Covered karst terrains between limestone elevations and ridges of NNW-SSE strike on Middle Cretaceous limestone on Mester-Hajag (Veress 2000a): *1* contour line, *2* semi-exhumed cone, *3* exhuming cone and ridge, *4* covered karst terrain formed through removal of cover, *5* depression of superficial deposit, *6* sediment reworking, *7* subsidence doline, *8* non-karstic pipe in doline, *9* alignment of the cross-section A-A′ (Fig. 4.63)

Mountain) or NNE to SSW direction (as on Mester-Hajag). Among the exhumed elevations of the bedrock covered karst terrains appear. The dimensions of such terrains are determined by the mutual arrangement of the ridges, while the strike of these terrains is determined by the direction of bedrock elevations and ridges. The outcrops of bedrock prevent the transport of cover sediment over the surface, and the cover sediment is conveyed into the karst and results in depressions of superficial deposit (Fig. 4.64).

- On blocks of medium elevation where partially permeable rock (marl) or poorly permeable rock (dolomite) occurs close (1–2 m) to the surface, subsurface cavity formation takes place. Through the collapse of thin ceilings, small-scale landforms similar in morphology to collapse dolines emerge on the surface of blocks, while remnant caves are observed in the sides epigenetic valleys (Fig. 4.61c).

- On blocks of medium elevation, valleys inherited over limestone are common. Runoff into them comes from either the impermeable cover sediment of the bearing block or from the cover of the neighbouring block. Particularly on antecedent sections of epigenetic valleys, water seepage feeds the karst water reserves of the mountains.
- The blocks in lower (below 300–400 m) positions covered with impermeable rocks are either buried karsts (Fig. 4.61d) or open karsts. Buried karst terrains do not affect the karstification of the surrounding blocks since the bearing blocks are either in marginal position or, if they are in the mountain interior, are at lower elevations than the neighbouring blocks. Epigenetic valleys of different sizes and in different stages of development are found on the blocks.

4.6.2.4 Covered Karst of Glaciokarsts (High-Mountain Karsts)

In the area of glaciokarsts, geomorphic evolution is due to karstification and glacial erosion (Smart 2004), but linear erosion and mass movements are also influential. Karstification could have happened before the Pleistocene glaciations or in some interglacials (Veress 2012b). Consequently, paleodolines could have come about before the Pleistocene (preglacial dolines) or in some interglacials (interglacial dolines). Cirques and glacial troughs could have formed in preglacial paleodolines or paleouvalas (Figs. 2.7 and 2.8). Interglacial dolines were generated on the floors of glacial valleys. Glacial valleys are dissected by rock basins, cuestas and hogbacks. Bögli (1964, 1976) determined two types of glaciokarsts: the schichtrippen karst (in case of horizontal beds, it is called schichtreppenkarst) and the rundhockerkarst. The former one develops on cuesta-like terrains, while the latter forms on roche moutonnée terrains. Both the open karren formation and the covered karren formation and thus the karstification happen mainly on the surfaces of above-mentioned karst types with different dipping.

While the interior parts of cirques are mostly uniformly covered, the cover of glacial troughs is partial and patchy. The sediment cover in glacial troughs is of moderate thickness. Depending on the shape and position of the valley depressions or to what extent they are filled or lined, karst terrains of variable morphology and pattern develop. The shape of the covered karst and the size of the patches depend on the shape, size and coveredness of paleokarst depressions and glacial depressions. It is common that the two landform assemblages of different origin jointly determine the pattern of the covered karst: interglacial dolines come about in glacially created landforms.

Under outcrops of the cuestas, paleodolines are often found which are elongated in the direction of the strike of outcropping strata faces. This strike may coincide with the direction of the valley axis, and the cuestas are aligned where the slope of the valley side is opposite to the dip direction of strata (Fig. 1.11). The result is rows of covered karst patches in strike direction. If the cuestas are perpendicular to the valley axis, the rows of covered karst patches are also aligned perpendicular to that.

The pattern of covered karst also depends on glacier type (Fig. 4.15), moraine type, the location of mountain failures and other mass movements and their intensity. It is common that mass movements result in a too thick cover for a uniform covered karst formation process (Figs. 4.17 and 4.19).

In general, the amount of glacial accumulations is decreasing in the glacial valley from bottom to top. In the cirque, however, both the glacial deposit and debris of other origin accumulate in considerable thickness. The valley floors dissected by preglacial dolines may be continuous in both cross-section and longitudinal section (Figs. 4.13 and 4.16c, d), only absent on the steepest slopes. In the valleys with interglacial dolines, glacial deposits primarily accumulate in dolines. The degree of weathering for the glacial deposit (its clay content) is increasing towards the lower end of the glacial trough. On the more weathered glacial deposit enriched in clay, the density of subsidence dolines (including an increased proportion of dropout dolines) is higher.

The rate of karstification for the glacial valleys is enhanced by the partial or complete lack of outflow, caused by glacial overdeepening or paleokarst formation. Particularly the karstification and covered karst formation of glacier valleys developed in preglacial dolines shows high intensity. Covered and open karst formation in glacial valleys produced a mosaical pattern, which is virtually impossible to classify. Even so, if we look for regularity in karst distribution, it is claimed that covered karst patches are concentrated on the floors of paleodolines, giant grikes and the forelands of cuestas dissected with giant grikes covered by debris. The location, size and shape of covered karst, however, are largely modified by mountain failures and other mass movements.

The cover of glaciokarst covered karst is not only morainic material or its reworked variety but also rock debris and blocks of frost shattering and mountain failure (Figs. 4.17 and 4.19). With the growing debris grain size, subsidence dolines are reducing in number and density. The varieties of glaciokarst covered karst are the cirque and the glacial valley covered karst.

4.6.2.4.1 Cirque Covered Karst

Covered karst is often restricted to the cirques of the glacial valleys (Figs. 4.65 and 4.66), either in lack of a glacial trough or the gradient of the glacial trough is so high that mass movements and fluvial erosion predominate in it. Cirques often formed in paleodolines, paleouvalas (Figs. 4.65 and 4.66), where present-day karstification is intensive. Cirques which are affected by uniform karstification and deepening are called cirque dolines (Barrère 1964). The karst pattern of cirque valleys shows three varieties.

In the case of a solitary cirque, a single independent island-like covered karst patch is generated (first variety). Along the major glacial troughs, cirques are aligned with short tributary glacial valleys connected to the main glacial valley. The cirques or their short glacial troughs are in a hanging position above the main valley (second variety, Fig. 4.67). The main valley without covered karst is bordered by

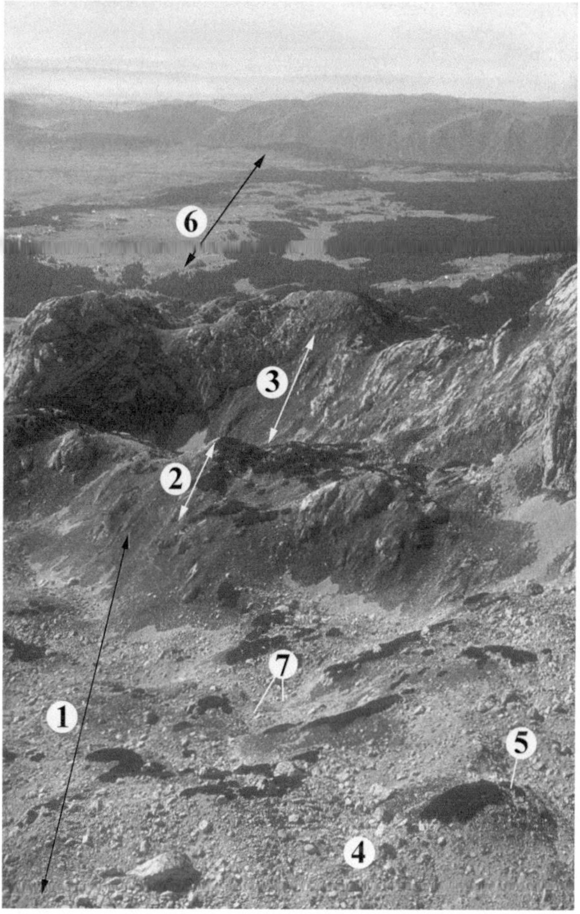

Fig. 4.65 Cirque dissected by glacial erosion (Valoviti do, Durmitor Mts): *1* closed cirque developed from a preglacial doline (Valoviti do), *2* ridge, *3* preglacial giant doline, *4* moraines, *5* roche moutonnée, *6* lower surface around the mountains covered with morainic deposit, *7* suffosion doline

covered karst patches of the cirque row (Fig. 4.15d). Finally, a group of cirques can be associated with a major glacial valley (third variety, Fig. 4.15c), creating a larger covered karst patch (Fig. 2.8).

The covered karst terrain of the cirque is uneven, dissected by roches moutonnées, cuestas, hogbacks, morainic mounds and rock basins (Figs. 4.65 and 4.66). Karstification in the covered karst area of the cirque could have begun quite recently, gradually developed to the trigger of the retreating snow and ice cover. In the area of the mentioned cirque below Triglav, there was a cirque glacier even in the 1950s (Gams 2002). Karst landforms developed in glacial deposits and debris often in very high density, above giant grikes and shafts (Figs. 4.68 and 4.69). The deposits on the floors of glacially overdeepened cirques was transported into the karst through the shafts and suffosion dolines and depressions of superficial deposit formed.

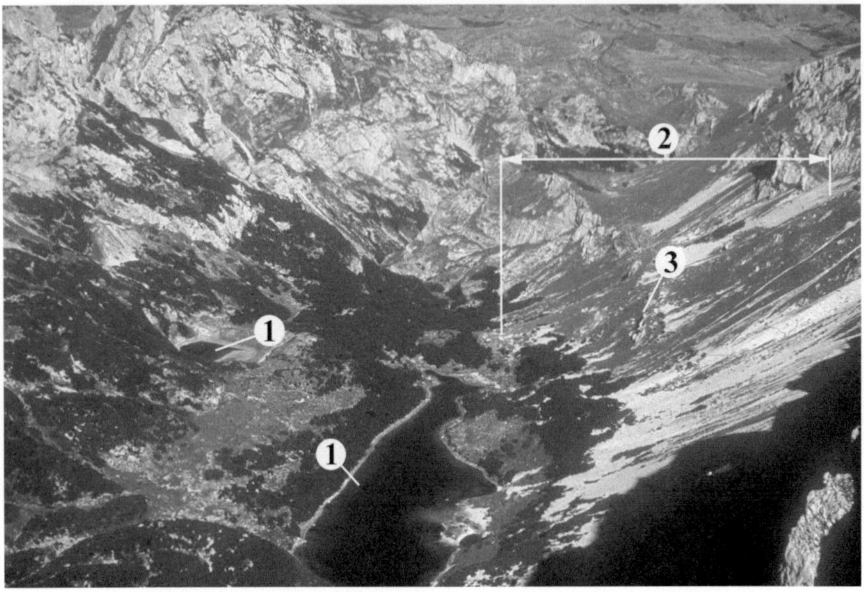

Fig. 4.66 Cirque formed from giant doline with rock basin (Durmitor Mountains): *1* rock basin, *2* debris slope, *3* margin of partial doline covered with debris

Fig. 4.67 Cirque above the Vrata Valley (Julian Alps): *1* horn, *2* cirque, *3* glacial trough, *4* valley formed through the action of meltwater and mass movements, *5* roche moutonnés, *6* shaft, *7* morainic deposit, *8* Vrata Valley

Fig. 4.68 Covered karst features of the cirque below Triglav (Julian Alps): *1* counterslope scarp (roche moutonnée), *2* covered karst in cirque, *3* debris fan of the cirque, *4* debris-covered giant grike with rows of subsidence dolines, *5* shafts, *6* subsidence dolines

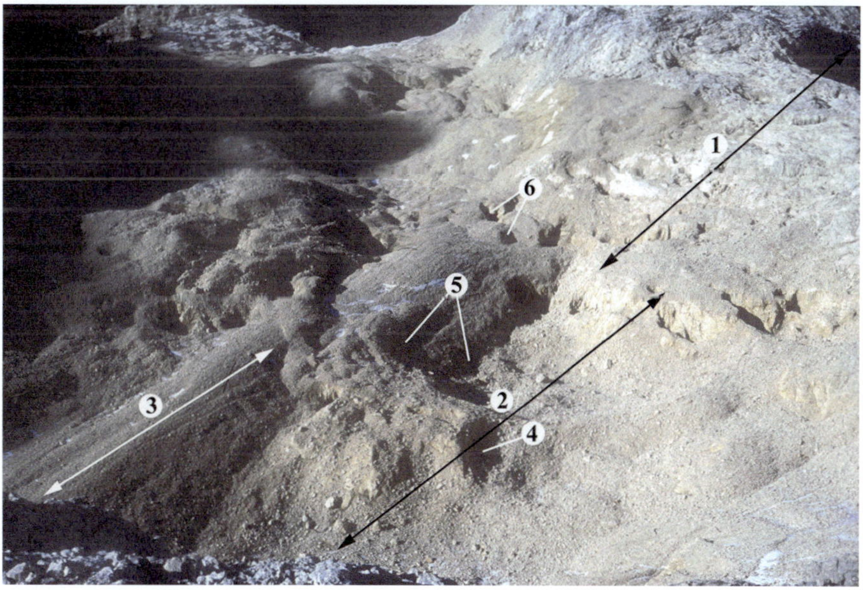

Fig. 4.69 Covered karst features of another detail of the cirque below Triglav: *1* uncovered karst detail of cirque, *2* covered karst detail of cirque, *3* debris fan, *4* subsidence dolines, *5* shafts, *6* unfilled shafts

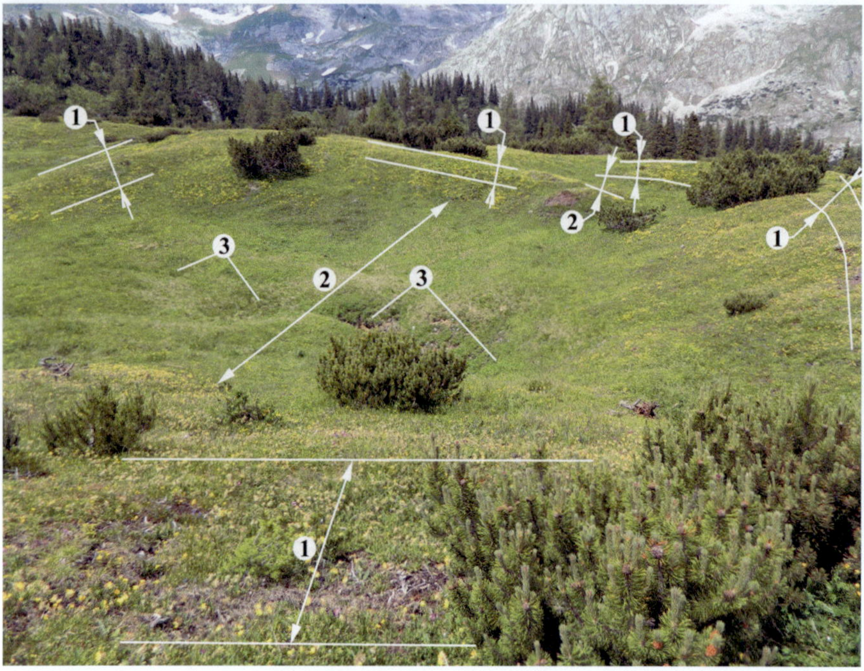

Fig. 4.70 Karstification of terrains enclosed by morainic hills (cirque of a glacial valley in the Hochschwab): *1* morainic hill, *2* covered karst terrain enclosed by morainic hills, *3* subsidence doline

In the cirques with morainic mounds and poor drainage, connected covered karst terrains come about where subsidence dolines occur with high frequency (Fig. 4.70). Because of high doline density, the uneven surface between dolines can erode into a plain after the morainic material has been transported into the dolines and from there into the depth of the karst (Barrère 1964).

4.6.2.4.2 Glacial Valley Covered Karst

The covered karst of the valley extends to both the cirque valley and the trough valley. In parallel valleys, covered karst stripes stretch parallelly on the glaciokarsts, but they are seldom continuous. It commonly occurs that, since the cover sediment primarily accumulates in dolines (or rock basins), the pattern is striped with patches (Figs. 4.71, 4.72, 4.73, and 4.74). The following modified versions are observed.

- It is common that the members of the glacier network are connected to large-scale outlet glaciers. Although on the floors of tributary glaciers, covered karsts of striped pattern with patches developed, there is no karstification on the floor of the main glacier since its floor reached the karst water table. This variety is exemplified by the glacial trough of Grundlsee in Totes Gebirge (Fig. 2.8).

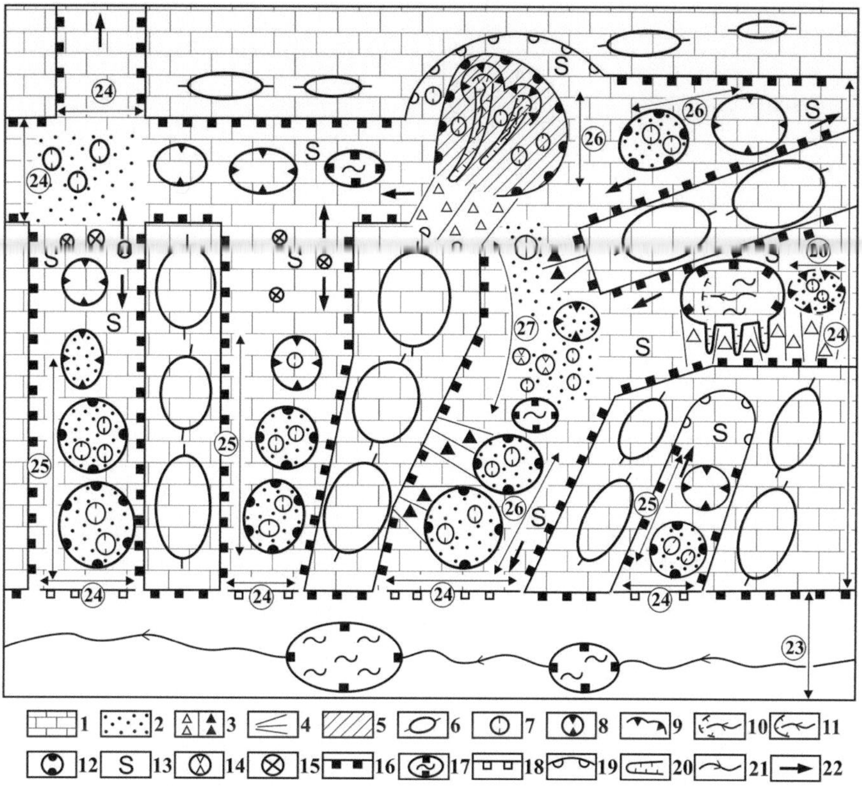

Fig. 4.71 Striped-patchy covered karst of high mountain with glacier network (example of Totes Gebirge, not to scale): *1* limestone; *2* moraine; *3* frost-shattered debris of siliceous limestone, blocks and debris of mountain failures; *4* debris fan; *5* reworked, transformed rock debris (with high clay content); *6* elevation (horn); *7* subsidence doline; *8* paleodoline; *9* paleouvala; *10* subaqueous ponor; *11* ponor; *12* depression of superficial deposit; *13* schichttrippenkarst; *14* recent solution doline; *15* shaft doline; *16* glacial trough; *17* rock basin with lake; *18* transverse scarp; *19* cirque; *20* gully, ravine; *21* watercourse; *22* direction of surface slope; *23* main glacial trough; *24* tributary glacial trough; *25* striped-patchy karst (concealed karst), patches are aligned in the inter glacial dolines of glacial valleys (with uncovered karst between patches); *26* covered karst (concealed or cryptokarst) patches on the heap of mountain failure or debris fan; *27* covered karst (concealed karst) terrain composed of moraine patch and local rock debris accumulation and interrupted by uncovered patches with solution dolines

- In karst regions, piedmont glaciers can also develop on the lower portions of the karst margin. Composite covered karst develops, and to its striped and patchy covered karst pattern, a covered karst zone is linked in the area of the former piedmont glaciers. The latter is bordered by an open karst zone (with old solution dolines without cover and still active) (in the Durmitor, between Žabljak and the Tara River (Figs. 4.75 and 4.76)). A mixed karst zone is found in the area of the former piedmont glacier: patches of open karst (higher surfaces) and covered karst (glacially moulded paleodolines) alternate (with subsidence dolines within

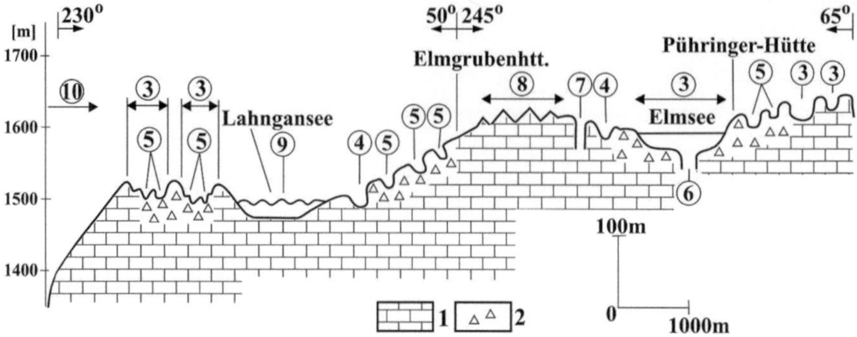

Fig. 4.72 Theoretical profile across the floor of a glacial valley of the Totes Gebirge (along tourist paths 214 and 201): *1* limestone; *2* morainic deposit, siliceous debris, blocks of mountain failure; *3* paleodoline; *4* recent solution doline; *5* suffosion doline; *6* ponor in lake; *7* shaft; *8* cuesta; *9* rock basin; *10* main glacial valley

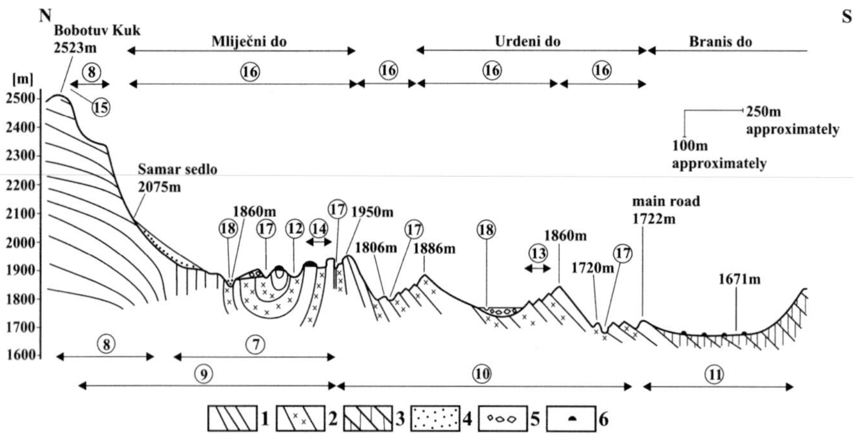

Fig. 4.73 Section of doline row remoulded by ice and less infilled with morainic deposit (Durmitor): *1* thick limestone strata poor in silica, *2* thin-bedded limestone series rich in silica, *3* flysch, *4* frost-shattered debris, *5* morainic deposit, *6* roche moutonnée, *7* root zone of the recumbent fold which created Šareni Pasovi mountain, *8* front of recumbent fold, *9* cirque, *10* glacial trough of N to S direction, *11* glacial trough of E to W direction, *12* rock basin, *13* cuesta, *14* hogback terrain, *15* horn, *16* paleodoline, *17* solution doline, *18* suffosion doline

the paleodolines). The area of the paleodolines represents depressions of superficial deposit today (Fig. 4.77).

- A special case of composite karst is when the merging of tributary glaciers created a major glacier, which terminated on the mountain slope in a patch elongated in the direction perpendicular to that of the feeding tributary glaciers. In this case, mixed karst forms in the area of the main glacier. The covered karst has developed both in the paleodolines and on the surfaces between them (Figs. 4.78

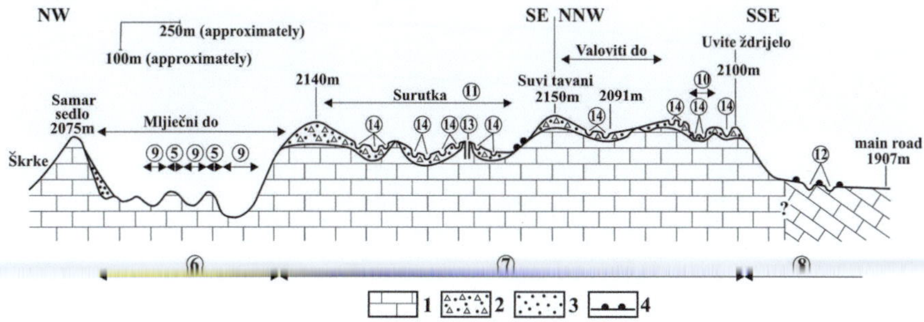

Fig. 4.74 Theoretical profile of the Durmitor Mountains between Mliječni do and Uvite ždrijelo: *1* limestone, *2* morainic deposit and clayey morainic deposit, *3* frost-shattered debris, *4, 5* roche moutonnée, *6* cirque, *7* glacial valley of NW to SE orientation, *8* glacial valley of W to E orientation, *9* rock basin, *10* paleodoline, *11* paleouvala, *12* recent solution doline, *13* shaft, *14* subsidence doline

and 4.79). Since the cover is thin, limestone outcrops even in the interior of paleodolines, and thus the covered karst of the paleodolines is interrupted by patches of open karst (Figs. 4.80, 4.81, and 4.82).

- From the narrow plateau (e.g. of the Hochschwab), glacial valleys radiate. If covered karst only appears in the cirques, the row of covered karst is aligned parallel to the covered karst patches of the plateau (as on the northern margin of the Hochschwab). If covered karst is also present in the glacial troughs, rows of covered karst patches perpendicular or at low angles to the covered karst row of the mountains dominate the covered karst pattern (as on the southern margin of the Hochschwab, Fig. 4.83).
- On plateaus (such as the Dachstein) where the glaciers formed plateau glaciers or filled up dolines or linked several dolines, glacial deposits accumulated in dolines. The pattern of the covered karst reflects that of paleodolines. Covered karst terrains are restricted to paleodolines. The patches of covered karst do not show any orientation.
- On lower and smaller karst plateaus, ice only accumulated in paleodolines and in their immediate environs. Although the dolines were moulded by ice, cover deposit did not accumulate on their bottom in every case. Covered karst patches developed in dolines which had sediment (glacial deposit) influx (Fig. 4.84) as exemplified in the Eastern Alps by the Rax, Schneeberg and probably the Schneealpe. In the Riss glacial, these plateaus were parts of the glacier network of the Eastern Alps (Van Husen 2000). In the Würm, there were only minor glaciers in the area. Covered karst patches could have developed in dolines which date back before the Würm and probably not in the post-Riss dolines since no glacial deposit reached them. These are still active as proved by their large size (Veress 2012b). The older paleodolines were remoulded by glacial erosion as centres of ice formation. Such covered karst patches occur in small number and sporadic distribution.

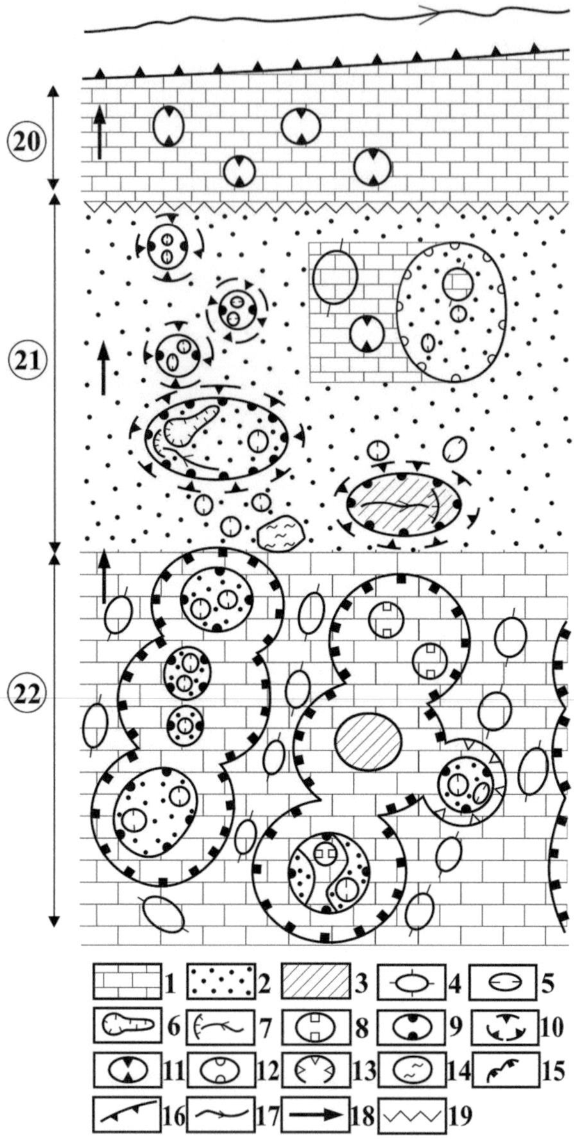

Fig. 4.75 Karst zones in high mountains bordered by plains (on the example of the Durmitor, not to scale): *1* limestone, *2* morainic deposit, *3* impermeable cover (clayey morainic deposit), *4* elevation, *5* subsidence doline, *6* subsidence doline with ravine, *7* ponor, *8* recent solution doline, *9* depression of superficial deposit, *10* buried infilled paleodoline, *11* old but still active solution doline, *12* paleokarst depression of plain floor, *13* cirque, *14* rock basin, *15* glacial trough formed of dolines, *16* river valley, *17* watercourse, *18* surface slope, *19* boundary of ice cover, *20* uncovered zone, *21* mixed concealed zone (largely covered with moraine), *22* striped-patchy covered karst

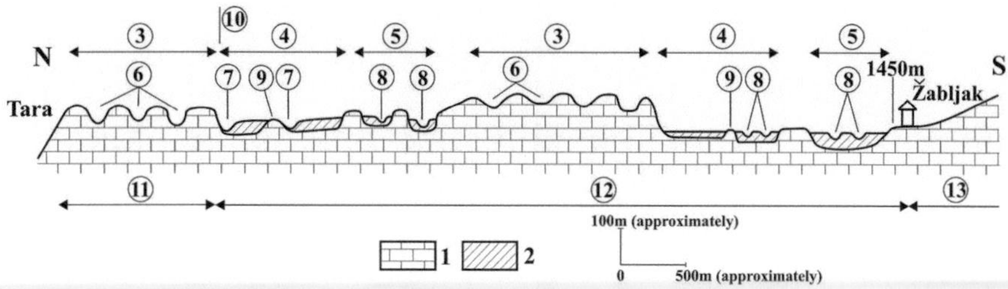

Fig. 4.76 Theoretical profile between Žabljak and the Tara Valley (northern foreland of the Durmitor Mountains): *1* limestone, *2* morainic deposit, *3* zone of solution dolines, *4* former glacially planated karst terrain with roches moutonnées, *5* half infilled paleodolines (depressions of superficial deposit), *6* old but still active dolines without fill, *7* half suffosion doline, *8* suffosion doline, *9* roche moutonnée, *10* probable termination of former glacier, *11* uncovered karst, *12* mixed concealed covered karst, *13* Durmitor

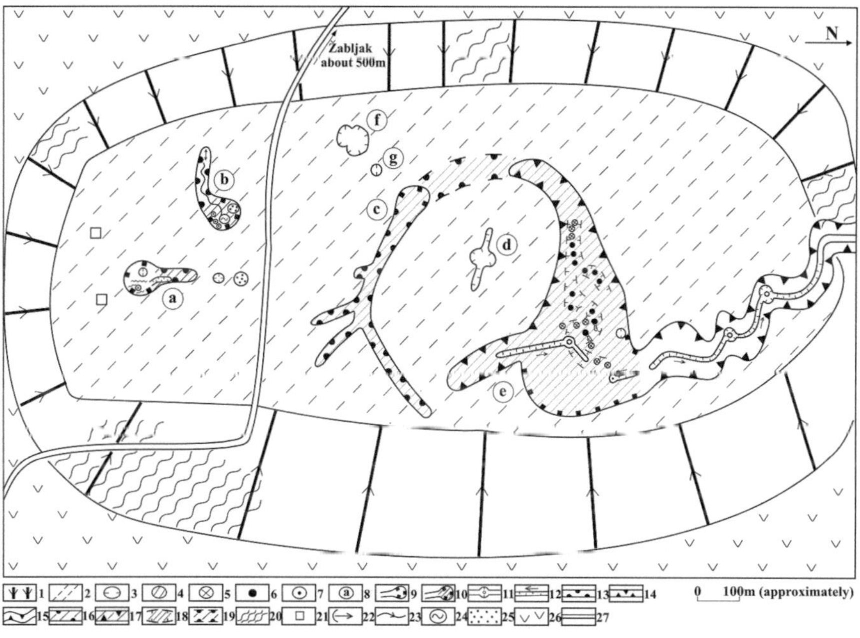

Fig. 4.77 Morphological map of the depression of superficial deposit (paleouvala) near Žabljak (sketch, not surveyed): *1* side slope of paleodoline, *2* infilled floor of paleodoline, *3* suffosion doline (diameter more than 1–2 m), *4* infilled suffosion doline, *5* suffosion doline (diameter less than 1–2 m), *6* suffosion doline with debris on floor (diameter less than 1–2 m), *7* dropout doline or opening in cover (diameter less than 1–2 m), *8* doline code, *9* ponor-like doline (covered karst ponor), *10* infilled covered karst ponor, *11* shallow solution doline on channel floor, *12* slope direction of gully and its floor, *13* ravine, *14* valley, *15* valley with incised meanders, *16* infilled ravine, *17* infilled valley, *18* buried gully, *19* buried ravine, *20* limestone outcrop, *21* limestone block (moraine), *22* artificial water inflow, *23* watercourse *24* lake, *25* bog in doline, *26* karst terrain bordering paleodoline, *27* road, *a–c*, covered karst ponor with creek (on the floor of types '*a*' and '*b*' there are subsidence dolines of different types), *e*, valley with a channel (gully) on its floor, there are a lot of small suffosion dolines on the valley floor, *f*, coalesced suffosion doline which is separated into part dolines, *g*, larger suffosion doline

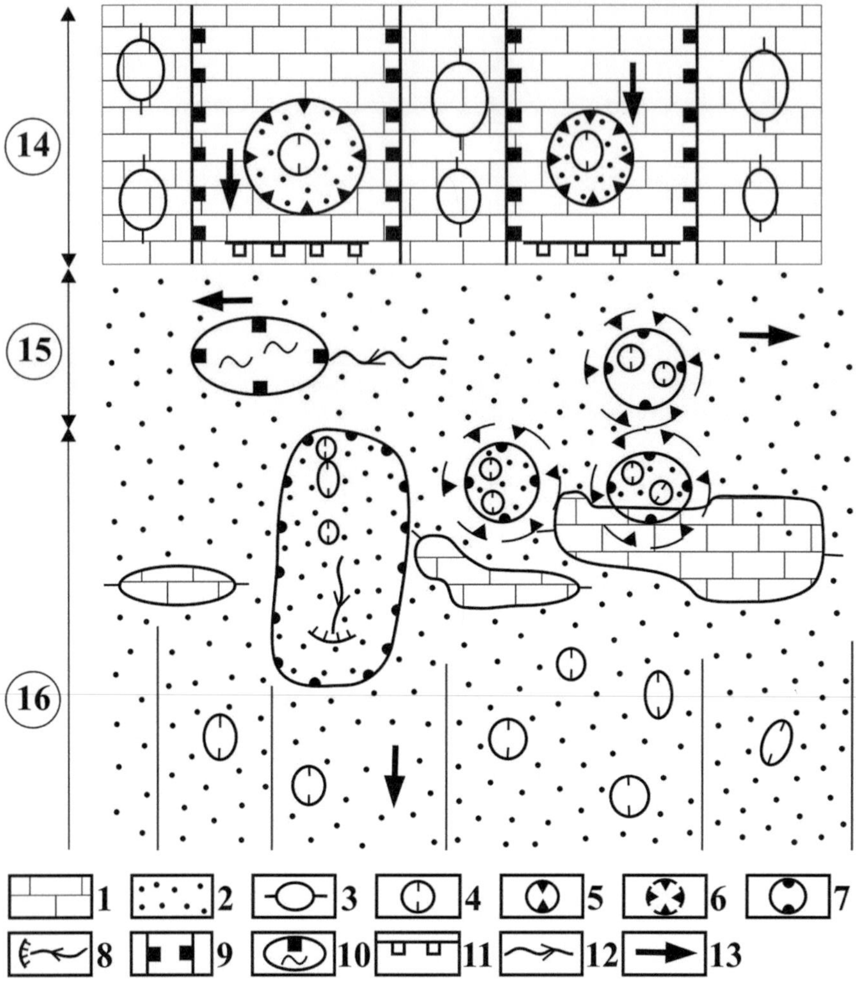

Fig. 4.78 Karst zones at main and tributary glaciers, example of Tauplitz alm (Totes Gebirge) (not to scale): *1* limestone, *2* morainic deposit, *3* elevation, horn, *4* subsidence doline, *5* paleodoline, *6* buried paleodoline, *7* depression of superficial deposit, *8* covered karst ponor, *9* glacial trough, *10* rock basin, *11* transverse scarp, *12* watercourse, *13* direction of surface slope, *14* tributary glaciers with paleodolines on their floor and paleodolines with covered karst patches on their floor, *15* continuous covered karst on the floor of the main glacial valley, *16* mixed (concealed karst) zone of the moraine of ice which overflowed the margin of the main glacier

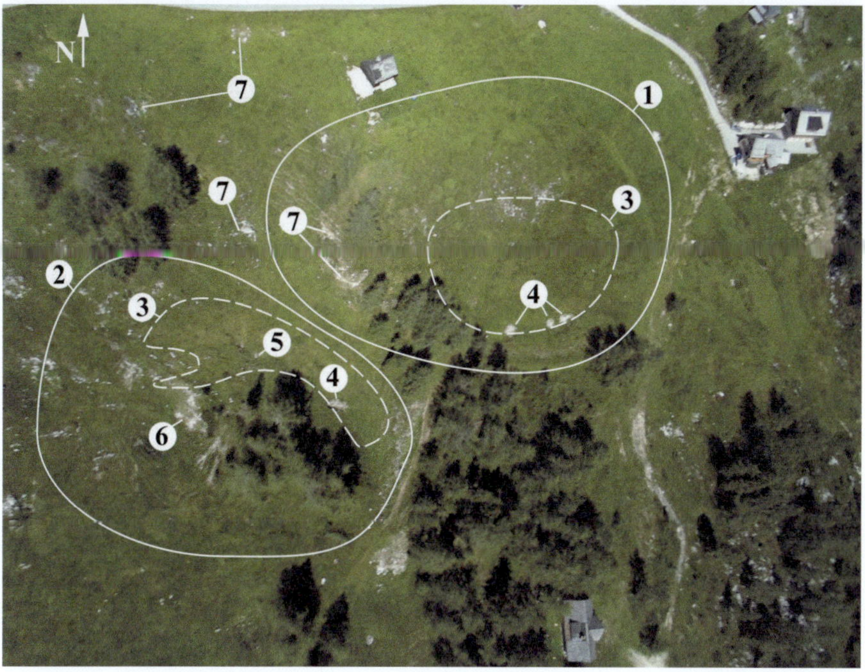

Fig. 4.79 Aerial photograph of the double paleodoline Tauplitz alm: *1* margin of the eastern paleodoline, *2* margin of the western paleodoline, *3* covered karst patch of paleodolines, *4* half-doline, *5* subsidence doline, *6* solution doline, *7* limestone outcrops from below cover sediment

- A covered karst pattern controlled by paleodolines has developed in high mountains of mediterranean environment (e.g. in the Biokovo Mountains). The surface of the Biokovo Mountains is dissected by a network of large-scale uvalas and dolines (Fig. 4.85), a polygonal karst pattern (Telbisz et al. 2005). In some dolines, snow and ice fill accumulated probably in the Pleistocene. In dolines filled (lined) to a limited degree, patches of covered karst occur. For the purposes of agricultural cultivation, the doline floors with cover sediment (solution residue) were planated, and the former subsidence dolines were destroyed. Today over the abandoned, not cultivated doline floor, embryonic subsidence dolines of maximum 0.5–1 m diameter and 1–2 m depth are beginning to reappear. On some of the covered karst patches of several square metres area, five to six subsidence dolines are counted now.
- In large-scale dolines, uvalas and cirque valleys, the covered karst patches are of zonal pattern (internal or local zonation). Here a debris slope is created without karstification (buried karst zone) bordered by a concealed karst zone or a cryptokarst zone followed by open karst. In the concealed karst zone, dolines appear, and, where the impermeable cover sediment wedges out, blind valleys

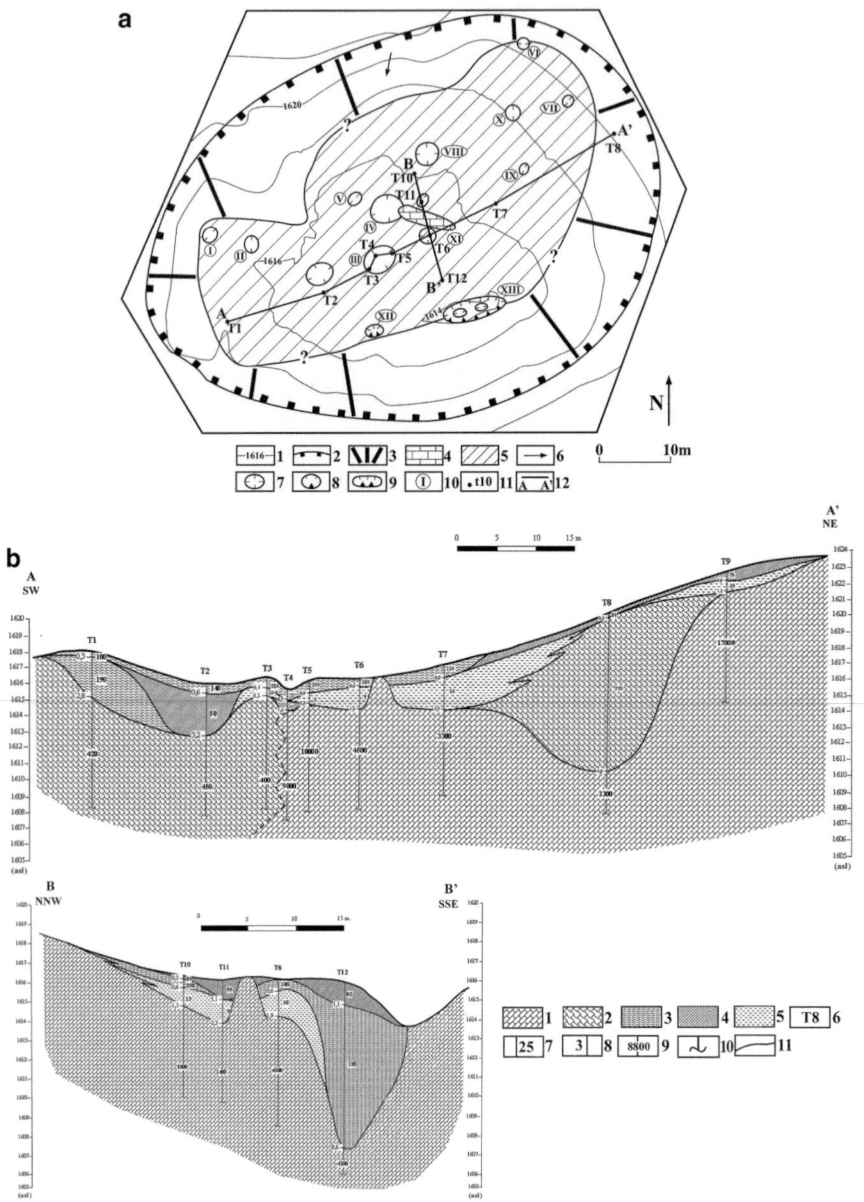

Fig. 4.80 (**a**) Map of the eastern of the paleodoline near Tauplitz alm: *1* contour line, *2* paleodoline, *3* uncovered side slope of paleodoline, *4* uncovered floor of paleodoline, *5* covered floor of paleodoline, *6* dip direction of layer, *7* subsidence doline, *8* half-doline, *9* elongated half-doline, *10* identification code of doline, *11* site and code of VES measurement, *12* alignment of geoelectric–geological profile. (**b**) Geoelectric–geological profiles of the doline shown in (**a**): *1* limestone, *2* limestone debris (fragmented limestone), *3* limestone debris (clayey), *4* clay with limestone debris, *5* clay, *6* identification code of VES measurement, *7* geoelectric resistivity of the series (Ohmm), *8* base depth of geoelectric series, *9* geoelectric resistivity of bedrock (Ohmm), *10* approx. penetration of measurement, *11* geoelectric series boundary

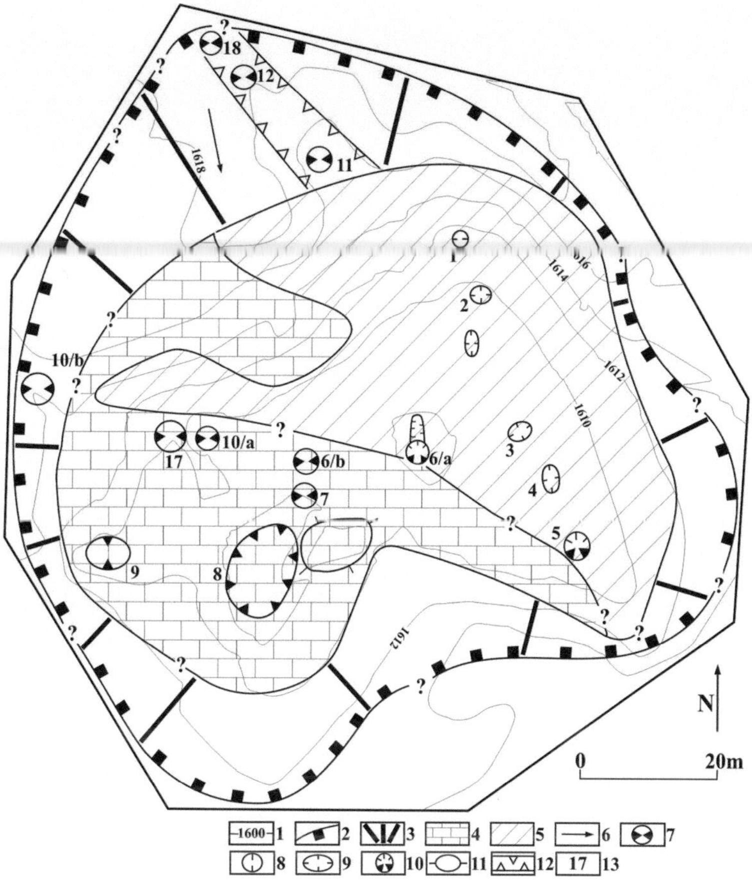

Fig. 4.81 Map of the western of the paleodoline near Tauplitz alm: *1* contour line, *2* paleodoline, *3* uncovered side slope of paleodoline, *4* uncovered floor of paleodoline, *5* covered floor of paleodoline, *6* dip direction of layer, *7* solution doline, *8* subsidence doline, *9* elongated subsidence doline, *10* half-doline, *11* roche moutonnée, *12* solution giant grike, *13* identification code of doline

with ponors (or only ponors) also occur. Covered karst patches reflecting internal zonation occur in the Totes Gebirge. The buried karst may be absent (e.g. in the paleodolines of the Hochschwab plateau).

4.6.2.5 Mediterranean Karst

According to climate, this type belongs – together with subtropical monsoon – to the subtropical climate (warm temperate), but characterised separately, since the karst of the subtropical monsoon shows similar properties to tropical karsts. On mediterranean karst, three patterns (varieties) of covered karst are distinguished: polje covered karst, karst hill covered karst and cuesta covered karst.

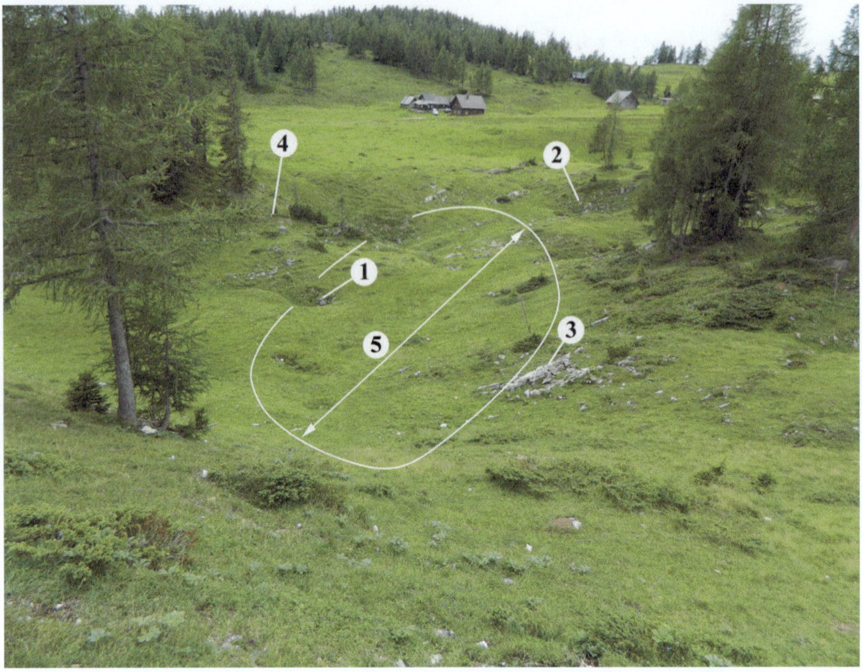

Fig. 4.82 View of the western of the Tauplitz alm double paleodoline: *1* half-doline, *2* solution doline, *3* side of paleodoline on bedding plane, *4* side of paleodoline on bedding heads, *5* covered karst floor of paleodoline

The polje covered karst is widespread on the mediterranean karst (e.g. on the lower surfaces of the Dinaric Mountains) as well as on tropical karst, where the poljes often constitute the external (noncontinuous) zone of the inselberg karst (Sweeting 1958).

Covered karst develops on the floors of poljes. The degree of coveredness of poljes shows remarkable variations: the floors of some poljes are completely covered, while others are only partially. The non-karstic sediments of poljes can be underlying rocks and cover sediments. The cover sediments can be fluvial and lacustrine deposits or solution residue. If the area of the polje is sufficiently large and the non-karstic sediment is regularly developed, the covered karst can form in zones on the polje floor and concealed karst zone and cryptokarst zone also emerges. In both zones, open karst patches can also occur. In other poljes, the pattern of cryptokarst and concealed karst terrains are patchy and mosaic-like. Since the polje groups are arranged along tectonic lines (Jakucs 1977; Ford and Williams 2007), the covered karst patches of the poljes are also adjusted to the structural directions of the polje-bearing mountains. On the floors of the temporarily or permanent waterlogged poljes, katavothra and associated complex features, according to Jennings (1985), alluvial streamsink dolines occur.

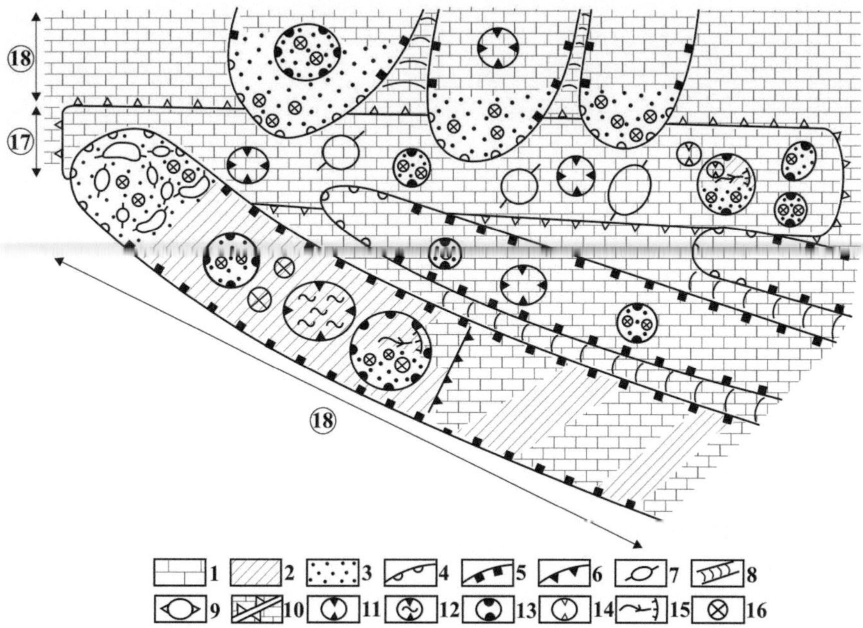

Fig. 4.83 Covered karst pattern of narrow plateau and glacial valleys radiating from the plateau (example of the Hochschwab, not to scale): *1* limestone, *2* impermeable non-karstic cover, *3* permeable non-karstic cover, *4* cirque, *5* glacial trough, *6* transverse scarp, *7* horn, *8* aréte, *9* morainic hill, *10* plateau, *11* paleodoline, *12* paleodoline with lake, *13* depression of superficial deposit, *14* recent solution doline, *15* ponor, *16* subsidence doline, *17* plateau with patches of concealed and cryptokarst, *18* glacial valley with concealed and cryptokarst patches on its floor

On concealed karst polje floors, subsidence dolines occur, while on cryptokarst terrains, ponors can be found. Subsidence dolines may develop independent from karst water (dry polje) or bound to karst water (temporarily waterlogged polje). The subsidence dolines and alluvial streamsink dolines often appear in groups, and their density is high.

On mediterranean karsts, karst hill morphology can develop in great extension (e.g. in SW part of Croatia, in the coastal zone as well as in the area of Bosnia and Macedonia). Karst hills occur in poljes and are widespread in large poljes and between dolines (Veress and Péntek 2010), in uvalas generated from the merging of dolines and in uvala systems (Fig. 4.86). Polje covered karst and karst hill covered karst often appear in conjunction (Fig. 4.86a). It is common that the latter develop in the interior of the former or on the margin (around Baćina, Croatia). On the floors of uvalas and uvala systems, covered karst may occur (solution residue, fluvial deposit). The fills on the floors of these landforms constitute the covered karst, which has a pattern of irregular patches and patch networks. The patches of cover sediment can be linked to each other where the ridges between neighbouring dolines lowered. On covered karst patches, limestone can outcrop and form internal karst hills. On such terrains, both concealed karst and cryptokarst terrains occur. In

Fig. 4.84 Row of subsidence dolines in a paleodoline on the Rax *1* subsidence doline, *2* paleodoline

Fig. 4.85 Covered karst patches in the Biokovo Mountains (Croatia)

Fig. 4.86 Karst hill covered karst: (**a**) (*top left*) karst hill covered karst on polje margin near Baćina, (**b**) (*top right*) covered karst depression enclosed by karst hills between Baćina and Ploče, (**c**) (*bottom left*) covered karst encircled by karst hills between Baćina and Ploče, (**d**) (*bottom right*) system of depressions dissected by karst hills at Ploče (Croatia, along road number 513): *1* karst hill, *2* permanent or intermittent lake (subsidence doline)

Fig. 4.87 Covered karst patches in front of cuesta (Crimean peninsula): *1* cuesta, *2* solution doline, *3* subsidence doline

Fig. 4.88 Elongated subsidence doline (Crimean peninsula): *1* cuesta, *2* partially infilled solution doline, *3* subsidence doline, *4* exposed bedding head

Croatia on the cover deposits of depressions (solution dolines and uvalas) between karst hills, intensive agricultural cultivation was performed, and, consequently, subsidence dolines have been destroyed here. Where cultivation ceased, embryonic dolines have developed on the cover sediment of depressions. Elsewhere only those subsidence dolines have been preserved in the cultivated areas, which contained lakes of permanent (Fig. 4.86) or temporary water cover (Fig. 4.86c), which formed where the elevation of the surface does not exceed the level of the high (permanent lake) and low karst water table (temporary lake). Cuesta covered karst is found on the Crimean peninsula, where on terrains in front of cuesta, asymmetric solution dolines are aligned parallel with the strike direction of the cuestas (Fig. 4.87). The asymmetry of the solution dolines is sometimes reduced by doline fills, which represent the covered karst. The covered karst patches of doline floors are aligned in rows in the direction of the cuesta. The covered karst patches are mostly elongated. In the fill suffosion, dolines dominate (Fig. 4.88).

Karst water has a predominant role in the control of karst processes (including covered karst formation) particularly on polje and karst hill covered karsts.

4.6.2.6 Tropical Covered Karsts

In tropical karst, the following karst types occur (Williams 1971, 1972a, b; Day 1979; Balázs 1973, 1986, 1990; Sweeting 1995; Zhang 1980; Zhu 1982; Zhu and Zhu 1988; Smart et al. 1986): fenglin, fengcong, polygonal karst, pinnacle karst.

Fig. 4.89 Covered karst among the mountain groups of a fengcong

On most of such karst types, covered karst is widespread. The individual types occur in the following varieties:

- Karst cones fenglin with conical hills and closed depressions and valleys between the hills.
- Karst tower fenglin, where the hills are divided by a network of narrow corridors.
- Isolated towers fenglin (fenglin with intermountain plains), where plains occur between the towers. Four forms of this variety are distinguished (Williams 1987): (1) residual hills rise above the carbonate plain covered with alluvium, (2) carbonate hills rise above the non-carbonate plain covered alluvium, (3) carbonate hills rise above the buried dissected carbonate surface and finally (4) carbonate towers rise from the non-karstic basement.
- Plain fengcong, where mountain groups are separated by plain surfaces (Fig. 4.89).
- Depression fengcong with major depressions (Fig. 4.90) and valleys between mountain groups.
- Fengcong canyon variety in the sides of deeply incised river valleys.

Between the mountains of the fengcong mountain groups, depressions of variable size occur: dolines similar in size and morphology to temperate belt dolines or cockpit dolines as well as poljes or polje-like landforms. There are several transitional varieties between these two types, distinguished both by size and morphology. Covered karst may occur in the various depressions (the bigger they are, the higher is their probability) as well as on plains between mountain groups.

Fig. 4.90 Covered karst formed in a doline of fengcong-type karst: *1* subsidence doline, *2* gully of doline

The individual depressions are of elongated form, of meandering shape, indicating inheritance from former valleys (Sweeting 1995). Any of them can be covered karst as exemplified from Yunnan Province, China. In the epigenetic valleys, dolines may occur in rows divided by walls (Ford and Williams 2007). In this case, the covered karst on the floor of epigenetic valleys is of patchy pattern. The patches are arranged in rows. On the covered karst patches of the valleys, depressions active as katavothra are typical (Zhang 1980). However, according to a widespread opinion, the floors of the depressions of the fengcong lie above the karst water table (Balázs 1986; Williams 1987; Ford and Williams 2007).

Polygonal karsts do not only occur on tropical karsts but also on subtropical (mediterranean) karsts (Ford and Williams 2007). On tropical karst, they are typical of the fengcong (on plain fengcong and depression fengcong). The polygonal karst is a network of ridges with elevations in the nodes. The cockpit karst is a variety of polygonal karst, where with the development of cockpit dolines the ridges are lowered (Williams 1969).

On polygonal karst (or on its cockpit karst variety), the density of depressions is high. Covered karst develops in the depressions (Fig. 4.30), the interior of which can be composite on tropical karsts and divided into compartments. The depressions of the various polygonal karsts are of variable size, which is governed by the density of shafts of the epikarst, eventually controlled by the density of faults in the bedrock and the distribution of water replenishment (Williams 2004). The size of depressions on the other hand controls the dimensions of the covered karst terrains. According to Williams (1971), karst depressions are linked if time allows. Therefore, the size of covered karst terrains depends on the age of the polygonal karst. On polygonal karst, covered karst terrains are of mosaical pattern.

In the polygonal karst of New Guinea, in the centre of the covered karst terrains of the depressions, there are dolines (active as ponors), to which ditches lead (Williams 1971, Fig. 4.30). With the increasing size of depressions, the ditch systems become more and more composite. From these depressions, the cover sediment (probably weathering product) is transported into the karst. The starlike cockpits (doline) of the cockpit karst of Jamaica are not uniform but are divided into open and covered partial depressions (Sweeting 1973). The latter can be concealed karsts and cryptokarsts. On these terrains, ponors and subsidence dolines occur. On the floors of cockpit karst depressions, the lakes formed in wet periods (Waltham et al. 2005) attest to the high karst water table reaching the level of the basement.

On pinnacle karst, the elevations show steep sides. In stone forests, the assemblage of towers is predominant, but ridges also appear. The stone forest of Lunan is described as a variety of covered karst (Chen et al. 1986; Maire et al. 1991; Sweeting 1995). The stone forest covered karst has at least two varieties according to the degree of exhumation: the partially exhumed stone forest and the covered stone forest. In the partially exhumed stone forest, the towers rise above the surface built of cover sediment. In the covered stone forest, the landforms do not rise above the sediment cover environment. The landforms are often covered by compact rock (basalt) (Song and Liang 2009), i.e. cryptokarst. With the weathering of the basalt cover, parts of the stone forest evolve into concealed karst. According to their elevation, valley, hilltop and hillside stone forests are distinguished (Song 1986). Larger exhumed stone forests mainly occur for the hilltop variety. For the latter, the extension of the covered karst is smaller, while for the valley floor variety, it is larger. Another variety of the pinnacle karst is the arête and pinnacle karst, where exhumed ridges are prevalent, and the pinnacles are less prevalent. In the tsingy, giant grikes dominate.

In conjunction the types and varieties of tropical karst constitute genetical and hydrological karst systems, best exemplified by the South China Karst due to the widespread distribution of carbonate rocks. The character and spatial manifestation of karst systems determine the pattern of covered karst. Several factors (rock structure, climate, surface elevation) decide which of the enumerated types develop next to each other. On higher-lying terrain, polje, embryonic fengcong, plain fengcong, depression fengcong, covered stone forest karst, karst cones fenglin and karst towers fenglin occur, while on lower ground plain, fengcong and isolated towers fenglin can be found.

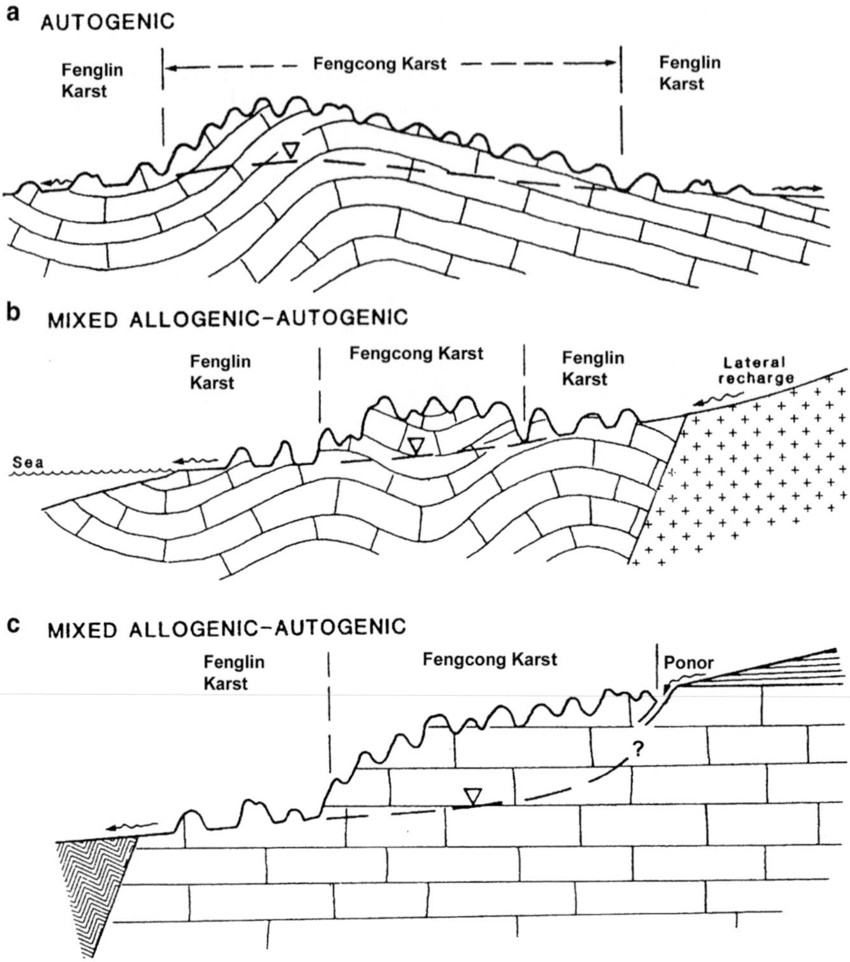

Fig. 4.91 Tropical autogenic and mixed autogenic–allogenic karst (Williams 1987) – modified

The properties of the non-karstic rocks of tropical karst are the following:

- Covered karst occurs where non-karstic rocks wedge in the karst. But covered karsts are also widespread where cover sediment accumulated.
- The bedrock can be dissected to a great extent (Wilford and Wall 1965). The bedrock is dissected by large-scale infilled dolines or grikes, pinnacles (Bergado and Selvanayagam 1987). Numerous transitional stages are possible between the varieties with dolines and pinnacles. Thus, the cover thicknesses are highly variable. But because of the dissected nature of the bedrock, the lateral extension of the covers of similar thickness can be variable too.

- The non-karstic cover of the covered karst is composed of thick autogenic solution residue (autochthonous deposit) or reworked weathering residue as well as fluvial deposit derived from the non-karstic terrain (allochthonous deposit). The solution residue can also be reworked or of fluvial origin, mixed with sediment from non-karstic terrain.
- On the fenglin karst of the inselberg karst, the karst water table lies at the level of intermountain plains – irrespective of autogenic, allogenic or mixed karst type. At the same time, on a fengcong-type karst, the karst water table is below that of the depressions (Williams 1987, Fig. 4.91).

According to its cover, at least four varieties of covered karst are identified:

The first variety occurs on tropical depression covered karst, where there are minor depressions (tropical depression covered karst) between inselbergs. Patches of covered karst appear in the depressions (Fig. 4.90). Such depressions do not yet receive cover sediment from the environs, their cover is composed of weathering residues of variable thickness and in their area, there is no wedged-in non-karstic rock. Because of the solution residue of variable thickness and composition, between the slightly covered and partially infilled depressions, there are numerous transitional stages. In such depressions, no watercourses or stream systems are found. Since the bottom of the depressions is above the karst water table, their covered karst features developed independent of karst water. They may occur in the depressions of cone groups of any fengcong-type karst and are widespread on depression fengcong and on the fengcong canyon variety.

The second variety develops on polygonal covered karst and plain fengcong. Cover sediments are not only weathering products but also partially reworked deposits. Covered karst formation is only independent from karst water in the depressions of the cone groups, not or not everywhere on plain terrains or epigenetic valley floors. Such covered karst terrains occur on fengcong karsts where the mountain groups are separated by plain surfaces (plain fengcong), there are valleys between inselbergs (karst cones fenglin), in depressions developed from epigenetic valleys as well as where there are (systems of) watercourses in the depressions (polygonal or cockpit sections of the fengcong karst).

The third variety of the covered karst is associated with intermountain plains (intermountain plain covered karst). This is typical of varieties 1 and 3 of the isolated towers fenglin. Since for variety 3 the karst bedrock is uneven, where the bedrock is in deeper position, the cover may become too thick. As a consequence, covered karst does not develop in every case, neither in this variety. Intermountain plain covered karst occurs, for instance, along rivers around Guilin (Fig. 4.29).

The intermountain plain covered karst shows the following properties:

- The covered karst terrains of neighbouring intermountain plains may merge.
- Covered karst formation is governed or influenced by karst water.
- The cover sediment is partly solution residue and partly fluvial deposit.

- The major intermountain plains themselves show covered karst pattern mostly shaped by the rivers flowing in intermountain plains. Thus, the cover is divided into stripes of different thickness and composition, the boundaries of which show similarity with the courses of rivers (Fig. 4.29).
- Where intercalated non-karstic rocks occur, tropical varieties of allogenic cryptokarst may also be present.

According to Williams (1987), the jointly developed fengcong and fenglin (the former with higher and the latter with lower elevation) can be of autogenic and mixed allogenic–autogenic type. In the latter case, the fengcong–fenglin assemblage is bordered by a terrain of non-karstic rocks. If the karst water table lies below the surface, ponors emerge on the margin of the non-karstic terrain. Sediments of the non-karstic terrain are transported into the karst interior through ponors (Fig. 4.91c). The character of the karst will be semi-allogenic covered karst (or its tropical variety) since the sediments eroded from the non-karstic terrain cannot reach the surface of the fenglin, but are transported into the caves through ponors. If the karst water table is at the level of the fenglin next to the non-karstic terrain, no ponor formation takes place. The transported sediment accumulates in the intermountain plains of the fenglin and creates the tropical intermountain plain covered karst (of zonal pattern) of the recent allogenic covered karst. Since the fengcong surface lies higher, it does not receive cover sediment of non-karstic origin. Neither can the fenglin on the opposite side receive cover sediment as the fengcong forms a barrier to sediment transport. In the area of the fengcong and the fenglin of opposite location, the material of the covered karst is supplied by solution residue or the denudation of the non-karstic cover (Fig. 4.91b).

The fourth variety develops on tropical karren (stone forest, arête and pinnacle karst, tsingy), occasionally under consolidated cover (cryptokarst), e.g. the stone forest of Lunan formed under basalt (Song and Liang 2009). The sections of the stone forest buried under laterite with depressions of variable size and shape between the pinnacles derived from this karst (Song and Liang 2009). The latter covered karst appeared on a partially exhumed stone forest. Although according to Salomon (2009) tsingy formed under cover sediment, in our opinion, the Bemaraha tsingy developed through the merging of caves and grikes (Veress et al. 2008), while the Ankarana tsingy on barren surface. The depressions dismembering the Ankarana tsingy resulted from solution along faults (Veress et al. 2009). Therefore, the presence of covered karst is not assumed for either tsingy – neither for the present nor for the time of their origin.

It is probable that terrains dissected by pinnacles and grikes are widespread in tropical regions (Wilford and Wall 1965; Waltham et al. 2005), which are covered varieties of the pinnacle karst. They are regarded covered karsts and also occur on the negative forms of the above-presented karst types (e.g. in intermountain plains). Dolines can develop above grikes on the surfaces of cover sediment when the cover sediment is transported into the grikes (Waltham et al. 2005).

4.6.3 *Structural Covered Karsts*

4.6.3.1 Platform Covered Karst

On platform karsts, the bedrock strata are horizontal or slightly tilted. The low elevation of the surface favoured the development of the cover. Therefore, covered karsts are common on platform karsts. The cover developed uniformly and is homogeneous, not or slightly variable in horizontal direction. The denudation and dissection of the cover is of limited extent and primarily happens at valleys.

If the covered karst surface is sloping and the cover is impermeable, recent allogenic karst may emerge in the area. If the surface is very slightly sloping or flat, but the cover is impermeable and consolidated rock, caprock dolines appear on the allogenic karst. With decreasing surface slope, the allogenic nature weakens or ceases and exclusively caprock dolines become present. If the cover sediment is unconsolidated permeable rock, subsidence dolines are typical of the karst (Beck and Sinclair 1986). In the former case, a single cryptokarst zone emerges on the karst, while in the latter case, a single concealed karst zone appears. If both impermeable (consolidated) and permeable rocks are present, the platform karst is divided into two zones: one cryptokarst zone and one concealed karst zone, as exemplified by the Central Kentucky Karst, where the sandstone terrain is buried karst, while the 'Sinkhole Plain' is concealed karst.

If the bedrock is highly prone to karstification, gypsum or rock salt, covered karst formation is caused by stopping (breccia pipe formation) (Waltham et al. 2005). The stoppings and the accompanying collapses are widespread all over the covered karst – irrespective of lithology (autogenic cryptokarst). The karstification of the cover will be homogeneous even if there are both consolidated and unconsolidated cover sediments. Consequently, in this case, a single zone develops on the platform karst. The extension of covered karst on rock salt and gypsum depends on the distribution of these rocks. It may happen that in anteclises, gypsum reaches positions close to the surface (in Ukraine). Then the extension of the covered karst is determined by the location and strike of the anteclise.

4.6.3.2 Salt Diapir

Covered karst on rock salt forms patches of smaller extension. The reason for this is that rock salt, even if it originally had a large spatial extension, reaches the surface in diapirs (Ford and Williams 2007).

4.7 Conclusion

Covered karsts are classified by the origin and type of their cover sediment, age of the covered karst, bearing landforms, type of bearing karst and the pattern of cover sediment.

The covered karst is cryptokarst if the cover is impermeable rock and concealed karst if the cover is not impermeable rock. The cover material can be intercalated into the karstic rock, derived from the bedrock (terra rossa, laterite, morainic deposit), rocks transported over the karst (fluvial deposits) and rocks originated on and alien to the karst (marine sediments, volcanic rocks). In the covered karst areas, karstification is either syngenetic (the evolution of landforms is driven by karstification on the cover) or postgenetic (the cause of landform generation on the cover is that an older landform of the bedrock loses its sediment fill).

Covered karst can form in surface landforms, through the covering of the depressions only or through the burial of such forms together with their environments. Following the denudation of the surface, the patches of cover are only preserved in depressions and create karstic patches. Patches of covered karst occur in depressions of structural origin (syncline, tectonic graben), of erosional origin (glacial valley, river valley, fluvial terrace, abrasional platform), in paleokarst depressions (doline, uvala, blind valley) and/or recent karst landforms (polje, depressions and lower terrains of fengcong and fenglin).

According to pattern, for block mountains, we distinguished recent allogenic, renewed, semi-allogenic, mantled allogenic, horst covered karsts, for glaciokarsts cirque valley and glacial valley covered karsts and platform and salt diapir covered karsts. For climatically controlled karst, tundra, taiga, polje, karst hill, cuesta, tropical depression, polygonal, intermountain plain and stone forest covered karst can be identified.

The recent allogenic covered karst receives sediment from the bordering non-karstic terrain. In the karst, in downslope direction, buried, cryptokarst, mixed composite covered karst and open karst zones alternate. The renewed allogenic karst has lost its non-karstic environs which provided its cover sediment. Therefore, its more elevated margin is a zone of open karst, which is followed by mixed composite covered karst and buried karst zones. In the area of a semi-allogenic karst, sediments from the bordering non-karstic terrain do not accumulate since the watercourses of the karst remove the sediment. It has a uniform cover of unconsolidated permeable sediment (in a single zone). The mantled allogenic karst has a consolidated impermeable cover with the following zones: buried karst, cryptokarst (transitional cryptokarst), concealed karst and open karst. It shows variable karstification processes if the bedrock is dissected by paleodepressions. On horst covered karst, the various blocks show different kinds of karstification depending on their previous evolution. Buried karst develops on blocks in lower position covered by impermeable rock. On higher-lying cryptokarst blocks, the streams of the epigenetic valleys incising into the impermeable cover open up cavities of the bedrock. Water from the watercourses seeps away and feeds the karst water reserves of the mountains and contributes to cavity formation under channels. Concealed karst forms on blocks where elevated position led to the partial or total removal of the impermeable cover and the block is covered by permeable material.

On glaciokarsts the cover sediment (mostly morainic deposit) accumulates in glacial valleys, primarily in paleodepressions. The pattern of the covered karst is primarily controlled by the type of the glacier. In glacial valley covered karst, parallel stripes of covered karst or parallel covered karst stripes with patches are typical patterns. On composite glaciokarst, striped pattern is replaced by zonal pattern, while for plateau glaciers or incipient glaciation, the pattern of the covered karst reflects that of the paleodolines. In cirque covered karst, the cover is of patchy pattern.

The tundra-covered karst mostly develops where rock salt and gypsum occur. This covered karst is of patchy pattern. The taiga-covered karst may be continuous between valleys, where the cover developed without interruption, while it is patchy on valley floors. On polje karst, covered karst can be continuous but also zonal (e.g. cryptokarst and concealed karsts alternate). The shape, size and pattern of covered karst reflect the shape, size and pattern of poljes. On karst hill covered karst, the pattern of the covered karst is patchy, and the size and shape of patches are adjusted to the size and shape of uvalas dissected by karst hills. The patches can also be linked. The cuesta covered karst appears in the asymmetric dolines of the cuestas, and the covered karst patches form rows parallel to the cuesta.

The tropical depression covered karst comes about in the smaller depressions of the fengcong karst. The covered karst patches reflect the pattern of the depressions. The tropical polygonal covered karst forms in larger depressions. Patches of covered karst are larger, interlinked and separated into concealed and cryptokarst. The intermountain plain covered karst appears on fenglin karst and shows internal zonation. The stone forest type of covered karst develops where the bedrock is dissected into pinnacles and ridges.

The platform covered karst is associated with karst of low elevation. The covered karst is not or only poorly dissected into zones. Varieties develop over different bedrock types. The salt diapir covered karst is typical of a rising salt diapir and has a patchy pattern.

References

Andrusov D, Borza K, Martiny E, Pospisil A (1958) O povode a dobe vzniku tzv. terra rossy juzného a stredného. Slovenska. Geol. Zbornik SAV 9:1, Bratislava

Balázs D (1970) Zsombolyok a Central Kentucky Karszton (Shafts on the Central Kentucky Karst). Karszt és Barlang 1:21–24 (in Hungarian)

Balázs D (1973) Relief types of tropical karst areas. In: Jakucs L (ed) IGU symposium on karst morphogenesis. József Attila University, Szeged, pp 16–32

Balázs D (1986) Kína karsztvidékei (Karst regions of China). Karszt és Barlang 2:123–132, in Hungarian

Balázs D (1990) A Dél-Kínai karsztvidék főbb barlangtípusai (Main cave types of the South China Karst). Karszt és Barlang 1:53–60 (in Hungarian)

Bárdossy Gy (1961) A Sümeg környéki bauxit (Bauxites around Sümeg). Bány Lapok, Budapest 7:457–463 (in Hungarian)

Bárdossy G (1977) Karsztbauxitok (Karst bauxites). Akadémia Kiadó, Budapest, 413 p (in Hungarian)

Barrére P (1964) Le relief karstique dans l'ouest des Pyrénées centrales. Rev Belge Géogr Ed Soc Roy Belge Géogr Special Publ Karst et Climats Froids 88(1–2):9–62

Beck BF, Sinclair WC (1986) Sinkholes in Florida: an introduction. Florida Sinkhole Research Institute Report, 85-86-4, 16 p
Bergado DT, Selvanayagam AN (1987) Pile foundation problems in Kuala Lumpur Limestone, Malaysia. Q J Eng Geol 20:159–175
Bögli A (1964) Le Schichttreppenkarst. Un exemple de complexe glaciokarstique. Revue Belge de Geographie 1(2):64–82
Bögli A (1976) Die wichtigsten Karrenformen der Kalkalpen. Karst processes and relevant landforms. International Speleological Union, Commission on Karst denudation. Department of Geography, Philosophical Faculty, University of Ljubljana, Ljubljana, pp 141–149
Brook GA, Ford DC (1980) Hydrology of the Nahanni karst, northern Canada the importance of extreme summer storms. J Hydrol 46:103–121
Businszkij GI (1964) Types of karst bauxite deposits and their Genesis. Symposium on bauxites, oxydes et hydroxydes d'aluminium. Zagreb 1:93–106
Chen Z, Song L, Sweeting MM (1986) The pinnacle karst of the stone forest, Lunan, Yunnan, China: an example of a subjacent karst. In: Paterson K, Sweeting MM (eds) New directions in karst. Geobooks, Norwich, pp 597–607
Combes (1969) Rechèrches sur la genèse des bauxites dans le Nord-Est de l'Espagne, le Languedoc et l'Ariège (France)- Mém. Centre d'Études et de Recherches Géologique et Hydrogéologie de, Montpellier, 342 p
Corbel J (1947) Observations sur le karst couvert en Belgique. Bull Soc Belg Études Géogr 17(1/2):95–105
Cvijič J (1893) 'Das Karstphänomen' Versuch einer morphologischen Monographie. Geogr Abh 5(3):217–329
Day MJ (1979) Surface roughness as a discriminator of tropical karst styles. Zeitschrift für Geomorphologie, Supplement-band 32:1–8
Dehao Z (1982) Evolution of peak cluster depressions in the Guilin area and morphometric measurement. Carsologica Sin 2(2):127–184 (in Chinese)
Delannoy JJ (1997) Recherches géomorphologiques sur les massifs karstiques du Vercors et de la transversal de Ronda (Andalousie). Les apports morphogiques du karst, tése de doctorat. Presses Universitaires du Septentrion, Lille, 678 p
Dongus H (1962) Alte Landoberflächen der Ostalb. Forschungen zur Deutschen Landeskunde, 134. Bundesanstalt für Landeskunde und Raumforschung, Bad Godesberg, 71 p
Fekete S (1988) Trópusi talajok (Tropical soil-types). Akadémiai Kiadó, Budapest, 503 p (in Hungarian)
Filippov AG (2004) Mineral deposits in karst. In: Gunn J (ed) Encyclopedia of caves and karst science. Fitzroy Dearborn, New York, pp 514–515
Ford DC (2004) Bear rock karst, Canada. In: Gunn J (ed) Encyclopedia of caves and karst science. Fitzroy Dearborn, New York, pp 137–138
Ford DC, Stanton WI (1968) Geomorphology of the south-central Mendip Hills. Proc Geol Assoc 79(4):401–427
Ford DC, Williams PW (1989) Karst geomorphology and hydrology. Unwin Hyman, London, 601 p
Ford DC, Williams PW (2007) Karst geomorphology and hydrology. Unwin Hyman, London, 561 p
Gams I (1994) Types of contact karst. Geografia Fisica e Dinamica Quateraria 17:17–46
Gams I (2002) Changes of the Triglav Glacier in 1945–94 period in the light of climatic indicators. http://ai.ijs.si/mezi/personal/triglav/
Grimes KG (2009) Laterite karst, unpublished poster displayed at the 7th international conference an geomorphology (ANZIAG), Melbourne, 6 p
Grimes KG, Spate AP (2008) Laterite karst (Andysez No 53). ACKMA J 73:49–52. International conference on geomorphology (ANZIAG), Melbourne, 6 p
Gunn J (2004) Fluviokarst. In: Gunn J (ed) Encyclopedia of caves and karst science. Taylor and Fitzroy Dearborn, London, pp 363–364

Gvozdetskiy NA (1965) Types of Karst in the U.S.S.R. Prob Speleol Res Prague:47–54

Gvozdetskiy NA (1981) Karst. Izd-vo Miszl, Moscow, 214 p

Hevesi A (1980) Adatok a Bükk hegység negyedidőszaki ősföldrajzi képéhez (Data to the quarternary paleogeographical features of the Bükk Mountains). Földtani Közlöny 110(3–4):540–550 (in Hungarian)

Hevesi A (1986) Hideg vizek létrehozta karsztok osztályozása (The classification of karsts created by cold waters). Földrajzi Értesítő XXXV(3–4):231–254, in Hungarian

Jakucs L (1977) Morphogenetics of karst regions. Adam Hilgar, Bristol, 284 p

Jaskó S (1957a) A Bicske, Szár, Tatabánya és Tarján közötti terület bauxitföldtani leírása (Bauxite geological description of the area included by Bicske, Szár, Tatabánya and Tarján, Hungary). MÁFI Évk, Budapest 46(3):505–523 (in Hungarian)

Jaskó S (1957b) Bauxit teleproncsok Veszprém és Nagyvázsony környékén (Rags of bauxite deposits near Veszprém and Nagyvázsony, Hungary). MÁFI Évk, Budapest, 46:525–530 (in Hungarian)

Jennings JN (1985) Karst geomorphology. Basil Blackwell, New York, 293 p

Kannan RC, Nettles NS (1999) Remedial measures for residential structures damaged by sinkhole activity. In: Beck BF, Petit AJ, Herring JG (eds) Hydrology and engineering geology of sinkholes and karst proceedings of the seventh multidisciplinary conference on sinkholes and karst. Balkema, Rotterdam, pp 135–139

Katzer F (1909) Karst und Karsthydrographie, Zur Kunde der Balkanhalbinsel, no. 8 Sarajevo, 94p

Knechtel MM (1963) Bauxitization of terra rossa in the Southern Appalachian retion. U.S. Geological Survey professional paper, 475-C, Washington, DC, pp 151–155

Knez M, Slabe T (1999) Unroofed caves and recognizing them in karst relief (discovered during expressway construction at Kozina, South Slovenia). Acta Carsologica 28(2):103–112

Knez M, Otoničar B, Slabe T (2003) Subcutaneous stone forest (Trebnje, central Slovenia). Acta Carsological 32(1):29–38

Korpás L (1981) A Dunántúli-középhegység oligocén-alsó-miocén képződményei (Oligocene-Lower Miocene Formations of the Transdanubian Central Mountains in Hungary). MÁFI Évkönyve, Budapest, 140 p

Korzhuev SS (1961) Merzlotnyi karst Srednego Prilen'ya i nekotorye osobennosti yego proyavleniya (The Middle-Lena frozen karst and its characteristics). In: Sokolov NI, Gvozdetskiy NA, Balashov LS (eds) Regionalnoe karstovedenie. Izdatelstvo AN SSSR, Moscow, pp 207–220

Korzhuev SS (1972) Drevnij karszt i cikli karsztoobrazovania Szibirszkoj plattormi (A paleokarst and karstification cycles on the Siberian platform) Trudi moszkovszkogo obsesztva iszpitatelej prirodi XLVII:141–151

Lauriol B, Gray JT (1990) Drainage karstique en Mileu de Pergélisol: le cas de l'ile d'Akpatok, T. N. O. Canada. Permafr Periglac Process 1:129–144

Lu Y, Cooper AH (1997) Gypsum karst geohazards in China. In: Beck BF, Stephenson JB (eds) Engineering geology and hydrogeology of karst terrains. Balkema, Rotterdam, pp 117–126

Lucas J (1872) Studies in Nidderdale, 319 p (Reprint publication) Forgotten Books, Hongkong

Maire R, Zhang S, Song S (1991) Génese des karsts subtropicaux de Chine du sud (Guizhou, Sichuan, Hubei). Grottes et karsts tropicaux de Chine Méridionale, Karstologia mémoires 4:162–186

Malott CA (1939) Karst valleys. Bull Geol Soc Am 50(12):1984

Malott CA (1945) Significant features of the Indiana karst. Proc Indiana Acad Sci 54:8–24

Móga J, Németh R (2005) The morphological research of the basalt and loess covered plateaus in the Bakony Mts (Transdanubian middle mts.-Hungary). Acta Carsologica 34(2):397–414

Németh R (2005) A Kab-hegyi bazalttakaró depresszióinak vizsgálata (Investigation of the depressions of the basalt cover of Kab Hill). Karszt és Barlang 2000–2001 évf.:33–41 (in Hungarian)

Nicod J (1976) Corrosion du type crypto-karstique dans les karsts Méditerranéens. In: Gams I (ed) Karst processes and relevant landforms. Department of Geography, Ljubljana University, Ljubljana, pp 171–180

Papastamatiou J (1964) Les gisements de bauxite en Grèce. Symposium on bauxites, oxydes et hydroxydes d'aluminium. Zagreb 1:285–293
Pataki A, Nyírő T (1983) A nyirádi-deáki bauxitbánya karsztos feküje és ennek bányászati vonatkozásai (The karstic bedrock of the Nyirád-Deák bauxite mine and its mining aspects). Földtani Kutatás 26:19 p (in Hungarian)
Peljtek EL (1971) Meslorozhdeniya Boksitov Eniseyskogo kryazha i Sibirskoy Platformy (Bauxite quarries from the Jenyiszej Mountain and Siberian platform) – V kn. In: Sapozhnikov DG (ed) Platformennye boksity S.S.S.R. Nauka, Moscow, pp 221–262
Pollard W, Omelon C, Andersen D, McKay C (1999) Perennial spring occurrence in the expedition Fiord area of western Axel Heiberg Island, Canadian High Arctic. Can J Earth Sci 36:105–120
Pulina M (2005) Le karst et les phenomenes karstiques similaires des regions froides. In: Salomon JN, Pulina M (eds) Les karsts des regions climatiques extremes. Karstologia Mémoires, 14, Presses Universitaires de Bordeaux, Bordeaux, pp 11–100
Quinlan JF (1967) Classification of karst types: a review and synthesis emphasizing the North American literature 1941–1966. Natl Speleol Soc Bull 29:107–109
Quinlan JF (1978) Types of karst, with emphasis on cover beds in their classification and development. PhD thesis, Geology, University of Texas, Austin
Quinlan JF, Pohl ER (1967) Vertical shafts actively promote slope retreat and dissection of the solution escarpment and the Chester Cuesta in the Central Kentucky Karst. Natl Speleol Soc Bull 29:109
Quinlan JF, Smith AR, Johnson KS (1986) Gypsum karst and salt karst of the United States of America. Le Grotte d'Italia 4(13):73–92
Racoviţă G, Moldovan O, Onac B (eds) (2002) Monografia carstului din Munţii Pădurea Craiului. Institul de Speleologie 'Emil Racoviţă, Cluj-Napoca, 264 p
Roglič J (1964) Karst valleys in the dinamic karst. Erdkunde 18(2):113–116
Roglič J (1965) The delimitation and morphological types of the Dinaric karst. Naše Jame 7(1–2):12–20
Salomon JN (2009) The tsingy karrenfields of Madagascar In: Ginés A, Knez M, Slabe T, Dreybrodt W (eds) Karst rock Features. Karren Sculpturing Zalozba ZRC. Carsologica, 9. Institut za raziskovanje krasa ZRC SAZU, Postojna-Ljubljana, pp 411–422
Salomon JN, Pomel S, Nicod J (1995) L'évolution des cryptokarsts: comparaison entre le Périgord-Quercy (France) et le Franken Alb (Allemagne). Z Geomorphol 39(4):381–409
Salvigsen O, Elgersma A (1985) Large-scale karst features and open taliks at Valdeborgsletta, outer Isfjorden, Svalbard. Polar Res 3(2):145–153
Sásdi L (1990) Az Aggtelek-Rudabányai-hegység karsztjának földtani fejlődéstörténete (Geological evolution of the Aggtelek-Rudabánya Mountains). Karszt és Barlang I:3–8 (in Hungarian)
Sawicki LS (1909) Ein Beitrag zum geographischen Zyklos im Karst. Geogr Z (Vienna) 15:185–204, 259–281
Slabe T (1999) Subcutaneous rock forms. Acta Carsologica 28(2):255–271
Slabe T, Liu H (2009) Significant subsoil rock forms. In: Gines A, Knez M, Slabe T, Dreybrodt W (eds) Karst rock features. Karren Sculpturing, Zalozba ZRC. Carsologica, 9. Institut za raziskovanje krasa ZRC SAZU, Postojna-Ljubljana, pp 123–137
Smart C (2004) Glacierized and glaciated karst. In: Gunn J (ed) Encyclopedia of caves and karst science. Fitzroy Dearborn, New York, pp 388–391
Smart P, Waltham AC, Mingde Y, Ying-jun Z (1986) Karst geomorphology of Western Guizhou. Cave Sci (BCRA) 13(3):89–113
Song LH (1986) Origination of stone forest in China. Int J Speleol 15(1–4):3–33
Song L, Liang F (2009) Two important evolution models of Lunanshilin karst. In: Ginés A, Knez M, Slabe T, Dreybrodt W (eds) Karst rock features. Karren Sculpturing Zalozba ZRC. Carsologica, 9. Institut za raziskovanje krasa ZRC SAZU, Postojna-Ljubljana, pp 453–459

Soriano MA, Simón JL (2001) Subsidence rates of alluvial dolines in the central Ebro basin, Northeastern Spain. In: Beck BF, Herring JG (eds) Geotechnical and environmental applications of karst geology and hydrology. Balkema, Lisse, pp 47–52

Sowers GF (1986) Correction and protection in limestone terrane. Environ Geol Water Sci 8:77–82

Sowers GF (1996) Building on sinkholes: design and construction of foundations on karst Terrain. ASCE Press, New York, 202 p

Stefanovits P (1976) Talajtan. (Soil science) Mezőgazdasági Kiadó, Budapest, 380 p (in Hungarian)

Sweeting MM (1958) The karstlands of Jamaica. Geogr J 124:184–199

Sweeting MM (1973) Karst landforms. Columbia University Press, New York, 362 p

Sweeting MM (1995) Karst in China. Its geomorphology and environment. Springer, Berlin, 265 p

Szabó PZ (1964) Neue daten und Beobachtungen zur Kenntnis der Paläokarsterscheinungen in Ungarn. Erdkunde 18(2):135–142

Székely K, Szentes G (1981) A Mammut-barlangrendszer földtani és geomorfológiai vázlata (Geological and geomorphological study of the Mammoth Cave System). Karszt és Barlang I-II:15–20 (in Hungarian)

Tan M (1992) Mathematical modelling of catchment morphology in the karst of Guizhou, China. Z Geomorphol N F 36(1):37–51

Telbisz T, Dragašice H, Nagy B (2005) A horvátországi Biokovo-hegység karsztmorfológiai jellemzése terepi megfigyelések és digitális domborzatelemzés alapján (Doline morphometric analysis and karst morphology of Biokovo Mt (Croatia) based on field observations and digital Terrain analysis). Karsztfejlődés X:229–243 (in Hungarian)

Thomas TM (1974) The South Wales interstratal karst. Trans Br Cave Res Assoc 1:131–152

Van Everdingen RO (1981) Morphology, hydrology and hydrochemistry of karst in permafrost near Great Bear Lake, Northwest Territories, Paper 11. National Hydrological Research Institute of Canada, Calgary, 53 p

Van Husen D (2000) Geological processes during the quaternary. Mitteilungen der Österreichischen Geologischen Gesellschaft 92:135–156

Veress M (1992) Karsztmorfológiai sajátosságok a Pádis fedett karsztjának példáján (Karstmorphological characteristics based on the example of the covered karst of Pádis). Földrajzi Közlemények CXVI(3–4):125–141 (in Hungarian)

Veress M (2000a) Covered karst evolution Northern Bakony mountains, W-Hungary. A Bakony Természettud. Kut. Eredményei, 23, Bakonyi Természettudományi Múzeum, Zirc, 167 p

Veress M (2000b) Középhegységi karsztok néhány típusa (Some types of middle mountain karsts). Földrajzi Közlemények XXIV (XLVIII) 1–2:1–28 (in Hungarian)

Veress M (2005) Adalékok a Tábla-völgyi dülő (Tési-fennsík) fedett karsztosodásához ((in Hungarian) Data on covered karstification of the doline in Tábla valley, Tési plateau). Karsztfejlődés X:267–291 (in Hungarian)

Veress M (2006) Adatok a Tési-fennsík két térszínrészletének fedett karsztosodásához (Survey/ data on covered karstification of two surface parts of the Tési plateau). Karsztfejlődés XI:171–184 (in Hungarian)

Veress M (2009) Investigation of covered karst form development using geophysical measurements. Z Geomorphol 53(4):469–486

Veress M (2010) Adalékok az Aggteleki-fennsík völgyeinek fejlődéséhez (Some data to the valley evolution of the Aggtelek Plateau). Karszt és Barlang, 2008 évf (I–II):3–12 (in Hungarian)

Veress M (2011) Adatok a Mecsek-hegység fedett karsztosodásához a Cigány földi mintaterületről vett példák felhasználásával (Data to the covered Karstification of the Mecsek Mountains by using investigation data of Cigany Föld)– Karszt és Barlang, 2010 évf (I–II):9–30 (in Hungarian)

Veress M (2012a) New data on the development of the Baradla Cave. Acta Carsologica 42(2–3):193–204

Veress M (2012b) Glacial erosion and karst evolution (Karren formation on the surfaces formed by glaciers). In: Veress B, Szigethy J (eds) Horizons in earth science research. Nova Science Publishers Inc, New York, pp 1–94
Veress M, Péntek K (1996) Theoretical model of surface karstic processes. Z Geomorphol 40(4):461–476
Veress M, Péntek K (2010) The development of hum slopes due to karren formation. Karst Dev 1(2):23–36
Veress M, Zentai Z (2009) Karsztjelenségek minősítése a Bükk-hegység néhány mintaterületén a mészkőfekü morfológiájának és a fedőüledékek szerkezetének értékelésével (Classification of Karst phenomena on a few sample areas of the Bükk Mountains by using morphology of the limestone floor and the structure of the sedimentary rock). Karszt és Barlang, 2007 I–II:37–54 (in Hungarian)
Veress M, Lóczy D, Zentai Z, Tóth G, Schläffer R (2008) The origin of the Bemaraha tsingy (Madagascar). Int J Speleol 37(2):131–142
Veress M, Tóth G, Zentai Z, Schläffer R (2009) The Ankarana tsingy and its development. Carpathian J Earth Environ Sci 4(1):95–108
Veress M, Németh I, Unger Z, Kéri P (2013) Predicting potential sites of covered Karstification. J Geogr Geol V(1):1–18
Veress M, Unger Z (2015) Kab mountain: Karst under a Basalt Cap. In: Lóczy D (ed) Landscapes and landforms of Hungary. Springer, Heidelberg/New York/Dordrecht/London, pp 55–62
Waltham W (2004) Construction on karst. In: Gunn J (ed) Encyclopedia of caves and karst science. Fitzroy Dearborn, New York, pp 247–249
Waltham T (2008) Fengcong, fenglin, cone karst and tower karst. Cave Karst Sci 35(3):77–88
Waltham AC, Fookes PG (2003) Engineering classification of karst ground conditions. Q J Eng G Hydrogeol 36:101–118
Waltham T, Bell F, Culshaw M (2005) Sinkholes and subsidence. Springer, Berlin, 382 p
White WB (1988) Geomorphology and hydrology of karst Terrains. Oxford University Press, New York, 464 p
White WB, Watson RA, Pohl ER, Brucker R (1970) The Central Kentucky karst. Geogr Rev 60(1):88–115
Wilford GE, Wall JRD (1965) Karst topography in Sarawak. J Trop Geogr 21:44–70
Williams PW (1966) Morphometric analysis of temperate karst landforms. Irish Speleol 1:23–31
Williams PW (1969) The geomorphic effects of groundwater. In: Chorley RJ (ed) Water, earth and man. Methuen, London, pp 269–284
Williams PW (1971) Illustrating morphometric analysis of karst with examples from New Guinea. Z Geomorphol 15(1):40–61
Williams PW (1972a) Morphometric analysis of polygonal karst in New Guinea. Geol Soc Am Bull 83:761–796
Williams PW (1972b) The analysis of spatial characteristics of karst terrains. In: Chorley RJ (ed) Spatial analysis in geomorphology. Methuen, London, pp 136–163
Williams PW (1987) Geomorphic inheritance and the development of tower karst. Earth Surf Process Landforms 12(5):453–465
Williams PW (2004) Polygonal karst and paleokarst of the King Country, North Island, New Zealand. Z Geomorphol 136(Suppl):45–67
Worthington (1984) The paleodrainage of an Appalachian Fluviokarst: Friars Hole, Went Virginio. McMaster University, MSc thesis, 218 p
Xeidakis GS, Torok A, Skias S, Kleb B (2004) Engineering geological problems associated with karst terrains: their investigation, monitoring, and mitigation and design of engineering structures on karst terrains, Bulletin of the geological Society of Greece vol. XXXVI, Proceedings of the 10th International Congress, Thessaloniki, April 2004, 1932–1941
Xuewen Z, Dehao Z (1988) Karst caves in Guilin. Geological Publishing House, Beijing, 249 p (in Chinese, English summary)

Zámbó L (1970) A vörösagyagok és a felszíni karsztosodás kapcsolata az Aggteleki-karszt délnyugati részén (The relationship between red clays and surface karstification at the southwestern part of Aggtelek karst). Földrajzi Közlemények 94(18):281–293 (in Hungarian)

Zámbó L (1986) Karsztvörösagyagok CO_2 termelés és a karsztkorrózió összefüggése (The connection between the CO_2 production of karst red clays and karst corrosion). A Nehézipari Műszaki Egyetem Közleményei I. sorozat Bányászat 33/1-4:125–138 (in Hungarian)

Zámbó L (1993) The soil cover as a factor of Karstification. Ann Univ Sci Rol Eötvös Nom Sect Geogr XXII:67–74

Zámbó L (1998) A talajtakaró (Soil cover). In: Baross G (ed) Az Aggteleki Nemzeti Park. Mezőgazda Kiadó, Budapest, pp 97–117 (in Hungarian)

Zhang Z (1980) Karst types in China. GeoJournal 4(6):541–570

Zuffardi P (1976) Karst and economic mineral deposits. In: Wolf KH (ed) Handbook of Strata – bound and stratiform ore deposits, vol 3. Elsevier, Amsterdam, pp 175–212

Chapter 5
Covered Karst Landforms

Abstract The landforms of covered karst, originated in cryptokarst or concealed karst environment, are presented. Cryptokarst landforms include caprock dolines, ponors, blind valleys, epigenetic valleys and remnant caves. Typical concealed karst landforms are subsidence dolines, closed gullies (blind gullies) and blind suffosion gullies. On both types of covered karst, karren, depressions of superficial deposit and covered karst ponors occur. Numerous varieties of covered karst landforms are distinguished. According to their morphology, karren represent 15 varieties. According to their origin, caprock dolines represent 3; according to size, dropout dolines represent 8; and according to size, morphology and environment, suffosion dolines represent 8 varieties. Subsidence dolines are characterised by morphological parameters (distinguishing between features of karstic and non-karstic origin), ground plan, cross-section and shape of slope. The patterns of doline groups and their morphological environments are demonstrated as well as some varieties of subsidence pseudokarst depressions. Ponors are grouped according to their positions occupied in the karst into karst marginal and karst interior ponors. Karst interior ponors show four varieties according to the position of the rock boundary, while covered karst ponors have three varieties. Depressions of superficial deposit are characterised and grouped according to their position, bedrock morphology, cover sediments and degree of coveredness. Where it is reasonable, covered karst landforms on evaporites are presented separately (karren, salt step, solution subsidence trough, solution-induced depositional basin). The pseudokarst-covered karst landforms are also described.

Keywords Cryptokarst • Concealed karst • Subsoil • Karren • Caprock doline • Subsidence doline • Ponor • Depression of superficial deposit • Epigenetic valley • Remnant cave

5.1 Introduction

In the classification of covered karst, landforms concealed and cryptokarst, pseudokarst and (partly) anthropogenic landforms are distinguished. Ascending hot waters also induce cavity formation on covered karst (e.g. in Hungary, the caves of the thermal karst of the Buda Mountains). Although this kind of cavity formation can

M. Veress, *Covered Karsts*, Springer Geology,
DOI 10.1007/978-94-017-7518-2_5

be regarded as a special way of covered karst formation, this phenomenon is not treated here.

Cryptokarst features develop on impermeable consolidated rocks, on structural rock boundary (inside the impermeable cover) or on true rock boundary (where the impermeable cover ends), at local or pointlike water influx into the karst. Even if no water reaches the karst, different features can form on the cover. They develop on rock salt and gypsum independent of water influx from the surface (landforms of autogenic cryptokarst). Such depressions form when the ceilings of deep-seated cavities cave in, while other landforms come about to the effect of rising karst water table (see below).

The landforms of concealed karst form on permeable (mostly non-cohesive) rocks, independent of true rock boundary. Water transfer into the karst is not necessarily of local nature. (If it is, it happens on hidden rock boundary or half rock boundary.) The characteristics of covered karst landforms and their formation are the following:

- The same form can develop both on cryptokarst and on concealed karst, such as karren and depressions of superficial deposits (DSD), while for other forms, only different varieties can develop on the two types of covered karst. There are other features which are restricted to cryptokarst (caprock doline) or exclusively to concealed karst (subsidence doline).
- On covered karsts, in addition to karst landforms, non-karstic landforms also emerge. The evolution of karstic and non-karstic features mutually presupposes each other. A non-karstic form may be transformed into a karstic feature (e.g. a valley into a blind valley).
- The karst landforms of bare surfaces have their counterparts as covered karst varieties. The diversity of covered karst landforms is higher. However, not every covered karst landform has an uncovered variety. Within a landform type (e.g. karren), on the other hand, uncovered karst landforms show a higher degree of variation.
- Most of the covered karst landforms are similar to uncovered karst landforms, but in some cases the differences are remarkable. It is common that there are transitional forms between covered and uncovered forms (e.g. in the case of dolines).
- Covered karst landforms originate from the karstification of the bedrock, but processes in the cover also play a part. Thus, the covered karst phenomena are due to the modifying influence or entirely to the effect of processes in the cover.
- Although complex processes take place on uncovered karsts too, this statement is even more valid for cover karst. To processes in the bedrock, those in the cover are added, such as material transport, erosion and mass movements associated with water motion in the cover. Therefore, the landforms spring from numerous influences and increase diversity in size and morphology.
- Landforms on covered and uncovered karst can transform into each other. Karstification often begins on covered karst. With the removal of the cover sediments, the landforms of covered karst are destroyed, become inactive or are

transformed. In other cases, uncovered karstification is followed by covered karst formation. The landforms of uncovered karst (dolines, uvalas) determine the site of covered karst formation and occasionally even the size and shape of landforms.

- As it has been already mentioned, landforms on cryptokarst are different from those on concealed karst. Therefore, and because non-karstic evolution may be prevalent at the beginning, there are different ways of landform evolution on the two types of covered karst.

The landforms of concealed karst are subsidence dolines, blind gullies and blind suffosion gullies. The landforms of cryptokarst are ponors on true rock boundary, caprock dolines, epigenetic valleys, remnant caves and blind valleys. Karren, covered karst ponors and DSDs occur on both types of karst (Fig. 5.1).

5.2 Subsoil Karren

5.2.1 General Description

Karren are often defined as solutional microforms (Veress 2010a). Defining the karren, however, is not easy since recently megakarren have also been distinguished (Ginés 2004, 2009; Grimes 2012).

According to surface coveredness, Bögli (1960, 1976) identified subsoil karren, karren on half-covered and bare terrains. In the development of karren on bare terrain, rainwater fallen on uncovered surfaces plays an important part. The karren of half-covered terrains come about on the margin of soil or sediment cover, and their development is partly or entirely due to runoff from the cover patch. The resulting karren are transitional between subsoil karren and bare surface karren. Subsoil karren are solutional features formed on the bedrock below cover sediments. The forms under the cover are rarely inherited over the cover. Forms only develop on the cover if the cover sediment is transported into the karren depressions of the bedrock (Curtis et al. 1976; Trudgill 1985). In this case depressions can emerge on the cover surface above the grikes of the bedrock (Trudgill 1986). Bedrock karren are inherited over the cover if the sediment cover is thin and the density of bedrock forms is low. On thicker sediment cover and with higher form density of the bedrock, the material deficit becomes cumulative and larger landforms (dolines) emerge on the surface. However, there is no sharp boundary between the two types of form.

Covered karren and subsoil karren cannot be distinguished from each other and no difference is made between them in literature. As exceptions, Ford and Williams (1989) describe solution pits, channels, karren shafts and wells developed below weathering debris. Gines (1990, 1996) considers that tube and cup karren originated under cover. The Lunan stone forest is regarded karren formed under debris and called subjacent karst (Chen et al. 1986), cryptokarst (Maire et al. 1991) or subcutaneous karst (Slabe 1999; Knez and Slabe 2002).

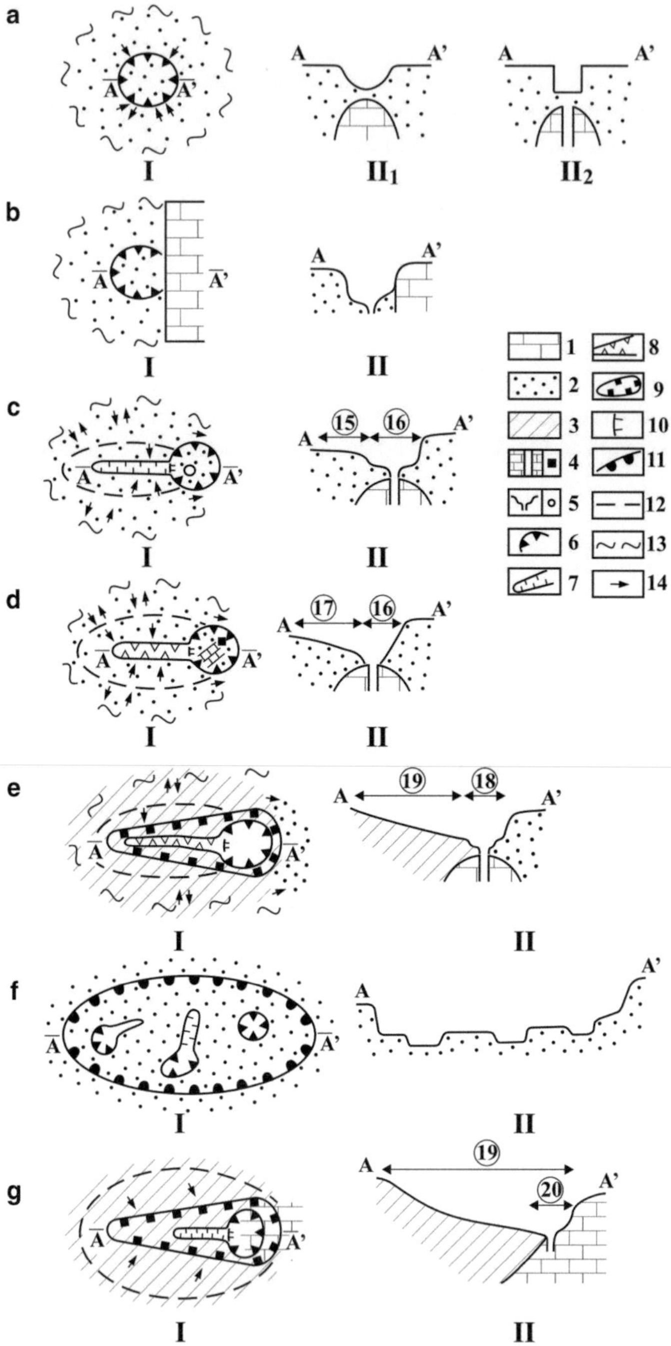

Fig. 5.1 Covered karst landforms (without caprock dolines and karren). Ground plan: *1* limestone, *2* permeable cover, *3* impermeable cover, *4* chimney, shaft, *5* non-karstic pipe, *6* doline, *7* gully, *8* ravine, *9* blind valley, *10* scarp on the side slope of doline, *11* DSD, *12* divide, *13* limit of bifurcation, *14* water flow; section: *15* gully, *16* doline, *17* ravine, *18* covered karst ponor, *19* blind valley, *20* ponor, *I* top view, *II* cross-section, (**a**) doline, (**b**) suffosion half-doline, (**c**) doline with gully, (**d**) doline with ravine, (**e**) covered karst ponor, (**f**) DSD, (**g**) blind valley ponor

Karren are produced by water flow and seepage (White 1988; Veress 2010a). On bare surfaces, karren formation is induced by water flow or seepage. Under soil or cover sediment, karren form to the effect of water seepage (through the infiltration of rainwater) or of slow water flow on the boundary between the cover and the rock. (For karren developed below the water table, both types of water motion are present: initial seepage is gradually replaced by water flow.)

If the terrain becomes covered, karren of flow origin are transformed or their morphology is modified. In the opposite situation, if the sediment cover is denuded, karren features are also transformed. A good example for the transformation of bare karren when covered is rundkarren. When covered, rinnenkarren transformed into channels with overhanging walls and their dividing ridges become rounded (Bögli 1960; Veress 2009a).

The classification of subsoil karren by origin was completed by Slabe and Liu (2009), who distinguished karren formed below soil and cover sediment and those developed where the soil wedge out. The former are due to seepage through the cover (such are cups and channels), water flow on the contact of the cover and rock (scallops, channels) as well as to water table fluctuation (channels, subsoil tubes and notches). Half-bells and notches form at the level of soil cover.

If the cover of the karstic rock is impermeable, cryptokarren result, and if it is permeable, concealed karren appear. On cryptokarst, karren formation is only possible if the cover sediment is fractured (karren formation is local) or the bedrock surface receives water laterally or from below. Covered karren develop under both continuous and discontinuous cover (soil and cover patches). In the latter case, bare and covered karren patches alternate on the karst. If the cover is patchy, where it ends, karren aligned in the same direction of the flow on bare karren. Thus, for example on carbonate rocks, rinnenkarren (channels) develop, and on evaporites, rinnenkarren, rillenkarren and meanderkarren develop. Patchy pattern is particularly common on glaciokarst. On glaciokarst covered karren patches occur in areas of paleodolines or in larger uncovered karren landforms. The solutional features created under sediment (fill) of the cave mentioned by Slabe (1992, 1995) are also regarded as covered karren. Covered karren are typical where solution residues abound (on tropical karsts) or where intensively soluble rocks (rock salt) are present.

Covered karren can only be observed if they are exhumed or exposed. Being of small size, they do not build surface landforms. Their total distribution under the cover cannot be reconstructed. Karren under cover create a varied and extensive morphology on the bedrock, ranging from simple to very complex (Tan 1987; Bennett 1997). The bedrock types of various karst regions (assemblages of grikes and pinnacles of different size and density) are presented by Waltham and Fookes (2003). The dissection of the bedrock by karren depends on groundwater: bedrock surfaces greatly dissected by karren are found under tropical climate, to the effect of aggressive water from the bordering non-karstic terrain or of groundwater (Waltham and Fookes 2003).

During covered karst formation, karren may be transformed into other karst features. Thus, subsoil karren further develop into dolines (Williams 1983; Veress and Péntek 1996). Shafts under cover sediment also come about through the merging of karren features (solution pipes or chimneys) (see Chap. 7). The resulting shafts generate certain varieties of subsidence dolines. If the cover sediment is transported away, surface evolution does not promote doline formation, but the destruction of the cover leads to the development of grikes filled with soil (Trudgill 1985) and eventually of uncovered terrain with transformed karren (see below).

The assemblage of large-scale karren, such as the Pinnacles karst, is constituted of secondary karren, which bear minor primary karren (channels, scallops, chimney remnants and others).

5.2.2 *Subsoil Karren on Carbonate Rocks*

Since the infiltrating water is dissipated in soil and cover sediment, solution is of areal character and the rock is less dissected by karren features. According to the investigations by Trudgill (1976, 1985), in North Yorkshire, aerial solution only caused moderate dissection of the rock surface. Elevations of various shapes and sizes are left behind of the original surface. Trudgill (1976, 1985) identified undulation surfaces (under turf mat) and arcuate forms (under wet, acid peat soils). In the case of karren formation under cover sediment, however, mostly intensive dissection is observed – particularly if the process lasts for a longer period. The resulting features are presented below.

5.2.2.1 Irregular Karren Features

- Subsoil cavernous karren (Bögli 1960, 1980; Gams 1976; Zseni 2004, 2009): assemblage of large circular cavities and passages. Such karren mostly develop along zones of weakness (joints) in the rock.
- Subcutaneous tubes (Slabe 1999; Zseni 2004, 2009; Gams et al. 2011): cavities of variable position, filled with sediment or soil, of irregular, circular, elliptical, semicircular or omega shape, which can merge to form cavity systems. The smaller cavities are 1–10 cm, and the larger are up to 1 m diameter.

5.2.2.2 Vertical Karren Features

- Karren wells and solution pipes (Ford and Williams 1989; Zseni 2004, 2009) are pipes of vertical position and circular cross-section in the rock. The sediment fill is rather variable. Solution pipes do not only develop in carbonate rocks but also in gypsum under cover (clay) (Penck 1924) as well as in rock salt. Such so-called Geologischen Orgeln are partly or totally filled with debris or in situ soil. Large syngenetic solution pipes also form in cover sediment, in the dune sand of coastal areas (Jennings 1973; Gardner 1983; McKee and Ward 1983; Grimes 2004, 2009a). Syngenetic solution pipes are generated through the solution of the carbonate content of unconsolidated deposits and cementation along roots (Grimes

Fig. 5.2 Group of solution pipes on the Madagascar coast

2009a). Pipes in chalk (Kirkaldy 1950; Rodet 1992) are paleoshafts of 1–3 m diameter and 10–30 m depth (Waltham et al. 2005), formed under Paleogene cover (Edmonds 2001).

On the coast of Madagascar, in large rock blocks, solution pipes of 1–2 m diameter and 5–10 m vertical extension occur (Veress et al. 2008a). The solution pipes mostly cross the rock but sometimes have dead ends, and they may fork in both downward and upward directions, often merging with each other (Figs. 5.2 and 5.3). They may appear in high density, at merely 1–2 m from each other.

Even on bare surfaces, solution pipes mostly form only if there is soil accumulation on their floors. These features are often totally filled with soil. In their formation, rock joints are of decisive importance (Slabe and Liu 2009).

In the tropics, intensive solution generates large-scale varieties of solution pipes under soil or cover sediment. In the sides of pinnacles of stone forest karsts, vertical half-chimney-like features are common (Fig. 4.5). They are chimney remnants formed through the merging of chimneys which developed under cover and were subsequently exhumed. Chimneys and chimney remnants can also form through solution on bare surfaces, such as in the sides of pinnacles of the Madagascar tsingy (Veress et al. 2008b).

Root karren, formed along roots and to the effect of root acids, resemble solution pipes (Jakucs 1971, 1980).

- Subsoil pinnacles (Trudgill 1985; Ford and Lundberg 1987; Ford and Williams 1989; Grimes 2004; Zseni 2009) are steep-walled elevations. Pinnacles include conical remnant landforms on coasts (Veress 2005; Gomez-Pujol and Fornós 2009) or on bare terrains of glaciokarsts (Bögli 1976; Veress 2010a) (solution spikes, spitzkarren). The typical subsoil pinnacle karren are composed of cones of several tens of centimetres height (Fig. 5.4), generated through the solutional broadening of subsoil grikes.

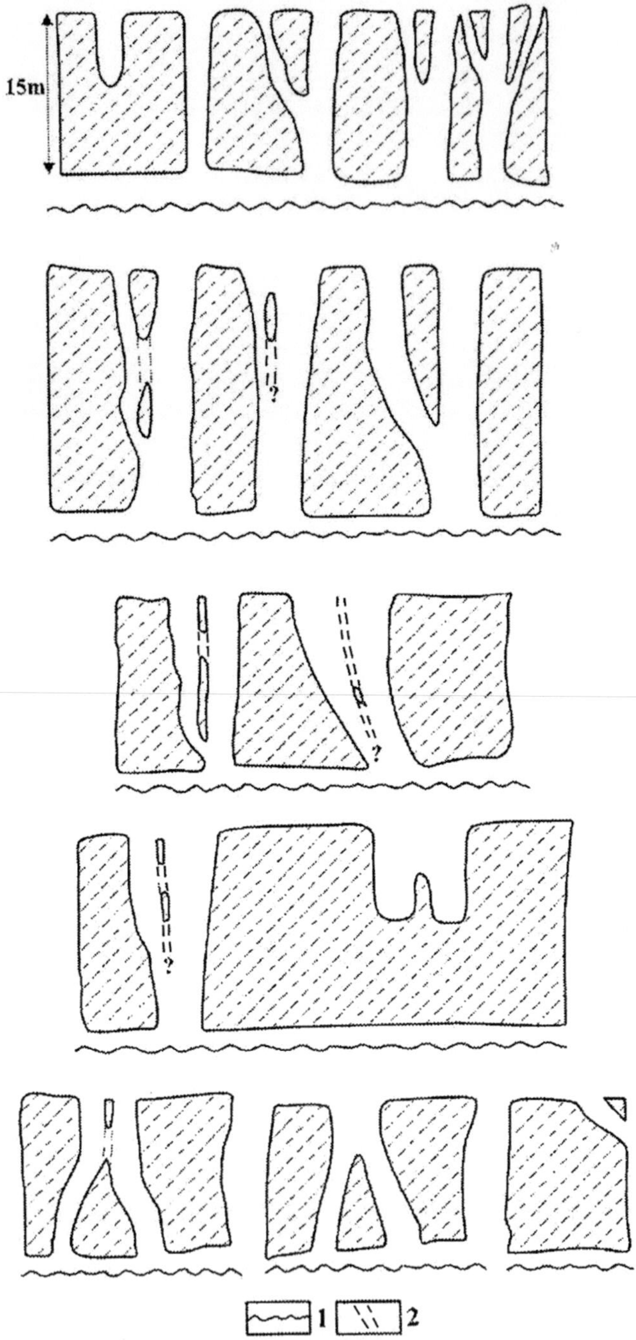

Fig. 5.3 Main patterns of the solution pipes of the Madagascar rocky coast along profiles: *1* sea level, *2* former dividing wall

Fig. 5.4 Subsoil pinnacle on the side of a subsidence doline of the Pádis plateau (Romania)

In the tropics pinnacles emerge under laterite in large extensions, similarly to the grikes of the temperate belt. This form assemblage is constituted of elevations bordered by large-scale grikes and/or chimneys (primary karst features). The laterite surrounds the elevations, fills the gaps between them and even buries the elevations. The laterite is underlain by a largely dissected surface, where the floor of the depressions lies at 55 m, while the dividing ridges rise up to 18 m depth below the ground (Bergado and Selvanayagam 1987, Fig. 5.5). The pinnacles formed on marble karst under laterite and gravel are 3–10 m high (Brook 2004).

Pinnacle karst is an assemblage of tropical karren with bare (uncovered) karst varieties, such as in Australia, on the Judbarra karst (Grimes 2012; Martini and Grimes 2012) or on the tsingy of Madagascar (Veress et al. 2008b, 2009) and varieties generated on covered karst and exhumed by now. The stone forest karren, such as in Lunan, are pinnacle karsts formed under cover and partially exhumed by the present time (Yu and Yang 1997).

5.2.2.3 Linear Karren Features

- Rundkarren (Sweeting 1955, 1973; Bögli 1960; Haserodt 1965; Perna and Sauro 1978; Veress 2010a): steep-walled, vertical channels, which developed from rinnenkarren channels through subsoil or subcutaneous solution if the rinnenkarren are buried (Bögli 1960; Haserodt 1965; Veress 2009a).
- Subsoil channels (Slabe and Liu 2009; Gams et al. 2011; Knez et al. 2011): developed on steep rock faces where the rock contacts with soil or cover sedi-

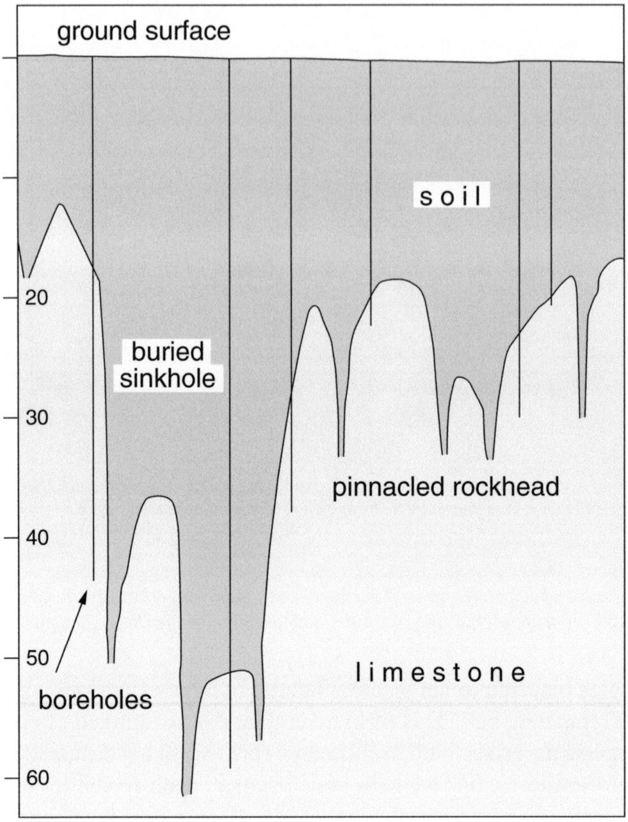

Fig. 5.5 Pinnacle karst under cover sediment (Malaysia, Kuala Lumpur, Waltham et al. 2005)

ment. Channels on bare surfaces and covered terrains are markedly different. In the channels of bare surfaces (rinnenkarren or runnels), there are embedded internal channels and rinnenkarren systems on gentle slopes (Veress 2010a), while in the channels on covered surfaces, benches and subsoil scallops occur (Slabe and Liu 2009), although scallops not formed under sediment cover may be also present in the rinnenkarren of bare terrains (Veress 2010a). Slabe and Liu (2009) identify a broader (more than 20 cm wide) and a narrower variety (less than 20 cm wide). The alignment of the latter may differ from the direction of the bearing slope. The channels of the Lunan stone forest are semicircular or of reverse omega shape in cross-section (Fig. 5.6). Channels also develop on gypsum and rock salt (for channels on rock salt, see below). Gypsum channels are described by Sásdi (1987) from Alsótelekes (Aggtelek Karst, Hungary), where 5–15 cm wide channels are formed on gypsum underlying Pannonian clay in the wall of an open-cast mine.

- Subsoil grikes (kluftkarren), cutters (Howard 1963; Ford and Williams 2007; Zseni 2009): vertical or oblique depressions of parallel walls or narrowing

Fig. 5.6 Exhumed subsoil karren (Lunan area): *1* channel, *2* scallops

downwards, with widths of several tens of centimetres to ca 1–2 m and depths of several metres.

On temperate (and mediterranean) karsts, series of depressions and dividing ridges are exposed (subsoil grikes) (Figs. 5.7, 5.8, 5.9 and 5.10). The cover is mostly terra rossa and sometimes yellow clay (Pigott 1965). With the cover sediment transported into them, the grikes are inherited onto the surface or only cover remnants are preserved or the cover is totally denuded because it is transported into the grikes (Trudgill 1985, 1986).

There are grikes widening downwards with inner grikes on their floors (Figs. 5.10 and 5.11IId), grikes wedging out downwards (Figs. 5.7 and 5.11Ib), grikes with parallel walls (Figs. 5.8 and 5.11Ia) and pocket-like grikes (Figs. 5.9 and 5.11IIc).

The grikes are often of oblique position and the neighbouring grikes may cross or merge with each other (Fig. 5.8), detaching large blocks from the bedrock mass. Where the bedrock is heavily jointed, grikes are deep and narrow (Figs. 5.7 and 5.8), while where it is less jointed, grikes are wider and shallower. Features of low depth, however, can also emerge through the truncation of dividing walls, when the bedrock is remarkably denuded between grikes (Fig. 5.9).

Fig. 5.7 Subsoil grike in doline side (doline 1 of Pádis-2 area, Fig. 4.43); the cover is reworked sandstone debris: *1* grike wedging out downwards, *2* fracture broadened by solution, *3* grike with parallel walls, *4* bedding plane grike

Fig. 5.8 Subsoil grike (China, near Lunan, the cover is solution residue): *1* grike broadening downwards, *2* grikes crossing each other, *3* grike with parallel walls, *4* bedding plane grike

Fig. 5.9 Subsoil grike (Croatia, mine wall next to main road number 1); the cover is solution residue: *1* grike, *2* kamenitza on the floor, *3* ridge between grikes, *4* detached section of bedrock

Fig. 5.10 Subsoil grike in a roadcut of the road to the Pádis plateau (the cover is terra rossa mixed with soil): *1* grike narrowing downwards, *2* grike with parallel walls, *3* grike broadening downwards, *4* interior grike, *5* detached section of bedrock

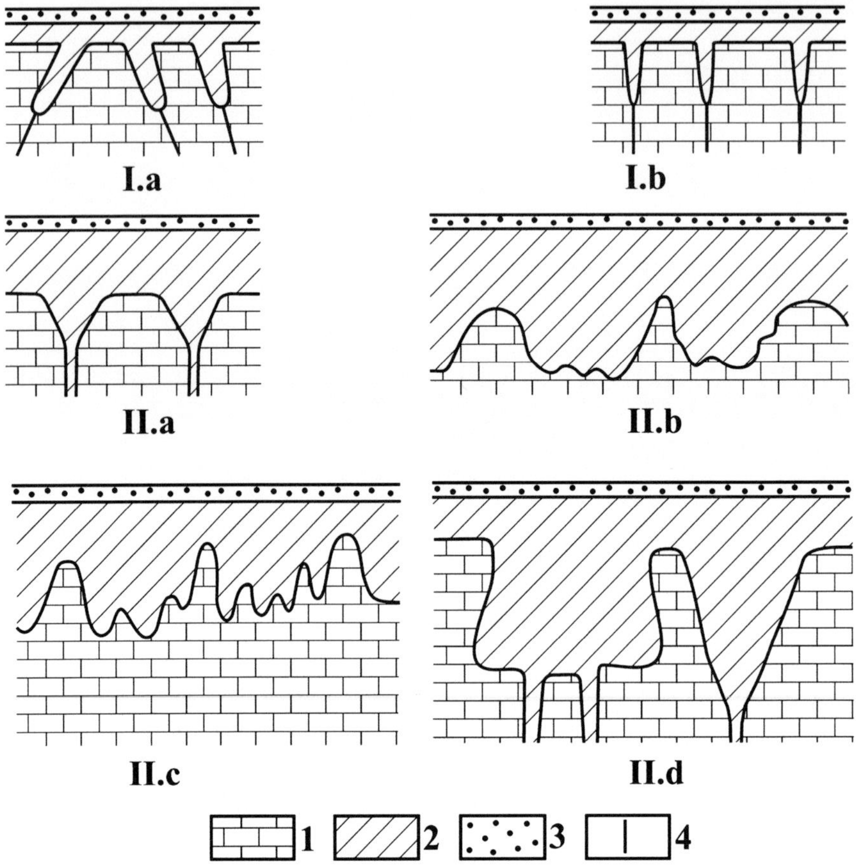

Fig. 5.11 Subsoil grike *I* Grikes in less denuded bedrock: grike with parallel walls (**a**), simple grike narrowing downwards (**b**). *II* grikes of more denuded bedrock: grikes narrowing downwards continued to depth in interior grikes with parallel walls (**a**), composite grikes (**b**, **c**), composite grike continued to depth in interior grike with overhanging and then parallel walls (**d**) *1* limestone, *2* solution residue, *3* soil, *4* fracture

Using data obtained from the interpretation of photographs, the parameters of karren were calculated (Table 5.1). Therefore, the received values are without dimension. For covered karren, the interpretation was made from photos of exposures, while for bare karren, from orthophotos. In the former case, the alignment of the profile of measurement followed the direction of exhumation, and in the latter, perpendicular to the direction of features. It is observed that specific solution (S_w) is more intensive below covered karren. Since the density of karren (K_d) is similar on covered and uncovered karsts, the mentioned variation can be explained by the lesser width of karren on uncovered karst. The bare karren do not show internal dissection (d_d).

Table 5.1 Some characteristics of subsoil and bare karren

Locality	Coveredness	Karren type	S_w	d_d	K_d
Side of doline Pádis-2	Reworked debris	Grike	0.82	1.8	0.19
Pádis (road cut)	Terra rossa	Grike	0.93	2.22	0.30
Croatia (road cut)	Terra rossa	Grike	0.74	0.83	0.32
Lunan (road cut)	Laterite	Grike	0.64	0.87	0.22
Totes Gebirge	–	Grike	0.61	0	0.23
Totes Gebirge	–	Grike	0.33	0	0.22
Totes Gebirge	Below soil patch with *Pinus mugo*, on bare slope	Rinnenkarren	0.4	0	0.19
Totes Gebirge	–	Rinnenkarren	0.38	0	0.17

Making comparisons within the karren of covered karst, it can be seen that specific solution and internal dissection for the Pádis are higher than for the Croatian and Lunan (South China) karren. Since the intensity of solution is lower on the Pádis than in the Croatian and Lunan study areas, the higher specific width in the Pádis study area is explained by the higher density of features, which is probably related to the higher density of joints in the rock. The higher density of features favours the merging of neighbouring grikes. This is evidenced by the higher degree of dissection for the Pádis. It is probable, however, that in Lunan the lower specific width is due to the younger age of karren formation. Figure 5.8 shows that the slope is exhumed, and consequently, grike formation has begun recently at the lowermost level.

The grikes may grow further to form giant grikes (Gams 1971). Ruiniform karst comes about if the grikes form surface landforms because of the removal of cover sediments (soil). The form assemblage that develops through the broadening of dividing walls is called corridor karst or labyrinth karst (Brook and Ford 1978; Brook and Fenney 1996).

Cutters are grikes in dolomite (Howard 1963), features of parallel walls with flat bedrock surface in between. On gypsum, grike formation is rapid (Pfeiffer and Hahn 1972), and on rock salt, they also form a dense network but their depth is moderate (see below). This is explained by the solution of the surfaces between grikes.

- Subsoil or subcutaneous half-bells (Slabe 1999; Slabe and Liu 2009; Zseni 2004, 2009; Grimes 2004): depressions of vertical cliffs, which develop under the channels of uncovered limestone cliffs, under the cover (Slabe and Liu 2009).

5.2.2.4 Horizontal Karren Features

- Subsoil or subcutaneous notches (Jennings 1985; Slabe 1999; Gams 1971, 1976; Gams et al. 2011): form at the level of soil and sediment cover. In cross-section, they are (almost) semicircular depressions of several metres length with sharp upper edges and rounded lower edges. They are mentioned with various names: grooves (Paton 1963), undercut notches (Waltham 1984), solution notches (Jennings 1973) or swamp notches (Ollier 1984). In particular, they can extend vertically if the height of soil or sediment cover is reducing due to denudation and they are exhumed. On their surface secondary minor non-subsoil features of solution origin appear. A variety formed during solution along bedding planes are bedding plane grikes (Slabe and Liu 2009; Veress et al. 2008b).

The notches at the base of tropical inselbergs are occasionally longer than 100 m, 20 m deep and 6 m high (Jennings 1985; Sweeting 1973). They develop in swamp environment (swamp slots) (Wilford and Wall 1965).

Notches are also typical for metamorphic rocks of carbonate content (Veress et al. 1998), where they occur in variable size and shape (developing varieties are located at the level of soil and weathering residue). Notches also occur in caves, large-scale channels filled with soil, where bench-like features or pockets appear (Szunyogh 1999).

5.2.2.5 Circular Karren Features

- Subsoil cups or subcutaneous cups (Gams 1971; Trudgill 1985; Slabe and Liu 2009): depressions of circular shape in plan view and semicircular in cross-section; diameters range from several centimetres to several tens of centimetres. They occur on gently sloping terrain.
- Subsoil or subcutaneous scallops (Slabe 1999; Zseni 2009): scallops (Fig. 5.6) of shell shape with depths less than the diameter. They occur in groups on steep rock surfaces (grike or cave walls).
- Tafoni-like features (Zseni 2009): groups of depressions on vertical cliffs, with diameters ranging from 1–2 to 10–20 cm. Their depth usually exceeds the diameter of the entrance. Their cross-section is mostly elongated in the strike direction of rock strata.

5.2.2.6 Karren Developed Below the Karst Water Table

A special group of covered karren is represented by karren generated by the influence of groundwater or karst water. Such forms also develop under impermeable cover, associated with two types of water motion (Slabe and Liu 2009). One is karst water rising to the level of cover or even higher, and the other is the solution effect of water laterally percolating in the cover sediment. If the cover is permeable, they emerge to the effect of the mentioned water motion or if the water level sinks below the bedrock surface and the infiltrating rainwater exerts a solution influence. During waterlogging, notches, subsoil spongy rock surfaces, subsoil tubes and vertical and horizontal subsoil channels come about. As a consequence of subaqueous solution, the rock mass is separated into blocks wrapped in solution residue (Slabe and Liu 2009).

Fig. 5.12 Karren generated by ascending karst water motion (Cerkniško polje, Slovenia). (**a**) Alluvial streamsink doline, (**b**) doline floor, (**c**) site of karst water issue. *1* grike, *2* bedding plane grike, *3* chimney, *4* blind chimney, *5* cavity formed on bedding plane, *6* doline margin, *7* margin of interior depression, *8* rainwater rill, *9* bedding plane

Karren of subaqueous solution develop above katavothra. At a katavothron of Cerkniško polje, where the bedrock outcrops from below the cover, the following karren are identified: bedding plane grikes, cavities, scallops, solution pipes, blind chimneys, natural arches and subsoil grikes (Fig. 5.12).

5.2.3 Evaporite Karren

The karren on evaporites resemble to those on carbonate karsts. They are classified by size (microforms, small forms, mesoforms) and coveredness (bare, half covered and covered) (Macaluso and Sauro 1996; Madonia and Sauro 2009). The covered karren of gypsum karsts on Sicily are runnels, shafts, grikes, small pinnacles between cutters and stone forests (Macaluso and Sauro 1996; Madonia and Sauro 2009). Small pinnacles are exhumed and stone forests are exposed by mining (Forti

Fig. 5.13 'Salt glaciers' of Korond Stream valley (Salt Hill, Parajd, Romania): *1* scar formed by salt fluidization, *2* tongue of fluidized material. *Left* (**a**) 'salt glacier' on the steeper valley side; *right* (**b**) 'salt glacier' of the gentler valley side

and Sauro 1996; Macaluso et al. 2001). On gypsum, in the environs of cover patch, micro-meanders, meandering runnels and rundkarren occur. On the salt karst of the Urals, covered features are the hallows, tubes and shafts (Andrejchuk and Eraso 1996). On rock salt at cover patches, micro-meanders and micro-loops are observed (Macaluso and Sauro 1996).

At Parajd, in the valley side of the Korond Stream, rock salt is exposed at several localities as a result of mass movements of the valley side deposits, and sometimes the salt moves together with the cover sediment. This latter phenomenon is typical when the salt is fluidized (Urai et al. 1986), as in the case of salt glaciers (Jennings 1985; Ford-Williams 2007). Thus, the salt is mostly exposed at the minor salt 'glaciers' of the valley side (Fig. 5.13). As a consequence of these movements, bowl-shaped scars appear on the upper sections of the valley sides and continue in debris fans of cover sediment or salt breccia. In salt breccia, the insoluble debris grains are cemented by salt as a result of mixing during glacier-like movement or the precipitation of dissolved salt. Precipitation happens on the grains of the cover sediment or on the contaminations of salt. (Salt can precipitate at some metres distance from the site of solution.) On the upper margin of exposures, the uneven salt surface is exposed with the following types of karren (Veress et al. 2011).

Grikes are closely spaced narrow features of salt with steep sides (Fig. 5.14). Where the salt is exposed, broader depressions with less steep slopes also occur (subsoil channels) (Fig. 5.15). In the continuation of both grikes and channels, rinnenkarren develop on the uncovered part of salt too. In the continuation of grikes, narrow varieties occur (Fig. 5.14), while those in the continuation of covered

Fig. 5.14 Subsoil grike on a slope formed in salt exposed by a 'salt glacier' (Korond Stream valley): *1* grike, *2* channels linked to grike

(buried) channels are broad (Fig. 5.15a). Channels can develop on salt breccia too (Figs. 5.16 and 5.17).

On rock salt, rillenkarren are common on the margins of sediment cover, along cover patches (Fig. 5.18a) or on salt breccia (Fig. 5.18b). Under salt breccia, major karren tables and in their sides micro karren tables between rillenkarren emerge (Fig. 5.18b). Chimneys are also common on salt as well as a series of transitional features between chimneys and subsidence dolines.

5.2.4 *Subsidence Pseudokarren*

In the cover sediment, features of karren size category also occur, for instance, in laterite (Grimes 2009b; Grimes and Spate 2008). White (1988) calls forms which are due to material deficit in the cover sediment suffosional pseudokarst. It occasionally happens that landforms in the cover are generated not only by piping but also through collapse. Therefore, we do not call them suffosional pseudokarren, but subsidence pseudokarren.

Pseudokarren can also develop in the fills of true karren. Such form group occurs on the Asiago Plateau, near the spring of Bosco Secco (Italy). The features

Fig. 5.15 Rinnenkarren in salt (Korond Stream valley). *Left* (**a**): large-size channels (rinnenkarren) in salt produced by runoff and outflow from the cover sediment; *right* (**b**) channels in salt fed by outflow from the erosional depressions of the cover *1* channel zone filled with cover sediment, *2* channel zone, *3* zone of rinnenkarren on salt breccia (?), *4* sheet erosion of soil and cover sediment, *5* rill erosion in the cover, *6* zone of rinnenkarren (channels primarily formed due to runoff from the cover)

Fig. 5.16 Channels in salt breccia (Korond Stream valley)

Fig. 5.17 Channels formed in salt breccia in the Atacama Desert (Chile)

Fig. 5.18 Rillenkarren in rock salt. *Left* (**a**), karren at the termination of sediment cover in the side of a ponor of the Salt Hill; *right* (**b**), karren table formed at salt breccia with rillenkarren, pinnacles (spitzkarren) and minor karren tables in its side (Atacama Desert, Valle de la Luna)

generated in the fill are subsidence pseudokarren grikes (with an elongated ground plan), subsidence pseudokarren pipes (with a circular ground plan) and pseudokarren caves.

The subsidence pseudokarren grikes developed in the fills of limestone grikes (Figs. 5.19 and 5.20). They show a pattern of mesh network directed parallel to the bearing slope and wedge out both upslope and downslope. The forms without outflow in the fill of true karren are some metres long, 10–20 m wide and very shallow, their depth is growing downslope, and it is some tens of centimetres. The grikes of the bedrock can be partially exposed by pluvial erosion (Fig. 5.19) or through the subsidence of the cover fill. Among the suffosion pseudokarren grikes deriving from the latter process, there are examples the edge of which developed in the cover or fill (Fig. 5.20), while in case of others the edges coincide with the edges of the exhumed grikes of the bedrock, probably because the cover has been removed from the edges of the grikes through pluvial erosion (Fig. 5.19). Two varieties are identified: suffosion pseudokarren grikes with gentle sides and dropout pseudokarren grikes with steep sides and uneven floors dissected by collapse heaps.

Subsidence pseudokarren pipes (Fig. 5.20) are vertical features of some tens of centimetres diameter and 1–2 m depth in the cover. They also have two varieties: those formed above the chimneys of the bedrock and those above the caves of the fill. In this case, earth bridges, ceiling remnants of pseudokarren caves, occur between the pseudokarren pipes. Pseudokarren caves are horizontal forms in the fill of the bedrock grikes, some tens of centimetres in diameter and some metres long (Fig. 5.20).

5.3 Caprock Dolines

Caprock dolines are typical of cryptokarsts with preferably (but not necessarily) consolidated sediment cover and/or if there is large-scale material deficit in the bedrock. The origin of caprock dolines is due to collapse(s). The dimensions and shape of the surface depression are largely determined by the dimensions and shape of the caved-in cavity in the bedrock sediment. Their generation mostly depends on the intensity of the solution process on the bedrock. Thus, this doline type is very common on gypsum and rock salt. On gypsum, form density (for caprock and subsidence dolines together) can be as high as 200 dolines per square kilometre (Waltham et al. 2005). It is favourable for caprock doline generation if the salt or gypsum is overlain by carbonate rocks, since water conduction is of large scale which induces cavity formation in salt and gypsum. They form in consolidated rocks (e.g. dolomite overlying gypsum), or in unconsolidated rock, it is underlain by consolidated rock (Waltham et al. 2005). In the latter case, inheritance happens from the consolidated to the unconsolidated rock.

Synonymous denominations include caprock collapse doline (Williams 2004), subjacent collapse doline (Jennings 1985) and solution subsidence (Sweeting 1973). Collapse and caprock dolines are treated together by Ford and Williams (2007) with

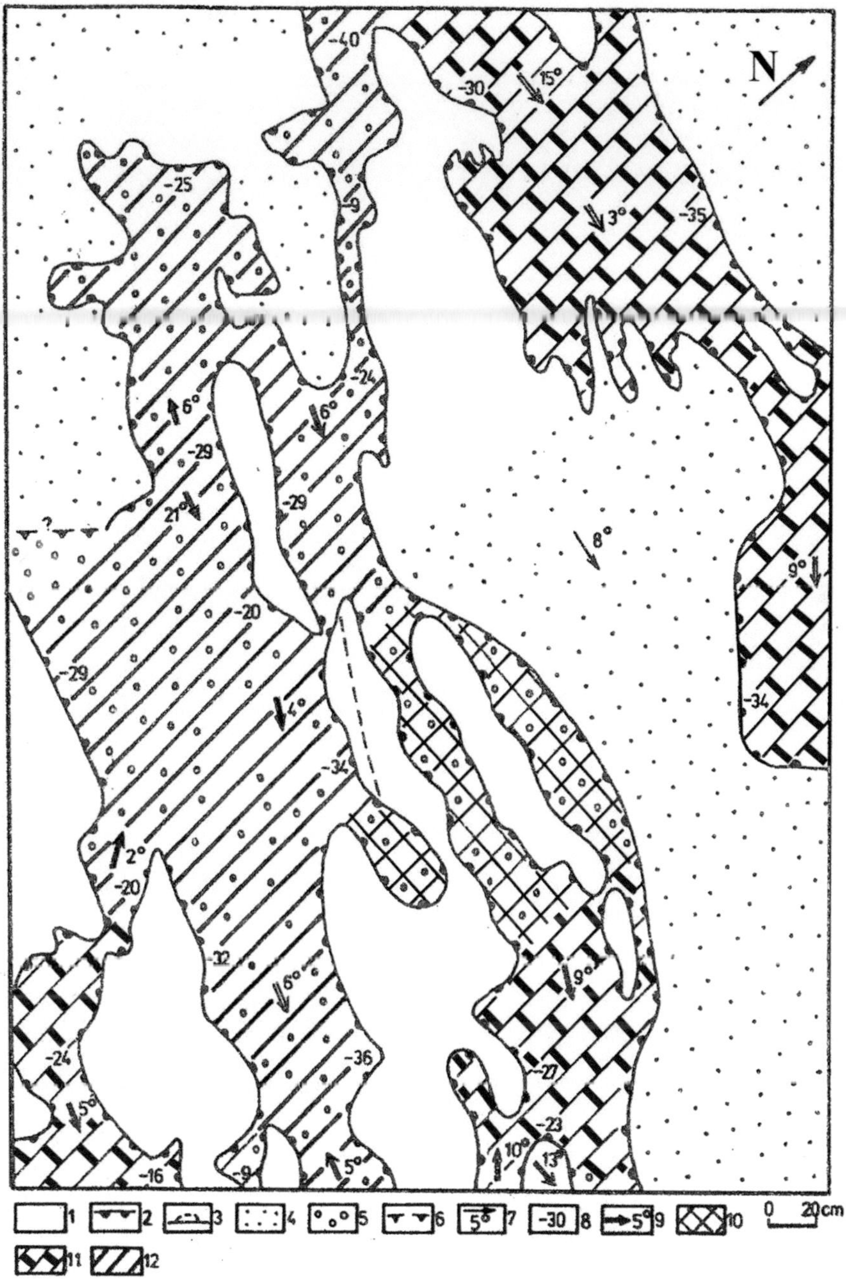

Fig. 5.19 Geomorphological map of subsidence pseudokarren, northern part of study area near the Bosco Secco spring (Asiago Plateau, Italy, Veress et al. 1999): *1* terrain without karren formation in the absence of soil and cover sediment, *2* vertical wall of grikes in bedrock, *3* overhanging walls of grikes in bedrock, *4* cover sediment on terrain free of grikes, *5* fill in karren, *6* gentle slope in cover sediment, *7* slope direction and angle on terrain without karren, *8* depth of grikes in bedrock, cm (from margin to floor), *9* direction and angle of slope on the floor of pseudokarren, *10* buried (infilled) grikes in bedrock, *11* grike partly exhumed by pluvial erosion, *12* suffosion pseudokarren grike

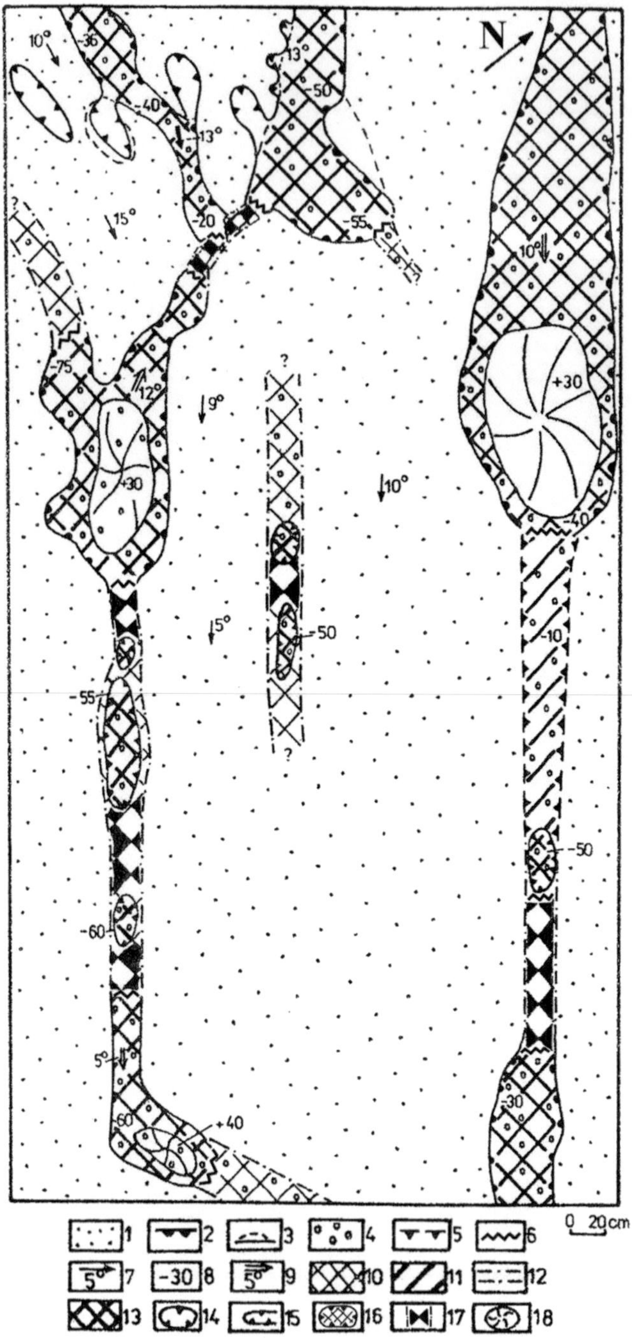

Fig. 5.20 Geomorphological map of subsidence pseudokarren, southern part of study area near the Bosco Secco spring (Asiago Plateau, Italy, Veress et al. 1999): *1* terrain covered with sediment, free of karren formation, *2* vertical wall of grikes in bedrock, *3* overhanging wall of grikes in bedrock, *4* fill in karren, *5* gentle slope on cover sediment, *6* vertical wall in fill, *7* direction and angle

only partial distinction. It may happen that the cavity in the carbonate rock is directly inherited over the surface where there is no cover (collapse doline), but the collapse of the large-scale and near-surface cavity can take place with cover on the surface over which the collapse is inherited also directly (caprock doline).

Caprock dolines show a great diversity both for size, morphology and origin. There are varieties transitional to collapse dolines and others to subsidence dolines. The following three types are identified.

Dolines of the first type (C_1) result from the caving-in of large-scale and near-surface cavities of the bedrock. C_1 caprock dolines are transitional towards subsidence dolines (dropout doline) if the cover is unconsolidated rock. Caprock dolines transitional towards subsidence dolines have been described from Florida (Davies and LeGrand 1972).

The characteristics of C_1 caprock dolines are the following:

- In top view, they are mostly circular.
- In lateral view, they are bordered by (sub)vertical walls.
- Their dimensions are highly variable. Large dolines are more common.
- The floor is uneven with collapse heaps (fragments of the caved-in ceiling), debris fans and debris slopes (fragments of the walls) and depressions or passages, created by the subsequent movements of the collapsed cover.
- On the sides of two cavities of opposite position, cave opens.
- Such dolines are mostly solitary.
- In the floor under the caved-in non-karstic material, bedrock (limestone, gypsum, rock salt) appears.

Part of the sandstone dolines of South Wales probably belong into this type (Thomas 1954, 1963, 1974; Bull 1977, 1980). Some of the dolines are above pipes or shafts with debris heaps accumulated in allogenic caves under the dolines (Bull 1977, 1980). They are C_1 caprock dolines because a precondition to pipe formation is the existence of an allogenic cave. The dolines under which only chimneys and no allogenic cave developed (Waltham et al. 2005) are C_3 caprock dolines (see below). However, dolines developed by the collapse of the cemented dune sand as described from SE Australia (Grimes 1994; Webb et al. 2010) and by the collapse of the dolomite above gypsum (Lybia) also belong to this type.

The dolines formed on cover above salt in Germany probably also belong to this type (Cramer 1941; Hundt 1950). Dolines of similar origin also occur above horizontal through caves formed in rock salt. The collapses of a cave ceiling in the Atacama Desert (Valle de la Luna) have been inherited over the cover salt breccia. These features, however, are transitional since the salt breccia is not impermeable. Thus, the forms are transitional from caprock to dropout dolines. Similar dolines

Fig. 5.20 (continued) of slope on terrain without karren formation, *8* depth of grikes and solution pipes in bedrock, cm (from margin to floor of the karren in bedrock), *9* direction and angle of slope on the pseudokarren floor, *10* buried (infilled) grikes, *11* suffosion pseudokarren grike, *12* pseudokarren cave, *13* dropout pseudokarren grike, *14* solution pipe in bedrock, *15* ruined solution pipe in bedrock, *16* pseudopipe, *17* deposit bridge, *18* collapse heap formed on the gypsum karst of Ukraine (Klimchouk and Andrejchuk 1996, 2002) as well as at Kungur caves in the Urals (Gorbunova 1979; Andrejchuk and Klimchouk 2002). At these places, the chimneys under the dolines formed on the ceilings of horizontal caves.

The second type (C_2) of inherited dolines (Fig. 5.21) is associated with large-scale pipes developed over large, deep-seated cavities or caves (Fig. 5.22, Waltham et al. 2005). Pipe development begins at the large cavities if there is water circulation between the evaporite and its bedrock (Klimchouk 2004). Pipes grow upwards, towards the ground surface. The resulting debris fills the pipe as well as the deep-seated cavity, where collapse heaps also accumulate, and generates breccia pipes. The pipes can be filled with the collapsed material of the rock crossed (Klimchouk and Andrejchuk 1996) and also with the cover of the evaporite (Walters 1977).

Depressions only develop if the breccia pipe is exposed to the ground surface; however, this does not happen at all places (Fig. 5.22). As a consequence of exposure, a depression of steep walls takes shape on the surface (Ford and Williams 2007). The stability of the walls of the resulting depression depends on the strength of the rock over which inheritance happened. Inheritance can happen over both consolidated (even over gypsum) and unconsolidated rock. The doline may further deepen through repeated collapses. If there is no deepening, the lower part or bottom of the doline fills up.

Some caprock dolines, such as Berezniki (Ural Mountains), are 150 m deep (Andrejchuk 2002). The diameter of the surface depression (doline) depends on the diameter of the breccia pipe, the cover rock type and thickness. The diameter of caprock dolines of consolidated rocks probably does not exceed the diameter of pipes. The diameter of caprock dolines in cohesive rocks is also equal to that of breccia pipes (Klimchouk and Andrejchuk 1996). The boundary between the side slope of the doline and the wall of the breccia pipe cannot be identified. The caprock doline is essentially the near-surface section of the breccia pipe. When developed, the walls of dolines in consolidated and cohesive rocks are vertical. The dolines are widening and, as mentioned, deepening through renewed collapses. The side slopes are getting gentler through mass movements and pluvial erosion. The process leads to doline broadening and infilling. In non-cohesive (or less cohesive) rocks, the diameter of the caprock doline exceeds the diameter of the breccia pipe even when it forms. It is possible since the side slope of the doline is less steep in the first place. Naturally, with the subsequent rapid erosion of side slopes, the dolines are further broadening. The doline can develop in superimposed consolidated and unconsolidated (non-cohesive) rocks. In this case, in the upper, non-cohesive section, the doline slope is gentle, while in the lower consolidated rock, the doline has vertical walls. The diameter of the doline is smaller in the lower part than in the upper (Andrejchuk 2002). Intensive broadening is particularly intensive in non-cohesive rock. Thus, the diameter of the mentioned Berezniki doline grew from 150 to 200 m within some years. On the doline floors, lakes are common. The floors are dissected by minor depressions (Andrejchuk 2002), and some dolines also have outflow channels (Ford and Williams 2007).

Fig. 5.21 C_1- and C_2-type caprock dolines (From Origo 2013)

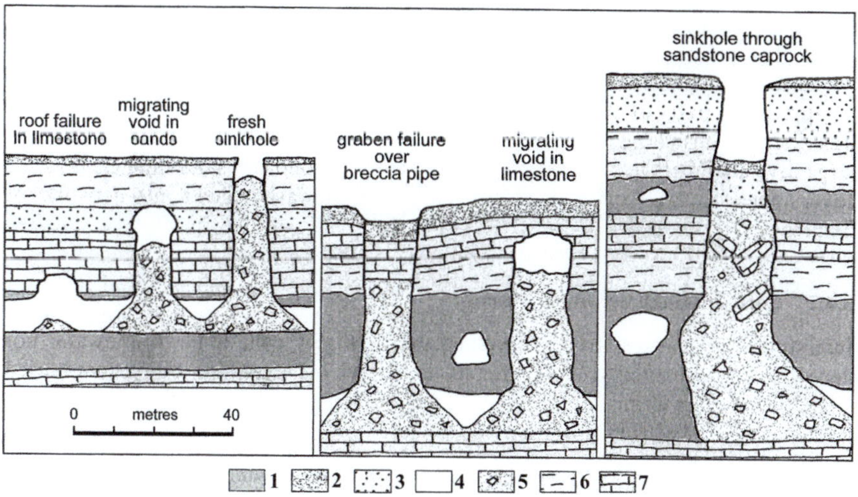

Fig. 5.22 C_2 caprock dolines and breccia pipes at various stages of their development (Waltham et al. 2005): *1* gypsum, *2* soil cover, *3* sand and sandstone, *4* cave, *5* breccia pipe, *6* clay and mudstone, *7* limestone and dolomite

Since the development of pipes presupposes the presence of large cavities (particularly, if the pipe reaches the surface from great depth), they can mostly originate on easily soluble evaporite. Therefore, the pipes and the associated C_2 dolines, as it was mentioned, emerge above gypsum and rock salt. During the formation of a pipe of several hundred metres long, large amounts of collapsed debris accumulate. To accommodate the accumulating debris in the cavities of gypsum and rock salt, cavity formation has to be very rapid in these rocks, which is proved that at Wink Sink (Texas), the breccia pipe which is found there today did not exist in 1928 (its development was triggered by water flow in a borehole), but by 1980 reached the surface (Johnson 1987, 1989; Johnson et al. 2003). Breccia pipe formation is favoured by the continuous dissolution of the easily soluble collapse material. There is no large-size cavity at Wink Sink, only a pipe totally filled by collapse material (Johnson 1987; Johnson et al. 2003). Over a mine in the Ural Mountains, even shorter time (merely some months) was necessary for cavity or pipe development and the resulting emergence of a doline (Andrejchuk 2002).

Breccia pipes rise to the surface from great depths. In North America, their lower end may lie as deep as 1200 m (Quinlan et al. 1986). From the Chinese coal mines, breccia pipes from similar depths were described (stoping from 700 m depth, Lu and Cooper 1997, Fig. 5.23). The latter are paleokarst formations above Ordovician gypsum and have diameters ranging from several 10 m to several 100 m.

On rock salt, where numerous breccia pipes are found next to each other, large-size depressions develop above them: solution subsidence troughs and solution-induced depositional basins (Martinez et al. 1998; Klimchouk 2004). Solution subsidence troughs are elongated in ground plan and deeper, while solution-induced depositional basins are less elongated and shallower. The former are filled up to a lesser degree, while the latter to a greater degree (Olive 1957; Klimchouk 2004). Where the salt terminates, the edge of salt retreats and salt slopes are generated (Olive 1957).

Breccia pipe generation can be triggered by mining activities too, such as the origin of doline Wink Sink 1 (Johnson et al. 2003) or a doline near Hutchinson (Kansas) (Walters 1977; Wassmann 1979). In limestone breccia pipes are rare, but occasionally form to the effect of ascending hot water. Breccia pipes in limestone are paleokarst features. Paleokarst breccia pipes are sometimes exposed by caves and cavities formed subsequently. Breccia pipes are described from above the limestone cavities in the wall of the Grand Canyon (Wenrich and Sutphin 1994) or from the Paiute Cave (Hose and Strong 1981).

C_2 caprock dolines were described by Quinlan et al. (1986), Walters (1977) from North America, Klimchouk and Andrejchuk (1996), Gorbunova (1979) from Ukraine, Cooper (1998) from England, Lu and Cooper (1997) from China and Andrejchuk (2002) from Russia.

The third caprock variety (C_3) comes about if there is no large-size cavity in the carbonate bedrock, but only chimney, shaft or their groups. The chimney or shaft develops to the cover rock and is inherited to the consolidated cover. The chimneys or chimney groups do not lead down to larger cavities but are gradually getting narrower, e.g. on Kab Mountain (South Bakony, Hungary).

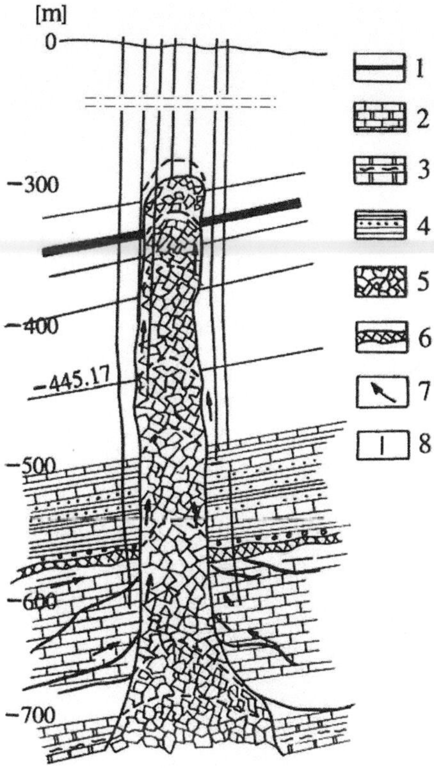

Fig. 5.23 Breccia pipe stoping from gypsum, China (Lu and Cooper 1997-modified): *1* Permian Shanxi Formation, *2* carbonate rocks, *3* carbonate rocks with gypsum, *4* sandstone and shale, *5* collapse column showing various stage of its upward progression, *6* paleokarst horizon between the Ordovician and Carboniferous sequences, *7* direction of water flow, *8* boreholes complied from borehole data of the Bureau of Kailuan Mine

The morphology of dolines of this variety resembles the morphology of solution or suffosion dolines (Fig. 5.24): side slopes are relatively gentle and there is no collapse heap on their floor, but they have a chimney (shaft) below or grike in the bedrock. The latter are bordered by collapse blocks. The morphological explanation of these properties is that in the bedrock (as mentioned) a vertical chimney of small diameter developed. In the lack of a large horizontal cave, the collapse is of limited area (the formation process is presented in detail in Chap. 7). Such dolines form above non-allogenic caves.

Probably dolines of this type occur near the Pecos River (Sweeting 1973). There are lakes in dolines lying near the water level of the river (Sweeting 1973). This circumstance excludes the existence of large horizontal cavities, typical of C_1 inherited dolines. Here dolines are formed in sandstone. This type of inherited dolines occurs on the sandstone of the Central Kentucky Karst (Dicken 1935; Balázs 1970; White et al. 1970) and on the karst of Transvaal (Brink and Partridge 1965).

The caprock dolines of Kab Mountain show a diverse morphology. They originate exclusively in basalt and in basalt and loess, where basalt is mantled by loess. In this case, the interior of the dolines may contain subsidence dolines. The following varieties occur on Kab Mountain:

- The doline is depression of the flat basalt terrain (Fig. 4.60).

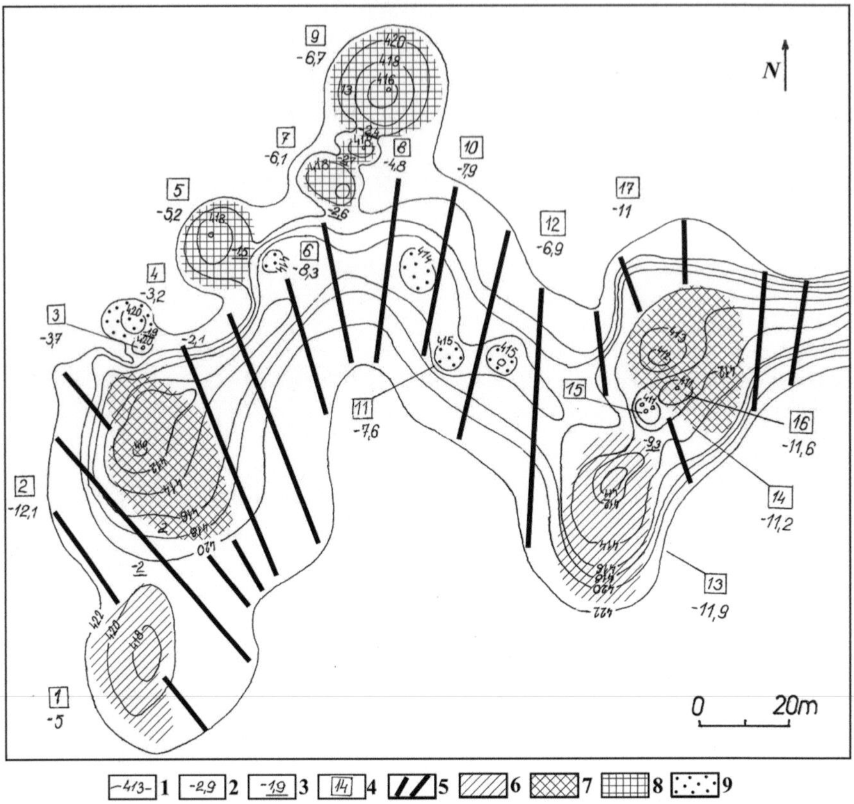

Fig. 5.24 Geomorphological map of the M-2 system (Kab Mountain, Hungary): *1* contour lines, *2* deepest point of a karstic depression (started from 422 m elevation), *3* depth of sill between subdepressions (from 422 m level), *4* code of depression, *5* epigenetic valley, *6* passive ponor (doline), *7* intermittently active ponor, *8* caprock doline, *9* subsidence doline

- Gully, ravine or valley leads to the doline (Fig. 5.25).
- The doline is situated on a valley floor formed on basalt (Figs. 4.59e, f and 5.24).
- Some dolines can be both horizontally and vertically composite (Fig. 5.25).

5.4 Subsidence Dolines

5.4.1 General Description

Subsidence dolines originate on unconsolidated (mostly non-cohesive) cover (Fig. 5.1a–d, Williams 2004). They form not only above limestone but also above gypsum and rock salt if not covered by unconsolidated cover sediment (Martinez et al. 1998; Waltham et al. 2005; Cooper and Waltham 1999). In the present terminology,

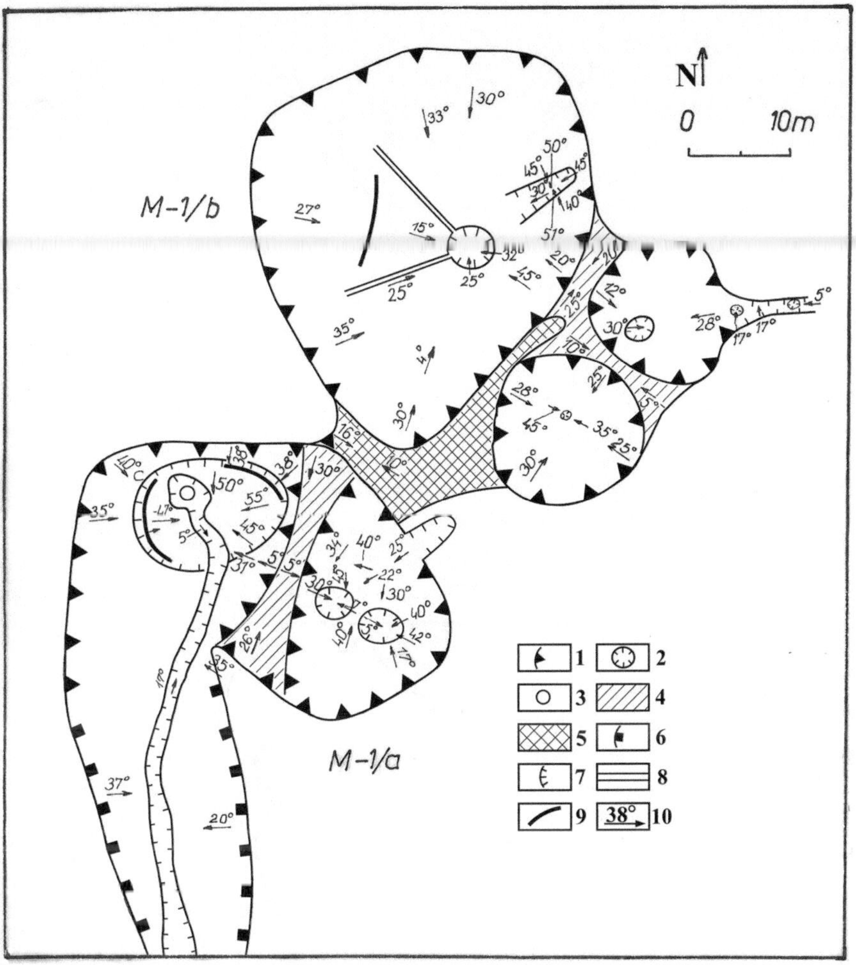

Fig. 5.25 Group of caprock dolines on Kab Mountain (the M-1 system): *1* caprock doline; *2* subsidence doline; *3* passage, cave; *4* ridge between partial dolines; *5* col between dolines; *6* valley; *7* gully; channel; *8* rill; *9* scar of mass movement; *10* angle and direction of slope

the literature identifies two types: dropout dolines and suffosion dolines (Drumm et al. 1990; Tharp 1999; Waltham and Fookes 2003; Williams 2004). They develop where chimneys, shafts and grikes occur in the bedrock (Jammal 1984; Beck 1986; Veress 2000; Waltham et al. 2005). The dropout doline variety forms when a cavity in the cover caves in, particularly if the cover is cohesive rock (Waltham et al. 2005) or the lower part of the cover is in motion by solifluction (Drumm et al. 1990; Williams 2004), while suffosion dolines develop on non-cohesive rock by suffosion (Williams 2004; Waltham et al. 2005).

The name subsidence doline was introduced by Jennings (1985), and other denominations for this type are also in use. White (1988) calls the dropout doline cover-collapse doline and the suffosion doline cover-subsidence doline. In literature various dolines are often named sinkholes. Beck and Sinclair (1986) mention the above landforms under the names of cover-collapse sinkhole and cover-subsidence sinkhole, resp. Other authors do not distinguish between the two types of subsidence dolines. Ford and Williams (1989) use the name of suffosion doline, Culshaw and Waltham (1987) subsidence and Veress (2000) doline with ponor for all subsidence dolines. Cvijič (1893), Sweeting (1973) and Bögli (1980) call the subsidence dolines alluvial dolines. Cvijič (1893) describes this doline type as funnel shaped. Sweeting (1973) emphasised that on its slopes, slides are common and their floor is often waterlogged or swampy. According to Sweeting (1973), the older the dolines of this type, the bigger they are. Sweeting (1973) mentions the suffosion dolines of morainic terrains as shake holes, a term introduced by Lucas (1872), who defined them as depressions generated by collapse.

Subsidence dolines are widespread in the karst regions of the Earth, not only on concealed karsts but also in the large depressions of bare karsts (or bare karst zones) (dolines, uvalas and poljes). A complete overview of their occurrence is impossible. Only some of them are cited: Yorkshire Dales Karst; Pennines; Pennsylvania (Waltham et al. 2005); Ingleborough (Sweeting 1973); Georgia (Hyatt et al. 1999); Ebro Basin (Soriano and Simón 2001); Central Kentucky Karst (White et al. 1970); Zambia (Brook 2004; Spooner 1971); South England (Sperling et al. 1977); Florida (Beck and Sinclair 1986); tundra karsts in Siberia (Gvozdetskiy 1981; Korzhuev 1961); evaporite karsts, e.g. Delaware Basin (Martinez et al. 1998); and Ukraine (Klimchouk and Andrejchuk 2002). Subsidence dolines of other regions investigated by the present author can be also mentioned: part of the Alps, Dinaric Mountains, the Pádis, Bakony, Mecsek, Crimean peninsula, etc. The number of subsidence dolines on individual covered karsts may be several thousands.

Subsidence dolines differ from solution dolines not only in their occurrence partly or totally in the cover but also in water inflow of the larger specimen from their environs conveyed into the karst. They, however, cannot be regarded ponors. It is not because they do not form on true rock boundary but also because they do not have a separate catchment bordered by divide. (Such a terrain is called the hinterland of the doline.)

5.4.2 Morphology of Subsidence Dolines

Subsidence dolines (primarily suffosion dolines) develop partly in the bedrock (Fig. 5.26a) or exclusively in the cover (Fig. 5.26b–d). If the doline floor is bedrock (Fig. 5.26b, c), the doline only has a chimney or a shaft and has no non-karstic pipe (formed in the cover). A similar situation arises if the doline is only partly developed in the cover (Fig. 5.26a). If the doline floor is also in the cover or the doline with limestone floor is partially infilled subsequently, there is a non-karstic pipe in the doline (Fig. 5.26d, e).

The morphology of subsidence dolines is simple (Fig. 5.27) and composite (Fig. 5.28). While the dolines of simple morphology have a non-karstic pipe and/or

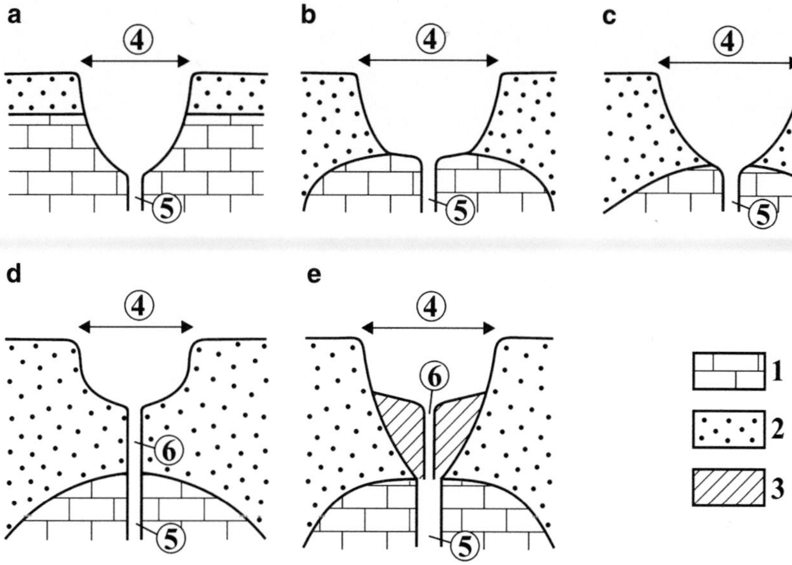

Fig. 5.26 Doline partly in the cover, partly in the bedrock (**a**), doline developed in the cover, but its floor is bedrock (**b**), only conduit of doline in the bedrock (**c**), conduit of doline partly in the cover (**d**, **e**). *1* limestone, *2* cover, *3* doline fill, *4* doline, *5* chimney, shaft, *6* non-karstic pipe

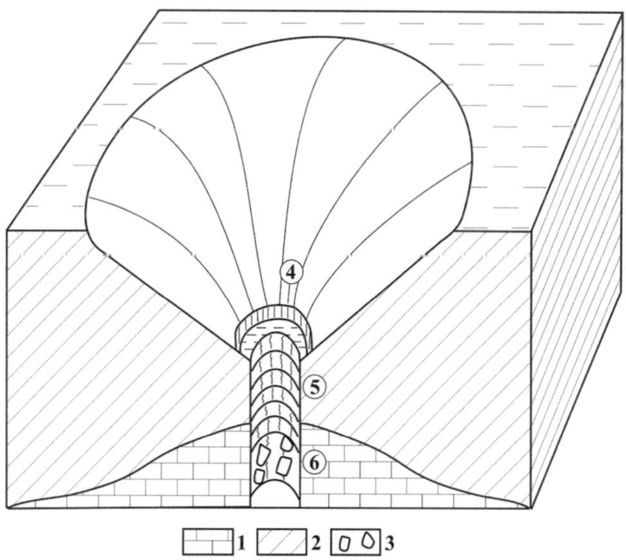

Fig. 5.27 Elements of a simple suffosion doline: *1* limestone, *2* cover sediment, *3* collapsed material, *4* interior subsidence doline, *5* passage, *6* chimney, shaft

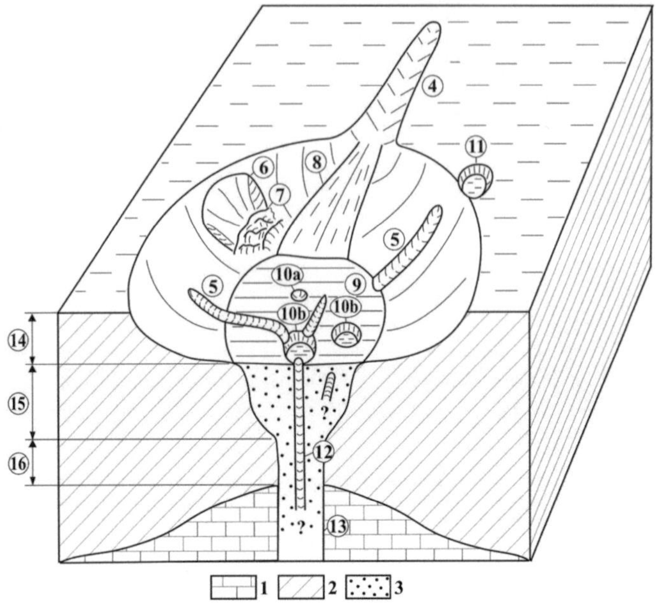

Fig. 5.28 Elements and morphology of a composite suffosion doline: *1* limestone, *2* cover sediment, *3* doline fill, *4* ravine, *5* gully, *6* slump, *7* slump heap, *8* alluvial fan, *9* flat floor, *10.a* interior suffosion doline, *10.b* interior dropout doline, *11* subsidence doline in marginal position, *12* passage in fill, *13* chimney, shaft, *14* upper part of doline without fill: with the erosion of slopes the doline was broadening and its slopes were getting gentler, *15* infilled lower part of the doline, *16* infilled non-karstic pipes in the doline

chimney, the interior of dolines of composite morphology is dissected by various smaller features of karstic origin or associated with them (non-karstic pipe, chimney, interior doline) or of non-karstic origin. Features of non-karstic origin can originate from destructive processes such as scars of mass movements and erosional landforms. Landforms of material transport and accumulation are collapse heaps, debris fans and alluvial fans.

In literature (Waltham et al. 2005), the water conduits of subsidence dolines on the surface are called shallow holes. Shallow holes are either open or clogged, but, in spite of being infilled, can conduct water. They are continued in non-karstic pipes (Figs. 5.28 and 5.29), which occasionally broaden towards the bedrock. Their widths are several centimetres or tens of centimetres. If broader, they are transitional to interior dropout dolines and originate from collapse (Fig. 5.29a). If narrower, they are generated by rainwater seepage (Fig. 5.29b). Non-karstic pipes are associated with the shafts in the bedrock and mostly occur above them. If this is not the case, non-karstic pipes are partially infilled subsequently or, as getting longer, have not yet cut across the cover. They are vertical, oblique or of variable direction. Non-karstic pipes occur on the floor of subsidence dolines as solitary features (Fig. 5.30a) or in groups. There are probably two ways of developing in groups: from the main

Fig. 5.29 *Left* (**a**) non-karstic pipe formed through collapse in the sediment cover above a katavothron (in the polje next to road number 513, Croatia), *right* (**b**) passage some centimetres across formed through infiltration (Tauplitz alm, Totes Gebirge). *1* embryonic suffosion doline, *2* non-karstic pipe

non-karstic pipe, secondary non-karstic pipes come about (Fig. 5.30b–d), or several chimneys form in the bedrock with individual non-karstic pipes above them (Fig. 5.30e, f). The combination of the two patterns is also possible (Fig. 5.30g). The density of (incipient) non-karstic pipes is sometimes very high in the environs of katavothra (4–5 of them in 1 m^2). This is only possible if the non-karstic pipes of the cover constitute an intricate (branching) network. Non-karstic pipes can change their direction in the cover. In this case or if the non-karstic pipe is long because the cover is thick, non-karstic pipe is getting clogged more rapidly. Non-karstic pipes can be infilled or buried under mass movement heaps. The clogged or buried non-karstic pipes often reopen.

On the floor of suffosion dolines (particularly if it is filled to a plain surface), interior subsidence dolines, most often dropout dolines, appear. In the sides of interior dropout dolines, collapses are common. They can be partially filled and their floor becomes plain. Neighbouring dropout dolines may merge with each other. Interior dropout dolines are of small dimensions (some tens of centimetres deep and 1–2 m in diameter). There are several specimens of them in a single doline. The transition between non-karstic pipes and dropout dolines is gradual. Dropout dolines can also appear on the side slopes or edges of dolines. Those in marginal positions broaden through repeated collapses and merge with the main doline. In the dolines, non-karstic pipes and chimneys (shafts) occasionally occur side by side. The possible varieties of such landforms are presented in Fig. 5.31.

Mass movements are common in all covered karst landforms, including subsidence dolines. Among them, collapses directly contribute to dropout doline formation. After the doline has formed, according to the site of mass movement, the process can be external or internal. In the former case, the material of the movement arrived into the doline from the outside buries both its floor and side slopes. Internal mass movements form on the side slopes of dolines. Then the cover rocks of the side slope or its gully are displaced. The mass movements of covered karst depressions are of small dimension. The scar of mass movement formed on the side slope and is a step or spoon-like depression of arcuate or, less often, of straight edge. The

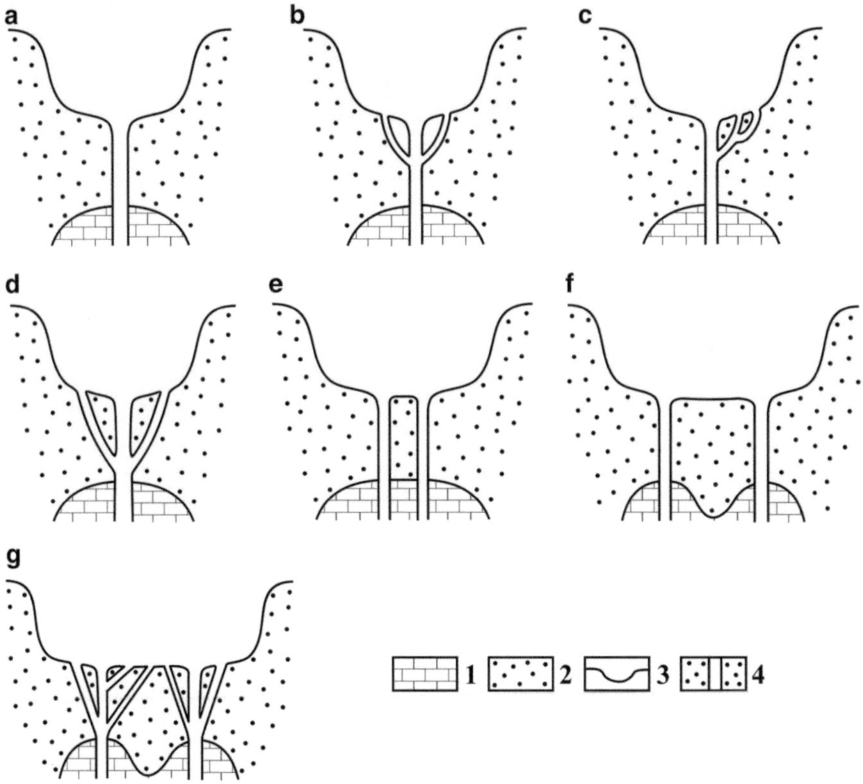

Fig. 5.30 Passage patterns in cover: *1* limestone, *2* cover, *3* doline, *4* passage, (**a**) solitary non-karstic pipe, (**b–d**) branching non-karstic pipe, (**e**, **f**) parallel non-karstic pipes, (**g**) non-karstic pipes with multiple branches

displaced material comes to rest on the doline floor and mostly forms an uneven surface with mounds there.

In a depression, mass movement occurs if:

- Slope inclination increases.
- The material of the floor is transported into the karst (increasing doline depth or reducing the support of sediments on side slopes).
- With the karstic subsidence of the floor, the depression is getting deeper.
- Pressure on the upper slope section increases (due to the accumulation of sediment, snow, ice, human structures or waste).
- The cover sediment receives much water (from rainwater runoff on slope, from the watercourse arriving at the margin of the depression or meltwater of a remnant snow patch).
- The cover sediment receives water from the intermittent lake in the depression or exerts a sucking effect on the deposits of the side slope.

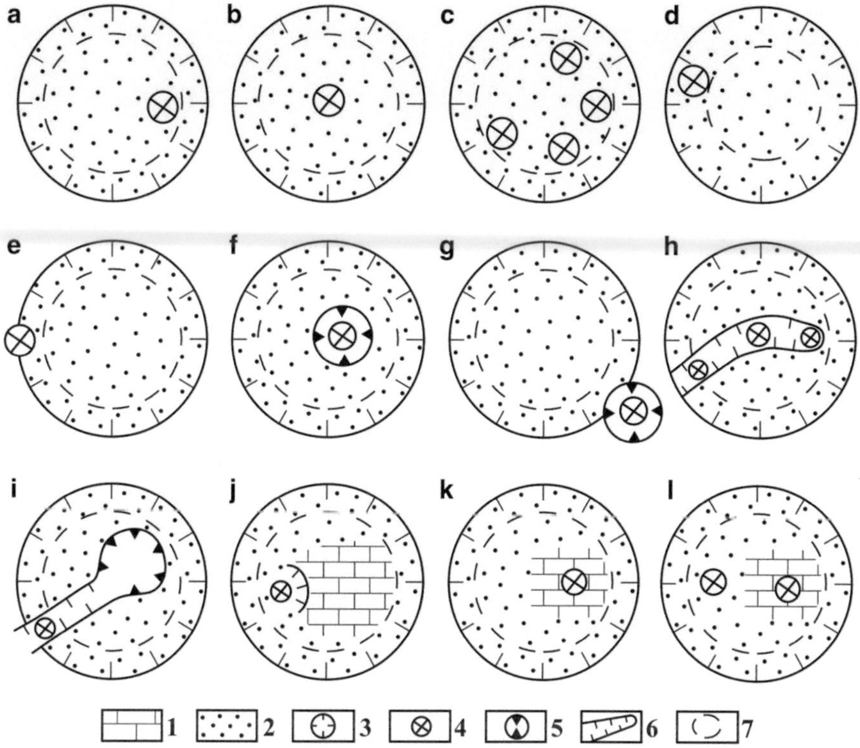

Fig. 5.31 Non-karstic pipes and shafts of dolines: *1* limestone; *2* cover sediment; *3* doline; *4* non-karstic pipe, chimney (shaft); *5* interior or marginal subsidence dolines; *6* gully, ravine; *7* edge of almost flat doline floor; non-karstic pipe on the edge of floor (**a**), non-karstic pipe in the interior of floor (**b**), non-karstic pipes arranged in wreathe form on the floor (**c**), non-karstic pipe on the side slope of the doline (**d**), non-karstic pipe on the margin of the doline (**e**), non-karstic pipe in the interior subsidence doline (**f**), non-karstic pipe in the marginal subsidence doline (**g**), non-karstic pipes in the gully of the doline (**h, i**), shaft of ponor on the margin of the limestone outcrop of the floor (**j**), shaft in the limestone floor (**k**), non-karstic pipe and shaft on the doline floor (**l**)

- The vegetation partly or totally dies.
- In the slope deposits freeze and thaw or drying and wetting alternates.

The mass movements of karstic depressions are one time or repeated, continuous or periodical. To the effect of mass movements, the depression is getting wider, the side slopes are getting gentler, and depth decreases.

Karstification and mass movements can strengthen each other's influence. Thus, karstic deepening increases slope inclination, favouring mass movement, while mass movements lead to the thinning of cover sediment, which may increase the intensity of karstification. Interior dolines occasionally form at the mass movement scar. The material derived from the mass movement modifies the development of the depression and often contributes to its getting inactive. The generated material accumulates on the floor and buries the non-karstic pipes of the depression.

Fig. 5.32 Local collapse of the marginal dropout doline of depression D-5 (Tés Plateau, Bakony, Hungary): *1* partial depression formed before 2010, *2* doline detail formed after July 2010, *3* wall of the landform due to collapse, *4* vegetation displaced to lower position during collapse, *5* collapse blocks of the cover

The mass movements of the covered karst depression include collapse, slide, solifluction, gelisolifluction and soil and regolith creep.

Collapses have several varieties. According to size, the collapse can be local (Fig. 5.32) or extending over the whole depression. In the former case, separate blocks are displaced on sections of various length of the doline side slope and become tilted or rotated. In the case of extended collapse, the whole area of the depression is affected by the collapse. For this variety, it is common that the collapsed cover moves in one mass but gets divided into blocks during the collapse (Fig. 5.33). In the above cases, the collapses also extend to the subsoil cover, but more frequently only affect the soil. Along the minor depressions or non-karstic pipes, the sediment is divided into blocks while it moves downslope. This is the continuation of the collapse that triggered landform development. This involves patterns similar to desiccation cracks (Fig. 5.29a). The blocks get into lower and lower position towards the non-karstic pipes.

On the slopes of depressions, slides are the most common processes (Fig. 5.34), which take place only on soils or also on cover sediment. In the latter case, the cover of smaller thickness is displaced as a layer slump. There is no sharp boundary

Fig. 5.33 Collapse extending over the whole doline (Pádis plateau, Romania)

Fig. 5.34 Slump scar (on the slope of the ravine of depression D-5 on the Tés Plateau)

Fig. 5.35 Rotational landslide on the slide slope of an alluvial streamsink doline (polje next to road number 513, Croatia)

between slides affecting the soil and also the cover sediment. During layer slumps, arcuate or wavy, steep steps several tens of centimetres high come about where the cover fails (Fig. 5.34). As during the process thicker and thicker cover sediment is displaced, the movement increasingly takes the form of block slide. As a result, a rotational landslide is created (Fig. 5.35). There are numerous transitions between slumps and rotational landslides (Fig. 5.36). Slumps take place over layers which are not impermeable. Layer slump can also happen as displacement over the bedrock.

Solifluction movements (Fig. 5.37a) also often take place in the sides of depressions. During solifluction, the sediment does not cover the whole floor of the depression. Gelisolifluctional movements happen on glacio-, tundra- and taiga-covered karsts. The surface of the displaced material becomes uneven with pillow-like bulges or isolated soil patches of various sizes (Fig. 5.37b).

Soil and regolith creep are less influential in shaping the slopes of depressions. Data on such mass movements are presented in Table 5.2 for the dolines of the Hárskút Basin (Bakony Mountains, Hungary). The bedrock is Cretaceous limestone and the cover is leached loess and soil of several tens of centimetres thickness. Material displacement was measured on individual side slopes using the method described in Chap. 3. Maximum displacement was 0.5–1 cm per year and probably was not uniform. The data from the rows of rods indicate the speed of displacements

Fig. 5.36 Slumped cover block in the side of doline I in the Pádis-2 area

Fig. 5.37 *Left* (**a**) solifluction on the side slope of a ponor on the Pádis plateau, *right* (**b**) gelisolifluction in the side of a subsidence doline along the Lena River (Solomonov et al. 2010) (soil is divided into patches by the movement of slope deposits of the doline over permafrost)

varying with sites. Maximum values were observed in the central part of the doline side slope selected for measurement.

For mass movements of this type, no scar is created on the slopes of the depression, but accumulation on the floor of the depression can be remarkable. Creep may bury the passages of the depression and modify the shape of interior dolines.

Mass movement may lead to the destruction of vegetation or its adjustment to the movements (see Chap. 6), where tree trunks bend (creep) or fall down (collapse).

Table 5.2 Mass movements in the cover sediment of the side slopes of the subsidence dolines G-5/a and G-9 (Hárskút Basin, Bakony Mountains) (displacements of rods, cm)

Subsidence doline G-5/a			
	Date of installation	Date of repeated measurement	
Rod identification number	24.08.1985	11.05.1990	28. 05.1995
1	0.0	0.0	+1.8
2	0.0	+1.4	+4.2
3	0.0	+1.3	+6.1
4	0.0	+2.7	+5.4
5	0.0	+1.4	+2.8
6	0.0	+3.6	+3.6
7	0.0	+1.2	+2.6
Subsidence doline G-9			
	Date of installation	Date of repeated measurement	
Rod identification number	02.05.1987	10.05.1991	28.05.1995
1	0.0	+0.8	+0.6
2	0.0	+2.0	+1.3
3	0.0	+0.2	+1.6
4	0.0	+1.5	+3.2
5	0.0	+0.4	+0.6

Erosion features in dolines are rainwater rills, gullies and ravines. Gullies and ravines begin to grow from the interior depressions of dolines and do not only occur on the doline floors but also extend to the side slopes through retreating erosion (Fig. 5.38). Such features also form on the neighbouring terrains or are associated with the doline margin (Fig. 5.38g, h) or from the interior of the doline extend over the margin (Fig. 5.38c, d). If the doline fills up, an outflow gully can emerge (Fig. 5.38i).

The landforms and dimensions of subsidence dolines are largely controlled by the fact whether they have a gully or ravine leading to them. (Such landforms are typical of suffosion dolines.) If they have it, they are able to grow (deepen) more intensively than in the lack of such a gully. On the other hand, if there is water inflow, the chance for infilling is greater. On the plain surface generated by accumulation, secondary landforms (interior subsidence doline, gullies, non-karstic pipes) develop. The depth and width of dolines are related to each other: the deeper the doline, the wider it is. A 2 m deep doline can reach maximum 8 m diameter (Waltham et al. 2005).

In ground plan, dolines are most often circular (Figs. 5.39 and 5.40Ia), but those of elongated or chasm-like shape also occur (Figs. 5.39, 5.40Ib, c, and 5.41). The side walls of chasm-like dolines are often subvertical, parallel to each other, and

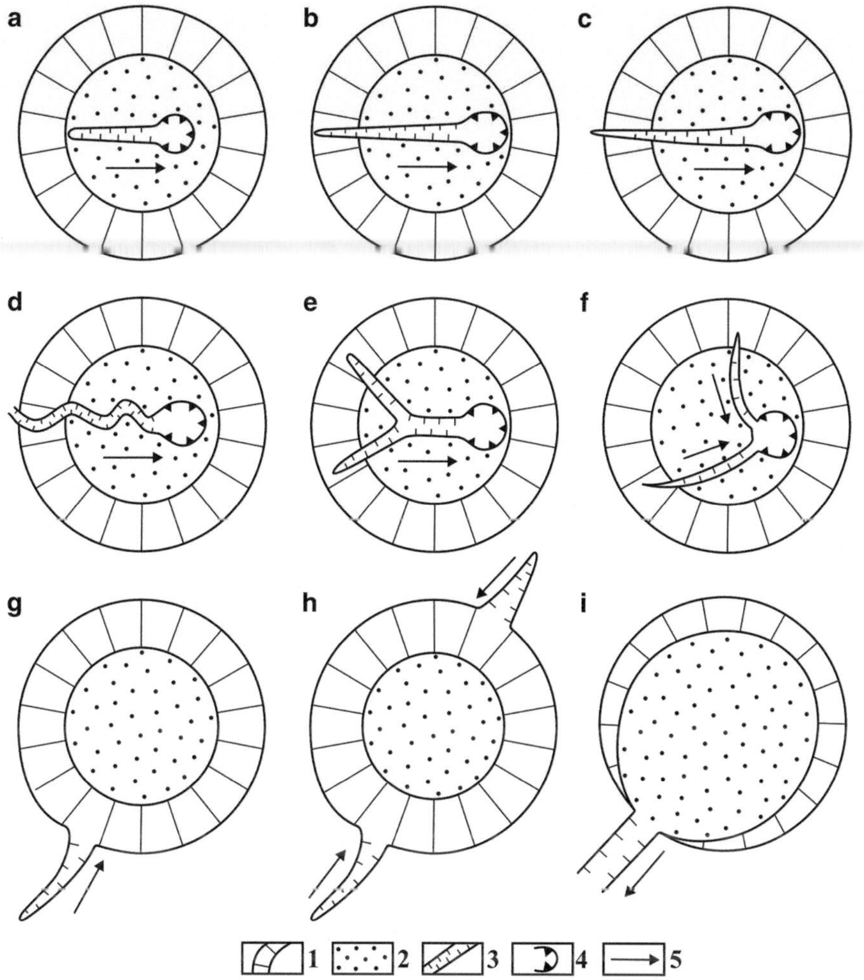

Fig. 5.38 Channel and gully patterns of subsidence dolines: *1* doline side; *2* doline floor; *3* gully, ravine; *4* non-karstic pipe or shaft or interior depression; *5* slope direction. Straight gully on doline floor (**a**), on floor and side slope (**b**), floor, side slope and bordering terrain (**c**), meandering gully (**d**), gully branching fork-like (**e**, **f**), gully, ravine linked to the doline (**g**), several gullies, ravines linked to the doline (**h**), outflow gully draining the infilled doline (**i**)

their lower sections are constituted by the bedrock (dropout dolines). The elongated dolines (as well as some of the chasm-like dolines) may be of suffosion type with gentle side slopes. (The elongated or chasm-like dolines usually form rows.) Less commonly they are of sinuate (Fig. 5.40Ie), narrowing (Fig. 5.40Id) and irregular shape (Fig. 5.40If). Dropout dolines, although mostly circular, include rectangular

(Figs. 4.43 and 5.40IIa) or triangular specimen (Fig. 5.40IIb). Composite, uvala-like dolines are also common (Figs. 5.39, 5.40III and 5.42). In this case two neighbouring (partial) dolines in close position form a single system with only a narrow dividing wall between them. Since partial dolines are mostly of small size, it is not probable that merging happened through doline growth as for the solution uvalas on uncovered karsts. Although according to Brook and Allison (1986), uvalas can be created from subsidence dolines – similarly to uncovered karst uvalas. In our opinion, the reason for their merging is that they developed very close to each other. The composite dolines are suffosion dolines, possibly twinned (Figs. 5.39 and 5.40IIIa) or triply composite (Figs. 5.39, 5.40IIIb and 5.42). Vertically composite dolines also occur. Elongated or chasm-like dolines are found on the floor of circular (Fig. 5.40IIIe, f) or elongated or occasionally chasm-like (Fig. 5.40IIId) main dolines.

By the elevation of dividing walls (thresholds), there are two varieties of uvala-like suffosion dolines identified. The elevation of the threshold level equals the margin of the doline system. Then the doline, as it has been mentioned, is divided into partial dolines (Figs. 5.40IIIa, b). Threshold elevation can be lower than the doline margin. In this case, the partial dolines are aligned in a larger depression. This way multiple composite uvala-like dolines (Fig. 5.40IIIg) are generated.

The subsidence dolines are bordered by straight (Figs. 5.43f and 5.44), normal (Fig. 5.43a–c), convex (Figs. 5.43d and 5.45) and concave (Figs. 5.43e and 5.46) slopes. For the shaping of slopes and floor, several factors, strengthening or weakening each others' influence, are responsible. As it has been mentioned above, the slopes have different inclinations, lengths and morphologies in the same doline. Slope length is controlled by deepening/accumulation, while slope shape and inclination by deepening and the erosion or reworking of slope sediments.

In general, relying on general slope evolution, it is claimed that convex slopes border doline deepening by suffosion. If the doline has a flat floor and moderate depth, it is at an initial stage of its evolution. The doline is similarly young if its side slope is convex and its floor is not flat (Fig. 5.43d_1–d_2). For such dolines, the rate of deepening still surpasses the rate of denudation of the side slope. Deep dolines with flat floor and convex slopes are rare. Such dolines probably develop when, although the doline floor receives abundant sediment, it is capable to transport it into the karst. The flat floor is not necessarily due to the accumulation of sediment which eroded from the slopes on the floor. A flat floor also develops if subsidence of suffosion origin is widespread on the doline floor or if there is an abundant influx of sediment from the hinterland of the doline. A reduction in the sediment transport of passages can also contribute to its development.

For dolines with normal slopes, the infilling of the floor has already started (Fig. 5.44a–c). In dolines with concave slopes (Fig. 5.44e), there is denudation on the

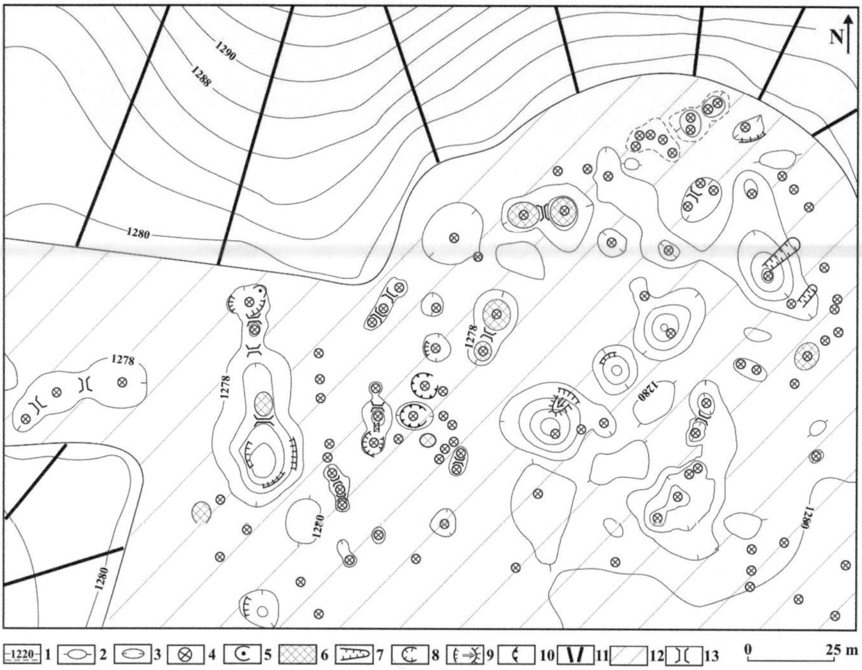

Fig. 5.39 Detail of karst morphological map of NW of Uvala Râchite (Pádis-1, Pádis plateau) (the boundary of major dolines is drawn along the first closed contour): *1* base and supplementary contour lines, *2* elevation, *3* subsidence doline, *4* minor subsidence doline or interior doline (which cannot be represented by contour lines), *5* non-closed subsidence doline, *6* flat-floored doline, *7* gully, *8* soil slide, *9* slump, *10* collapse, *11* side slope of DSD on limestone, *12* covered karst (concealed karst) floor of DSD, *13* col, dividing wall

upper slope section too. With the accompanying infilling, the slope becomes shorter. The floor of dolines with concave slopes is mostly flat. Straight slopes emerge when dropout dolines develop or mass movements (collapses, slides) take place. Therefore, straight zones on slopes occasionally also develop in suffosion dolines (Fig. 5.43f).

Composite slopes result from interior doline formation on the flat floor (Fig. 5.43e$_4$). Similar slopes come about if the cover is thin and the layer heads or blocks of the bedrock generate zones of different inclination on the slope (Fig. 5.43g).

In asymmetric dolines, opposite slopes are of different inclination (see below). With the infilling of the floor, asymmetric dolines of flat floor develop (Fig. 5.43h$_2$). The infilled, fossilised dolines are shallow and flat or the floor is of gentle inclination (Fig. 5.43i).

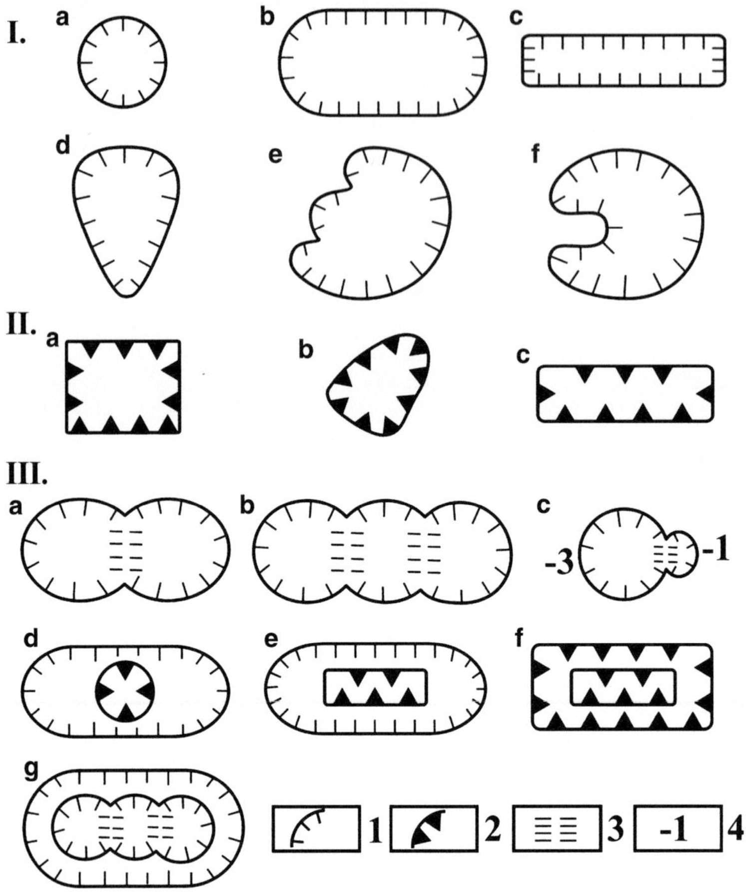

Fig. 5.40 Patterns of doline ground plans. *I* mainly suffosion dolines: dolines of circular (**a**), elongated (**b**), crevice-like (**c**), narrowing (**d**), sinuate (**e**), irregular (**f**), ground plan. *II* dropout dolines: *square* (**a**), *triangular* (**b**), crevice-like (**c**), doline ground plans. *III* composite (uvala-like) dolines: twinned (**a**), triple composite (**b**), marginal hanging doline (**c**), doline elongated on *top* and circular at *bottom* (**d**), elongated on *top* and crevice-like at *bottom* (**e**), crevice-like both on *top* and at *bottom* (**f**), elongated on *top* and composite at *bottom* (**g**), *1* doline margin (mostly suffosion doline), *2* margin of mostly dropout doline, *3* dividing wall between partial dolines, *4* doline depth

As it has been mentioned, the dropout dolines differ from suffosion dolines in respect of slope angle and cross-section. The side slopes of dropout dolines are vertical (Fig. 5.47a, b) or occasionally overhanging (Fig. 5.47c). The vertical slope is of composite shape if the dropout process takes place periodically (Fig. 5.47d). The floor of dropout dolines is often flat and shows collapse heaps (Fig. 5.47e).

The inclination of the doline slope also depends on the lithology of the slope: on sand a slope of 35° is typical (Waltham et al. 2005). If the dolines are predominantly formed in debris of some centimetres diameter, slope inclination is low (Fig. 5.48Ia),

Fig. 5.41 Crevice-like subsidence dolines (A, B, Pádis plateau)

Fig. 5.42 Uvala-like subsidence dolines (Uvala Râchite)

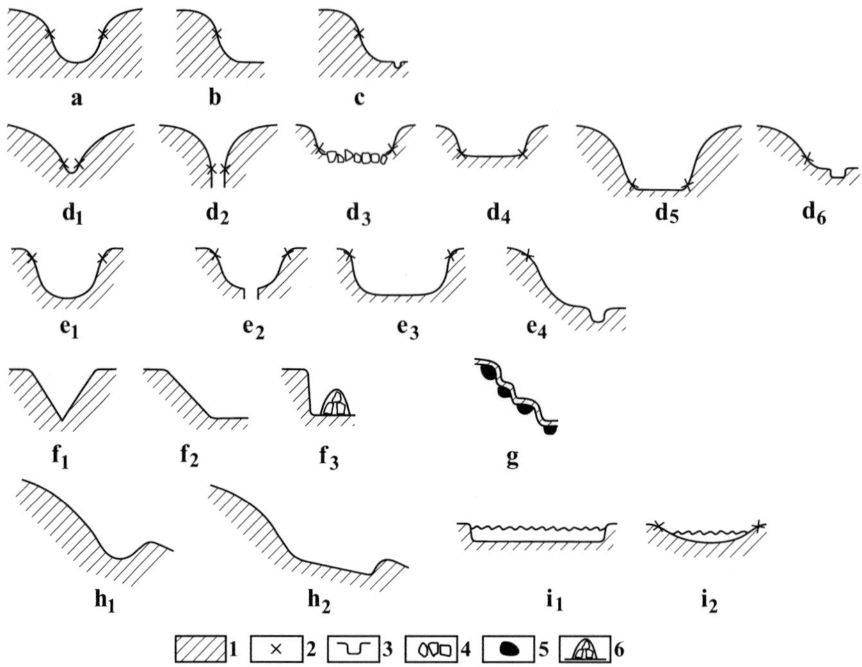

Fig. 5.43 Cross-sections and slope shapes of subsidence dolines: (**a**) doline with normal slopes and narrow floor; (**b**) doline with normal slopes and flat floor; (**c**) doline with normal slopes and flat floor with interior depression; (**d_1**) doline with convex slopes and narrow floor; (**d_2**) doline continued in passage with convex slopes; (**d_3**) doline with convex slopes and flat floor with rock debris and blocks; (**d_4**) doline with convex slopes, small depth and flat floor; (**d_5**) doline with convex slopes, larger depth and flat floor; (**d_6**) doline with convex slopes, flat floor and interior depression; (**e_1**) doline with concave slopes; (**e_2**) doline with concave slopes, continued in a passage; (**e_3**) doline with concave slopes and flat floor; (**e_4**) doline with concave slopes, flat floor and interior depression; (**f_1**) doline with straight slopes and narrow floor; (**f_2**) doline with gentle straight slopes and flat floor; (**f_3**) doline with steep straight slopes and collapse heaps on its floor; (**g**) doline with composite slopes; (**h_1**) asymmetric doline; (**h_2**) symmetric doline with flat floor; (**i_1**) fossil, infilled doline with straight slopes and flat floor; (**i_2**) fossil, infilled doline with concave slopes; *1* cover sediment, soil; *2* inflection point; *3* interior doline; *4* rock debris and blocks; *5* rock blocks detached from the bedrock but remained in situ; *6* collapse heap

and if formed in collapsed material, the inclination is higher (Fig. 5.48Ib). The inclination is even higher if the doline is formed in fine sediment, partly or totally cohesive rock (clay) or its structure ensures high strength (Fig. 5.48Ic). Such a rock is loess, the calcareous micropipes of which increase the strength of this sediment.

In the same rock type, slopes of different inclination can arise if the origin or the stage of development differs. On the other hand, slopes on different rocks can have similar inclinations.

Doline asymmetry (Fig. 5.49) has various reasons. If the suffosion doline is found on sloping terrain, the downhill slope is steeper and shorter than the opposite slope (Fig. 5.50a_1). Therefore, major subsidence dolines or dolines on slopes of

Fig. 5.44 Subsidence doline with straight side slopes (Hárskút Basin, Bakony Mountains)

depressions of superficial deposit are always asymmetric. For subsidence dolines on rock salt, the uphill slope is steeper and longer (Fig. 5.50a_2). Sweeting (1973) explains the origin of the gentler side with the downwash of the cover from the upper slope section, which reduces slope inclination of the doline (Fig. 5.50b_1). The doline becomes asymmetric if the cover from the doline slope is removed (Fig. 5.50b_2) or if karstification is more intensive on one side than on the opposite side (Fig. 5.50c). The half-subsidence dolines have an asymmetric cross-section too. In this case, the doline develops at the termination of the cover sediment. The slope developed in the cover sediment is gentle with small inclination, while the opposite that is built up of the bedrock is steep. The half-subsidence dolines mainly develop at cuestas where the steep side slope is built up of the front of the cuesta (Fig. 5.50d). If the doline formed in the fill of solution crevice, its side slope at the termination of the crevice is also steeper than on the opposite side (Fig. 5.50e).

Dolines occur as solitary features or in groups. Solitary dolines lie in the interior of paleodolines (Fig. 5.51a), on its side slope (Fig. 5.51b), on its margin (Fig. 5.51c) or in head valleys (Fig. 5.51d). Suffosion dolines particularly occur in groups. Group arrangement is irregular in cirques (Fig. 4.6), paleodolines, paleouvalas (Fig. 5.51f) and cockpit dolines (Fig. 4.23) and between morainic hills (Fig. 4.70). Dolines are aligned in rows in paleodolines (Figs. 4.79, 4.80a, 4.81 and 5.51e), on valley floors (Figs. 4.34 and 5.51h, i), on valley sides (Fig. 5.51j), above fractures or buried cracks (Figs. 4.68, 5.51k and 5.52), along bedding planes (Fig. 5.51l), in channel (Figs. 4.33, 5.51m_1 and 5.53) and in gullies (Figs. 4.77, 5.54, 5.55 and 5.56). It is possible that following the track of the meandering channel, the alignment of the doline row is not straight (Fig. 5.51m_1).

Fig. 5.45 Suffosion doline with convex side slopes (Hochschwab, Austria)

Fig. 5.46 Doline with concave side slopes (Uvala Râchite)

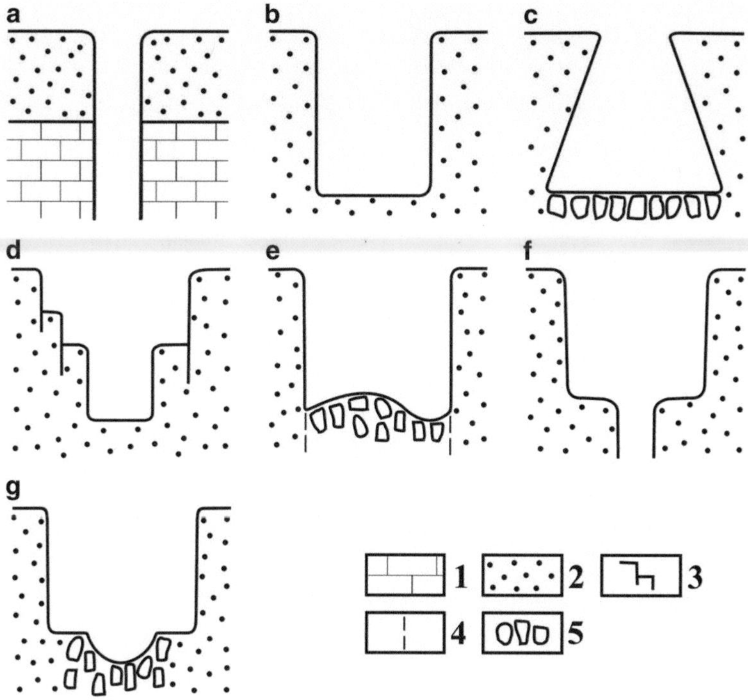

Fig. 5.47 Dropout dolines in cross-section: (**a**) dolines of small diameter, vertical walls, continued in a shaft, (**b**) doline with vertical walls, (**c**) doline with overhanging walls, (**d**) doline with stepped sides, (**e**) doline with collapse heap on its floor, (**f**) doline with a non-karstic pipe on its floor, (**g**) doline with a depression on its floor, *1* limestone, *2* cover, *3* failure front in the cover, *4* buried failure front, *5* collapse blocks of the cover

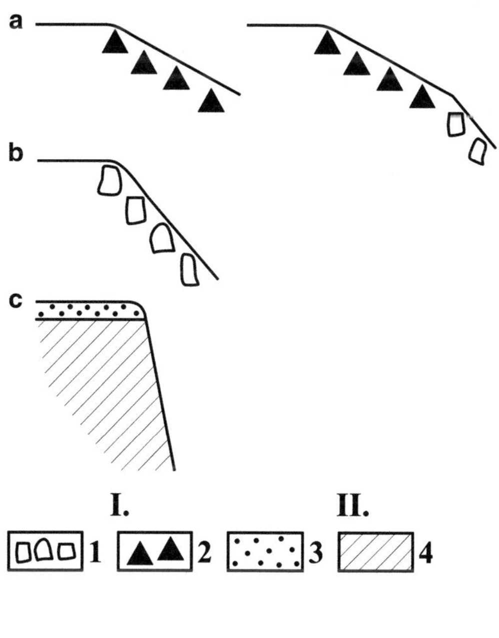

Fig. 5.48 Doline slopes. *I* simple slope: (**a**) gentle slope (below ca 40°) in frost-shattered debris; (**b**) steep slope (ca 40°–60° slope) in moraine, collapsed material; (**c**) steep (subvertical) slope in fine-grained, cohesive deposits (clay). *II* composite slope: *1* collapsed material, moraine, *2* frost-shattered debris, *3* soil, *4* fine-grained sediment (clay)

Fig. 5.49 Asymmetric subsidence doline (Tauplitz alm, Totes Gebirge, Austria)

The dolines transform the morphology of gullies or ravines in which they form in high density. Only 1–2 m long, col-like remnants of the original gully floor are preserved with counterslope sections and local embayments on the margin (Figs. 5.51m_2, 5.54 and 5.55). Gullies develop into blind gullies with dolines on their floor. In other cases, the gully has no outflow in the first place, and the elongated form is either due to covered karst formation processes or incised in this way at some stage of gully development (suffosion gully, Fig. 5.55). The piping origin of gullies is evidenced by the lack of passages in their minor depressions (Fig. 5.55); thus, the cover cannot be transported by runoff on the channel floor. The gullies either have no catchment (Figs. 5.55 and 5.56) or the catchment is too small to accommodate a watercourse. The lack of watercourse is proved by the vegetation on their floor.

5.4.3 Classification of Subsidence Depressions

Subsidence landforms are either produced by karstic processes (subsidence dolines) or by non-karstic processes (subsidence pseudokarst depression).

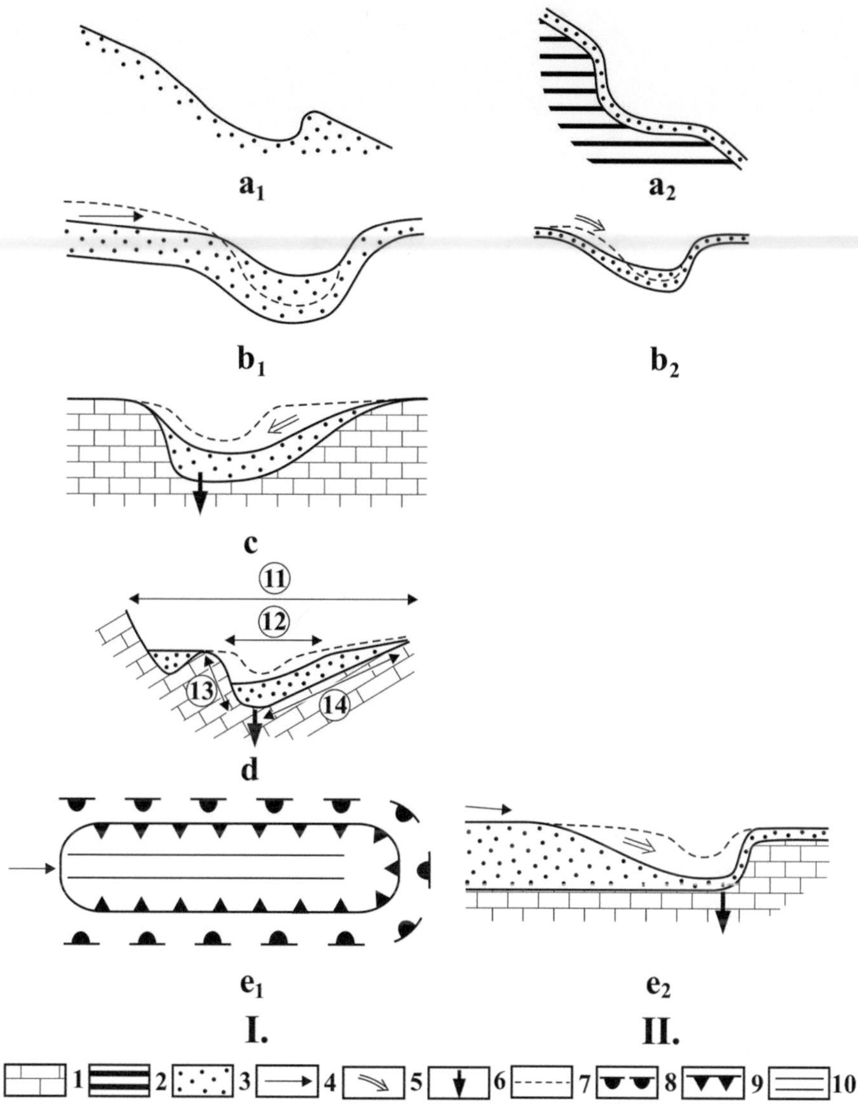

Fig. 5.50 Asymmetric subsidence doline features: (**a**) asymmetric subsidence doline elongated in slope direction: for suffosion doline in limestone, the downslope doline side is steeper and shorter; (**a**$_1$), for dropout doline in rock salt, the upslope side is steeper and shorter (**a**$_2$); (**b**) subsidence doline asymmetric because of reworking of cover, inflow by watercourse outside the doline (**b**$_1$); doline asymmetric because of reworking within the depression (**b**$_2$); (**c**) subsidence doline asymmetric because of more intensive karstification of the bedrock; (**d**) subsidence dolines formed at bedding head; (**e**) asymmetric subsidence doline formed at the termination of solution crevice (I, *top* view, II, lateral view); *1* limestone, *2* rock salt, *3* cover, *4* inward transport of the cover from the hinterland, *5* reworking of cover in the doline, *6* sediment transport into the karst, *7* former ground surface, *8* buried solution crevice, *9* steep side slope of doline, *10* gentle side slope of doline, *11* asymmetric solution doline, *12* asymmetric suffosion doline (half-doline) formed at the escarpments on the floor of solution doline, *13* bedding head, *14* bedding plane

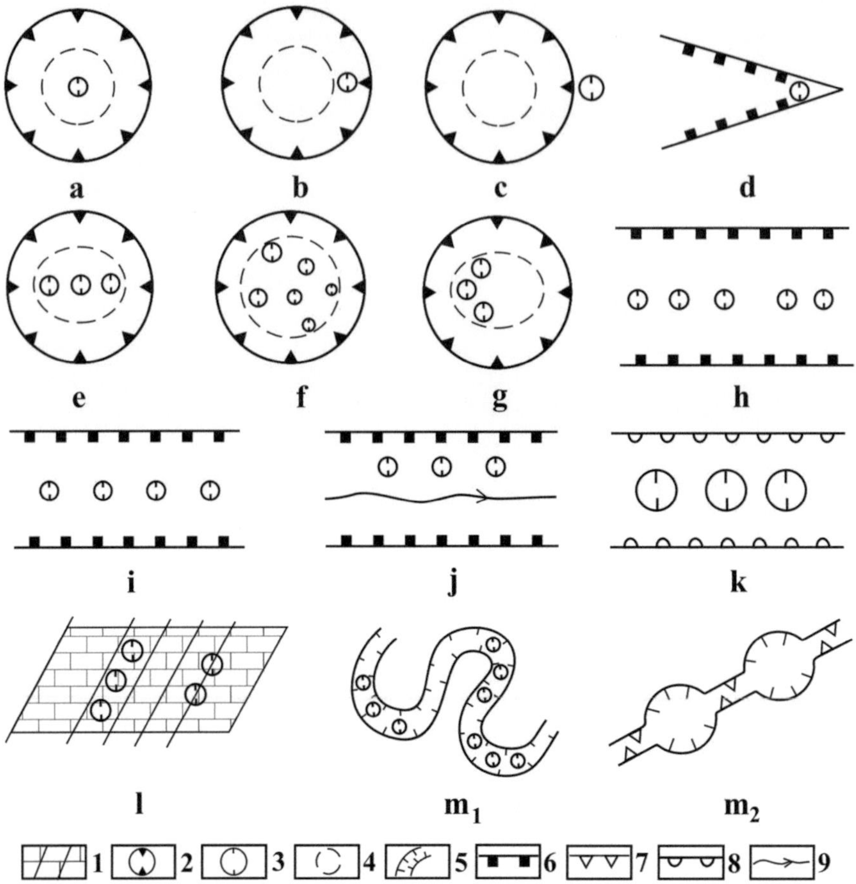

Fig. 5.51 Doline patterns: *1* limestone and beds, *2* solution paleodoline (DSD), *3* subsidence doline, *4* edge of floor of paleodoline (DSD), *5* channel, *6* valley, *7* gully, *8* giant grike (bogaz), *9* watercourse; solitary doline in the centre of paleodoline (**a**), on its side slope (**b**), on its margin (**c**), in headwaters (**d**), doline row in paleodoline (**e**), doline group in paleodoline (**f**), dolines in arcuate, wreathe-like arrangement on the floor of plaeodoline (**g**), on valley floor (**h**, **i**), in valley side (**j**), above crevice (**k**), on bedding boundary (**l**), in channel ($\mathbf{m_1}$), in gully ($\mathbf{m_2}$)

5.4.3.1 Types of Subsidence Dolines

The two types of subsidence dolines, dropout and suffosion dolines, cannot be clearly distinguished. On the one hand, dropout dolines can be transformed into suffosion dolines, when their side slopes are getting gentler. On the other hand, the morphology and development of the dropout doline changes: the small dropout dolines of high mountains begin to develop further through suffosion instead of collapses. The distinction is not sharp since there are transitional forms between the two types (Sperling et al. 1977). Kinship is also indicated by the occurrence of dolines of both type at the same site, next to each other. Dolines of both types appear

Fig. 5.52 Subsidence dolines formed on moraine along fracture (crack) (Durmitor): *1* active subsidence doline, *2* inactive subsidence doline, *3* rills produced by runoff from rainfall or snowmelt, *4* Sušica canyon

Fig. 5.53 Subsidence dolines of a flood channel (1) of Ponor polje (Pádis plateau)

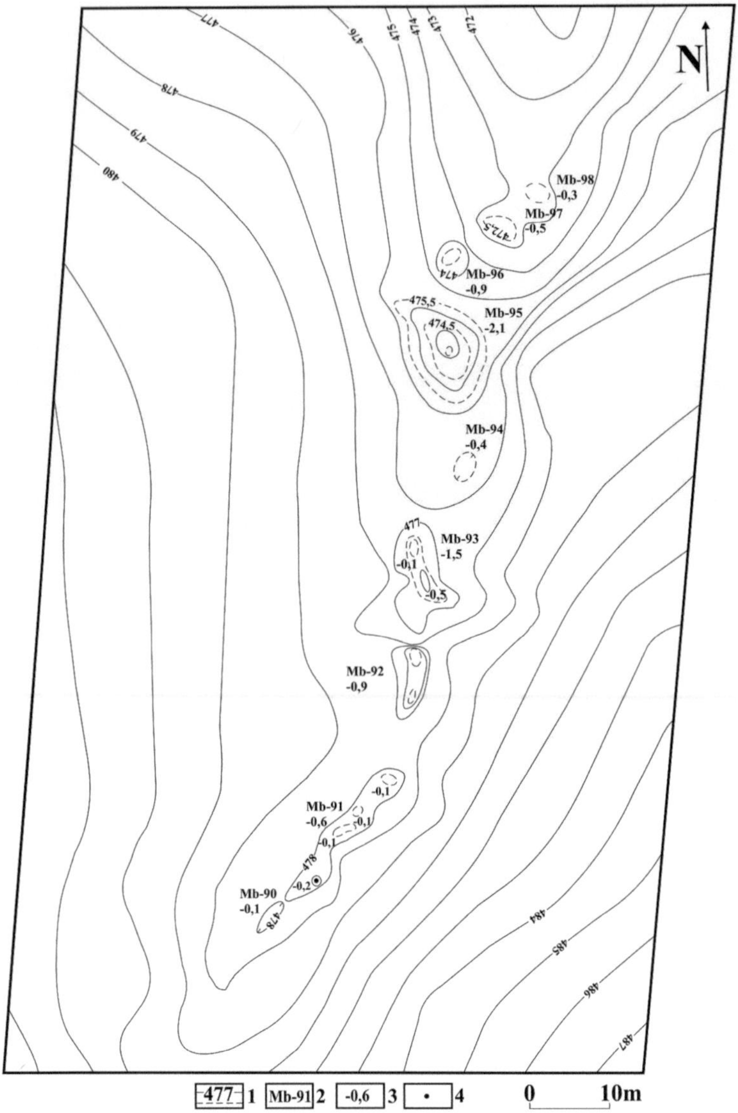

Fig. 5.54 Subsidence dolines developed in a ravine (Mester-Hajag, Bakony): *1* base and supplementary contour lines, *2* identification code of doline, *3* doline depth, *4* passage

mixed even on the gypsum karst of Ukraine (Klimchouk and Andrejchuk 2002), where, because of the intensive solution of gypsum, the formation of dropout dolines is rapid.

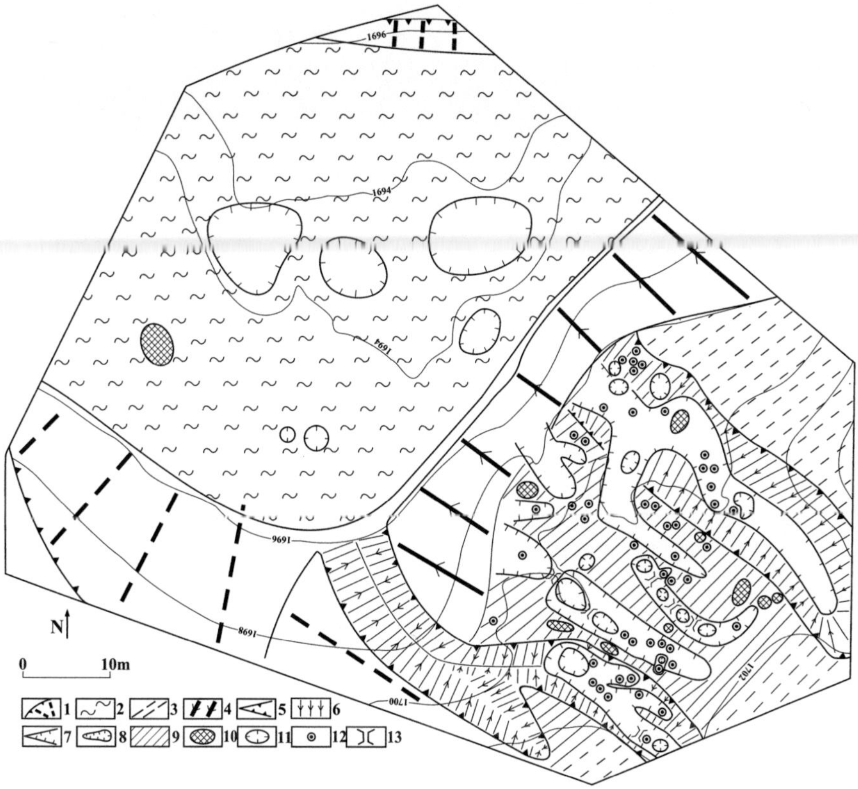

Fig. 5.55 Karstification on alluvial fan (Dachstein): *1* paleodoline and its side slope, *2* floor of paleodoline, *3* surface of alluvial fan, *4* front of alluvial fan, *5* ravine, *6* side slope of ravine, *7* gully, *8* blind suffosion trough, *9* ridge between ravines, *10* elevation, *11* suffosion doline with diameter more than 2 m, *12* suffosion doline with 0.5–2.0 m diameter, *13* col

In literature a third type, the compaction doline, is also distinguished (Waltham and Fookes 2003). This type of doline develops through the compaction of the fills of buried dolines (paleodolines) (Waltham et al. 2005). It cannot be regarded an independent type neither morphologically, genetically nor for its occurrence. Its side slopes equally show collapse (Waltham et al. 2005) and subsidence character and origin (Waltham and Fookes 2003).

A joint property of dropout dolines is their side slopes that are constituted by failure front(s). These surfaces cut across the cover and are without soil and vegetation. On the side walls, collapses follow almost continuously or are renewed regularly and their floor is uneven. The passage and the related gully in them only occur in interior dropout dolines and not in all cases even there.

The cover may be of two kinds: the lower part of the dropout dolines (e.g. Winter Park sinkhole 1) is in clay, while the upper is in sand (Beck and Sinclair 1986). The lower part in clay has vertical walls, while the upper slopes are less steep.

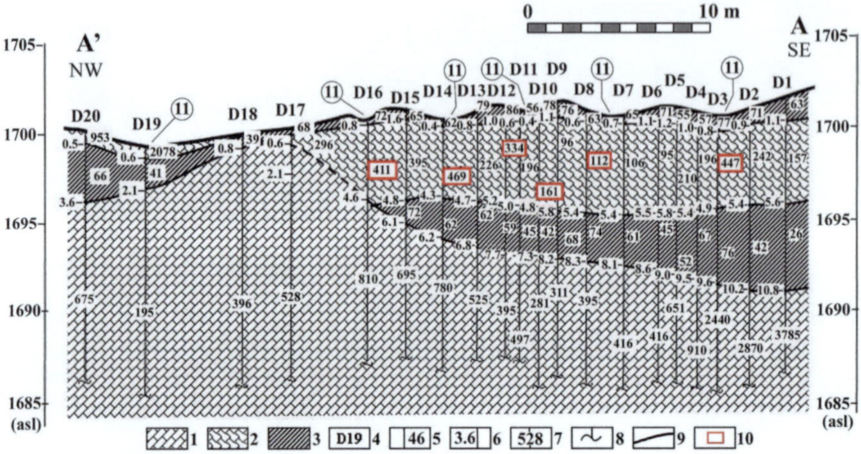

Fig. 5.56 Geoelectric geological profile across the alluvial fan shown in Fig. 5.55: *1* limestone, *2* fractured limestone, *3* clay with limestone debris, *4* identification code of VES measurement, *5* geoelectric resistivity of series (Ohmm), *6* base depth of geoelectric series, *7* geoelectric resistivity of the bedrock (Ohmm), *8* approximate penetration depth of VES measurement, *9* boundary of geoelectric series, *10* series of higher resistivity compared to its environment, *11* doline

The varieties of dropout dolines are the following:

- Large-size dropout dolines (D_1) have diameters and depths of several tens of metres and the largest may reach or even surpass 50 m in diameter (Waltham and Fookes 2003). They resemble and represent transition to caprock dolines. Dolines of this type primarily form above easily soluble rocks (gypsum, rock salt) if the cover on the surface is cohesive, e.g. in the Salt Hill of Parajd. In Parajd the cover is not thick, and in the sides of dropout dolines, rock salt outcrops (Fig. 5.57). They also appear in non-cohesive cover if below the cover there are large cavities close to the surface, as described from Florida (Beck and Sinclair 1986; Jammal 1984) and Zambia (Spooner 1971).
- Medium-size dropout dolines (D_2) are less than several tens of metres in diameter and depth. They occur on carbonate karsts (Figs. 5.33 and 5.58). Among the karsts investigated by author, such dolines occur in the Mecsek Mountains and on the Pádis plateau.
- Breccia-pipe dropout dolines (D_3): the chimney developed in rock salt or the pipe (shaft) above the cavity in salt extends upwards and inherits over the cover (Fig. 5.59). The depth of the doline exceeds its diameter. In the wall of the pipe rock, salt is exposed and the chimney ruins become visible on the wall. Although there are similarities with caprock doline variety C_2, there are also differences in many respect: the cover is thin above the rock salt and the cover is exclusively unconsolidated rock.
- Small-size dropout doline in three varieties. The variety rock block dropout doline (D_4) is not deeper than several tens of centimetres. On the floor and sides, the limestone bedrock outcrops in blocks or as intact base rock (Fig. 5.60). It

Fig. 5.57 D_1 dropout doline developed above salt (Parajd, Salt Hill)

probably occurs on covered karst where the cover sediment is thin and constituted of coarse debris. This dropout doline is widespread first of all on glaciokarsts. The dropout doline with soil (D_5) is also smaller in diameter than several metres, and on the floor blocks of collapsed soil occur (Fig. 5.61). The bedrock is not exposed on the floor. It is a formation of short lifetime and widespread on covered karsts of fine sediment and soil cover. It grows into suffosion doline. The flat-floored dropout doline (D_6) is of several metres diameter and 10–20 cm depth. The floors of dolines are mostly flat and not dissected (Fig. 5.62). It occurs on glaciokarsts on clayey fine-grained cover. In ground plan the steep side slopes of the doline are not arcuate but strikingly straight. Figure 5.62 shows an exception: the slopes are composed of shorter arcuate sections. The bedrock is not exposed on the doline floor. It is probably also a young formation.

- Asymmetric dropout doline (Fig. 5.63) (D_7): side slopes are not vertical but cut across the soil and bedrock. The lengths of side slopes are variable and their downhill slopes are shorter or even missing (then the feature is not closed) and the opposite slopes are longer. The dolines of this type are usually small (some metres diameter) and quickly become inactive – probably because they are formed in fine-grained, clayey cover (clayey moraine). On the floor small intermittent or permanent lakes come about. A dropout doline variety is typical of glaciokarsts.
- Interior dropout doline (D_8) is one of the doline types of the interior of suffosion dolines (or occasionally of their margins) (the other type is suffosion doline). It can be deeper in diameter, and blocks of the bedrock may be exposed on its floor.

Fig. 5.58 D_2 dropout dolines (Doline 'A' Mecsek Mountains, Hungary, Doline 'B' Pádis)

Fig. 5.59 D_3 dropout doline (Parajd, Salt Hill) *1* chimney ruins

Fig. 5.60 D_4 doline (doline 'b' in the area Pádis-2, shown in Fig. 4.43)

The dolines of this variety have vertical walls (see Fig. 5.32), and their side slopes only become gentler during erosion.

- Suffosion dolines are much more widespread than dropout dolines on covered karsts. They are similar in size to dropout dolines (Waltham and Fookes 2003). The side slopes of suffosion dolines, although often steep, are covered with soil and vegetation. As opposed to the slopes of dropout dolines, the slopes of suffosion dolines do not cut across the cover sediment where they are formed.
- They are stable formations, either developed from dropout dolines or had been suffosion dolines even at the beginning of their evolution. While dropout dolines show a simple morphology, suffosion dolines are more variable. Elongated or crevice-like ground plan and variable slope conditions are typical of suffosion dolines. The side slopes of younger dolines are steeper, but later, as they are broadened through the erosion of slopes, slope inclination decreases (Sweeting 1973). Their interior is filled up, and on the resulting floor, microforms of various origins emerge. Their margins are often bordered by a terrain of moderate slope, some metres width, sloping towards the centre of the doline. Such terrains are not uniformly developed around the dolines, but they are locally narrow or missing. They probably result from pluvial erosion, where the cover is reworked from the neighbouring terrain into the doline. The varieties of suffosion dolines are classified according to size (age), morphology and environment.

Fig. 5.61 D_5 dropout doline (Hárskút Basin, Bakony Mountains), observed on 20 May 1998, formed some days before

Fig. 5.62 D_6 dropout doline (Durmitor)

Fig. 5.63 Active (**a**) and inactive (**b**) d_7 dropout dolines (Hochschwab, Austria)

- Large-size suffosion dolines (S_1) range from some metres to ca 60 m in diameter (Waltham and Fookes 2003, Fig. 5.42). Their present-day shape and size are reached through the transformation of older dropout dolines or the growth of smaller suffosion dolines. Examples occur where bedrock or limestone blocks appear on the floor or in the sides. Primarily the large suffosion dolines are associated with gullies and ravines. Their interior is dissected by minor karstic and non-karstic features. The dolines of this type develop above the extensive cavities of the bedrock, as exemplified on sand of the Florida karst (Beck and Sinclair 1986) and also common on other covered karsts (e.g. on the Pádis plateau).
- Embryonic suffosion dolines (S_2) are young and have small size, 10–20 cm deep features (Figs. 5.29b and 5.64a) and less than 1 m in diameter. They occur both inside and outside larger suffosion dolines.
- Interior suffosion dolines (S_3) occur first of all on the floor of large suffosion dolines (Fig. 5.64b) and occasionally on its side slope or margin, sometimes in groups. They are often found in alluvial streamsink dolines (see below). Both their diameter and depth can exceed 1 m. They mostly continue in a passage. Gullies may lead to interior suffosion dolines too (mostly from the floor of the bearing doline).
- Suffosion dolines with debris (S_4) are some tens of centimetres deep. Some varieties deepen exclusively into debris, while in others soil fills the gaps between debris fragments and blocks or buries them. In the first case, side slopes are steep. This variety is most widespread on glaciokarsts (Figs. 4.20, 4.68 and 5.65).
- Suffosion dolines between morainic hills (S_5) are very shallow, irregular-shaped dolines with gentle slopes (Fig. 5.66a). They are also found on glaciokarsts.
- Half-suffosion dolines (S_6) have limestone outcrops on one side. Occurring mainly at the base of heads of beds, they are small dolines of asymmetric cross-section (Fig. 5.66b).
- Suffosion gullies (S_7) are closed features elongated in downslope direction, occasionally more than 10 m long and 1–2 m across, some tens of centimetres deep. They are similar to gullies both in shape and size (Fig. 5.55).

Fig. 5.64 (**a**) S_2 doline, from the polje next to road number 513 (Croatia), (**b**) S_3 suffosion doline (Mecsek Mountains). *1* S_3 doline, *2* the paleodoline where the doline developed in fill

Fig. 5.65 (**a**) S_4 suffosion dolines in the debris of a paleouvala (Julian Alps, near Triglav), (**b**) S_4 suffosion dolines from the floor of a glacial valley (Totes Gebirge). *1* suffosion doline, *2* paleodoline

- Fossil dolines (S_8) form during the infilling of former subsidence dolines (mostly through suffosion). They are features of some tens of centimetres depth with a permanent or intermittent lake (Fig. 5.67). The erosional channel in the doline is produced by overflow rainwater.
- In the case of small cover thickness, the covered karst landforms of evaporites are varieties of dropout and suffosion dolines (Figs. 5.50a_2, 5.57 and 5.59), ponors and, on rock salt, salt scarps. Salt scarps are rows of gentle terrains of some metres width on the margin of salt diapirs (in Parajd on the slope of Salt Hill) divided by steeper slopes of some metres height (Fig. 5.68). The gentle steps smoothe into the slope of Salt Hill.

Fig. 5.66 (**a**) S_5 suffosion doline (Tauplitz alm, Totes Gebirge), (**b**) S_6 half-doline (Tauplitz alm, Totes Gebirge). *1* suffosion doline, *2* morainic mound, *3* doline slope bordered by steep scarp with bedding heads of limestone, *4* more gentle doline slope

5.4.3.2 Subsidence Pseudokarst Depressions

Doline-like landforms also develop to the effect of non-karstic processes. Such formations are similar to suffosion dolines with some metre-long convex slopes, but some specimen resemble dropout dolines. They are called subsidence pseudokarst depressions. White's (1988) denomination, suffosional pseudokarst, is only appropriate in certain cases since they are not always of suffosion origin. Subsidence pseudokarst features occur on the surface of lava flows covered with unconsolidated cover sediment, as exemplified by one of the lava flows on Hekla (Iceland), where funnel-like suffosion pseudokarst dolines of circular ground plan are found. They are of limited extension: both depth and diameter do not exceed some metres. Twinned (uvala-like) varieties also occur (Fig. 5.69), often in groups. The patterns of groups are either irregular (Fig. 5.70) or aligned in rows (Fig. 5.71). Elsewhere in volcanic pyroclast, depressions similar to dropout dolines are numerous (e.g. on Iceland, in the Askja caldera), and these are called dropout pseudokarst dolines. Their dimensions are small: 1–2 m in diameter and depth. Suffosion pseudokarst dolines of large diameter, small depth and gentle slopes also develop in loess (Jakucs 1977; Zeeden et al. 2007). In loess dropout-type pseudokarst features are also found, above loess pipes (Derbyshire et al. 2000). Similar forms are described from laterite too (Mendonca et al. 1993; Grimes 2009b).

Subsidence pseudokarst landforms also result from human activities (consequent pseudokarsts) (Halliday 2004). Collapse features induced by mining occur above galleries in chalk in northern Kent (Bell et al. 1999). Collapses above the galleries of coal mines on Dohányos Mountain (South Bakony, Hungary) lead to the formation of such depressions. They vary greatly in ground plan and cross-section (Fig. 5.72). Occasionally they are elongated (trenches) or circular (subsidence pseudokarst doline) in ground plan (Fig. 5.72). Trench-like forms are variable for width, length and inclination of side slope. Their cross-sections are symmetric, asymmetric or stepped asymmetric. Pseudokarst dolines resemble dropout dolines (Fig. 5.73). In composite trenches, trenches of smaller width and doline-like depressions occur

Fig. 5.67 S_8 suffosion doline (Hárskút Basin, Bakony Mountains)

Fig. 5.68 Steps on the Parajd Salt Hill: *1* step body, *2* step front

Fig. 5.69 Uvala-like suffosion pseudokarst dolines (valley near Hekla, Iceland)

(Fig. 5.74). On the floor of dolines, the bedrock is exposed with the bedrock chimney, which is also of non-solutional origin. The trench varieties are not restricted to loess, but they may occur on uncovered bare limestone terrains too.

5.5 Ponors

5.5.1 General Description

Ponors are karstic features developed on rock boundaries (Fig. 5.1g). On one side a valley with permanent or intermittent watercourse is attached to them, while on the opposite side or on the floor, a water conduit (chimney, shaft) appears in the outcropping limestone. The typical properties of ponors are the following:

- They are located in blind valleys or on valley floors.
- They have a catchment with clearly identifiable divide.
- They are mostly continued in an erosional cave (ponor cave).

Fig. 5.70 Irregular group of suffosion pseudokarst dolines (valley near Hekla)

Fig. 5.71 Suffosion pseudokarst dolines aligned in a row (valley near Hekla)

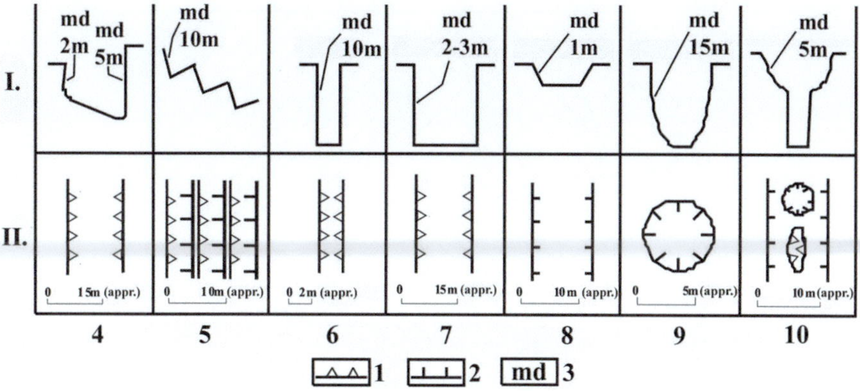

Fig. 5.72 Subsidence pseudokarst features above mine galleries (Dohányos Mountain, South Bakony, Hungary) *I* cross-section, *II* top view, *1* (sub)vertical side slope, *2* gentle side slope, *3* maximum depth, *4* asymmetric trench, *5* stepped asymmetric trench, *6* narrow trench, *7* broad trench, *8* shallow trench, *9* dropout pseudokarst doline, *10* composite trench

Fig. 5.73 Dropout pseudokarst doline formed on non-cohesive cover (loess) above the mine gallery of Dohányos Mountain

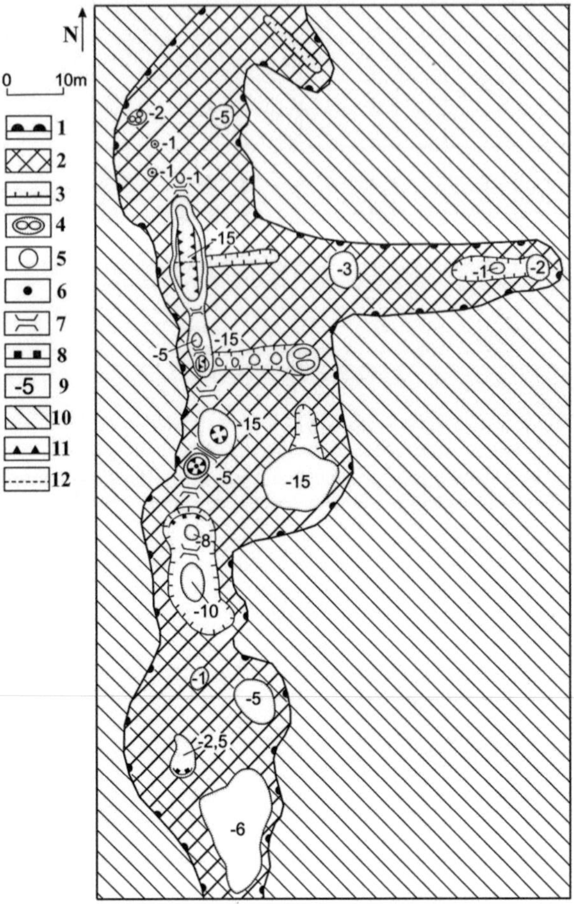

Fig. 5.74 Composite trench on loess from Dohányos Mountain: *1* edge of composite trench, *2* trench floor, *3* interior trench, *4* composite pseudokarst doline, *5* pseudokarst doline, *6* chimney, *7* col, *8* mass movement, *9* depth of landform, *10* neighbouring terrain, *11* (section of) landform in limestone, *12* fracture

Ponors were classified by Cvijič (1893), distinguishing between 'jama' (a narrow shaft getting narrower downwards) and 'trebič' (continued in cave). Ford and Williams (2007) also identify two types of ponor: one is with vertical passage and conduit, and the other is with lateral passage and conduit.

The ponor can be variable: funnel-like (with convex slope) or infilled (with concave slope). It is common that one side of the depression is a rocky cliff. On the floor of the infilled ponor, water conduction takes place through numerous passages (nonkarstic pipes) of several tens of centimetres diameter, which often form small

Fig. 5.75 Water conduit sites formed on the margin of channel of ponor (ponor of Ponor polje, Pádis plateau): *1* water conduit sites resembling subsidence dolines

depressions of water conduction, similar to subsidence dolines (Fig. 5.75). On the flat floor, not only sites of water conduction develop but subsidence dolines too, which often appear in groups and can be either dropout dolines or suffosion dolines.

The development stage of ponors depends on several factors: on rock structure (the presence of joints and faults enhances the growth of these landforms), the age of water conduction, climatic conditions and catchment properties (area and maximum relief of catchment, slope of valley leading to the ponor).

With surface erosion, the divides shift and catchment area changes with time. Catchment area can be increasing, decreasing or stable (not changing). It is increasing if on the superficial deposit the stream and thus its valley retreat. It is decreasing if the margin of the superficial deposit becomes denuded. The catchment area is stable if the superficial deposit does not become denuded at its margin because of its great thickness. (It cannot increase either because the superficial deposit is surrounded by non-karstic rock.)

5.5.2 *Classification of Ponors*

Ponors form on true rock boundaries (further: rock boundaries), while covered karst ponors form on not true rock boundaries.

5.5.2.1 Classification of Ponors According to Rock Boundaries

The rock boundaries and thus ponors can be located on the karst margin and in the karst interior. In the case of karst marginal position, the non-karstic rock is found on the margin of the bare karst. This situation results in the formation of a ponor row at the termination of blind valleys. The classification of karst marginal ponors is shown in Fig. 5.76 (Ford and Williams 1989, 2007). If non-karstic and karstic rocks alternate, several true rock boundaries are superimposed (or juxtaposed) (Fig. 5.76a). As a consequence, parallel rows of ponors emerge. The uppermost cover could be syngenetic with the karstic rock (consolidated rock) or postgenetic superposition (unconsolidated rock). Since the bedrock under the non-karstic rock is karstic, new sites of capture develop in the blind valleys (retreating capture), and with shifting rock boundary, the area of bare karst is extending. If the strata are tilted, in oblique position, the non-karstic bedrock outcrops and a unique ponor row comes about (Figs. 5.76b and 5.77I). Rock boundary with ponor row also forms if the karst margin is in contact with non-karstic rock due to tectonic movements (faulting) (Fig. 5.76c).

In all these cases, because of denudation and tectonic movements, an escarpment of the karstic rock develops on the karst margin, sloping towards the non-karstic terrain, while the non-karstic terrain gently slopes towards the rock boundary. The depressions of the ponors and blind valleys are aligned at the base of the escarpment, along the rock boundary. As mentioned above, the rock boundary may be shifted if a non-karstic rock mantles the bedrock (Fig. 5.76a).

Rock boundaries (along them ponors) can appear in the karst interior too in four different ways:

- The non-karstic permeable cover has a considerable extension on the karst. Where incising valleys cut through the cover, valley rock boundary comes about (Fig. 5.77IIa). If the karst water table lies below the channel floor, a ponor develops. With valley incision, the valley rock boundary is retreating. This involves the generation of newer and newer ponors (Jakucs 1956; Hevesi 1978, 1980; Ford and Williams 2007). Ponors on the floor are transformed into dolines (Jakucs 1956; Hevesi 1980). Active ponors are associated with gully or ravine at most. The ponor is seldom coupled with a blind valley.
- The rock boundary forms at the outcrop of limestone (Fig. 5.77IIc), where the cover does not overlie the karstic rock (Figs. 4.59a and 4.60). Such karst windows are found on the Pádis plateau, where the remnants of older karstification rise as elevations and the ponors are aligned at their base (Veress 1992). Tropical inselbergs can also be regarded karst windows if their environment is covered. On fenglin-type karst, ponors at the base of inselbergs are also common (Wilford and Wall 1965). Ponors are also present in the sides of exhumed tropical paleokarst hills described from Brazil (Balázs 1984). At the karstic rock patch (karst window) a short, arcuate rock boundary develops. However, the karst window may have such a small extension that only one ponor or one group of ponors develops on its area (the ponors are not aligned in a row). Depending on how

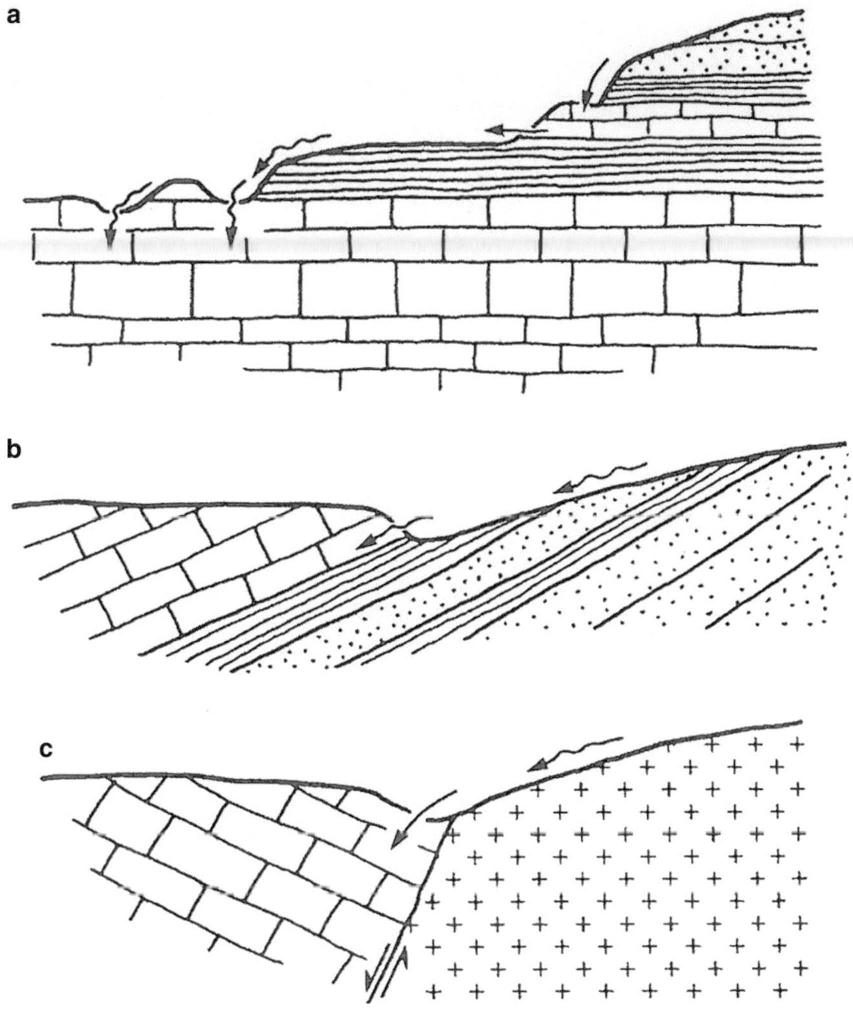

Fig. 5.76 Recharge by allogenic streams flowing (**a**) from overlying beds, (**b**) from underlying beds exposed upslope and (**c**) across a faulted contact with impervious rocks (Ford and Williams 2007)

advanced ponor formation and how large the catchment area and what its slope is, gullies, ravines or valleys lead to the ponors. In the case of several ponors, watercourses converging towards the karst window come about. In the isolated limestone patch, through caves may also appear (karst type 2 in the classification by Gams (1994)). The karst window can also be covered if the cover is thin above the limestone (either because the cover accumulated in small thickness or because it was denuded subsequently). In this case, the ponor or its predecessor is generated on buried rock boundary (or on structural rock boundary) and the

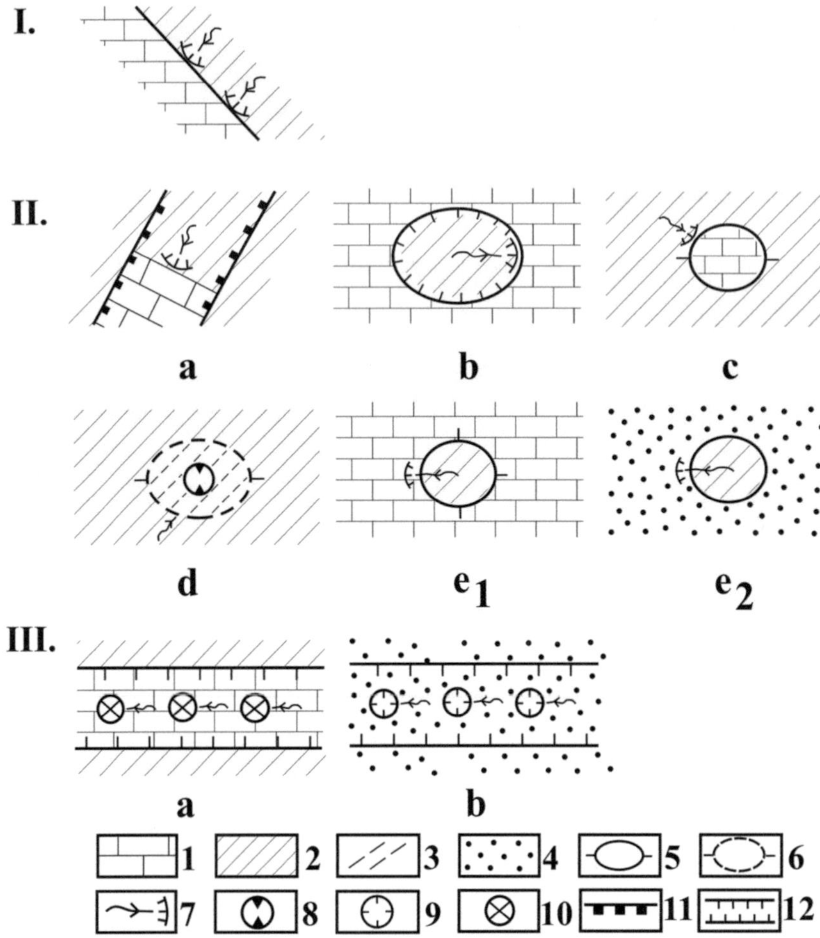

Fig. 5.77 Ponor types by rock boundary: *I* rock boundary on karst margin. *II* rock boundary in karst interior: rock boundary in valley (*IIa*), in karst depression (*IIb*), at elevation (*IIc*), at elevation covered in small part (*IId*), at impermeable cover patch (*IIe$_1$*), at impermeable cover bordered by permeable cover (e_2). *III* ponors of gully with ravine not on rock boundary: no cover on ravine floor (*IIIa*), cover on ravine floor (*IIIb*), *1* limestone, *2* impermeable cover, *3* thinning section of impermeable cover, *4* permeable rock, *5* elevation, *6* buried elevation, *7* ponor, *8* caprock doline transforming to ponor, *9* subsidence doline, *10* chimney, shaft, *11* epigenetic valley, *12* gully, ravine

true rock boundary developed only gradually, an example is Kab Mountain (Hungary), where above elevations of the karstic bedrock covered by basalt in small thickness caprock dolines emerge (Figs. 4.59b, c, f and 5.77IId). Starting from the caprock dolines, gullies and ravines extend through retreating erosion (Fig. 4.59b, c, e, f). The dolines gradually develop into ponors or covered karst ponors (see below) and often dissected into partial depressions.

- If impermeable rock patches occur on karstic rock, also local or short, arcuate rock boundary develops with a solitary ponor or group of a few ponors (Fig. 5.77IIe$_1$). The catchment of the ponors and their valleys is stable. With several ponors, a network of branching ravines forms on the cover patch. Occasionally the impermeable cover patch is surrounded by a permeable cover (Fig. 5.77IIe$_2$). In this case covered karst ponors (see below) emerge.
- If the rock boundary and ponor form in the fill of a paleodoline, paleouvala fill or in non-karstic cover karst of interior polje (Fig. 4.39), the rock boundary can be short (with a single ponor) or longer and arcuate. An escarpment invariably develops on the rock boundary and constitutes the side slope of the paleodepression (Figs. 5.77IIb and 5.78). In the case of a longer rock boundary, the ponors are aligned in an arcuate row. The catchment area is shrinking as the cover is being removed.

Along the rock boundary within the paleodepressions, the features of cryptokarst and concealed karst can occur together too. Thus, in a paleodoline of Pádis plateau, both ponors and subsidence dolines occur along the rock boundary (Fig. 4.41).

It can be experienced too that the rock boundary is dissected and consists of parts with different directions and has a wavy pattern. A good example for this is one of the paleouvalas in Totes Gebirge (Figs. 5.79 and 5.80). Here, the rock boundary is straight where its direction is perpendicular to the strike direction of the beds of the bedrock (Fig. 5.80a), and it is dissected where it is parallel with the strike direction of the beds (Fig. 5.80b). Mainly subsidence dolines, but also covered karstic ponors, occur along the dissected rock boundary of the paleouvala.

5.5.2.2 Covered Karst Ponors

Covered karst ponors are located in the interior of the covered karst. Therefore, the bedrock does not outcrop at the ponor and does not form an escarpment. Covered karst ponors do not have a well-identifiable catchment. They are transitional forms towards subsidence dolines and caprock dolines. The covered karst ponors in concealed karst environment are infilled to a great extent, and on their floor, interior subsidence dolines frequently occur. According to the rock boundary, the following varieties can be distinguished:

- The ponor formed on autogenic cryptokarst or transitional cryptokarst, i.e. not on true rock boundary, occasionally from C_3-type caprock doline (on transitional cryptokarst). Such features are transitional between ponors and caprock dolines. The catchment area of these covered karst ponors is extending, but mostly it is not well separated from other catchments (not independent), and the divide can be established only partially or not at all. The ponors of autogenic cryptokarsts are caprock dolines formed above the breccia pipes of evaporites. The ponors of transitional cryptokarsts developed in consolidated rocks above limestone, following structural or buried rock boundaries. An example for the former is a depression in the area of the Holbrook anticline (Arizona, USA) (Martinez et al.

Fig. 5.78 Ponor formed on rock salt (Parajd, Salt Hill): *1* margin of the doline bearing the ponor, *2* ponor, *3* doline floor with cover sediment (feeding area of ponor)

1998) and for the latter the depressions of C_3 type on the thin basalt cover of Kab Mountain (Fig. 5.77IId). Occasionally a subsidence doline is transformed into a covered karst ponor, as described from the Bükk Mountains, where a stream flowed into a subsidence doline formed in the vicinity of the channel of the Berva Stream (Kozma and Holló 2010). The subsidence doline developed on ca 25 m thick cohesive deposits (clay and clayey sand).

- Covered karst ponors develop close to true rock boundaries where the cover sediment is thinned out to the extent that waters from intermittent or permanent watercourses can percolate through it (Figs. 5.79 and 5.80). At the place where the giant grikes, shafts and chimneys of the cover had already developed before accumulation. In such cases the sediment accumulates in the environs of the chimneys and shafts.

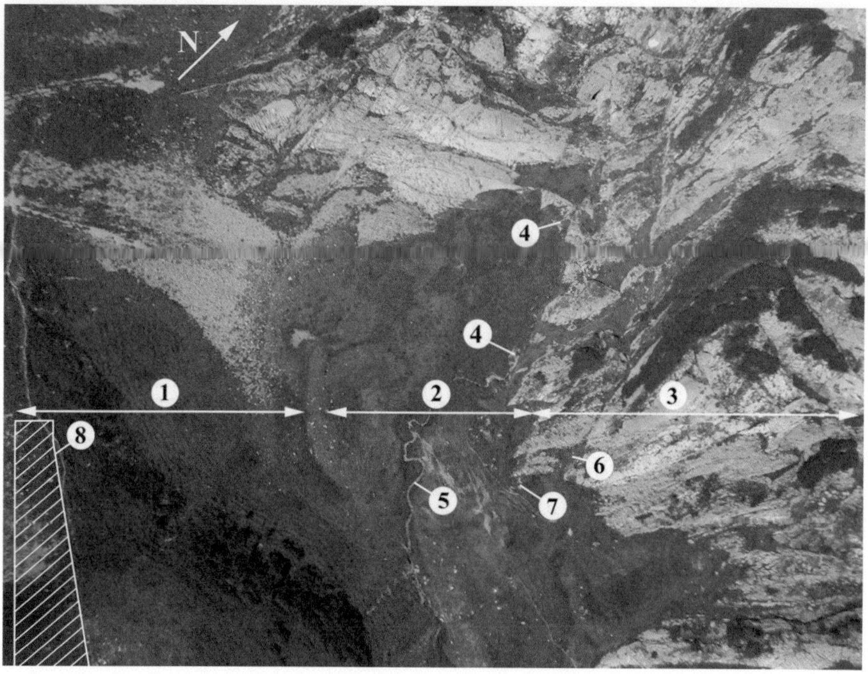

Fig. 5.79 Aerial photograph of the area IV under Wildgössl (cirque formed from paleouvala, Totes Gebirge): *1* side of paleouvala with debris fans, *2* floor of paleouvala with cover sediment, *3* slope of paleouvala with glacially eroded cuesta, *4* covered karst ponor, *5* gully, *6* covered karst terrain intruding into the depression of the cuesta, *7* a subsidence doline in the zone of thin cover sediment next to rock boundary, *8* zone of springs

- Another possible way of covered karst ponor development, as it was mentioned (Fig. 5.77IIe$_2$), if the impermeable rock patch is bordered by permeable cover. (This covered karst ponor variety is essentially generated in concealed karst environment. The impermeable cover sediment patch which is the feeding area often does not outcrop on the surface, but remains under permeable cover.) The covered karst ponor is bordered by a terrain of impermeable cover one side and of permeable cover on the other side (Fig. 5.81). The ponor is found at the end of a blind watercourse or blind gully. The catchment area of the ponor is small and stable. The impermeable cover sediment patches in the Bakony Mountains are mostly small in extension. For instance, the covered karst ponor in Fig. 5.81 is bordered by uncovered limestone in the E and loess-mantled limestone in the N. Similarly bare limestone surface is found ca 100 m to the W and ca 50 m to the S of the end of the ravine attached to the ponor. Karren passages, chimneys and shafts of gullies (Fig. 5.77IIIa) or subsidence dolines developed in the superficial deposit of gullies can operate as ponors.

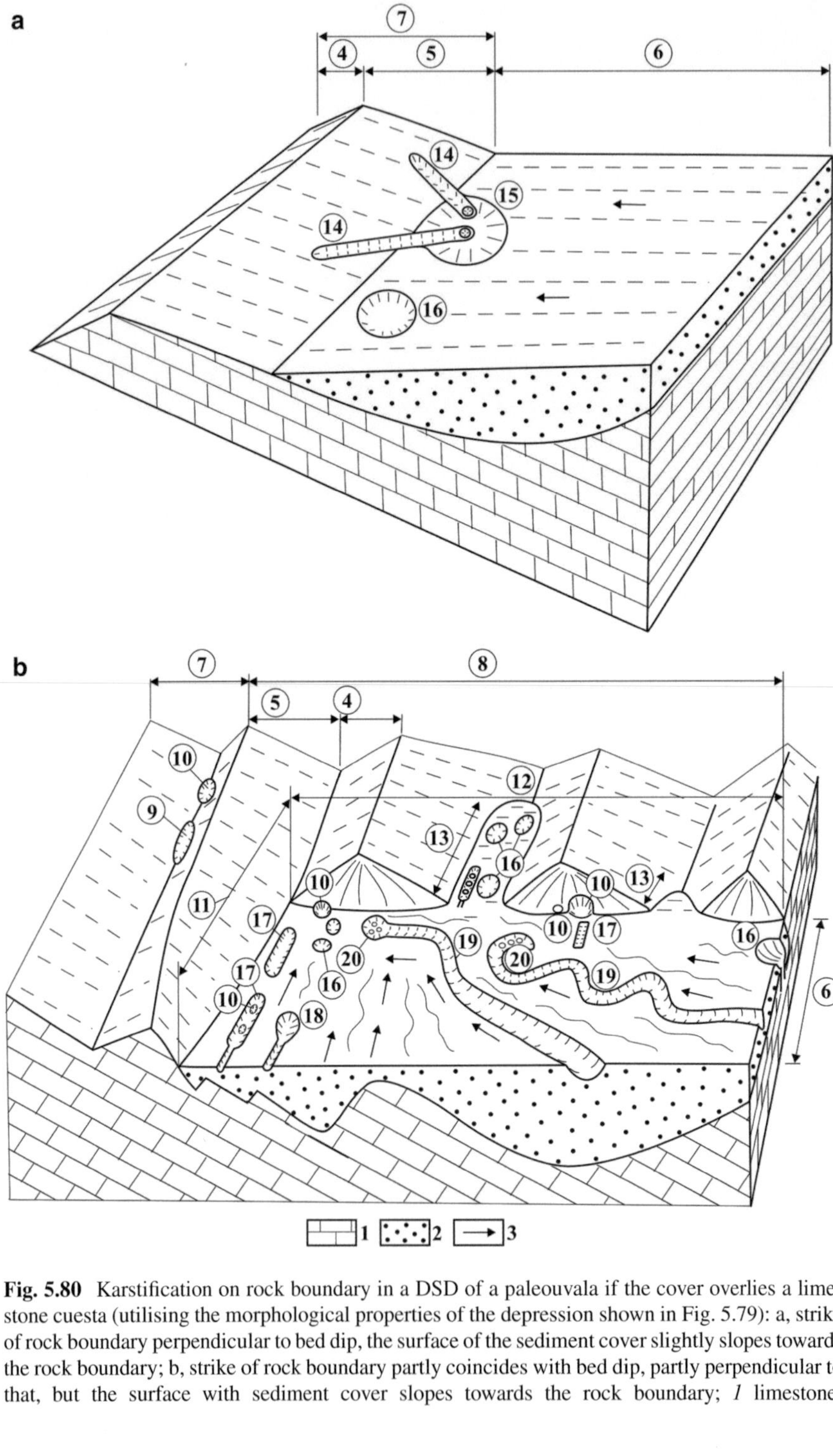

Fig. 5.80 Karstification on rock boundary in a DSD of a paleouvala if the cover overlies a limestone cuesta (utilising the morphological properties of the depression shown in Fig. 5.79): a, strike of rock boundary perpendicular to bed dip, the surface of the sediment cover slightly slopes towards the rock boundary; b, strike of rock boundary partly coincides with bed dip, partly perpendicular to that, but the surface with sediment cover slopes towards the rock boundary; *1* limestone;

5.6 Depressions of Superficial Deposit (DSD)

Veress (2000, 2008, 2009b) identified a further landform on covered karst, the depression of superficial deposit (Fig. 5.1f). These landforms originate from the erosion of cover sediment if the cover sediment is only transported into the cavities of the karst. If there is no closed and infilled depression on the bedrock, the DSDs develop exclusively in the cover sediment (Fig. 5.82a). If there is an infilled paleokarst landform on the bedrock, with the denudation of the cover sediment, they come about through the partial exhumation of the original karst depression (Figs. 5.82b, c and 5.83).

Various authors use different names for the depressions which are probably DSDs. Some are mentioned below. The landforms developed in cover sediment at katavothra are called 'alluvial streamsink dolines' by Jennings (1985). Korzhuev (1961) used the denomination 'kotlovina' for a large-size depression on the taiga karst, Károlyi and Ford (1983) mention the DSDs in moraines as 'bedrock closed depressions', while Láng (1971) describes 'ponor basins', closed depressions of hundreds of metres diameter in the blind valleys of the Aggtelek Karst and in the vicinity of the Postojna and Skocijanska caves (Slovenia). Davies and LeGrand (1972) use the name 'coves' for the large (300–1500 m diameter) depressions of the Appalache Mountains with minor dolines and ponors on cover sediment. Móga (2001) mentions 'basins drained underground' from the Aggtelek Karst. These are also large depressions with ponors and watercourses. Probably the features called 'karst valleys' by Crawford (1984) and the 'turloughs' (glacially moulded depressions in Ireland, which are filled with intermittent lakes) (Coxon 1986; Gunn 2006) are also DSDs. According to Beck and Sinclair (1986), large basins form through the broadening and then merging of subsidence dolines, and they are also DSDs. The landforms generated by the merging of blind valleys in the cover sediment are called 'small poljes' or 'border poljes' with alluviated floors (Ford and Williams 2007). This denomination is also used by others: Plan and Decker (2006) mentions the large depressions of the Hochschwab with sandstone floor as small poljes remoulded by glacial erosion. Goeppert et al. (2011) distinguish between two types of poljes developed on calcareous conglomerate, one of them occurs in cirques. Barrére (1964), however, calls the cirques further deepened by solution cirque dolines. Their deepening is explained by Goeppert et al. (2011) by erosion and solution. The other type forms in synclines also through the erosion of the cover. Consequently, these latter landforms are also DSDs.

Fig. 5.80 (continued) *2* cover; *3* slope direction of the covered karst terrain; *4* bedding head; *5* bedding plane; *6* gently sloping surface of sediment cover; *7* uncovered cuestas; *8* partly covered cuestas; *9* solution crevice; *10* shaft; *11* the dip of limestone beds is perpendicular to the strike of the rock boundary; *12* the dip of limestone beds coincides with the strike of the rock boundary; *13* covered karst terrain intruding into the depression of the cuesta; *14* solution crevice with partly uncovered, partly covered environment; *15* suffosion doline developed in the environs of crevices; *16* suffosion doline; *17* covered karst crevice; *18* subsidence doline with gully, *19* blind gully, *20* covered karst ponors and chimneys

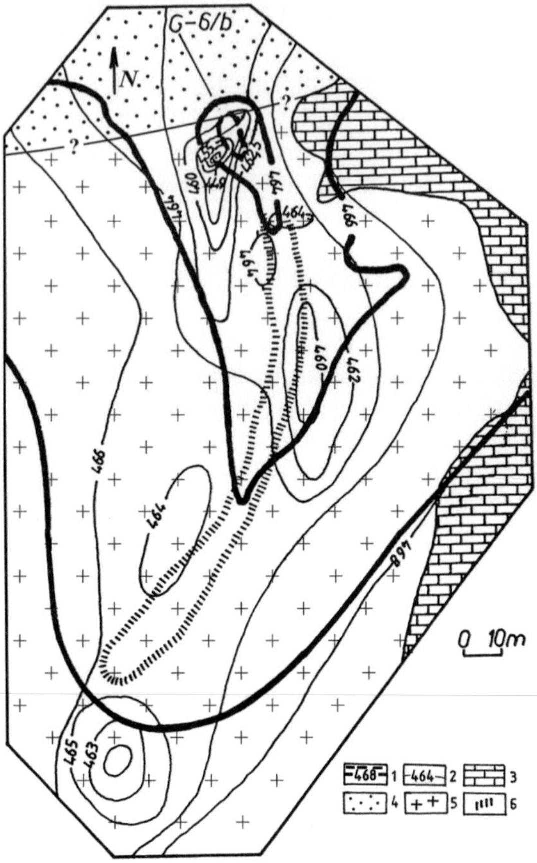

Fig. 5.81 Covered karst ponor marked G-6/b (Hárskút Basin, Veress 2000): *1* contour line of ground surface, *2* contour lines of limestone bedrock, *3* limestone, *4* loess, *5* grey and yellow clay with intercalated quartzite and sand gravel strips, *6* channel

5.6.1 Characteristics of the DSDs

The DSDs are composite landforms with subsidence dolines (e.g. dropout dolines, suffosion dolines) and/or ponors in their interior. Their origin is due to dolines and ponors. The cover sediment is transported into the cavities of the karst by pluvial and fluvial erosion through the passages of these karst features. Therefore, in the environs of dolines and ponors, with local denudation of the surface, closed forms develop in the cover sediment or paleokarst depressions are exhumed.

Their morphological properties are the following:

- Although at the beginning of their development they can be open, with further development they become closed formations without outflow.
- Their diameters and depths are variable. Their diameter ranges from some tens of metres to several hundreds of metres or occasionally even to 1–2 km, while their depth is smaller compared to their diameter: those with several times ten metres diameter is mostly several metres, and only the depressions with hundreds of metres diameter are several 10 m deep.

- They are often asymmetric in cross-section: one side is gentle and not distinct from the floor, while the opposite slopes are steep.
- Their floor is gently sloping and often dissected into parts of different slope.
- On their floor or margin, subsidence dolines (dropout doline, suffosion doline or their infilled varieties) occur as well as covered karst ponors, blind valley ponors and their combinations. On the floor of some depressions, minor interior DSDs are also found on the floor of the larger DSDs.
- Their floors and side slopes are dissected by rills, gullies and (blind) valleys linked to karstic features.

5.6.2 Classification of DSDs

DSDs are classified by Veress (2000, 2010b) according to their position on the karst, bedrock morphology, cover sediment and extent of coveredness into the following types:

According to their position, the depressions are either karst interior or karst marginal. Karst interior depressions (Fig. 4.41) are located on inflow side of the karst. They are common on recent allogenic-covered karsts. Karst interior depressions are found on autogenic karst or in the uncovered zone of renewed covered karst as well as in the covered karst zone of equally recent allogenic-covered karst, mantled allogenic karst and renewed covered karst and in the paleodepressions of glaciokarsts.

According to bedrock morphology, the depressions are either pseudodepressions (without a closed landform on the bedrock, Fig. 5.82a) or true depressions (with a closed depression on the bedrock, Fig. 5.82b, c). Pseudodepressions originate from karst marginal blind valleys if they are broadening (Fig. 5.84a). They can also develop from subsidence dolines if they are growing, i.e. broadening (Fig. 5.84b) or merging with erosional landforms (Figs. 5.84c, d, 5.85 and 5.86) or if the bedrock is dissected by ridges between the exhumed ridges (Fig. 5.84e, f). They occur on epigenetic valley floors if the floor subsequently comes under cover (Fig. 5.84g) or over katavothra (Fig. 5.84h). True depressions take shape in exhuming paleodolines (Figs. 5.83 and 5.87a–c), paleouvalas, paleo blind valleys (Figs. 4.26, 4.27 and 5.87d), poljes or polje sections (Fig. 4.28), depressions of fengcong karsts (Fig. 5.87e, f) and on rock salt (Figs. 5.87g and 5.88).

The depressions are also classified by the character of the cover sediment on their floor, which is either permeable (concealed karstdepression) or impermeable (cryptokarst depression). If the cover sediment is permeable, concealed karst with subsidence dolines forms on the floor (Figs. 5.39 and 5.82a). The subsidence dolines are mostly interior karst depressions. If the cover sediment is increasingly impermeable, although the karstic features develop on concealed rock boundary, an ever better developed gully network appears on the floor. If the cover is (partially) impermeable, cryptokarst with ponors, blind valley ponors and blind valleys develop on the floor of the depressions (Figs. 4.25, 4.48, 4.50 and 5.82b, c). Cryptokarst depressions are primarily in karst marginal position – although those in the karst interior may also occur.

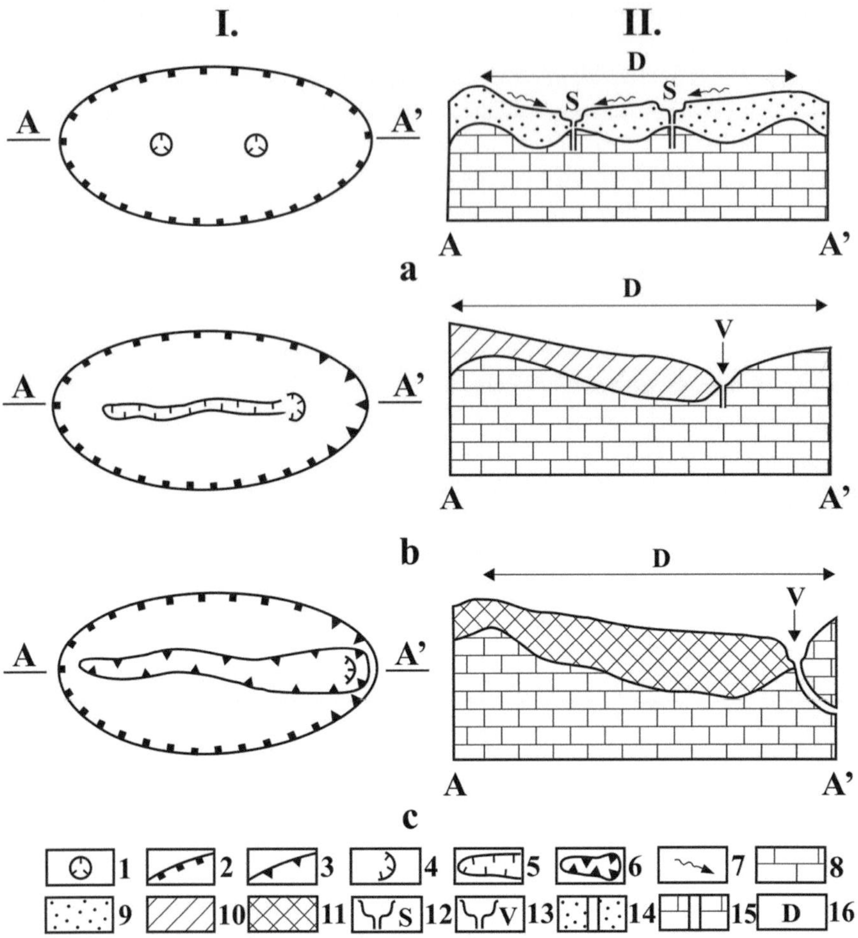

Fig. 5.82 Depressions of the superficial deposit (Veress 2012, modified): *I* planimetric representation: *1* subsidence doline, *2* the margin of the DSD, *3* the limestone margin of the DSD, *4* ponor, *5* gully, *6* blind valley, *7* sheet erosion, *II* cross-section, *8* limestone, *9* permeable cover sediment, *10* partly permeable cover sediment, *11* impermeable cover sediment, *12* subsidence doline, *13* ponor, *14* burrow, *15* shaft, *16* DSD, (**a**) there is no closed depression on the bedding head, (**b**, **c**) there is a closed depression on the bedding head

On the side slopes of the pseudodepressions, the bedrock does not outcrop, while in the true depressions, with the increasing removal of the sediment infilling the paleoform, this happens. According to the extent of exhumation, the true depressions are either covered or semi-exhumed (partly covered depressions). The side slopes of covered depressions are mantled by cover sediment. On the side slopes of the semi-exhumed depressions, the bedrock outcrops. The fill of paleodoline depressions of glaciokarsts is partly or totally due to glacial accumulation. However,

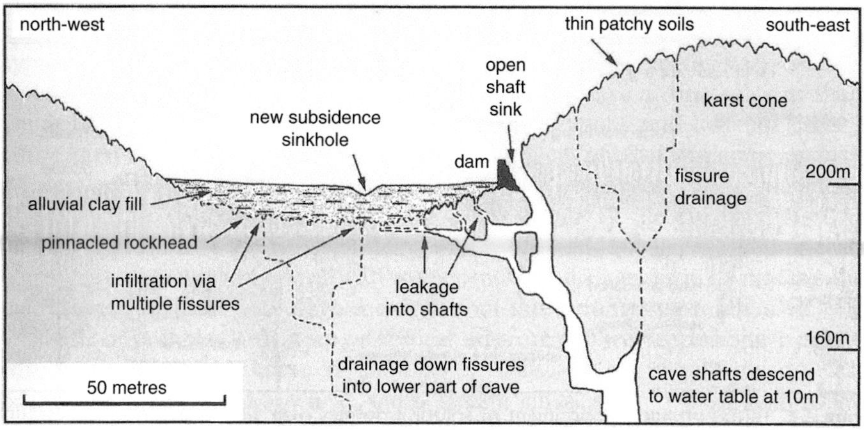

Fig. 5.83 True DSD on tropical karst (Waltham et al. 2005)

morainic deposits often did not fill in the dolines completely, but only covered them as a veneer. The side slopes were only partially covered or not covered at all. In the latter case, the depression must have existed at the time of glacial retreat. Though its initial depth increased with the transport of the cover sediment into the karst.

The depressions are either simple or composite. The simple depressions are only concealed karst (Figs. 4.64 and 5.89) or cryptokarst depressions (Figs. 4.48 and 4.50), while some parts of the composite type are concealed karst depressions and other are cryptokarst depressions (Figs. 5.90 and 5.91). A variety of concealed karst depressions is represented by those specimen in which ponors have already developed because the cover sediment on their floor is partly impermeable, e.g. clayey loess (Figs. 5.85 and 5.86).

As it has been mentioned, a variety of DSDs is the landform named alluvial streamsink doline by Jennings (1985), katavothra or alluvial streamsink-type DSD. Jennings (1985) introduced of this doline type with reference to the work of Cramer (1941).

They occur on polje floors or in depressions encircled by karst hills at katavothra. Sweeting (1973) describes these formations as a ponor type. The katavothron-type activity (and the alluvial streamsink dolina-type DSDs) are not restricted to the Dinaric Karst but widespread in other karst regions too (Japan, France, Tasmania, Jamaica etc.). These are karst landforms on terrains temporarily inundated by karst water. Karstic elements are shafts in the bedrock, non-karstic pipes, various subsidence dolines, D_8 dropout dolines and S_3 suffosion dolines (Figs. 5.92 and 5.93). Both of the latter landform types occur on the floor of this type of depression and its environs. Their growth is not only promoted by rills, gullies and ravines formed on its slopes and environs, but also by the non-karstic pipes and subsidence dolines on their margins in their vicinity (Fig. 5.92a). The latter are merged with alluvial streamsink-type DSDs.

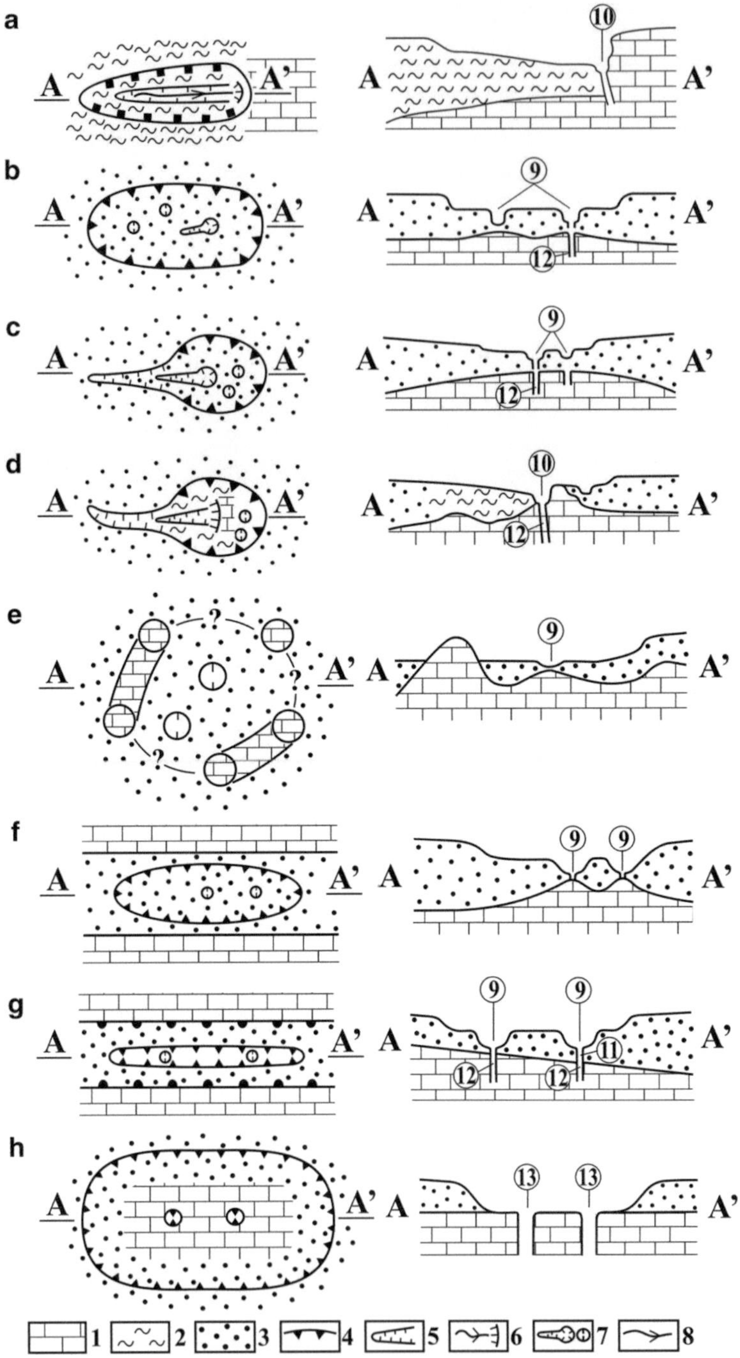

Fig. 5.84 Pseudodepressions formed in the following environments: through the broadening of blind valley on allogenic cryptokarst (**a**), through the denudation of the terrain bordering covered karst dolines (**b**), on concealed karst through the formation of ravine linked to doline (**c**), on concealed karst through ponor and valley formation (**d**), on concealed karst dissected by elevations and ridges (**e**, **f**), on floor of epigenetic valley (**g**), above katavothra (**h**). In *top* view: *1* limestone, *2* impermeable cover, *3* permeable cover, *4* DSD, *5* valley, ravine, *6* ponor, *7* subsidence doline, *8* watercourse; in cross-section: *9* subsidence doline, *10* ponor, *11* non-karstic pipe, *12* shaft, *13* katavothron

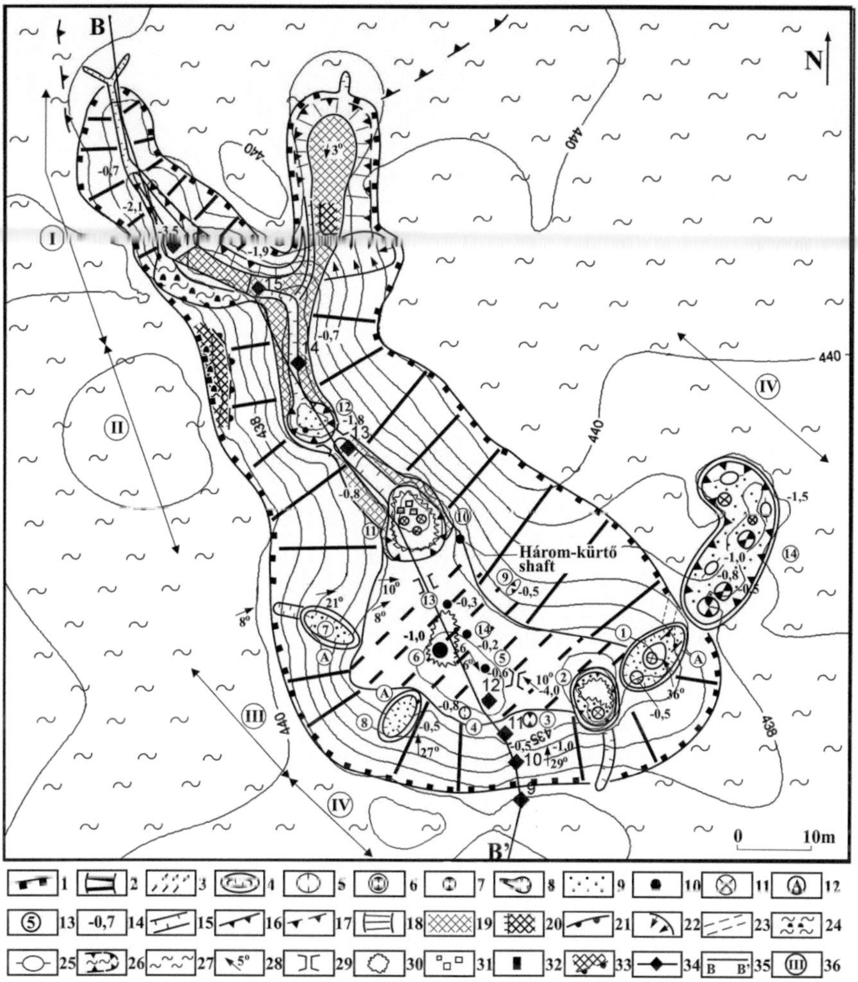

Fig. 5.85 Geomorphological map of DSD D-5: *1* margin of DSD; *2* side slope of DSD; *3* floor of depression; *4* suffosion doline (diameter above 2 m); *5* suffosion doline (diameter below 2 m); *6* dropout doline (diameter above 2 m); *7* dropout doline (diameter below 2 m); *8* ponor; *9* floor of karstic landform (diameter above 2 m); *10* chimney, shaft; *11* non-karstic pipe; *12* asymmetric doline; *13* identification code of karst depression; *14* depth of landform; *15* gully, channel; *16* ravine; *17* ravine transformed by pluvial erosion; *18* side slope of ravine; *19* floor of ravine; *20* step on the floor of ravine; *21* collapse scar; *22* slump; *23* horizontal non-karstic pipe in cover; *24* tongue of collapsed material; *25* collapse heaps; *26* valley leading to depression; *27* neighbouring terrain; *28* slope direction and angle; *29* col; *30* rock outcrop; *31* rock debris; *32* badger burrows; *33* edge of step created by badgers; *34* site of VES measurement; *35* profiles of VES measurements; *36* morphological zones of the depression. *I* zone of erosion and mass movements, *II* accumulation zone, *III* detail of floor of DSD without cover sediment, *IV* southern side slope of the DSD dissected by subsidence dolines

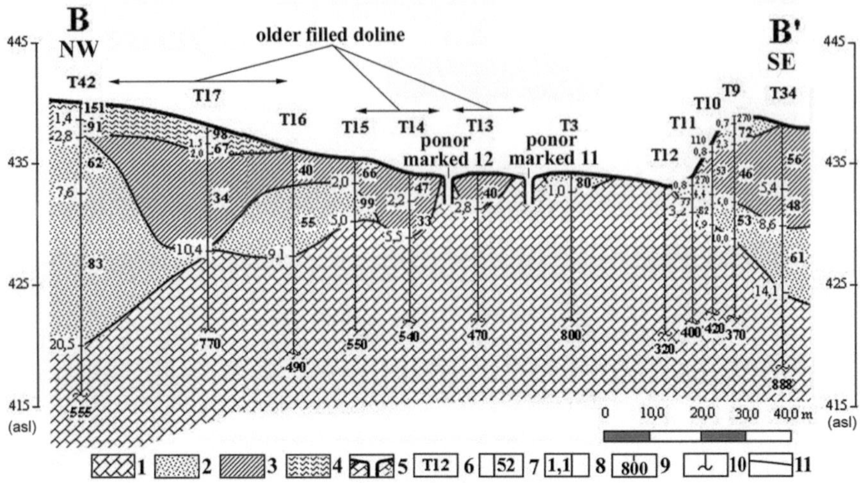

Fig. 5.86 Geoelectric geological profile of DSD D-5: *1* limestone, *2* (sandy) loess (with limestone debris), *3* loess (clayey or silty) or clay with limestone debris, *4* (clayey) limestone debris, *5* chimney, *6* code of VES measurement, *7* geoelectric resistivity of series (Ohmm), *8* base depth of geoelectric series (m), *9* geoelectric resistivity of bedrock (Ohmm), *10* approximate penetration depth of VES measurement, *11* boundary of geoelectric series

5.7 Karst Valleys

In karst areas valleys form on structural boundaries, on non-karstic rock wedged in into karstic rock, through cave collapse and inheritance. According to their position and morphology, the valleys are allogenic or through valleys, blind valleys and half-blind valleys or pocket valleys (Sweeting 1973; Jennings 1985; Ford and Williams 2007).

5.7.1 *Allogenic Valley*

The watercourses of allogenic valleys are permanent and are fed by watercourses originating from non-karstic terrains. The watercourses have considerable discharges and, consequently, are able to cut through the karstic rock patch. Their karstic section is narrow and canyon-like, while the non-karstic sections are wider (Sweeting 1973).

5.7.2 *Epigenetic Valley*

Epigenetic valleys develop if the valley formed on non-karstic cover sediment is inherited over the karstic bedrock. There may be permanent and intermittent streams in the valley and the valley may not contain any streams. It often occurs that because

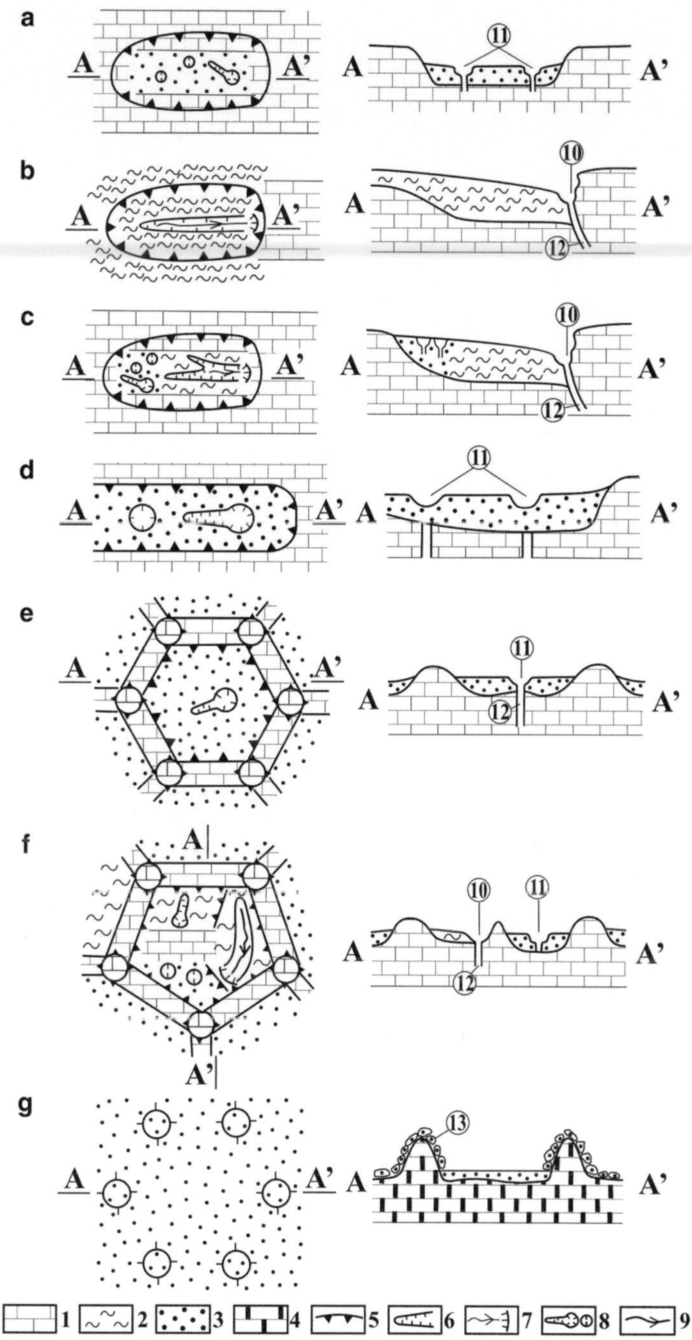

Fig. 5.87 True depressions formed in the following environments: (**a**) in paleodolines (concealed karst), (**b**, **c**) in paleodoline (cryptokarst), (**d**) in blind valley (concealed karst), (**e**, **f**) in doline of fengcong karst, (**g**) on rock salt. In top view: *1* limestone, *2* impermeable cover, *3* permeable cover, *4* rock salt, *5* DSD, *6* valley, ravine, *7* ponor, *8* doline, *9* watercourse; in cross-section: *10* ponor, *11* doline, *12* chimney, *13* blown sand blocks cemented by salt and displaced on slope

Fig. 5.88 DSD on rock salt (Atacama Desert)

Fig. 5.89 Detail of a DSD on the Pádis plateau (see map detail in Fig. 5.39) (Uvala Râchite): *1* margin of depression, *2* elongated subsidence doline, *3* uvala-like subsidence doline, *4* inactive subsidence doline

Fig. 5.90 DSD of exhuming paleodepression (Hochschwab): *1* ponor, *2* lake, *3* infilled ponor, *4* end of blind valley, *5* probable former level of fill, *6* valley, *7* gully, *8* cryptokarstic part of depression floor, *9* concealed karst part of depression floor, *10* margin of DSD (paleouvala)

of the different position of the karstic rock, the inheritance happens on one part of the valley only. In this case the stream of the valley which gets its water from the area covered with impermeable bed situated above the inherited section seeps on this section. (The place of seepage can vary depending on the discharge.) Such valley sections (epigenetic valley section) are steep and gorge-like. On this valley sections, there are no ponors and opened-up caves (remnant caves) often occur in the sides of the valley.

The valley turns into a dry valley if it has no intermittent watercourse since the karst lost its sediment cover. The incision of the valley floor ceases and the karst water table is gradually sinking below the level of the valley floor. The valley is increasingly drained since the rainwater fallen on the floor infiltrates. If the valley watercourse has a spring, its water seeps away within a short distance. Dry valleys were described by Fagg (1923), Lehmann (1936), Roglič (1964), Warwick (1964), Sweeting (1973), Pinchemel (1954) and Gunn (2004).

Two types of dry valleys are distinguished: one is the section of an allogenic river valley beyond its ponor (such valleys have a watercourse up to the ponor), and the other is a dry valley within the karst (Jennings 1985), where the watercourse is missing from the entire length of the valley (only active during intensive rainfall at most). Dry valleys in the continuation of allogenic valleys are demonstrated by Smith et al. (1969), Ollier and Tratman (1969), Jennings (1985), Clozier (1940) and

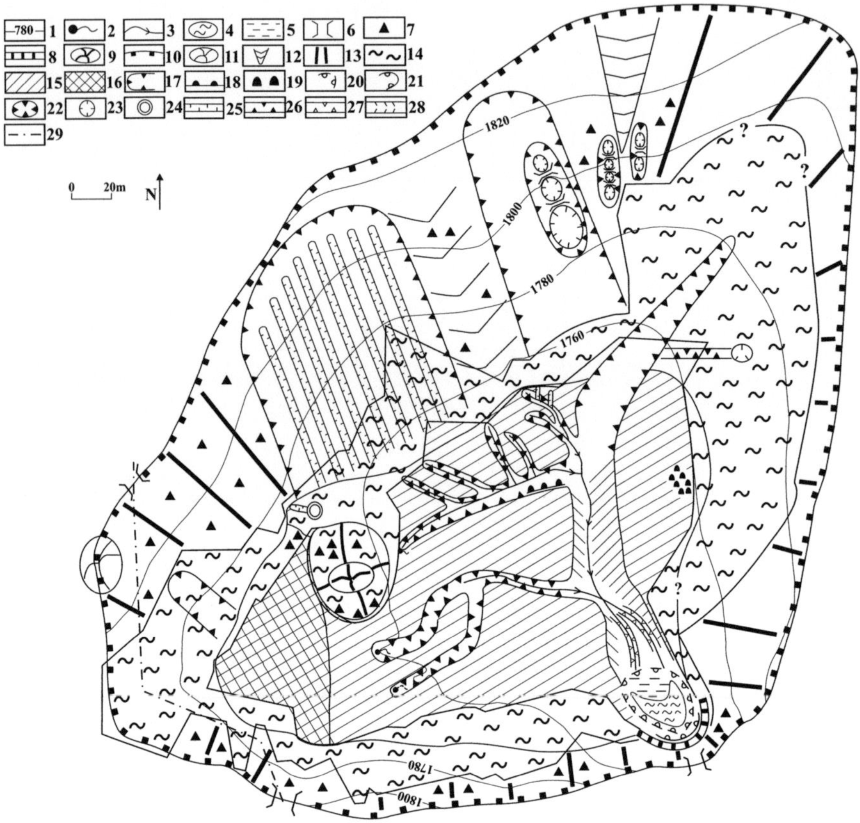

Fig. 5.91 Geomorphological map of the DSD of Hochschwab: *1* contour line, *2* spring (permanent and intermittent), *3* stream, *4* lake, *5* waterlogged area, *6* col, *7* exposed limestone, *8* cliff, *9* roche moutonnée or karstic residual feature, *10* edge of the DSD, *11* mound on the margin of the depression system, *12* limestone ridge, *13* denuded side slope of the depression, *14* accumulational slide slope of depression, *15* cryptokarstic part of the depression floor, *16* concealed karstic part of the depression floor, *17* nivation and mass movement features of the side slope, *18* slide, *19* thufur field, *20* infilled ponor, *21* active ponor, *22* recent uvala, *23* solution doline, *24* shaft, *25* rain furrow, *26* gully, *27* channel on valley floor, *28* valley, *29* tourist track

Trudgill (1985). The latter, for instance, in the Peak District, England, often show bowl-shaped headwater areas followed downvalley by a section between rock walls. The tributary dry valleys often hang above the main dry valley. According to Warwick (1964), dry valleys are of epigenetic origin. A condition to their development is sinking karst water table (Fagg 1954; Spark and Lewis 1957), caused by the subsidence of the base level of erosion (Pinchemel 1954; Jennings 1985; Williams 1983).

The eroded slopes of dry valleys are getting gentler and the floors are infilled (a valley with wide or bowl-shaped floor evolves). On the valley floor, small intermittent watercourses form channels, ravines and gullies. In dry valleys, karst features

Fig. 5.92 Remote (**A**) and close view (**B**) of alluvial streamsink doline-type DSD (the polje next to road number 513, Croatia): *1* alluvial streamsink doline, *2* interior doline, *3* interior doline on side slope, *4* exterior doline, *5* ravines

Fig. 5.93 Interior of alluvial streamsink doline-type DSD (Cerkniško polje, Slovenia): *1* DSD, *2* interior doline on floor, *3* valley of DSD, *4* chimney of katavothron, *5* non-karstic pipe, *6* water level of impounded lake, *7* channels ending at the water level of the former lake in the cover, *8* channels ending below the water level of the former lake in the cover, *9* channels crossing the water level of the former lake in the cover, *10* ring-like accumulation

can be absent, but they can occur too. On their floor there are ponors and dolines of various types. Lehmann (1936) also mentions depressions with lake from the floor of dry valleys. Dry valleys can originate on permafrost karst without inheritance.

5.7.3 *Blind Valley*

Blind valley forms if the valley or part of it represents a closed landform. Blind valleys were described by Cvijič (1893), Pohl (1955), Williams (1970), Jennings et al. (1980), Sweeting (1973), Gunn (2004) and Ford and Williams (2007). Cvijič (1893) distinguished between primary and secondary blind valleys. The primary blind valley is found on rock boundary (at the termination of the non-karstic rock), where ponors originate. The secondary blind valley is found on the floor of a normal karst valley where ponor forms in the channel. Ford and Williams (2007) also identified half-blind valleys, which develop in an early stage of blind valley formation when surface water conduction is less developed (karstic passages are small) or at a more advanced stage when conduits are infilled. In both cases the depression of the ponor is at least partially infilled and the water flowing here overflows the blind valley (or the ponor).

The size of blind valley depends on several factors. The length of the valley in the karst depends on the catchment area of the watercourse (on water discharge): the higher the discharge, the longer the karst section of the valley. If water discharge is sufficiently high, the watercourse can cut through the karst (allogenic valley forms). The length and width of the blind valley also depend on the carbonate content of the watercourse (Gams 1962). If carbonate concentration is high, the section of the blind valley developed in karstic rock is short and narrow, while where the concentration is lower, the blind valley is long and wide. As a consequence, blind valleys also grow through solution.

Blind valleys develop under variable environmental conditions: on karst margins, in karst interiors, karst depressions, rock basins or valleys. In the latter case, they develop in allogenic valleys, dry valleys or as independent blind valleys.

According to Ford and Williams (2007), their morphology changes during their evolution. Initially, if the water discharge of the watercourse drops, the valley floor (channel) is dissected, sink points appear and develop into 'stream sink depressions' with overflow. With further deepening overflow ceases and 'channel downstreams' result. The blind valley occurs at this point of evolution. The valley section beyond the ponor becomes dry. Occasionally, with the preservation of the cover patch, a row of 'sink points' takes place as the site of capture of the watercourse retreats.

The blind valleys on the karst margin are elongated in ground plan, while the karst interior blind valleys, the centripetal blind valleys are of circular shape. According to Ford and Williams (2007), the latter are transformed into 'point recharge dolines' with the removal of the cover. Eventually, with the complete denudation of the cover, dolines are left behind. Karst marginal blind valleys can deepen and broaden. With the merging of neighbouring blind valleys, larger closed depressions and boundary poljes come about (Ford and Williams 2007).

5.8 Remnant Caves

Remnant caves (Veress 2000) originate from the closed cavities of the phreatic zone through opening up to the surface. Remnant caves of uncovered karst and covered karst origin are identified. For those of uncovered karst origin, the formation and opening of the cavity is not associated with covered karst formation processes. This group includes coastal opened cavities. Both the first stage of covered karst remnant cave formation (the development of the cave) and the second (opening of the closed cavity) are bound to covered karst processes. Cavity formation takes place through solution in the phreatic zone (below the karst water table). Occasionally, both stages are linked to karstic influences but sometimes only the first or the second. Such cavities most often form during epigenetic valley formation. The characteristic properties of covered karst remnant caves are the following:

- They may occur on the sides or surface of blocks (Fig. 5.94), side slopes of karstic remnants (karst inselberg) or in valley sides (Fig. 5.95).
- Remnant caves are mostly cavities of minor dimensions (1–2 m high, several tens of metres long).
- In remnant caves features occur (or occasionally dominate) which cannot develop in the vadose zone (blind chimneys, spherical cavities). Such features indicate an environment of karst water (phreatic zone).
- Remnant caves are of horizontal extension, terminate blindly and are often found in hanging position in the sides of valley and karst inselbergs.
- Remnant caves occur in groups.
- Their altitudinal positions vary greatly within a short distance (e.g. in a valley) (Fig. 5.95).
- In the vicinity of remnant caves' numerous cavities, spherical cavity (or their ruins) is found on rock walls. Around them natural arches are also common as well as corridors (cave remnants) of several metres length with vertical cliffs (Figs. 5.94 and 5.96). Here the ceilings of former remnant caves are either missing or only their ruins are preserved. These remnant features could develop as the caves of this type are often in summit position and, consequently, has a thin roof.
- There is no tufa accumulation in the foreland of remnant caves and this indicates that they are not former outflow sites. Occasionally, there are deposits of non-karstic (e.g. fluvial) origin in the caves.
- In the remnant caves, reverse delta formation is often observed. This means that while the spring caves show corridors of single entrance branching with distance from the entrance, remnant caves may have several entrances, i.e. they often fork into branches towards the entrance.

The following varieties of remnant caves are identified (Fig. 5.97):

- Truncated remnant cave (Fig. 5.98a, b): Almost horizontal cavity, opened up during valley incision and shortening with the erosion of the valley side (of variable length but less than some metres).

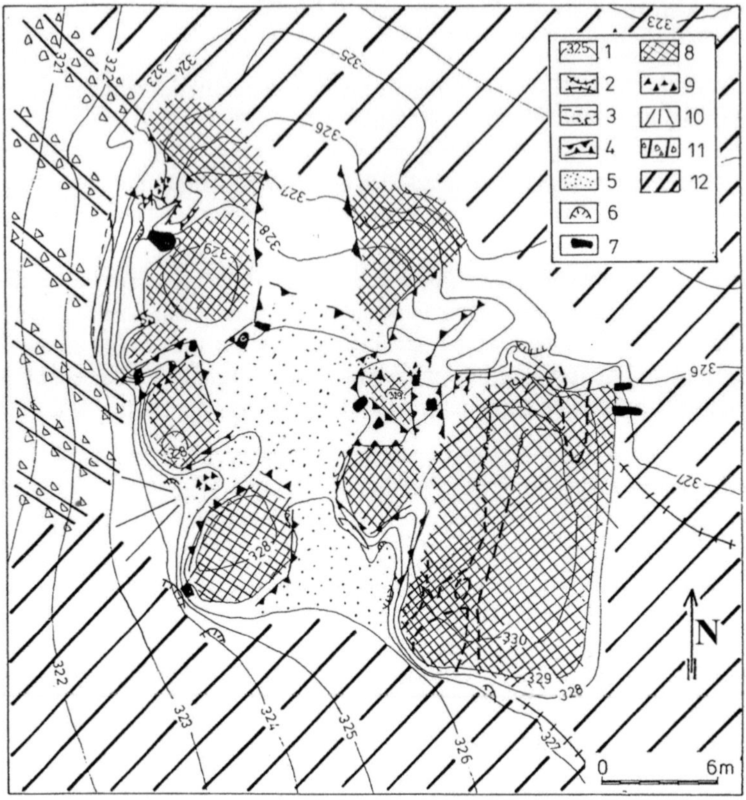

Fig. 5.94 Geomorphological map of Likas-kő of Hódos-ér (Likas-kő near the valley of Hódos stream, Bakony Mountains) (Veress and Futó 1987): *1* contour line, *2* col, *3* ruined remnant cave, *4* cave remnant, *5* fill, *6* remnant of spherical cavity, *7* chimney remnant, *8* block of Eocene limestone, *9* collapsed material, *10* debris fan, *11* debris slope, *12* terrain with unconsolidated, reworked sediment

- Ruined remnant cave (Fig. 5.98c, d): The ceiling of the remnant cave is destroyed on a shorter section. Above the destroyed part, a truncated remnant cave forms, while below it a ruined remnant cave originates. The ruined remnant cave is a through cave with two entrances.
- Cave remnant results from the destruction of the ceiling of the opened-up cave (Fig. 5.99). This may happen in two ways: The ceiling is either destroyed along the whole length of the former cave or only along some sections. In the first case, a depression bordered by steep rock walls and with the same width, depth and length as the former cave is left behind in the place of the cave. In the latter case, the cave remnant section is ruined remnant cave, while above it a truncated remnant cave forms (Fig. 5.98e, f).

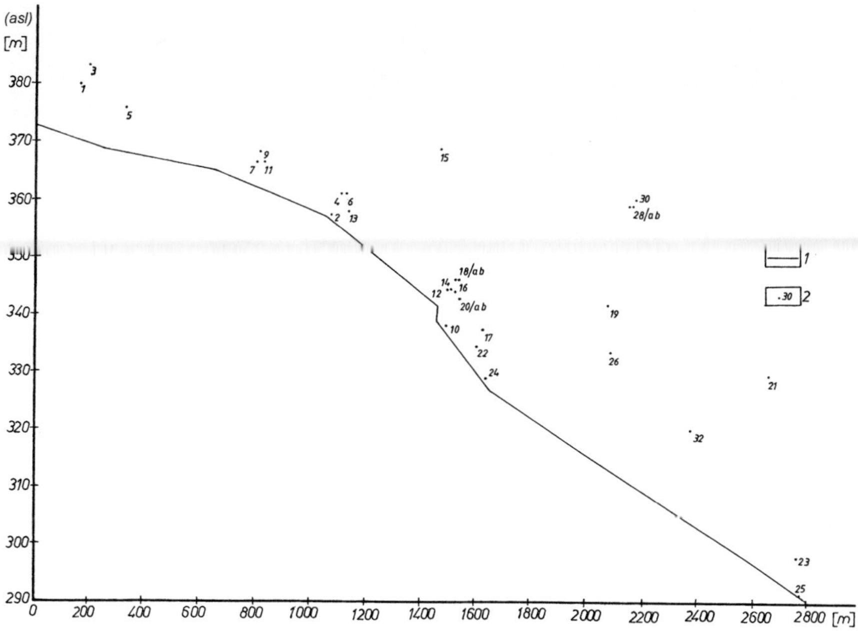

Fig. 5.95 Positions of remnant caves in the Ördög Valley (Bakony Mountains, Veress 2000): *1* channel floor, *2* remnant cave with identification number

Fig. 5.96 Cave remnants of Likas-kő of Hódos-ér: *1* cave corridor which lost its roof, *2* cave chamber which lost its roof

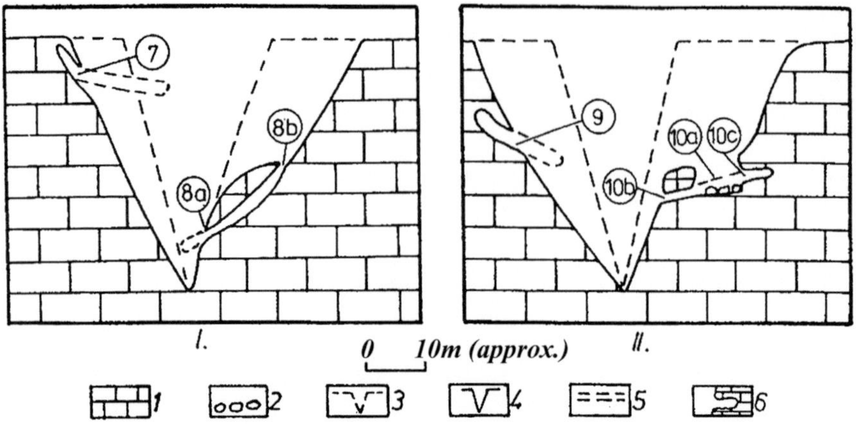

Fig. 5.97 Cave remnants exposed by erosion and further evolved by denudation of the valley side (*I*) and exposed and further evolved by the denudation of the valley side (*II*) (After Veress 2000): *1* karstic rock, *2* collapsed roof, *3* older valley, *4* present valley, *5* destroyed cavity section, *6* remnant cave, *7* ruined remnant cave exposed by stream erosion and truncated by the denudation of the valley side, *8* remnant cave exposed by linear stream erosion (*8a*) and further developed by the denudation of valley side into a ruined remnant cave (*8b*), *9* ruined remnant cave exposed and truncated by the denudation of the valley side, *10* ruined cave remnant exposed by the denudation of the valley side and roof collapse (*10a*) ruined remnant cave isolated by collapse (*10b*) and truncated cave (*10c*)

5.9 Conclusions

Among karren formed under cover are subsoil grikes and pinnacles are most widespread. The former are primarily typical of temperate karsts, while the latter are of tropical covered karsts. The subsoil grikes are particularly variable both for size and morphology. Among karren, subsoil channels and solution pipes are also common. Diverse karren features (bedding plane grike, notches, cavities, solution pipes, subcutaneous tubes) develop under covers which contain groundwater or karst water. Due to the intensive solution of rock salt, on the cover of this rock grikes, karren tables and solution pipes of salt breccia are common.

The three varieties of caprock dolines are the steep-walled C_1 and C_2 dolines both with large diameter and the C_3 caprock dolines of more gentle slopes. The C_1 variety is a large-size cavity, and the C_2 variety forms above a breccia pipe and the C_3 variety above a shaft.

The two types of subsidence dolines are dropout and suffosion dolines. Particularly the latter has variable slope shapes, cross-sections and morphology: gullies, mass movement landforms (scars originated from displacements and their accumulation features) as well as interior covered karst features (non-karstic pipes, shafts, interior suffosion and dropout dolines). Slope shape and cross-section of suffosion dolines are indicative of the processes taking place in them.

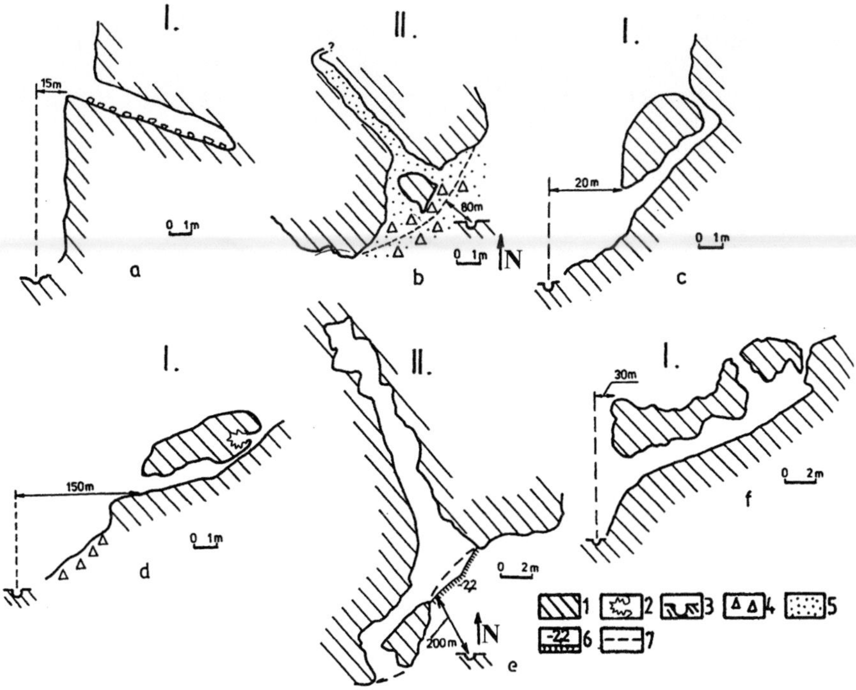

Fig. 5.98 Some typical remnant caves from the Bakony Mountains in longitudinal profile (*I*) and ground plan (*II*) (Veress 2000): *1* bearing rock, *2* spherical cavity, *3* channel, *4* debris, *5* soil, solution residue, *6* rock wall with depth, *7* margin of the cave entrance, (**a**) remnant cave opened up by linear erosion (C-3 or Cuhavölgyi Rejtett-fülke, Cuha Valley), (**b**) ruined remnant cave with several entrances opened up by the denudation of the valley side (K-8 or Kő-völgy rock hollow, Kő Valley), (**c**) remnant cave opened up by the denudation of the valley side and then turned to a ruined remnant cave (Km-1 or Átjáró Cave of Csesznek, Kőmosó Valley), (**d**) remnant cave opened up by the denudation of the valley side and then turned to a ruined remnant cave (M-4 or Likas-kő of Magos Mountain, Dudar Stream), (**e**) remnant cave opened up by the destruction of escarpment and then separated into truncated cave remnants and ruined cave remnants (M-5 or Csapóné-konyhája Cave, at the valley of Dudar Stream), (**f**) remnant cave opened up by the denudation of the valley side and then separated into ruined remnant caves (C-4 or Remete-lik, Cuha Valley)

The varieties of dropout dolines are large-size (D_1), medium-size (D_2), breccia pipe (D_3), small-size rock block (D_4), small-size soil (D_5), flat-floored (D_6), asymmetric (D_7) and interior dropout (D_8) dolines.

The varieties of suffosion dolines are the large-size (S_1), embryonic (S_2), interior suffosion (S_3), debris (S_4), inter-morainic hill (S_5), half suffosional (S_6) and fossil (S_8) dolines. Blind suffosion gullies (S_7) can also be regarded a variety of suffosion dolines.

Subsidence pseudokarst depressions form in variable environments. In the development of some of them, human activities have also played a part.

Ponors are either karst marginal or karst interior. Karst marginal ponors are sometimes aligned in a row with an escarpment at the karst terrain. The ponors are

Fig. 5.99 Remnant caves and cave remnants around Cave Ö-32 (Ördög Valley) (Veress 2000): *1* direction and angle of bed dip, *2* rock wall, *3* height of rock wall (m), *4* direction and angle of slope, *5* entrance to passable remnant cave, *6* impassable remnant cave, *7* impassable ruined remnant cave (through cave), *8* height of karstic passage in rock wall (m), *9* remnant cave, *10* cave remnant, *11* debris, *12* material of collapsed roof

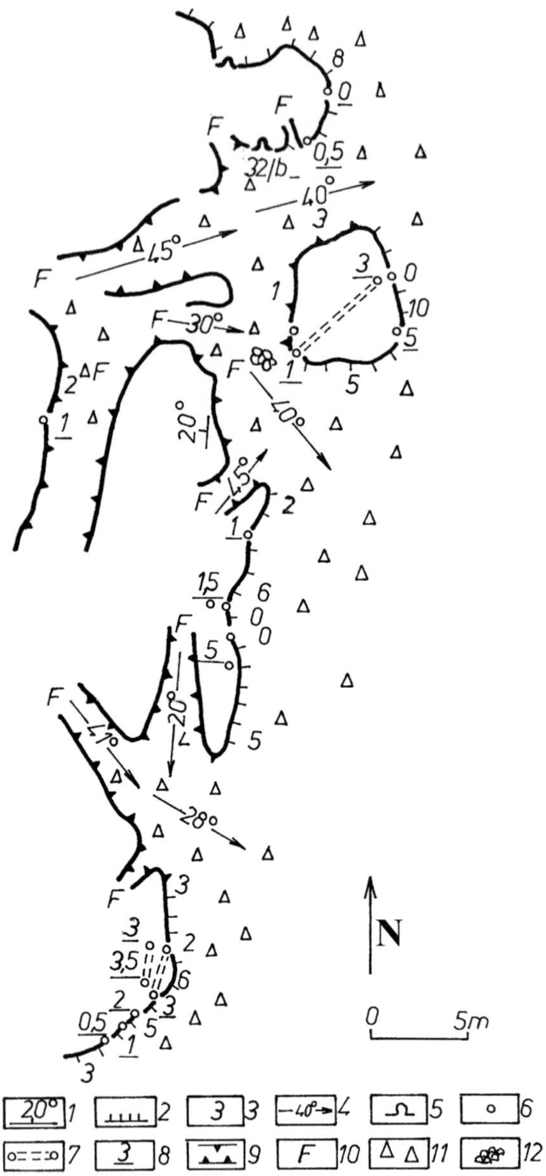

located in the end of blind valleys. Karst interior ponors develop on the floor of epigenetic valleys, at limestone outcrops, at impermeable cover patches overlying the karst, in karst interior poljes and paleodepressions. On covered karst, covered karst ponors are widespread. They do not have a well-definable, independent catchment, the bedrock does not outcrop on their margin, and there is no escarpment on the bedrock. Covered karst ponors may emerge through the further development of caprock dolines (C_2, C_3 types) on autogenic and transitional cryptokarsts, close to rock boundary on concealed karst and if the impermeable cover patch is bordered by permeable cover. Covered karst ponors are transitional between ponors and subsidence dolines. Three further types of closed gullies are presented: those further developed through suffosion or those of exclusively suffosion origin (S_7 subsidence doline variety). The third type is due to the linkage of the gully to an existent chimney. The latter has a ponor, while the former two have none.

The depressions of superficial deposit (DSDs) developed in cover sediment are landforms of large diameter compared to their depth, and they have ponors, covered karst ponors and subsidence dolines in their interior. By the morphology of the bedrock, pseudodepressions and true depressions are identified; by their cover, covered and half-covered depressions; by their position, karst marginal and karst interior; and by the lithology of their cover, sediment, concealed, cryptokarst and composite karst depressions are identified.

The epigenetic valleys on karst are valleys with permanent water flow (with V cross-section) or with intermittent water flow or dry valleys (with infilled floor). Remnants caves often occur in epigenetic valley sections (epigenetic gorges).

Remnant caves are opened-up cavities with solution features. During erosion, different varieties come about: truncated and ruined remnant caves and cave remnants.

References

Andrejchuk V (2002) Collapse above the world's largest potash mine (Ural, Russia). Int J Speleol 31(4):137–158

Andrejchuk V, Eraso A (1996) Karren landform on the artificial salt massives in the Ural area. In: Fornós IJ, Ginés Á (eds) Karren landforms. Universitat de les Illes Balears, Palma de Mallorca, pp 243–252

Andrejchuk V, Klimchouk A (2002) Mechanism of karst of the fore-Ural region, Russia (from observations in the Kungurskaja Cave). Int J Speleol N 31(1–4):89–114

Balázs D (1970) Zsombolyok a Central Kentucky Karszton (Shafts on the Central Kentucky Karst). Karszt és Barlang I:21–24 (in Hungarian)

Balázs D (1984) Exhumált trópusi őskarszt Lapinha vidékén (Minas Gerais, Brazília) (Exhumed tropical paleokarst in the Lapinha area) (Minas Gerais, Brazil). Karszt és Barlang II:87–92 (in Hungarian)

Barrére P (1964) Le relief karstique dans l'ouest des Pyrénées centrales. Rev. Belge Géogr. Ed. Soc. Roy. Belge Géogr. Special Publ., Karst et Climats Froids 88(1–2):9–62

Beck BF (1986) A generalized genetic framework for the development of sinkholes and karst in Florida, USA. Environ Geol Water Sci 8:5–18

Beck BF, Sinclair WC (1986) Sinkholes in Florida: an introduction. Florida Sinkhole Research Institute Report 85–86-4, 16 p

Bell FG, Culshaw MG, Cripps JC (1999) A review of selected engineering geological characteristics of English chalk. Eng Geol 54:237–269

Bennett D (1997) Finding a foothold. New Civil Engineer (4 Dec) pp 24–25

Bergado DT, Selvanayagam AN (1987) Pile foundation problems in Kuala Lumpur Limestone, Malaysia. Q J Eng Geol 20:159–175

Bögli A (1960) Kalklösung und Karrenbildung. Zeitschrift für Geomorphologie, Suppl. 2:4–21 (an English translation by E. Werner was published in Cave Geology 1(1):3–28, 1975)

Bögli A (1976) Die wichtigsten Karrenformen der Kalkalpen. In: Karst processes and relevant landforms. Department of Geography, Philosophical Faculty, Ljubljana, pp 141–149
Bögli A (1980) Karst hydrology and physical speleology. Springer, Berlin, 284 p
Brink ABA, Partridge TC (1965) Transvaal karst: some considerations of development and morphology, with special reference to sinkholes and subsidence in the Far West Rand. S Afr Geogr J 47:11–34
Brook GA (2004) Africa, Sub-Sahara. In: Gunn J (ed) Encyclopedia of caves and karst science. Fitzroy Dearborn, New York, pp 20–23
Brook GA, Allison TL (1986) Fracture mapping and ground subsidence susceptibility modelling in covered karst terrain: the example of Dougherty County, Georgia. Int Assoc Hydrol Sci 151:595–606
Brook GA, Fenney TP (1996) Morphology and denudation of quartzite and limestone pavements in Southern Africa and North America: are they small scale versions of labyrinth karst? In: Fornós IJ, Ginés Á (eds) Karren landforms. Universitat de les Illes Balears, Palma de Mallorca, pp 25–39
Brook GA, Ford DC (1978) The origin of labyrinth and tower karst and the climatic conditions necessary for their development. Nature 275:493–496
Bull PA (1977) Cave boulder chokes and doline relationships. In: Proceedings of the 7th international congress speleology, pp 93–96
Bull PA (1980) The antiquity of caves and dolines in the British Isles. Z Geomorphol 36:217–232
Chen Z, Song LH, Sweeting MM (1986) The pinnacle karst of the Stone Forest, Lunan, Yunnan. China: an example of a subjacent karst. In: Paterson K, Sweeting MM (eds) New directions in karst. Geo Books, Norwich, pp 597–607
Clozier R (1940) Les Causses du Quency. Librairie J.-B. Baillère et Fils, Paris, 183 p
Cooper AH (1998) Subsidence hazards caused by the dissolution of Permian gypsum in England: geology, investigation and remediation. Geol Soc Eng Group Spec Publ 15:265–275
Cooper AH, Waltham AC (1999) Subsidence caused by gypsum dissolution at Ripon North Yorkshire. Q J Eng Geol 32:305–310
Coxon C (1986) A study of the hydrology and geomorphology of turloughs. Unpublished PhD thesis, University of Dublin
Cramer H (1941) Die Systematik der Karsdolinen. N Jb Miner Geol Paleontol 85:293–382
Crawford N (1984) Karst landform development along the Cumberland Plateau escarpment of Tennessee. In: LaFleur RS (ed) Groundwater as a geomorphic agent. Allen and Unwin, Boston, pp 294–339
Culshaw MG, Waltham AC (1987) Natural and artificial cavities as ground engineering hazards. Q Journal Eng Geol 20:139–150
Curtis LF, Courtney FM, Trudgill ST (1976) Soils in the British Isles. Longman, London, 364 p
Cvijič J (1893) 'Das Karstphänomen'. Versuch einer morphologishen Monographie. Geogr Abh A Penck 5(3):217–329
Davies WE, LeGrand H (1972) Karst of the United States. In: Herak M, Stringfield VT (eds) Important karst regions of the northern hemisphere. Elsevier, Amsterdam, pp 467–505
Derbyshire E, Meng X, Dijkstra TA (eds) (2000) In the thick loess terrain of north-west China. Wiley, New York, 288 p
Dicken S (1935) Kentucky karst landscapes. J Geol 43:708–728
Drumm EC, Kane WF, Yoon CJ (1990) Application of limit plasticity of the stability of sinkholes. Eng Geol 29:213–225
Edmonds CN (2001) Predicting natural cavities in chalk. Geol Soc Eng Geol Spec Publ 18:283–293
Fagg CC (1923) The recession of the chalk escarpment. Proc Croydon Nat Hist Sci Soc 9:93–112
Fagg CC (1954) The coombes and embayments of the chalk escarpment. Proc Croydon Nat Hist Sci Soc 12:117–131

Ford DC, Lundbeg J (1987) A review of dissolutional rills in limestone and other soluble rocks. Catena Suppl 8:119–140
Ford DC, Williams PW (1989) Karst geomorphology and hydrology. Unwin Hyman, London, 601 p
Ford DC, Williams PW (2007) Karst hydrogeology and geomorphology. Wiley, Chichester, 561 p
Forti P, Sauro M (1996) Gypsum karst of Italy. Int J Speleol 25(3–4):239–250
Gams I (1962) Slope doline slovenije v primerjalni metodi (Blind valleys in comparative method). Zbornik v. kongresa geografov FLRJ v Ljubljani, Ljubljana, pp 185–190
Gams I (1971) Podtalne kraške oblike (Subsoil karst forms). Geografski Vestn 43:27–45
Gams I (1976) Forms of subsoil karst. In: Proceedings of the 6th international congress of speleology, Olomouc. Academia, Prague, pp 169–179
Gams I (1994) Types of contact karst. Geogr Fis Din Quateraria 17:37–46
Gams I, Otoničar B, Slabe T (2011) Development of slope and related subsoil karren: a case study from Bela Krajina, SE-Slovenia. Acta Cartologica 40(2):329–340
Gardner RAM (1983) Aelianite. In: Goudie AS, Pye K (eds) Chemical sediments and geomorphology. Academic, London, pp 265–300
Ginés Á (1990) Utilización de las morfologías de lapiaz como geoindicadores ecológicos en la Serra de Tramuntana (Mallorca). Endins 16:27–39
Ginés Á (1996) Quantitative data as a base for the morphometrical definition of rillenkarren features found on limestones. In: Fornós IJ, Ginés Á (eds) Karren landforms. Universitat de les Illes Balears, Palma de Mallorca, pp 177–191
Ginés Á (2004) Karren. In: Gunn J (ed) Encyclopedia of caves and karst science. Fitzroy Dearborn, New York, pp 470–473
Ginés Á (2009) Karrenfield landscapes and karren landforms. In: Ginés Á, Knez M, Slabe T, Dreybrodt W (eds) Karst rock features. Karren sculpturing. Zalozba ZRC. Institut za raziskovanje krasa ZRC SAZU, Postojna, Carsologica 9, pp 13–24
Goeppert N, Goldscheider N, Scholz H (2011) Karst geomorphology of carbanatic conglomerates in the Folded Molasse zone of the Northern Alps (Austria/Germany). Geomorphology 130:289–298
Gómez-Pujol L, Fornós JJ (2009) Coastal karren in the Balearic Islands. In: Ginés Á, Knez M, Slabe T, Dreybrodt W (eds) Karst rock features. Karren sculpturing. Zalozba ZRC. Institut za raziskovanje krasa ZRC SAZU, Postojna-Ljubljana., Carsologica 9, pp 487–502
Gorbunova KA (1979) Morphology and hydrogeology of gypsum karst. All-Union Karst and Speleology Institute, Perm, 93 p (In Russian)
Grimes KG (1994) The south-east karst province of South Australia. Environ Geol 23:134–138
Grimes KG (2004) Solution pipes of petrified forests? Drifting sands and drifting opinions! Vict Nat 121(1):14–22
Grimes KG (2009a) Solution pipes and pinnacles in syngenetic karst. In: Ginés Á, Knez M, Slabe T, Dreybrodt W (eds) Karst rock features, karren sculpturing. Zalozba ZRC. Institut za raziskovanje krasa ZRC SAZU, Postojna, Carsologica 9, pp 513–523
Grimes KG (2009b) Laterite Karst. Unpublished poster displayed at the 7th international conference on geomorphology (ANZIAG), Melbourne, 6 p
Grimes KG (2012) Surface karst features of the Judbarra/Gregory National Park, Northern Territory, Australia. Helictite 41:15–36
Grimes KG, Spate AP (2008) Laterite karst (Andysez No 53). ACKMA J 73:49–52
Gunn J (2004) Fluviokarst. In: Gunn J (ed) Encyclopedia of caves and karst science. Taylor and Fitzroy Dearborn, London, pp 363–364
Gunn J (2006) Turloughs and tiankengs: distinctive doline forms. Speleogenesis Evol Karst Aquifers 4(1):1–4
Gvozdetskiy NA (1981) Karst. Izd-vo Myszl', Moscow, 214 p
Halliday WR (2004) Pseudokarst. In: Gunn J (ed) Encyclopedia of caves and Karst science. Fitzroy Dearborn, New York, pp 604–608
Haserodt K (1965) Untersuchungen zur Hohen- und Altersgliederung der Karstformen in den Nördlichen Kalkalpen. Münchner Geogr. H. 27. Regensburg, 114p

Hevesi A (1978) A Bükk szerkezet- és felszínfejlődése. (An outline of structural and geomorphological development of Bükk Mts). Földr Ért 27(2):169–203 (in Hungarian)
Hevesi A (1980) Adatok a Bükk hegység negyedidőszaki ősföldrajzi képéhez (Data to the quaternary paleogeographical features of the Bükk Mountains) Földtani Közlöny 110(3–4):540–550 (in Hungarian)
Hose LD, Strong TR (1981) The genetic relationship between breccia pipes and caves in non-karstic terranes in northern Arizona. In: Proceedings 8th international congress speleology. National Speleological Society, Huntsville, pp 136–138
Howard AD (1963) The development of karst features. Bull Nat Spel Soc 25:45–65
Hundt R (1950) Erdfalltektonik. Knapp, Halle, 145 p
Hyatt JA, Wilkes HP, Jacobs PM (1999) Spatial relationships between new and old sinkholes in covered karst, Albany, Georgia, USA. In: Beck BF, Pettit AJ, Herring JG (eds) Hydrogeology and engineering geology of sinkholes and karsts. Balkema, Rotterdam, pp 37–44
Jakucs L (1956) Adatok az Aggteleki-hegység és barlangjainak morfogenetikájához (Some data to the morphogenetics of the mountains and caves of Aggtelek). Földrajzi Közlemények 80(1):25–35 (in Hungarian)
Jakucs L (1971) A karsztok morfogenetikája (Morphogenetics of karsts). A karsztfejlődés varienciái. Akadémiai Kiadó, Budapest, 310 p (in Hungarian)
Jakucs L (1977) Morphogenetics of karst regions. Adam Hilger, Bristol, 283 p
Jakucs L (1980) A karszt biológiai produktum! (Karst is a biological product). Földrajzi Közlemények 104(4):331–344 (in Hungarian)
Jammal SE (1984) Maturation of the Winter Park sinkhole. In: Beck BF (ed) Sinkholes: their geology, engineering and environmental impact. Balkema, Rotterdam, pp 363–369
Jennings JN (1973) Karst. The M.I.T Press, Cambridge, MA, 253 p
Jennings JN (1985) Karst geomorphology. Basil Blackwell, New York, 293 p
Jennings JN, Bao H, Spate AP (1980) Equilibrium versus events in river behaviour and blind valleys Yarrango billy, New South Wales. Helectite 18:39–54
Johnson KS (1987) Development of the Wink Sink in West Texas due to salt dissolution and collapse. In: Beck BF, Wilson WL (eds) Karst hydrogeology: engineering and environmental implication. Balkema, Brookfield, pp 127–136
Johnson KS (1989) Development of the Wink Sink in West Texas, USA, due to salt dissolution and collapse. Environ Geol Water Sci 14:81–92
Johnson KS, Collins EW, Seni SJ (2003) Sinkholes and land subsidence due to salt dissolution near wink, west Texas and other sites in western Texas and New Mexico. Okla Geol Surv Circ 109:163–195
Károlyi MS, Ford DC (1983) The goose arm karst, Newfoundland. J Hydrol 61(1–3):181–186
Kirkaldy JF (1950) Solution in the calk in the Mimms Valley. Herts Proc Geol Ass 61:219–224
Klimchouk A (2004) Evaporite karst. In: Gunn J (ed) Encyclopedia of caves and karst science. Fitzroy Dearborn, New York, pp 343–347
Klimchouk A, Andrejchuk V (1996) Breakdown development in cover beds and landscape features induced by interstratal gypsum karst. Int J Speleol 24(3–4):127–144
Klimchouk A, Andrejchuk V (2002) Karst breakdown mechanisms from observations in the gypsum caves of the western Ukraine: implications for subsidence hazard assessment. Speleogenesis Evol Karst Aquifers (www.speleogenesis.info), 1(1):20 p
Knez M, Slabe T (2002) Lithologic and morphological properties and rock relief of the Lunan stone forests. In: Gabrovšek F (ed) Evolution of karst: from prekarst to cessation. Karst Research Institute ZRC SAZU, Postojna, pp 259–266
Knez M, Slabe T, Travassos LEP (2011) Karren on laminar calcarenitic rock of Lagoa Santa (Minas Gerais, Brazil). Acta Carstologica 40(2):357–367
Korzhuev SS (1961) Merzlotnyi karst Srednego Prilen'ya i nekotorye osobennosti yego proyavleniya. (The Middle-Lena frozen karst and its characteristics). In: Sokolov NI, Gvozdetskiy NA, Balashov LS (eds) Regionalnoe karstovedenie. Izdatelstvo AN SSSR, Moscow, pp 207–220

Kozma K, Holló S (2010) A Berva-pataki víznyelő kialakulása és pusztulása az egri Berva-bérc lábánál (The development and denudation of the ponor of Berva stream near the Berva mount in Eger) . Karsztfejlődés XV:103–112 (in Hungarian)
Láng S (1971) A hazai karsztok és környékük lepusztulásának egyes kérdései (Some questions of the denudation of Hungarian karsts and their environments). Karszt és Barlang I:1–4 (in Hungarian)
Lehmann H (1936) Morphologische Studien auf Java. Engelhorn, Stuttgart, 114 p
Lu Y, Cooper AH (1997) Gypsum karst geohazards in China. In: Beck BF, Stephenson JB (eds) Engineering geology and hydrogeology of karst terrains. Balkema, Rotterdam, pp 117–126
Lucas J (1872) Studies in Nidderdale. (Reprint Publication) Forgotten Books Hongkong, 319 p
Macaluso T, Sauro U (1996) The karren in evaporite rocks: a proposal of classification. In: Fornós IJ, Ginés Á (eds) Karren landforms. Universitat de les Illes Balears, Palma de Mallorca, pp 277–291
Macaluso T, Madonia G, Palmer A, Sauro U (2001) Atlante dei karren nelle evaporiti della Sicilia (Atlas of the karren in the evaporitic rocks of Sicily). Quaderni del Museo Geologico G. G. Gemmellaro 5. Dipartimento di Geologia e Geodesia, Universitá degli Studi di Palermo. 143 p
Madonia G, Sauro U (2009) The karren landscapes in the evaporitic rocks of Sicily. In: Ginés Á, Knez M, Slabe T, Dreybrodt W (eds) Karst rock features. Karren sculpturing. Zalozba ZRC. Institut za raziskovanje krasa ZRC SAZU, Postojna-Ljubljana, Carsologica 9, pp 525–533
Maire R, Zhang S, Song S (1991) Génese des karsts subtropicaux de Chine du sud (Guizhou, Sichuan, Hubei). Grottes Karsts Trop Chin Meridionale Karstologia Mém 4:162–186
Martinez JD, Johnson KS, Neal JT (1998) Sinkholes in evaporite rocks. Am Sci 86:38–51
Martini JEJ, Grimes KG (2012) Epikarstic Maze cave development: Bullita cave system, Judbarra/ Gregory Karst, tropical Australia. Helictite 41:37–66
Mckee ED, Ward WC (1983) Eolian environment. In: Scholle PA, Bebout DE, Moore CH (eds) Carbonate depositional environments. American Association of Petroleum Geologists Memoir 33. American Association of Petroleum Geologists, Tulsa, pp 131–170
Mendonca AF, Pires ACB, Barros JGC (1993) Pseudo-sinkholes in lateritic terrains, Brasilia, Brazil. In: Beck BF (ed) Applied karst geology. Balkema, Rotterdam, pp 43–49
Móga J (2001) A szerkezet és kőzetfelépítés szerepe a Szilicei-fennsík karsztos felszínformáinak kialakításában. Karsztfejlődés VI:143–159 (in Hungarian)
Olive WW (1957) Solution subsidence troughs, castile formation of gypsum plain, Texas and New Mexico. Geol Soc Am Bull 68(B693):351–358
Ollier C (1984) Weathering. Longman, London, 270 p
Ollier CD, Tratman EK (1969) Geomorphology of the caves. In: Tratman EL (ed) The caves of northwest Clare, Ireland. David and Charles, Newton Abbot, pp 59–95
Origo (2013) Gyilkos víznyelők (Killer Ponors) http://www.origo.hu/archivum/20130924-oriasi-karokat-okozhatnak-a-viznyelok.html (in Hungarian)
Paton JR (1963) The origin of the limestone hills of Malaya. J Trop Geogr 18:137–147
Penck A (1924) Das unterirdische Karstphänomen. Receuil de Travaux offert à J. Cvijič, Belgrade, pp 175–197
Perna G, Sauro U (1978) Atlante delle microforme di dissoluzione carsica superficiale del Trentino e del Veneto. Museo Tridentino, Trento, 176 p
Pfeiffer D, Hahn J (1972) Karst of Germany. In: Herak M, Stringfield VT (eds) Karst, important karst regions of the northern hemisphere. Elsevier, Amsterdam, pp 189–223
Pigott CD (1965) The structure of limestone surfaces in Derbyshire. Geogr J 131:41–44
Pinchemel P (1954) Les plaines de craie du Nord-Ouest du bassin parisien et du Sud-Est du bassin de londres, et leurs bordures. Armand Colin, Paris, 502 p
Plan L, Decker K (2006) Quantitative karst morphology of the Hochschwab plateau Eastern Alps, Austria. Z Geomorphol N F 147:29–54
Pohl ER (1955) Vertical shafts in limestone caves. Natl Speleol Soc Occas Pap 2:24
Quinlan JF, Smith AR, Johnson KS (1986) Gypsum karst and salt karst of the United States of America. Le Grotte Ital 4(13):73–92

Rodet J (1992) La Craie et ses Karsts Centre de Géomorphologie du Centre. National de la Recherche Scientifique, Caen, 560 p
Roglič J (1964) Karst valleys in the Dinaric Karst. Erdkunde 18:113–116
Sásdi L (1987) Gipszkarszt jelenségek Alsótelekesen (Gypsum karst phenomena at Alsótelekes). Karszt és Barlang I-II:17–22 (in Hungarian)
Slabe T (1992) Naravni in poskusni obnaplavinski jamski skalni relief (Natural and experimental cave rocky relief on the contact of water and sediments). Acta Carsologica 21:7–34
Slabe T (1995) Cave rocky relief. Znanstvenaraziskovalni Center Sazu, Llubljana, 128 p
Slabe T (1999) Subcutaneous rock forms. Acta Carsologica 28(2):255–271
Slabe T, Liu H (2009) Significant subsoil rock forms. In: Knez M, Slabe T, Dreybrodt W (eds) Karst rock features. Karren sculpturing. Zalozba ZRC. Institut za raziskovanje krasa ZRC SAZU, Postojna, Carsologica 9, pp 123–137
Smith DI, High C, Nicholson FH (1969) Limestone solution and the caves. In: Tratman EK (ed) The caves of northwest Clare, Ireland. David and Charles, Newton Abbot, pp 96–123
Solomonov N, Kolosov P, Kipriyanova L, Knapp HD, Zhuravlev A, Trofimova E, Maksakovskiy N, Butorin A, Petrovskaya E (2010) Nominaciya Prirodiny Park 'Lenskie Stolby' (Rossiyskaya Federaciya). http://www.nhpfund.ru/files/lenapillars-nature park-nomination-ru.pdf
Soriano MA, Simón JL (2001) Subsidence rates of alluvial dolines in the central Ebro basin, Northeastern Spain. In: Beck BF, Herring JG (eds) Geotechnical and environmental applications of karst geology and hydrology. Balkema, Lisse, pp 47–52
Sparks BW, Lewis WV (1957) Escarpment dry valleys near Pegsdon, Hertfordshire. Proc Geol Ass 68:26–38
Sperling CHB, Goudie AS, Stoddart DR, Poole GG (1977) Dolines of the Dorset chalklands and other areas in southern Britain. Trans Ins Br Geogr 2:205–223
Spooner J (1971) Mufulira interim report. Min J 276:122
Sweeting MM (1955) Landforms in North-West Country Clare, Ireland. Trans Inst Br Geogr 21:218–249
Sweeting MM (1973) Karst landforms. Columbia University Press, New York, 362 p
Szunyogh G (1999) A talajelborítás hatása a karros formakincs fejlődésre (The effect of soil cover on the development of karren features). Karsztfejlődés III:31–42 (in Hungarian)
Tan BK (1987) Some geotechnical of urban development over limestone terrain in Malaysia. Bull Int Assoc Eng Geol 35:57–63
Tharp TM (1999) Mechanics of upward propagation of cover-collapse sinkholes. Eng Geol 52:23–33
Thomas TM (1954) Swallow holes on the millstone grit and carboniferous limestone of the south Wales coalfield. Geogr J 120:468–475
Thomas TM (1963) Solution subsidence in southeast Carmarthenshire and southwest Breconshire. Trans Inst Br Geogr 33:45–60
Thomas TM (1974) The South Wales interstratal karst. Trans Br Cave Res Assoc 1:131–152
Trudgill ST (1976) Limestone erosion under soil. In: Panos V (ed) Proceedings of the 6th international congress of speleology II. Ba. Academia, Prague, pp 409–422
Trudgill ST (1985) Limestone geomorphology. Longman, New York, 196 p
Trudgill ST (1986) Limestone weathering under a soil cover and the evolution of limestone pavements, Malham district, North Yorkshire, UK. In: Paterson K, Sweeting MM (eds) New directions in karst. Geobooks, Norwich, pp 461–471
Urai JL, Spiers CJ, Zwart HJ, Lister GS (1986) Weakening of rock salt by water during long-term creep. Nature 324:554–557
Veress M (1992) Karsztmorfológiai sajátosságok a Pádis fedett karsztjának példáján (Karstmorphological characteristics using an example of the covered karst of Pádis). Földrajzi Közlemények 114(3–4):125–141 (in Hungarian)
Veress M (2000) Covered karst evolution Northern Bakony mountains, W-Hungary. A Bakony Természettud. Kut. Eredményei 23, Bakonyi Természettudományi Múzeum, Zirc, 167 p
Veress M (2005) A horvátországi Locrum szigetének tengerparti karsztjai (Coastal karren of the Locrum island, Croatia). Karsztfejlődés X:207–219 (in Hungarian)

Veress M (2008) Covered karstification on the karsts of Hungary. In: Kertész Á, Kovács Z (eds) Dimensions and trends in Hungarian geography. Geographical Research Institute Hungarian Academy of Sciences, Budapest, pp 69–90
Veress M (2009a) Rinnenkarren. In: Ginés Á, Knez M, Slabe T, Dreybrodt W (eds) Karst rock features, karren sculpturing. Zalozba ZRC. Institut za raziskovanje krasa ZRC SAZU, Postojna-Ljubljana, Carsologica 9, pp 151–159
Veress M (2009b) Investigation of covered karst form development using geophysical measurements. Z Geomorphol 53(4):469–486
Veress M (2010a) Karst environments. Karren formation in high mountains. Springer, Dordrecht, 230 p
Veress M (2010b) A magyarországi eltemetett és rejtett karsztos térszínek felszínfejlődése (Development of cryptokarstic and latent karstic surfaces in Hungary). Földrajzi Közlemények 134(4):373–391 (in Hungarian)
Veress M (2012) Glacial erosion and karst evolution (Karren on the surfaces formed by glaciers). In: Veress B, Szigethy J (eds) Horizons in earth science research. Nova Science, New York, pp 1–94
Veress M, Futó J (1987) Adatok a Hódos-éri Likas-kő morfogenetikájához (Contributions to the morphogenesis of the Likas-kő of the Hódos-ér). Karszt és Barlang I–II:9–16 (in Hungarian)
Veress M, Péntek K (1996) Theoretical model of surface karstic processes. Z Geomorphol 40(4):461–476
Veress M, Szabó L, Zentai Z (1998) Mésztartalomhoz köthető felszínfejlődés a Kőszegi-hegységben (Surface development based on calcareous content in the Kőszeg Mts). Földrajzi Értesítő 47(4):495–514 (in Hungarian)
Veress M, Zentai Z, Kovács Gy (1999) Álkarros formák a Bosco Seccoi forrás (Asiagói-fennsík) környékén (Pseudokarren features near the Bosco Secco spring, Asiago plateau). Karsztfejlődés II:19–29 (in Hungarian)
Veress M, Tóth G, Zentai Z, Schläffer R (2008a) Kürtőképződés egy madagaszkári sziklás parton (Chimney development on a rocky coast of Madagascar). Karsztfejlődés XIII:135–149 (in Hungarian)
Veress M, Lóczy D, Zentai Z, Tóth G, Schläffer R (2008b) The origin of the Bemaraha tsingy (Madagascar). Int J Speleol 37(2):131–142
Veress M, Tóth G, Zentai Z, Schläffer R (2009) The Ankarana tsingy and its development. Carpathian J Earth and Environ Sci 4(1):95–108
Veress M, Puskás J, Zentai Z, Zs B (2011) Development of karren formation on the Saltic Hill of Praid (Transylvanian Basin, Romania). Carpathian J Earth and Environ Sci 6(2):183–194
Walters RF (1977) Land subsidence in central Kansas related to salt dissolution. Bull Kansas Geol Surv 214:82 p
Waltham AC (1984) Some features of karst geomorphology in South China. Cave Sci 11:185–199
Waltham AC, Fookes PG (2003) Engineering classification of karst ground conditions. Q J Eng Geol Hydrogeol 36:101–118
Waltham T, Bell F, Culshaw M (2005) Sinkholes and subsidence. Springer, Berlin, 382 p
Warwick GT (1964) Dry valleys in the southern Pennines, England. Erdkunde 18:116–123
Wassmann TH (1979) Mining subsidence in the East Netherlands. In: Saxena SK (ed) Evaluation and prediction of subsidence. American Society Civil Engineers, New York, pp 283–302
Webb JA, Grimes KG, Lewis ID (2010) Volcanogenic origin of cenotes near Mt. Gambier, south-eastern Australia. Geomorphology 119(112):40–52
Wenrich KJ, Sutphin HB (1994) Grand Canyon caves, breccia pipes and mineral deposits. Geol Today 10:97–104
White WB (1988) Geomorphology and hydrology of karst terrains. Oxford University Press, New York, 464 p
White WB, Watson RA, Pohl ER, Brucker R (1970) The central Kentucky karst. Geogr Rev 60(1):88–115
Wilford GE, Wall JRD (1965) Karst topography in Sarawak. J Trop Geogr 21:44–70

Williams PW (1970) Limestone morphology in Ireland. Irish Geographical Studies, Department of Geography, Queen's University, Belfast, pp 105–124
Williams PW (1983) The role of the subcutaneous zone in karst hydrology. J Hydrol 61:45–67
Williams PW (2004) Dolines. In: Gunn J (ed) Encyclopedia of caves and karst science. Fitzroy Dearborn, New York, pp 304–310
Yu Y, Yang B (1997) Paleoenvironment during formation of Lunan Stone Forest. In: Song L, Waltham AC, Cao N, Wang F (eds) Stone forest: a treasure of natural heritage. Proceedings of the international symposium for Lunan Shilin to apply for World Natural Heritage Status, China Environmental, Science Press, Beijing, pp 63–67
Zeeden C, Hark M, Hambach U, Markovič SB, Zöller L (2007) Depression on the Titel loess Plateau: form – pattern – genesis. Geogr Pannonica 11:4–8
Zseni A (2004) Talaj alatti karrformák (Subsoil karren features). Karsztfejlődés IX:157–175 (in Hungarian)
Zseni A (2009) Subsoil shaping. In: Knez M, Slabe T, Dreybrodt W (eds) Karst rock features. Karren sculpturing. Zalozba ZRC. Institut za raziskovanje krasa ZRC SAZU, Postojna-Ljubljana, Carsologica 9, pp 103–121

Chapter 6
Covered Karst Processes

Abstract In this chapter the activity and processes related to covered karst depressions, particularly during floods, the associated phenomena and deposition are demonstrated. The reasons for activity are taken into account. Activities are typified by character and modes of water inflow. The phenomena related to activity (flood lake, overflow, throughflow, intermittent springs, onfilling) and types of activity (surface inflow, seepage, latent activity, composite activity) are identified. A typology of flood lakes, the deposits of lakes of various durations (series with plant waste, laminite), is proposed with explanations of their origin, from which conclusions are drawn to the conditions of depressions at the time of deposition. The deposits (charcoal, limonite, etc.) of fossil dolines are presented and the environments of their origin are analysed. Sedimentation from various types of suspensions in flood lakes was modelled in the laboratory. Landform evolution in dolines was classified according to its duration, and factors associated with the date and duration of landform evolution are considered. The changes in the depth of depressions over several years have been measured and the resultant landforms described. Comparing these two data series, the depressions were classified according to their material budget.

Keywords Activity • Character of activity • Activity phenomenon • Activity type • Series with plant waste • Laminite • Fossil doline • Material budget of depression

6.1 Introduction

The term activity denotes the situation that water flows into the karstic passages (shafts, chimneys) of caprock and subsidence dolines and ponors from the bordering surfaces, from cover deposits and bedrocks on the slopes of depressions. This can also happen at open karst depressions (e.g. at solution dolines) (particularly if the karst shaft is situated under a glacier or if a karst lake has an overflow channel). It is also regarded activity if the inflowing water seeps away at a point (or points) of the depression. There are normal and flood-time activities. Normal activity is either permanent or intermittent. Permanent activity is typical of ponors which are fed by a permanent watercourse. Most of the ponors and particularly the dolines are of intermittent activity. Intermittent activity can be normal if the amount of water

M. Veress, *Covered Karsts*, Springer Geology,
DOI 10.1007/978-94-017-7518-2_6

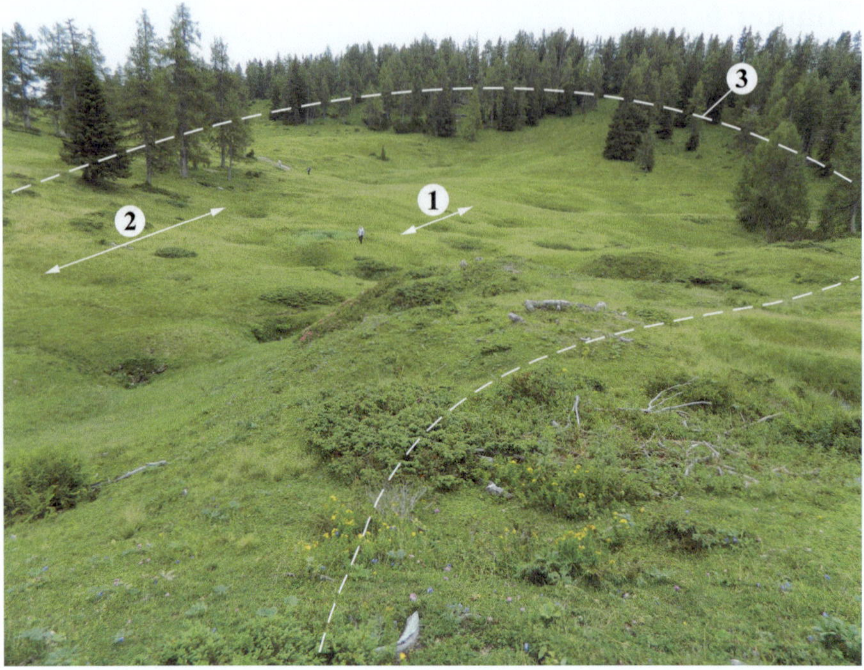

Fig. 6.1 Suffosion dolines with catchments (hinterlands) of different size in a DSD (Totes Gebirge, Tauplitz alm). *1* Suffosion dolines with small catchments, *2* Suffosion dolines with larger catchments, *3* Margin of DSD

inflow does not grow to an extreme extent. This means that the doline is able to convey the inflowing water into the karst without impoundment. During flood activity, however, so much water flows into the depression that impoundment results. Flood activity may result from numerous causes but primarily from intense rainfall or snowmelt. Flood activity occurs at depressions of both permanent and intermittent character. For numerous dolines, however, intermittent activity is restricted to flood activity only. In lack of flood there is no inflow into the doline from the surface. No activity is observed where the doline density is high (Fig. 6.1). In this case the doline has no or only limited catchment (hinterground). The gentle slope of the hinterground does not favour doline activity either (Fig. 5.89). The periods of activity, particularly of flood activity, lead to the operation of the following processes:

- Activity phenomena
- Accompanying deposition
- Development of partial (internal) landforms
- Intensive changes in the dimensions of landforms

Processes on covered karsts (particularly on concealed karsts) are more diverse than on uncovered karst. On covered karst the various processes often trigger each other and result in complex systems. The processes are closely linked to the development stage of landforms and the system of karst passages. A landform of a certain development stage induces a certain process (mainly activity phenomenon) and the opposite is also true: a given phenomenon influences the actual stage in the development of the landform.

The intensity and type of a process depends on the type of covered karst, thickness of cover, type of landform and climate on the covered karst.

6.2 Activity

6.2.1 Conditions of Activity

The starting date and duration of intermittent activity differs with karst areas and can even be different for the various depressions of the same karst area. It is common that (flood) activity in dolines next to each other due to precipitation does not start at the same time. It also happens, however, that out of two neighbouring dolines, one starts the activity before the other, while in a subsequent precipitation period, the order reverses. For the individual depressions, both the duration of activity (inflow) and its intensity can vary within a single precipitation period. These variations are due to numerous factors which influence water inflow (see below). The activities of individual dolines in the same karst area are quite different – particularly if the amount of precipitation is low or the precipitation extends over a longer time span. At the same time, with increasing intensity of precipitation and more lasting or repeated events of precipitation, the variation between the activities of dolines reduces. Consequently, in a given karst area, the activities of depressions (its start and intensity) tend to be more similar to each other.

The liability to activity depends – primarily, however, not exclusively – on runoff coefficient. The occurrence and duration of activity is influenced by the following factors:

- Rainfall intensity, duration and amount
- Rate of snowmelt, site and extent of snow accumulation
- Size and slope angle of hinterland (catchment area)
- Size, pattern and density of dolines
- Composition, properties of the cover sediments and soil (Fig. 6.2)
- Number and size of fossil dolines in the hinterland
- Water saturation of the cover (Fig. 6.2), presence or absence of ground ice, extent of freezing in the soil and extent of icing

The probability of activity is increased by prolonged precipitation (lasting for several days), when the cover sediments are saturated with infiltrating waters.

The effect of increasing precipitation amount and repeated rainfall events is confirmed by author's observations at the dolines of the Tés Plateau in 2010 (Veress et al. 2013; Veress 2013). Although no activity was observed then, during the field

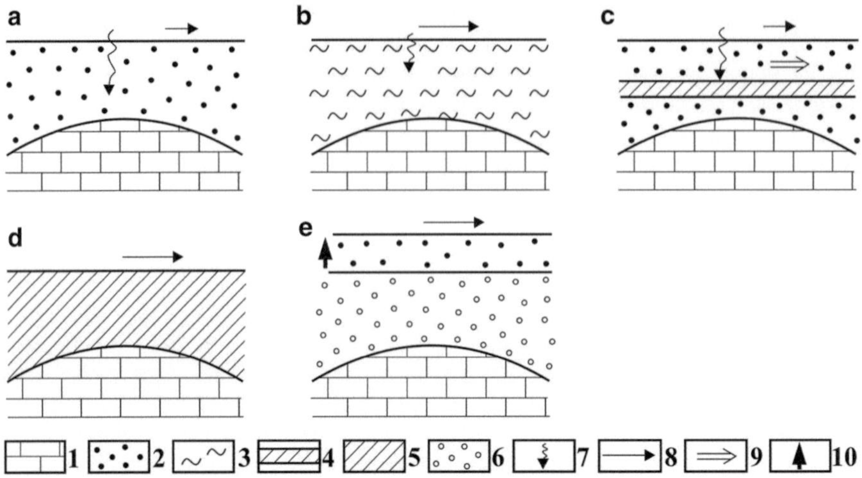

Fig. 6.2 Ways of water conduction on covered karst with different loessic materials. *1* Limestone, *2* Loess, *3* Decalcified loess, *4* Paleosol in loess, *5* Redeposited loess, *6* Clayey gravel, *7* Infiltration, *8* Surface runoff, *9* Seepage above the paleosol, *10* Rising groundwater level in loess, (**a**) most of rainwater infiltrates, (**b**) less water infiltrates into decalcified loess, (**c**) most of rainwater infiltrates, but little percolates through the paleosol, (**d**) increased impermeability of redeposited loess and surface runoff, (**e**) rising groundwater level in loess and therefore most of rainwater runs off on the surface

Table 6.1 Monthly amount of precipitation measured daily between 1 January 2010 and 8 October (using the amounts larger than 5 mm)

Month	Precipitation (mm)
January	25
February	48
March	11
April	107
May	237
June	148
July	55
August	107
September	158
October (until 8 October)	18
Total	914

work on 8 October 2010, numerous changes in the dolines of the plateau were detected which point to intensive (flood) activity in the recent past. In May 2010, the amount of rainfall was 237 mm (ignoring events of less than 5 mm), while it was almost 200 mm both in June and September (Table 6.1). Particularly large amounts were recorded between 12 and 16 May (152 mm) and on 25 and 26 September (68 mm).

Especially in high mountains, the role of snow can also be significant in inducing activity. At the time of snowmelt mostly (but not always), activity of low intensity but long duration is typical. At 1800–2200 m elevations, the large amounts of snow are added to this effect, for instance, in the Alps, where the density of covered karst-dolines is particularly high. Enduring activity is promoted by slow snowmelt (there are local snow patches at the mentioned elevations even in July) as well as snow compaction. A similar situation is expected in the autumn months, when the snow fallen in September or October often melts before autumn is over. As a consequence, the dolines with snow patches preserved in their hinterland (primarily in the gullies and ravines leading to the dolines and on the side slopes or bottom of the bearing paleodolines) are active for several weeks or months. Activity is not continuous but restricted to days and hours with sunshine and may last for several hours a day. Activity is promoted by frozen ground and if the formation of a snowmelt rivulet takes place always at the same site. For this reason, the soil is locally saturated with water and water from the rivulets formed on subsequent days cannot infiltrate.

On covered karst of moderate elevation (in middle mountains), snow acquires more significance on treeless terrain. The wind blows snow from such terrains into dolines and fills them with snow to their rim. (Similar accumulations of snow are naturally also observed in the zones with *Pinus mugo* or Alpine meadows in high mountains.) In the dolines snow undergoes compaction or sometimes firnification. For the mentioned reasons, in the dolines of the Bakony Mountains at 300–400 m elevation, snow patches occur even in May. Slow snowmelt ensures lasting activity in snow-filled dolines, amounting to even 240 h as estimated from our experience (counting with snow fill for 60 days and melting for 4 h a day). This probably means contact between water and rock over similar lengths of time (meaning solution of the same duration). (Contact with the rock can persist for an even longer period if karst passages are filled with water and water flow is reduced to slow seepage.) Therefore, for some dolines meltwater represents the overwhelming part of activity (not regarding flood activity). There are years when there is no activity at all induced by rainfall in the dolines of the Bakony karst or it occurs on few occasions in a year. Even in the latter case, activity is restricted to some hours at most. As it has been mentioned above, snowmelt rarely induces intensive activity, although during rapid snowmelt it is possible.

The start, duration and intensity of activity are also influenced by the size of the hinterland area. Thus, covered karst ponor K-1 in the Hárskút Basin, with the largest catchment area (ca 1 km^2), already received water when there was no water inflow into the dolines of small hinterland (some 100 m^2) at some hundreds of metres' distance. It is a general observation that with increasing slope inclination, the amount of runoff grows. Therefore, the dolines with hinterland which has a steeper slope also receive water earlier.

Where doline density is high (Figs. 5.39 and 6.1), there is probably no activity at all since dolines located in the interior of a doline cluster have no hinterland from where inflow could come. With growing doline size, the liability to (or the probability of) activity increases as the precipitation fallen on the slopes of the doline can be sufficient to induce the formation of rivulets. Moreover, the chance of removal of cover sediment is higher in the environs of a larger doline. Thus, the slope of the bordering terrain towards the doline is increasing and more and more water flows into the doline.

Activity is also influenced by the character of the cover. If the cover sediment has a high clay content (terra rossa, decalcified loess), infiltrating water leads to clay swelling and an increased impermeability of the cover (Fig. 6.2). The clay content of a sample from a doline of the studied paleodepression in the Hochschwab (Fig. 5.91) was 33.5 %. There some dolines were already linked with gullies. This indicates that the impermeability of the cover at this clay content is suitable to allow sufficient surface runoff to produce an erosional gully. Since clay content in the cover sediment can change substantially within short distances, it seems probable that the extent of impermeability and its temporal changes show major variations between the hinterlands of dolines close to one another – particularly in the case of low precipitation. On the other hand, if precipitation is abundant, even the cover sediment with a lower clay content can become impermeable. Therefore, it is probable that with higher amounts of precipitation, more and more dolines become active.

Rainwater will not seep away from the fossil dolines of the catchments of active dolines if they are filled with clay or argillified deposits. In the areas of fossil dolines, lakes with intermittent (rainfall) or permanent water surface form (Figs. 5.67 and 6.3). The rainfall is added to the water of permanent lakes or creates intermittent lakes, whose water overflows the rims of shallow depressions and reaches the active dolines. Essentially the fossil dolines are the catchments for active dolines (Fig. 6.3). Impermeable patches may occur in gullies and ravines on clayey covers of fluvial deposits. Therefore, in the hinterlands of dolines, there are major variations in the rate of runoff.

The extent of vegetation cover in the hinterland and its type and density also influence the occurrence and intensity of doline activity. In the above-mentioned Hárskút Basin, 13.4 mm of precipitation fell within 2 h on 9 May 1980. As the arable fields were not yet covered densely by crops, its consequence was the generation intensive inflow (flood activity). At the same time, in the dolines on grasslands, there was either no activity at all or less intensive activity. On 7 November 1982, when 13 mm fell (i.e. equal to the above-mentioned amount), activity was observed for the dolines on grasslands. At that time the soil and the cover sediment could have been highly saturated with water as the precipitation between 18 July and 7 August amounted to 159.7 mm. In the dolines of arable lands (covered with developed crops), there was no activity. All in all, with regard to vegetation cover in hinterland, dolines are prone to activity – if other conditions are similar – in decreasing order: arable (when free of crops), grassland, arable (in the growing season) and forest.

The significance of water saturation of the cover (soil) or its absence is underlined by a further observation. Following a dry spell, on 13 June 1978, the amount of 24-h precipitation was 30 mm in the Hárskút Basin. In spite of this considerable precipitation amount – in lack of previous rainfalls – no activity was observed in the dolines of the Hárskút Basin. Although the rainfall had a low intensity (obviously a factor of missing activity), we regard it probable that this amount was not sufficient to saturate

Fig. 6.3 Topographic map of the Eleven-Förtés doline group (Kőris Mountain, Bakony Mountains, Veress and Puskás 2007). *1* Contour line, *2* Wetland, waterlogged area with identification mark (fossil doline), *3* Limestone outcrop with identification code, *4* Shaft, *5* Excavated shaft, *6* Water flow, *7* Margin of DSD

the soil or the cover sediment – particularly, if we consider that part of the precipitation fallen in 24 h immediately evaporated or intercepted on the arable crops.

Jakucs (1956) associated activity with surface runoff, estimated in an empirical way. In the covered section of the Aggtelek Karst (on the catchment of a ponor of Baradla Cave), tin tubes were deepened into the soil to depths of 25–30 cm and 50 mm of water was poured into the tube. The height of the water column was measured at 3-min intervals. Plotting the height of the water column (which decreased because of seepage) against time (on the horizontal axis, 20 min of time equalled 5 mm of precipitation on the vertical axis), various seepage curves were received according to the initial moisture content of the cover sediment (Fig. 6.4).

The intensity of precipitation responsible for runoff was read in this coordinate system. Since the actual moisture content of the cover controls the infiltration curve, at a given time, runoff can be established where the artificial infiltration was measured. The runoff amount of the fallen precipitation equals the surplus water above the amount of seepage. Thus, if moisture content in the cover is determined by infiltration curve no. III, runoff appears if the amount of precipitation reaches or

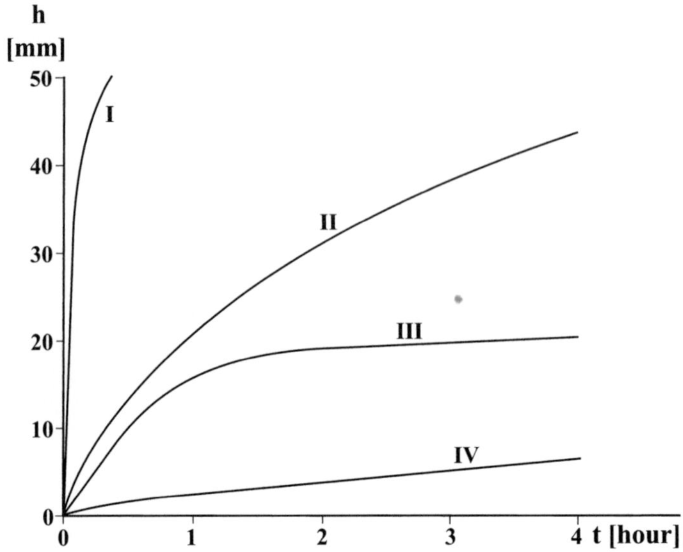

Fig. 6.4 Jakucs' infiltration curves for the covered karst at Aggtelek, on the catchment of the Baradla Cave (Jakucs 1956). *I* Dry soil desiccated in enduring drought and hot weather, *II* Uniformly dry soil, *III* Dry topsoil and moist subsoil, *IV* Very moist or frozen soil, *h* the height of the water column in the tin pipe, *t* time

exceeds 15 mm an hour or 18 mm (or more) in 3 h. If the moisture content of the cover follows the infiltration curve no. II, 20 mm per hour or 30 mm (or more) in 3 h has to fall at the site of measurement to generate runoff (Fig. 6.4).

Ground ice or surface ice could overwrite the impact of all the above-mentioned factors. In the presence of ground ice, all water runs off on the surface. Activity can take place independent from the amount of precipitation. Snowmelt can generate activity partly because of the existence of ground ice. At the same time, there are complex interactions between meltwater and ground ice. The higher the temperature is, the more snow meltwater is generated, and the soil warms up to cause more and more meltwater to seepage.

Eventually the activity of a doline will show very individual traits, only characteristic of the doline in question since the size of the catchment area (hinterland); the extent, nature and density of vegetation cover; as well as the pattern of the composition (degree of argillization) of cover sediment are different.

6.2.2 *The Characteristics of the Activity*

As mentioned, the activity is either continuous (for ponors with permanent watercourses) or intermittent. Intermittent activity is either normal or flood activity. During flood activity the amount inflow into the depression several-fold exceeds

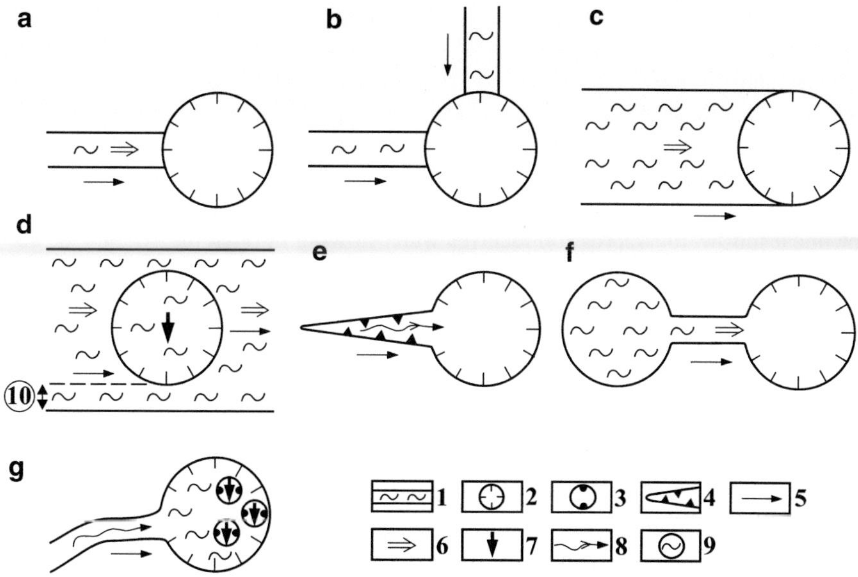

Fig. 6.5 Water inflow types inducing activity. (**a**, **b**) Water replenishment from rivulet, (**c**, **d**) water replenishment from sheet water, (**e**) water replenishment from intermittent ravines, (**f**) water replenishment from rivulet of lake overflow, (**g**) water conduction from intermittent lake through non-karstic pipes, *1* Rivulet and sheet water, *2* Doline, *3* Non-karstic pipe, *4* Gully, ravine, *5* Surface slope, *6* Water flow, *7* Drainage, *8* Channel flow, *9* Lake, *10* Water flow beside the doline

that of normal activity. The proportions of flood and normal activity vary with the individual depressions. During normal activity, the doline is able to conduct inflowing water, while during flood activity, it is unable to do that, and as a result, intermittent lakes and overflows are generated (see below). The intermittent activity is divided into active spells of various durations interrupted by inactive spells without inflow. The activity is called total period of activity when the water inflows (active spells) are interrupted by water inflow-free (inactive) spells. The reason for the separation on spells lies in fluctuating rates of snowmelt and the occurrence of repeated and/or high-intensity rainfalls. Within the same episode of intermittent activity, normal and flood activities may alternate.

Intermittent activity is caused by rainfall events, snowmelt, lake overflow (lakes in paleodolines, rock basins) or intermittent katavothron springs.

The types of water inflow are the following: superficial, rivulet, linear and valley-floor inflow (Fig. 6.5). In the case of superficial inflow, runoff from the bordering slopes arrives at the depression in the form of a broad sheet water (Fig. 6.5c, d). This kind of inflow is typical of covered karsts under temperate, mediterranean, tropical and arid to semiarid climate during intensive rainfalls if the hinterland is free of vegetation (e.g. arable land). It occurs at depressions in valley-floor position and without gully (ravine). Rivulet type of inflow involves water motion towards the depression without erosion (at most in rills) (Fig. 6.5a, b); it is of short duration unless it is associated with a snow patch. Similarly to sheet flow, rivulet inflow takes place on vegetation-free terrain preceding or succeeding sheetlike inflow. Occasionally, however, it emerges on the slopes of alluvial streamsink dolines (Fig. 5.93) and on covered karst of glaciokarst type (Fig. 5.52). On the grassed surfaces of glaciokarsts, the rivulets are preserved and visually detectable as the grass is greener there than elsewhere.

Inflow is linear into depressions to which gully, ravine or valley leads. In the case of linear inflow, the depression receives water flowing through the bottoms of gullies, ravines and valleys (Fig. 6.5e). Valley-floor inflow is characteristic of depressions located on the floor of valleys which are not blind valleys. The runoff water collecting on the floors of dry valleys in humid periods flows into the depressions.

The above types of inflow also occur in combination (double inflow), such as sheetlike and linear inflow combined at the covered karst ponor K-1 on 9 May 1980 (Veress 1987). Particularly on glaciokarsts, linear and rivulet inflow are typical. Flood activity is most probable during valley-floor or double (sheetlike + linear) inflow.

Water conduction happens anywhere on the floor of the depression and, thus, in the internal doline of the depression, on its margin or on the bottom of the gully with water inflow. Water inflow is found at non-karstic pipe, shaft or animal borrow (open water conduction) at a single or at several sites. It is common that with increasing discharge, additional water conduction sites are activated. Activity can take place simultaneously through seepage and open water conduction.

By character the activity is either summer or winter activity. If the ground is frozen, it is winter activity; if not, it is summer activity. As a consequence of the above, in some karst areas, summer activity is possible during the astronomical winter, and in the opposite way, winter activity is possible in the summer half-year. Vegetation-free terrains (e.g. arable fields) are more prone to summer activity. Due to snow cover, the liability to winter activity is higher in forested areas too.

Winter activity is mainly caused by snowmelt but can also be due to rainfall. On glaciokarsts and particularly on tundra and taiga karsts, winter activity is typical for most of the year. It is possible that certain depressions only receive water this way. Winter activity lasts for a long time but is divided into many shorter active spells. The activity is less typically of flood character. Limited amounts of debris reach the depression and the water mostly flows into the non-karstic pipes and shafts of the depression. The water is in contact with the rock wall for a long time (several hours a day). This contact is probably realised in the form of rivulets since the water inflow only seldom fully fills in the shaft if at all. For winter activity, with frozen ground, solution is more efficient than erosion, and thus, there is no sediment transport.

During summer activity, normal and flood activities are distinct. During summer activity, a remarkable proportion of water inflow into the depression often reaches the karst through seepage, which is pointlike and involves suffosional sediment transport from the dolines and thus the deepening of depressions through suffosion. During open water conduction, the debris is carried into non-karstic pipes or shafts and erosion of limited temporal and spatial extent can take place. Parallel to erosion, however, solution also happens. The effect of erosion is reducing with depths in the karst passage system (since the sediment deposition decreases), while solution is intensifying. The depth where erosion ceases is highly variable, depending on the size, position and morphology of chimneys.

During flood activity, intermittent lakes occasionally emerge in the depressions. They are primarily typical of summer activity (flood lakes can also develop during winter activity, particularly if the snow cover is deep and melts rapidly or if there is ice on the surface or in the cover sediment). The impact of flood lakes developed during winter and summer activities are different. In the flood lakes of winter activity, there is little sediment, and the depressions fill up at a slower rate. With frozen ground, seepage from the lakes is not possible. The water flows into the karst passages through open water conduction and exerts solution and, by way of the transported sediment, mechanical erosional influence there. In the flood lakes of summer activity, open water conduction is coupled with seepage. For flood lakes seepage is not pointlike but occurs over the entire surface of the depression. Therefore, material transport by suffosion intensifies. This, however, does not automatically involve the deepening of the depression as part of the lake sediments are deposited in the depression. Well-developed water conduction, however, may induce or lead to the formation of further non-karstic pipes and shafts.

All this may create favourable conditions partly for the emergence of further internal depressions and partly for karstic solution not limited to the existing shaft but more extended in space under the depression. The flood lake also contributes to the erosional evolution of the karst passage system. Water inflow into the shaft remobilises the previously deposited debris. Reworking is attenuated as the shaft is totally filled in with water, i.e. stagnant water conditions result. As long as the depression has no non-karstic pipe and shaft, the sediment input is deposited and further transported by suffosion into narrow cracks.

However, additional conditions are also necessary for a major filling: frequent and intensive activity, a vegetation-free hinterland, steep slopes and a non-cohesive cover sediment. These conditions increase the amount and grain size of the sediment transported into the depressions and deposited there.

Winter activity and summer non-flood activity are climate dependent, but summer flood activity is independent of climate, although it may be influenced by the rock properties of the catchment or the extent and nature of vegetation cover (vegetation cover is only partly climate dependent).

While on tundra and taiga covered karsts winter activity exclusively occurs, on temperate karsts both winter and summer activities are typical. On glaciokarsts winter activity is overwhelming, and on mediterranean and tropical karsts summer activity prevails.

6.2.3 Activity Phenomena

For flood activity the following phenomena occur: flood lake, overflow, throughflow, intermittent spring and onfilling (Veress 1987). Flood lakes can also form due to rising karstwater table, particularly if there is simultaneous water inflow into the depression.

Fig. 6.6 Ephemeral lake in a ponor of the Pádis Plateau. (**a**) The developing lake, **b** sediment transported by water flow (the material forms a mound because the water flow incised into the sediment transported by itself)

Flood lakes are intermittent standing waterbodies in depressions and occasionally in their environs (Figs. 6.6, 6.7 and 6.8). They are different from karstic lakes, which are permanent or, if intermittent, exist for several years, although the origin of both lake types is similar in certain cases: the presence of sediment fill or the rise of karstwater table. They can develop in the internal part or detail of depressions or over its entire area (Fig. 6.7b). They result from the impoundment of water as the amount of water inflow surpasses that of water seepage and open water conduction. The indicator of water input is the water flow developing on the surface of the lake (Fig. 6.7a). With the further rise of water table, the lakes may extend over the edge of the depression (Figs. 6.7c and 6.8c). In this case overflow happens: part of the lake water flows out of the depression as surface runoff (Figs. 6.5d and 6.7c). If the dolines are in valley-floor position and experience water inflow on the valley floor, repeated overflow phenomena lead to the formation of flood lakes in additional dolines. Consequently, a series of flood lakes comes about.

Flood lakes are found in depressions of various type and size (Fig. 6.9). Sometimes the water of the flood lake cannot be drained on the surface. Overflow cannot develop in blind valleys, poljes, cockpit dolines and true depressions of superficial deposit. In this case, however, the flood lake may inundate the environs of the depression. The lakes extend over (parts of) poljes (Komac and Zorn 2013), parts of blind valleys or cockpit dolines and DSDs. The inundation may extend over

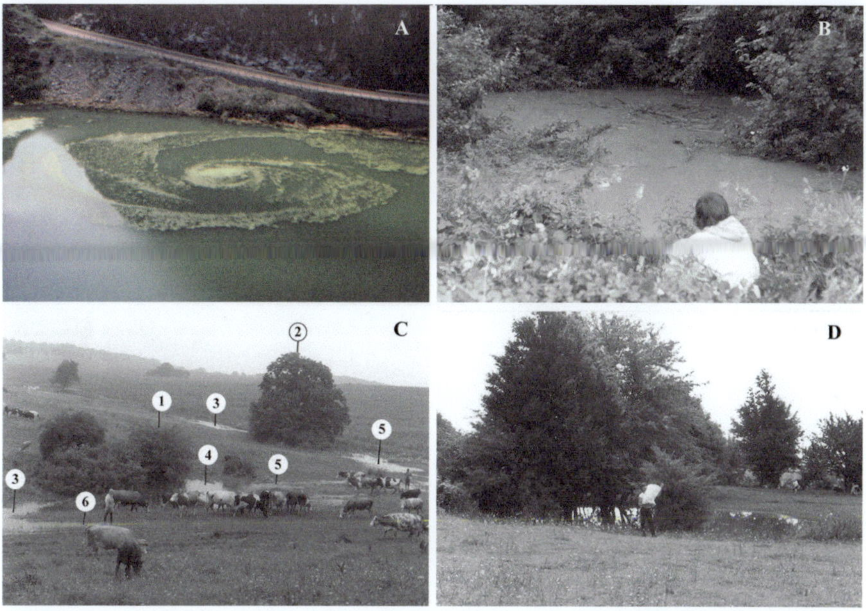

Fig. 6.7 (**a**) *Upper left*: a longer-duration lake with vegetation on its surface (South China karst; the pattern of vegetation refers to water flow and thus to water input), (**b**) *upper right*: longer-duration lake of covered karst ponor K-1 (Hárskút Basin) (with plant waste on its surface), (**c**) *lower left*: longer-duration lake of covered karst ponor G-9 (Hárskút Basin), (**d**) *lower right*: enduring lake in a suffosion doline (Hárskút Basin). (**a**) *1* Lake, *2* Water inflow, (**c**), *1* Covered karst ponor G-9, *2* Suffosion dolines K-2 and K-3, *3* Water inflow, *4* Intermittent lake, *5* Overflow, *6* Water conduction beside covered karst ponor G-9. Shooting Date: (**a**) 2013, (**b**) 1984, (**c**) 1984, (**d**) 1984

the areas of other depressions (e.g. dolines) or the lakes of the depressions merge into a single extensive lake (Figs. 6.8c and 6.9g). In tropical areas floods occasionally reach such great dimensions that the flood lake extends beyond the boundaries of the karst. Such a large lake formed on a marble karst in Zambia in 1926 and again in 1978–1979 and inundated the capital Lusaka (Brook 2004).

The flood lakes are widespread on karsts. Thus they occur on tundra karst (Korzhuev 1961; Pulina 2005), on temperate karst (Veress 2000), on glaciokarst (Veress et al. 2013) and on tropical karst (Zhang 1980).

The development of flood lakes has various causes (Fig. 6.10). They most often form at intensive inflow or if the period of inflow increases. They also emerge if the depression is filling up intensively or if the shaft or non-karstic pipe has filled up to a great extent because the discharge of water drainage decreases. In both cases the discharge of the water supply exceeds the discharge of water drainage (Figs. 6.10a, b). If the non-karstic pipe or the shaft loses its sediment, the rate of water conduction increases, and flood lakes rarely or not at all develop. The rate of water conduction is reduced if the fill is clayey deposit, which swells to the effect of the inflowing water. This way, during activity impermeability increases. Water conduction of

Fig. 6.8 Lakes in the DSD (alluvial streamsink doline) of the Cerkniško Polje. (**a**) *upper left*: State without lake, (**b**) *upper right*: partially water-filled DSD in 2002, (**c**) *lower left*: lake extended over the margin of the depression in 2013, (**d**) *lower right*: plant waste accumulated at the stagnant water level of the lake (with sinking level) in 2013

lower intensity may also be due to the underdeveloped karst passages. Flood lake can also develop over the snow filling of the depression (Fig. 6.11) and to the effect of an ice plug in the conduit. The weight of water in the flood lake can destroy the ice plug (Ford 2004; Ford and Williams 2007) and the lake is rapidly drained. In depressions above permafrost, even water inflow of low discharge produces flood lakes.

Flood lakes can also develop if the discharge of the water inflow does not exceed the discharge of water drainage. It may happen because of the impoundment of the inflowing water or the rise of the karstwater table. Impoundment may occur if the water seeping on the surface gets into the karstic passage system (Fig. 6.10c) or an onfilling happens to the impounded water of the karstic passage system. Because of the onfilling, the water table of the karstic passage system that had been filled up by water rises and creates a lake in the depression. The water inflow place that causes onfilling can be an inner doline (Fig. 6.10d); an inflow place outside the doline, for example, a chimney on the floor of the channel (Fig. 6.10e); or another doline (Fig. 6.10f). It may happen that the overflowing water of a lake developing in one doline creates the intermittent lake in another doline above the karstic passage filled up with water (Fig. 6.10g).

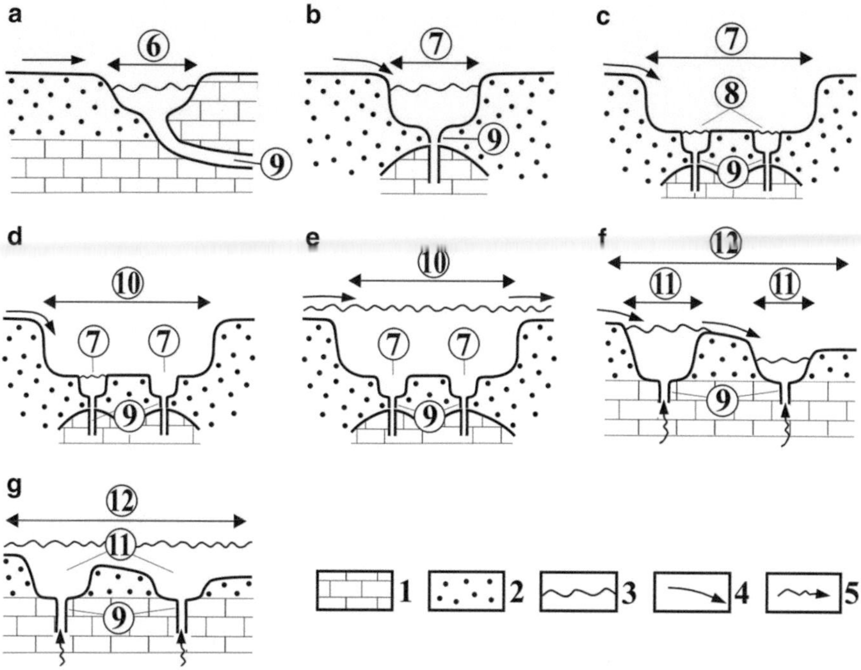

Fig. 6.9 Flood lakes in various karst landforms. *1* Limestone, *2* Cover, *3* Lake, *4* Water flow, *5* Karstwater flow, *6* Ponor, *7* Doline, *8* Internal doline, *9* Non-karstic pipe, shaft, *10* DSD, *11* Alluvial streamsink doline, *12* Polje. Lake formed in ponor (**a**), doline (**b**), internal doline within doline (**c**), doline of DSD (**d**), DSD (**e**), alluvial streamsink doline (**f**), detail of the polje (**g**); (**a–e**) due to water inflow, (**f**, **g**) due to water inflow and rising karstwater table

Karstwater can also be responsible for the formation of flood lakes. The water of the lake in the depressions derives partly or completely from karstwater with rising table. Such lakes do not only occur in the poljes of the Dinaric karst but, for instance, in Japan, France, Tasmania and on the tropical karsts of Jamaica and the South China Karst (Sweeting 1973).

Flood lakes of karstwater origin develop if the altitude of the high karstwater table exceeds the altitude of the floor of the depression. (A karstwater lake is regarded a flood lake if the rise of the karstwater table is triggered by intensive rainfall. Thus, depending on the intensity of rainfall, both flood lakes and non-flood lakes can develop at the same place.)

Onfilling can also contribute to the development of the intermittent lakes of karstwater origin. Thus, inner dolines of the DSDs which are in higher position operate as water drainage, and in this way they increase the altitude of the karstwater table, while in the case of inner dolines of lower position of the same DSDs, water outflow can be experienced.

Flood lakes of karstwater origin can develop by onfilling where the high karstwater table does not reach the depression floor but rainwater is added to the rising

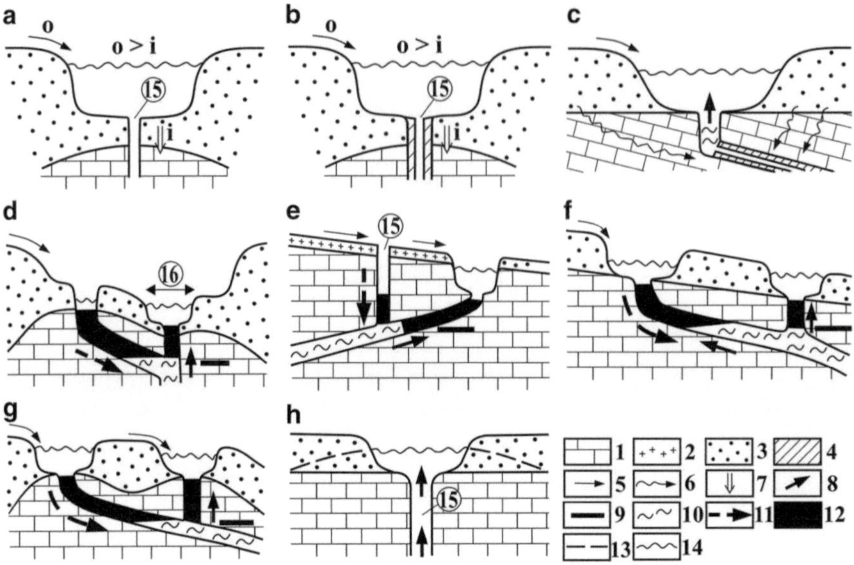

Fig. 6.10 Genetic types of flood lakes: inflow exceeds outflow (**a**, **b**), impoundment (**c**–**g**), outflow from karst (**h**), *1* Limestone, *2* Impermeable cover (basalt), *3* Permeable cover, *4* Fill of shaft, corridor, *5* Water inflow from the surface, *6* water replenishment from seepaging water, *7* Water inflow into the karst, *8* Water backflow because of the infilling of karst passages (clogging, passage bottlenecks and rising karstwater table), *9* Level of impoundment in karstic passages, *10* Water-filled karst hollows, corridors, shafts, *11* Onfilling to the water fill of the karstic passage, *12* water infill of the passage due to onfilling, *13* Karst water table, *14* Lake water level, *15* Non-karstic pipe and shaft, *16* Internal doline, 'o', discharge of water inflow, 'i', discharge of outflow

karstwater table. Such a phenomenon is expected in the Bakony Mountains, where, according to Böcker (1972), the rise of the karstwater table amounts to 100 m in some places (at Hárskút) in humid periods. On the Tés Plateau, the karstwater table is at 190 m elevation above sea level, while the floors of shafts below dolines are at 200 m elevation. Consequently, the shafts of the plateau are inundated by karstwater even at a level rise exceeding 10 m.

If the shafts under depressions are partially flooded by karstwater, the inflowing surface waters, which cannot flow further impounded because of the karstwater fill, may fully fill the shafts. Subsequently, the water filling the shafts is further impounded and reaches the floor of the doline.

The water conduction from flood lakes also takes diverse forms. It is possible that water is conducted away through one or more passages. If the conduits are found at different heights, with rising lake levels, further passages are activated. Water conduction can extend over the marginal depressions or to other depressions.

According to their duration, flood lakes fall into three classes (Veress 2000): ephemeral, moderate-duration and enduring lakes.

Fig. 6.11 Sediments deposited in the intermittent flood lake of a subsidence doline above compacted snow near Dudar (Bakony Mountains) in 1982. *1* Compacted snow or ice, *2* Sediment deposited on snow or ice (no sediment on the doline margins or obliquely deposited due to the subsequent melting and collapse of snow)

Ephemeral lakes only exist during a single period of activity and do not survive until the next activity (water inflow). After reaching its maximum, the water level of such lakes is rapidly dropping. They are mostly generated in such depressions on the bottom of which a shaft opens or a non-karstic pipe which is not clogged. The lakes of depressions with shafts develop through the complete filling of the shafts with water. After the shafts have been filled with water, inflow into the doline forms a lake. This is indicated by the accumulations of plant waste in the niches of the shaft walls (Fig. 6.12) as well as direct observations. Thus, the passage system of one of the dolines (G-5/b in the Hárskút Basin) formed along a bedding plane was filled after an event of intensive water inflow to the degree that the cavers who worked there could hardly escape. Filling with water is due to the lowering and narrowing of the passage system along the bedding plane, which is only some tens of centimetres high at ca 10–15 m distance from the entrance (less than 1 m even at the entrance). In the low passage the rate of water conduction is reduced, and as a result, the passage is filled with water. Subsequently, the water flowing into the doline is impounded on the floor.

The existence of the moderate-duration lake can last longer than the meteorological event (e.g. a spell of frontal rainfalls, which lasts for some hours or 1 day) during which the water inflow happened. Such lakes exist continuously over several periods of activity and survive several periods of inactivity. It is possible that weather fronts follow one another with short breaks between them. In this case a

Fig. 6.12 Plant waste from ephemeral lake on the wall of the shaft in the dropout doline Gy-12, Hárskút Basin, in 1981. *1* Grike in shaft wall, *2* Quartz gravel half buried under plant waste, *3* Plant waste

lake can last until the next meteorological event. (It is even more difficult to determine the time span for the existence of moderate-duration lakes if they are due to snowmelt.) The discharge of water inflow may fluctuate or there are interruptions between inflow periods. The reason for this is that rainfall intensity and duration may be variable within an event of one or several days' length. If rainfall intensity is decreasing (or the rain stops), the water level of the lake stabilizes for a period of various length (a standing level ensues) or drops but the lake is not drained or even rises after another inflow. For this reason, the moderate-duration lakes show oscillating or dropping level trends, but these trends of falling water level are interrupted by standing levels. Standing levels also occur in lakes with oscillating water table. From moderate-duration lakes, water conduction is manifested through seepage. The non-karstic pipes or shafts of the depressions are filled or the shafts are less developed. At the sites of seepage, water flow is laminar and it is only turbulent locally and at times in the shaft of the depression.

The time span of enduring lakes is longer not only compared to one active period but also to the total period of activity and, thus, to the triggering meteorological event. While moderate-duration lakes do not exist for more than 1 or 2 days, enduring lakes may last for several weeks. They lose their water through evaporation, although there are depressions where lakes also show seepage. The seeping water of such lakes does not carry sediment from the depression into the karst.

The time span of existence for enduring lakes may be equal or similar to the time span of karst lakes. The transition between flood lakes and karst lakes is represented by lakes formed in fossil depressions (S_8), which can be regarded karst lakes, but since many of them only exist during heavy rainfalls, they are also flood lakes.

The rates of water level subsidence (or rise) can be different for ephemeral and moderate-duration lakes reflecting either the discharge of water conduction or of water inflow or both of these parameters. The changes in the rate and extent of water level, however, exclusively depend on the rate of water inflow.

Depending on the nature of water conduction, ephemeral and moderate-duration lakes sometimes alternately form in the same depression. This happens in depressions where water conduction is by way of seepage. Thus, it is expected that in depressions bearing moderate-duration lakes, water inflow of low discharge often produces ephemeral lakes, but in depressions where ephemeral lakes develop, moderate-duration lakes only rarely form, with low probability, only during extreme rainfalls.

Throughflow (Fig. 6.7c) is observed for water inflow on the valley floor, which generates a lake in the depression and then the lake water overflows the depression.

Onfilling takes place if the risen karstwater table pours into the shaft of the depression from below and is impounded to generate a flood lake.

An intermittent spring (Fig. 6.13b) comes about if water seeping away from the depression and moving over an impermeable or partially impermeable sequence issues to the surface on the side slope of the depression. Intermittent springs are common, for instance, in the depressions of the Hárskút Basin, where during snowmelt, meltwater from snow patches seeps away on doline rims or in some distance from there and covering several metres issue as springs in the cover sediment or in the soil.

The springs evidence water storage or motion in the cover sediment and also indicate (partial) impermeability in the permeable cover. Springs on the doline slopes contribute to the prolongation of the duration of activity and the reduction of the intensity of flood activity.

6.2.4 *Types of Activity*

The following types of activity are identified: superficial water inflow, seepage, latent activity and complex activity. During superficial water inflow, water flows from the depression into the non-karstic pipe or shaft (Fig. 6.13a). Seepage occurs if below the site of seepage a non-karstic pipe is found and seeping over the cover sediment into the non-karstic pipe it reaches the bedrock. If there is no non-karstic pipe under the site of seepage, covering a longer route within the cover occasionally

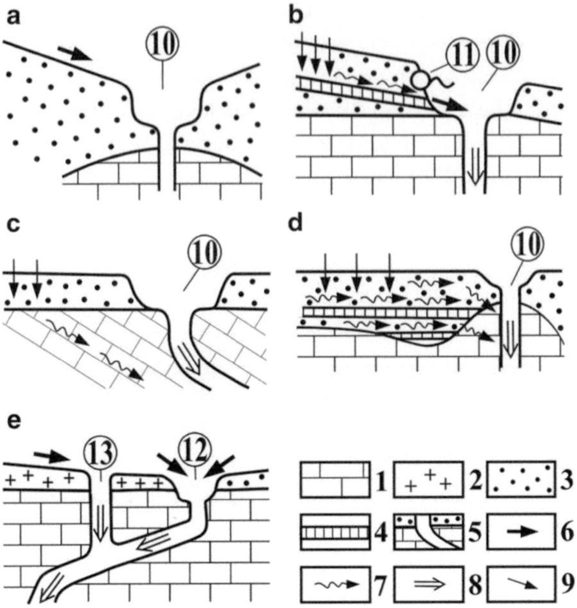

Fig. 6.13 Activity types. (**a**) By surface inflow, (**b**) by the water of an intermittent spring formed in the cover sediment, (**c**) by seeping water from cover to limestone, (**d**) water seepage over the impermeable intercalations of the cover, (**e**) complex activity (e.g. on Kab Mountain): water inflow through a shaft on the valley floor, by water flow from the valley floor (part of water overflows the shaft) and by water inflow from rivulet or rill, *1* Limestone, *2* Basalt, *3* Permeable cover, *4* Impermeable intercalation in the cover, *5* Non-karstic pipe, shaft, corridor, *6* Surface water flow, *7* Water seepage in the cover and bedrock, *8* Water flow in non-karstic pipe and shaft, *9* Infiltration, *10* Doline, *11* Intermittent spring, *12* Ponor, *13* Shaft on valley floor

arrives at the karst at several sites, through the shafts of the cover and the joint and grike system.

In the case of latent activity (Fig. 6.13b–d), the water does not reach the non-karstic pipe or shaft from the floor of the depression but seeps away outside the depression or derives from groundwater in the cover sediment. During latent activity, the shaft receives water even in lack of superficial water inflow. (This was observed at the passage system of the mentioned doline G-5/b, where water motion could be recorded even after surficial water inflow.)

Groundwater also contributes to latent activity. Groundwater in the cover was found in the catchment area of the covered karst ponor G-6/b (Fig. 5.81), where the boreholes in clayey cover were filled with water and indicated the presence of groundwater. The groundwater was also observed in the well deepened in the vicinity of the depression and was present for a longer period with a minimum fluctuation of level. Since no intermittent groundwater springs were observed, the groundwater seeped over the bedrock too. In the cover beds of yellow and grey clay alternate (Veress and Futó 1990). The grey beds indicate a reductive environment due to water saturation. Consequently, such beds are impermeable. The yellow beds point to an oxidative environment and water conduction (the water with oxygen content derives from the surface). Therefore, there is water motion in the yellow beds towards the covered karst ponor over the grey beds. Since no grey clay is found in the boreholes in the vicinity of the ponor, groundwater from the cover reaches the karst here. The relatively large extension of the cover patch and the stable water level in the well make it probable that in the ponor area water conduction into the karst is continuous and uninterrupted.

Latent activity indicates that water transfer from the cover sediment into the karst can last very long even after surficial water inflow ceased. It also shows that the karst passage system remains to be active longer than the duration of water inflow. During latent activity more water can reach the passage system of the doline (ponor) from the cover sediment than from the surface. The underground catchment of the doline is often more extensive than the surface catchment.

In the case of complex activity, water from the surface reaches the shaft through the non-karstic pipe of the depression and through a shaft outside the depression. The water which derives from two sources jointly flow into the passage system under the depression (Fig. 6.13e).

6.3 Sedimentation in Flood Lakes

Sedimentation in flood lakes was investigated in the field, in karst depressions and under artificial conditions, in the laboratory.

6.3.1 Sedimentation in the Laboratory

It was investigated in laboratory whether sedimentation depends on the rate of water level lowering and if it does, on what way, and the features developing in the experimental basin were also studied.

6.3.1.1 Rate of Deposition

6.3.1.1.1 In Measuring Cylinder

The matter of 63–125 μm grain size was found, conforming to literature data (Hiemenz 1986; Rohrsetzer 1991), to settle immediately or within 1–2 s from standing water. This time interval is enough for the sediment of such grain size to be able to settle down from standing water. The matter of 1–63 μm grain size also settles but slowly, according to our experiments as well. The colloids finer than 1 μm do not settle but adhere to surfaces (both in nature and in the experimental pool) and may be densed through evaporation or coagulated (and subsequently settles). For this reason, the rate of settling was measured for suspensions with matter below 63 μm grain size.

The bulk of dry matter in suspensions of 5 cm^3 volume taken from water depths of 5, 7.5 and 10 cm was measured at different times (with exponential time difference). A function was sought to describe the relationship between settling time and the corresponding bulk of matter for the measurement sites. Similarly, relationships were found between total amount of material (the initial concentration) and material amounts at various points of time (momentary concentration) and settling time. The crossing point of the two curves determines the point of time when the amount of suspended matter is halved (Figs. 6.14 and 6.15). A more precise determination of halving time is also possible (not presented here) if the natural logarithms of the functions is calculated. The shorter time interval corresponds to the halved mass, the more rapid the settling velocity of the grains is. Settling rate is determined if water depth is divided by halving time. The rate of sedimentation from the suspension increased by the limestone powder, plant waste and $CaCL_{2.}$ To the effect of these materials, coagulation took place, and thus, the colloids got into a greater grain size class with a greater rate of sedimentation.

6.3.1.1.2 In Experimental Basin

The relationship between the rate of water level subsidence and that of settling was investigated in the experimental basin preparing clay and clay/plant waste suspensions. The amount of settled substance at 5, 7.5 and 10 cm depths were measured. The selection of these depths was motivated by allowing comparison between settling rates in the measuring cylinders and the experimental basin. In Table 6.2 materials settled from clay and clay/plant water suspensions at the mentioned water depths, and water level dropping rates were quantified in the experimental basin. The Table also shows the settling rates of suspensions identified in measuring cylinders identical with the suspensions in the experimental basin.

The data in Table 6.2 prove that no substance settled from the clay suspension at high rates of water level dropping (0.56 cm/min, 0.62 cm/min and 1.13 cm/min). If water level dropping rate was reduced, settling happened from the same suspension. The amount of the settled substance increased by water depth. (This could be observed as the thickness of settled material grew towards the deeper parts of the basin.)

With high rates of water level sinking, no settling could happen from the clay suspension because the suspended particles, which have lower settling rates than the rate of water level sinking, were removed together with lake water. However, if the rate of water level sinking is reduced, particles larger than 1 μm settle, particularly if there is plant waste in the system. Colloidal particles smaller than 1 μm can also settle since there is sufficient time available for their coagulation, which increases grain size. The material of larger grains can settle because its settling rate exceeds that of water level sinking. (The non-coagulated colloidal residue adheres to surfaces or to the plant waste.)

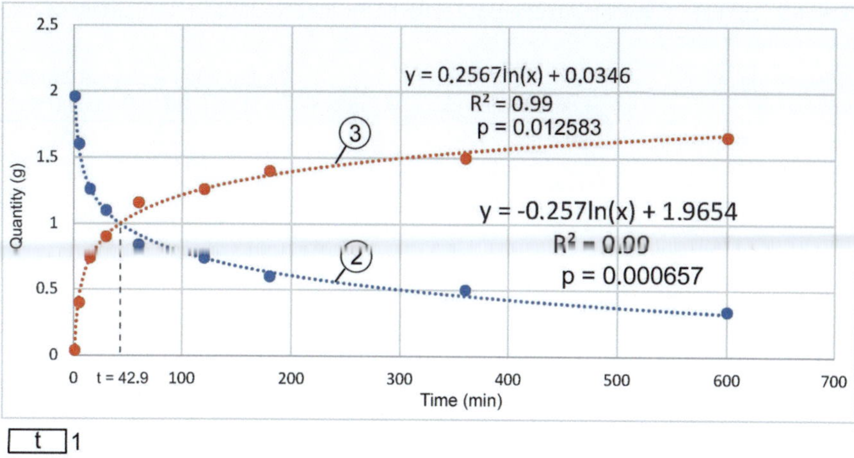

Fig. 6.14 Determination of the half-time of suspended material for clay suspension (Veress et al. 2015) *1* Half-time of concentration (bulk), *2* Concentration of suspended material (bulk) at the sample sites in the function of time, *3* Difference of initial and actual concentrations of suspended material (bulk) at the sample site in the function of time

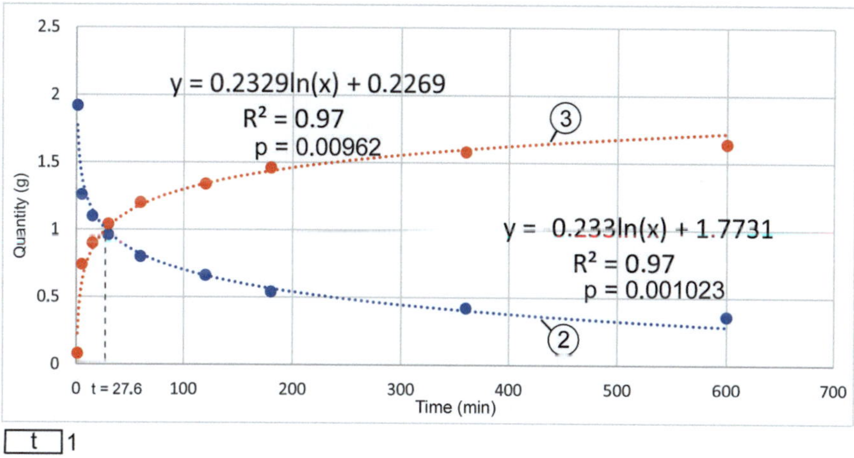

Fig. 6.15 Determination of the half-time of suspended material for clay-plant waste suspension. *1* Half-time of concentration (bulk), *2* Concentration of suspended material (bulk) at the sample sites in the function of time, *3* Difference of initial and actual concentrations of suspended material (bulk) at the sample site in the function of time

It can be stated that only that material can settle from the substance of the suspension whose rate of sedimentation is larger than the rate of water level lowering (Veress et al. 2015).

Table 6.2 The rates of water level subsidence of the lake in the experimental basin and the amount of the settled substance

Composition of the suspension		1 dm^3 Water + clay ≈ 0.2 m/m %			1 dm^3 water + clay ≈0.2 m/m% + plant waste ≈ 0.05 m/m %		
Distance of the settling place from the water level [cm]		5.0	7.5	10.0	5.0	7.5	10.0
Halving time of the settling of the setling substance [min]		23.9	39.9	62.9	15.7	21.1	27.6
Settling rate [cm/min]		0.209	0.187	0.158	0.318	0.355	0.362
Rate of water level decrease is large	Rate of water level decrease [cm/min]	5.6×10^{-2}	6.2×10^{-2}	1.13×10^{-1}	5.2×10^{-2}	6.9×10^{-2}	8.9×10^{-2}
	Time of subsidence [h, min, sec]	1 min 29 s	2 min 9 s	2 min 31 s	1 min 35 s	2 min 11 s	2 min 39 s
	Amount of settled substance [mg/cm^2]	No	No	No	6.4	2.4	8.3
Rate of water level decrease is small	Rate of water level decrease [cm/min]	1.4×10^{-5}	2.5×10^{-4}	2.8×10^{-4}	2.0×10^{-4}	3.6×10^{-4}	3.9×10^{-4}
	Time of subsidence [h, min, sec]	9 h 40 min	12 h 25 min	14 h 54 min	6 h 54 min	8 h 53 min	10 h 35 min
	Amount of settled substance [mg/cm^2]	3.1	7.5	12.8	10.0	13.0	51.2

Notice: Settling rate can be calculated if the settling distance (water depth) is divided by halving time

The amount of settled material is determined, with given grain size and suspension concentration, by the rate of water level sinking, the lifetime of the lake (with a given rate of level sinking) and the composition of suspension.

6.3.1.2 Depositional Forms in the Experimental Basin

In the experimental basin, a precipitation and a settling form were generated during settling. The latter one is adhered (colloid) or deposited (material with greater grain size than the colloid). Precipitation happens at a stagnant water level and its width grows with the salinity of the solution. The precipitation is linear or mosaical on the wall of the basin (Fig. 6.16) and continuous on the floor of the basin.

Fig. 6.16 Precipitation forms on the wall of the experimental basin. *1* Linear precipitation, *2* Mosaical surface precipitation

6.3.1.2.1 Coating and Settling Forms

Settling was performed with continuously sinking water levels or sinking interrupted with stagnation periods. The spatial extension of colloid coating forms adjusted to the position and extension of the bearing surface. Thus, on the side walls of the experimental basin in case of stagnant water level, a longitudinally extended linear settling form developed, while on objects submerged in the suspension, an annular settling form came about (Fig. 6.17), while no coating form developed on the floor with a dip angle of 9° of the experimental basin.

After a 24-h break, before level lowering started, linear and annular coating features are thicker. In case of various stagnant water levels which occurred at 20-min intervals, the linear and annular features developed below one another are ever thinner and paler.

The sediment with a grain size coarser than the colloid could deposit in case of continuous water level sinking or if water level sinking was interrupted by stagnant water levels.

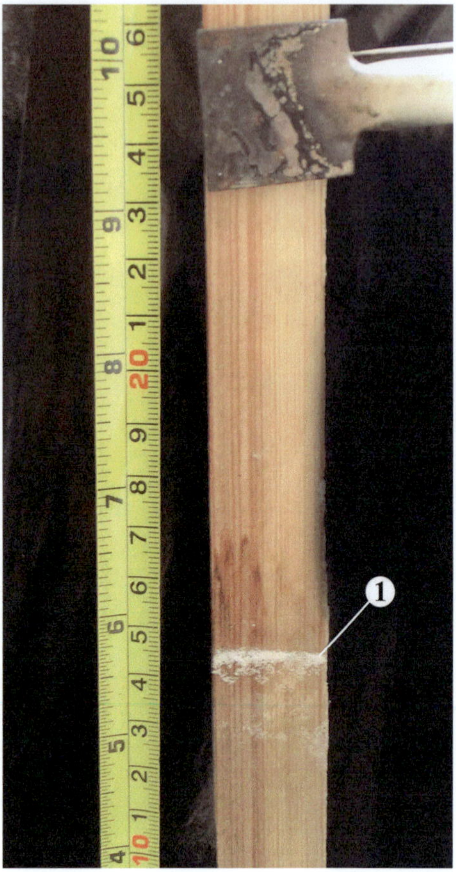

Fig. 6.17 Annular adherence on an object submerged in the experimental basin. *1* Colloid ring

In case of slow water level sinking, when the rate of the water level sinking of the suspension marked 'B' alternated between 1.1×10^{-3} and 3.2×10^{-3} cm/min, while that of the suspension "F" alternated between 1.9×10^{-3} and 3.4×10^{-3} cm/min, there was a continuous sedimentation (f) in both cases. (The data of suspensions 'A–G' are given in Table 3.3.) The following sedimentation zones developed progressing towards the inner side of the experimental basin in both cases (Fig. 6.18): outside a colloid belt (it was 1.7 cm wide in case of the suspension marked 'B' and it was 3.0 cm wide in case of the suspension marked 'F'). In the middle a sediment of lighter colour was deposited which probably had a grain size of 1–63 μm which was wider in case of the suspension marked 'B' (73.5 cm), while it was narrower in case of the suspension marked 'F' (71.5 cm). The third most inner sedimentation zone was wider in case of the suspension marked 'F' (8.5 cm) and it was less wide in case of the suspension marked 'B' (7 cm). The thickness of sediment increased progressing towards the inner side of the experimental basin. In case of the suspension marked 'B', it was 0.023 mm on the object slide (as compared to the stagnant water level between 7 and 14.5 cm), while it was 0.069 mm between 59.0 and 66.5 cm.

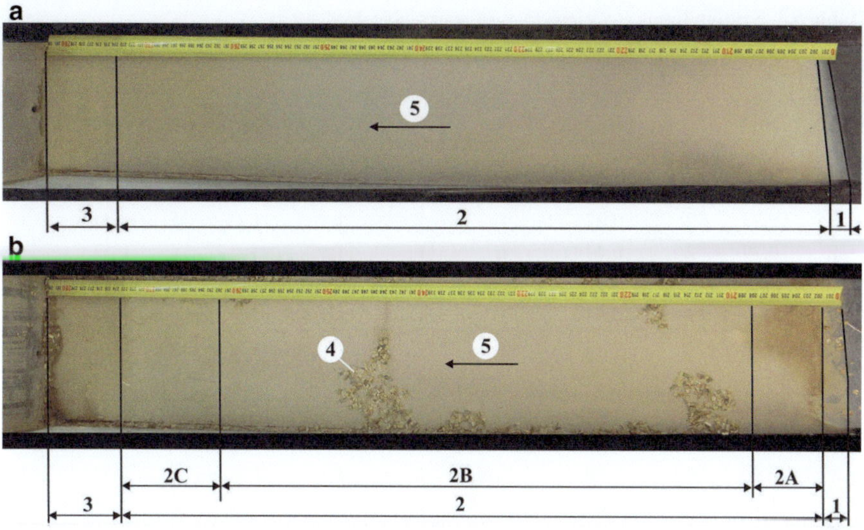

Fig. 6.18 Continuous settling form of the experimental basin in case of continuous water level sinking from suspensions marked *B* (**a**) and F (**b**). *1* colloid, *2* sediment with a grain size of 1–63 μm (subzones marked *2a–c*), *3* sediment with a grain size coarser than 63 μm, *4* plant waste, *5* dip direction of the floor of the experimental basin

In case of suspension marked 'F', the thickness was 0.042 mm between 7 and 14.5 cm and 0.091 mm between 59.0 and 66.5 cm.

The larger width of the colloid zone in case of suspension 'F' can be explained by the growing extent of the adherence of colloid grains. Because of this, the colloid grains sank, but preserving their charge, they could adhere to the floor in a greater width.

The thickening of the settling series towards the outflow of the experimental basin can be explained by greater settling time, however, not only by this. The fact that the darkest sediment (supposed to be built up of grater-sized grains) in case of suspension 'F' is the widest can be explained by the fact that because of coagulation, the sediment of coarser material (coarser than 63 μm) redeposited in a greater width. Another evidence for this is that the sediment thickness was larger at the measurement places on the floor at this suspension than at suspension marked 'B'.

When the water table sinking has been interrupted by stagnant water levels, thickening continuous settling forms developed. The thickening continuous settling form was divided into a thicker upper and a lower thinner part. The thicker part of this form developed close to the prevailing stagnant water level. The upper thicker part of thickening continuous settling forms is getting wider and thinner downwards towards the deeper parts of the basin (with gradually lower water levels). The lower parts of this settling form, however, are ever thinner towards the deeper parts of the basin and, to the effect of water seepage, are dissected by features similar to rainwater rills (Deák et al. 2013).

Table 6.3 Settling forms at various rates of water table decrease

Measured and calculated data				Settling forms developed in suspensions of various composition during different rates of water level decrease						
ΔV	m_{1-6}	t	v							
dm³	cm	min	cm/min	A	B	C	D	E	F	G
−1.	0.74	2	0.37	f	f	f	f	f	f	f
−2.	0.78	2	0.39	f	f	f	f	f	f	f
−3.	0.83	2	0.42	f	f	f	f	f	f	f
−4.	0.89	2	0.44	f_k	f	f	f	f	f	f
−5.	0.98	2	0.49	f_k	f_k	f	f	f	f	f
−6.	1.09	2	0.54	f_k	f_k	f_k	f	f	f	f
−1.	0.74	1	0.74	f_k	f	f	f	f	f	f
−2.	0.78	1	0.78	f_k	f	f	f	f	f	f
−3.	0.83	1	0.83	f_k	f	f	f	f	f	f
−4.	0.89	1	0.89	f_k	f_k	f	f	f	f	f
−5.	0.98	1	0.98	f_k	f_k	f_k	f	f	f	f
−6.	1.09	1	1.09	X	f_k	f_k	f	f	f	f
−1.	0.74	0.5	1.48	f_k	f	f	f	f	f	f
−2.	0.78	0.5	1.56	f_k	f	f	f	f	f	f
−3.	0.83	0.5	1.66	X	f_k	f	f	f	f	f
−4.	0.89	0.5	1.78	X	f_k	f_k	f	f	f	f
−5.	0.98	0.5	1.98	X	X	f_k	f	f	f	f
−6.	1.09	0.5	2.18	X	X	X	f	f	f	f

Deák et al. (2013)
ΔV volume of decrease in dm³ dm³:−1., −2., etc
m_{1-6} distance between surfaces before decrease (F_f) and after decrease (F_a) in cm
t: drainage time of a suspension of a volume of 1 dm³ 2 min, 1 min, 0,5 min
v: rate of water level decrease cm/min
A, B, C, D, E, F, G signs of the suspension types presented in Table 3.3 (the composition of the suspensions was given in Chap. 3)
f: continuous settling form
f_k: thickening continuous settling form
X: no settling form
Rate of water level subsidence: it was calculated based on the deepest point of the basin, rate of water level subsidence is accelerating during water loss (because of the shape of the basin); thus the given rates are average rates

The features formed during settling from suspensions of different composition on the settling floor at different rates of water level sinking are summarized in Table 6.3.

6.3.1.2.2 Settling Forms at Different Rates of Water Level Sinking

In Table 6.3 it can be seen that in case of suspensions A, B and C at a rate of water level sinking (0.34–0.54 cm/min), the development of a continuous settling form is dominant. In case of a medium rate of water level sinking (0.74–1.09 cm/min), the thickening continuous way of settling was characteristic. In case of a large rate of water level sinking (1.48–2.18 cm/min), the development of this settling form ceases in these suspensions, while in suspensions D, E, F and G (which are of higher water hardness and containing organic matter) continuous settling forms developed at every rate of water level sinking.

6.3.1.2.3 A Model for Settling Form Development on the Floor

During the process transforming the clay into suspension, the clay got into a domain of grain size smaller than 1 μm and between 1 and 63 μm. This is proved by the fact that a colloidal coating developed on the walls of the experimental basin and on the objects dipped into water. Sedimentation happened from the water which is another evidence for the presence of a material with a grain size of 1–63 μm. The deposited material, as we have mentioned above, created continuous and thickening continuous settling forms on the floor of the experimental basin (Figs. 6.18 and 6.19). During a decreasing, thus between two stagnant water levels, the water table of the lake of the experimental basin can be of three types:

- Stagnant
- Of a small, initial rate of water table sinking (V_S)
- Later of a great rate of water table sinking (V_l)

Starting from lower water tables, the rate of water table sinking is gradually increasing; the rates of water table sinking are accelerating because of the dipping of the floor of the experimental basin. Thus, both the experimental basin and the current water level downwards have a smaller extension. There should be a similar acceleration between two stagnant water levels too, also in case of the initial and the later water level sinking since the dipping of the floor is the same at each place. The difference of the two (initial and later water level sinking) is smaller or equal in case of smaller water level sinking, while it is greater in case of greater water level sinking.

In case of suspensions, D, E, F and G, the additives ($CaCl_2$, $KHCO_3$, dispergated plant additive) increased the coagulation of the colloid. It is probable that during coagulation, grains not only larger than 1 μm but also larger than 63 μm developed. There will be two consequences of coagulation:

- The coagulated material will not adhere, but it will deposit; therefore no colloidal coating will be created in case of stagnant water level.
- Because of coagulation, mainly the sedimentation rate of the grains of 1–3.9 μm will be so high that sediment will be created even in case of the greatest rate of water level sinking. The series will have a uniform thickness since the coarser sediment (coagulated colloid) will be deposited not only in case of smaller, initial water level sinking (or in case of stagnant water level) but also in case of large or larger sedimentation rate. The sediment will not have a thickening part in case of stagnant water level since because of coagulation there will not be any

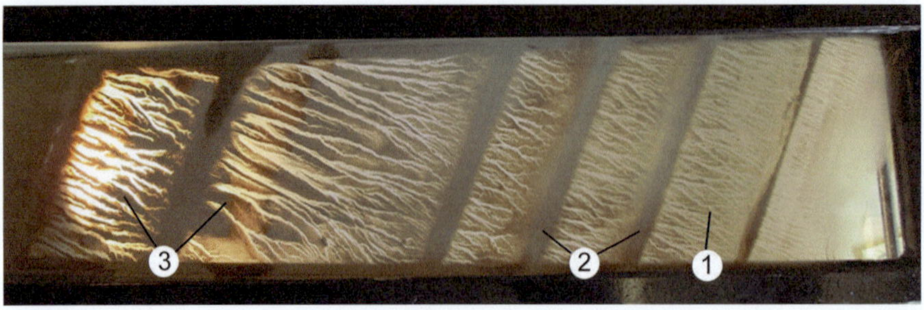

Fig. 6.19 Thickening continuous deposition form in the experimental basin. *1* thin bottom part of the deposition form of thickening continuous coating, *2* thick top part of the deposition form of thickening continuous coating, *3* rill-like form

fine-grained sediment in the suspension (or only of small amount) which could settle (Figs.6.19 and 6.20II).

As regards suspensions marked 'A' and 'B', in case of the development of a continuous settling form (the rate of water level sinking is small), the initial rate of water table sinking (V_S) will be slightly different from the later water table sinking (V_l). Because of this, sediments of similar size and thus similar thickness will be created everywhere, which are of diverse grain size (6.20Ia) because of slower water level sinking, but because of the lack of coagulation, the grain size coarser than 63 μm is absent or subordinated.

In case of medium rates of water table sinking, the initial (V_S) and the later (V_l) rates show a significant difference. Therefore the coarsest sediment but also the sediment of the finest grain do not only deposit in case of stagnant water level but even in case of initial water table sinking. In this case the upper part of the thickening continuous settling form develops. The reason for thickening is that because of slow water table sinking not only the coarse but also the fine-grained sediment deposits. However the thickening part will be even wider in case of water levels becoming lower and lower since the initial water level sinking is increasing. Thus, an even coarser (and thus even less) material will be able to deposit in a zone becoming even wider. The lower part of the thickening continuous sediment develops when the rate of water table sinking is large (thus in case of V_l). In this case only the coarsest (and thus small amount of) sediment is able to deposit (Fig. 6.20Ib). This zone can later become thinner which is proved by the rain furrows that developed on it (Fig. 6.19). By this the thickening continuous settling form develops.

Finally, in case of high rates of water level sinking (mainly in the case of the suspension marked 'A'), even the rate of sedimentation of the coarsest sediment will be lower than the rate of water level sinking. Therefore no sedimentation will happen.

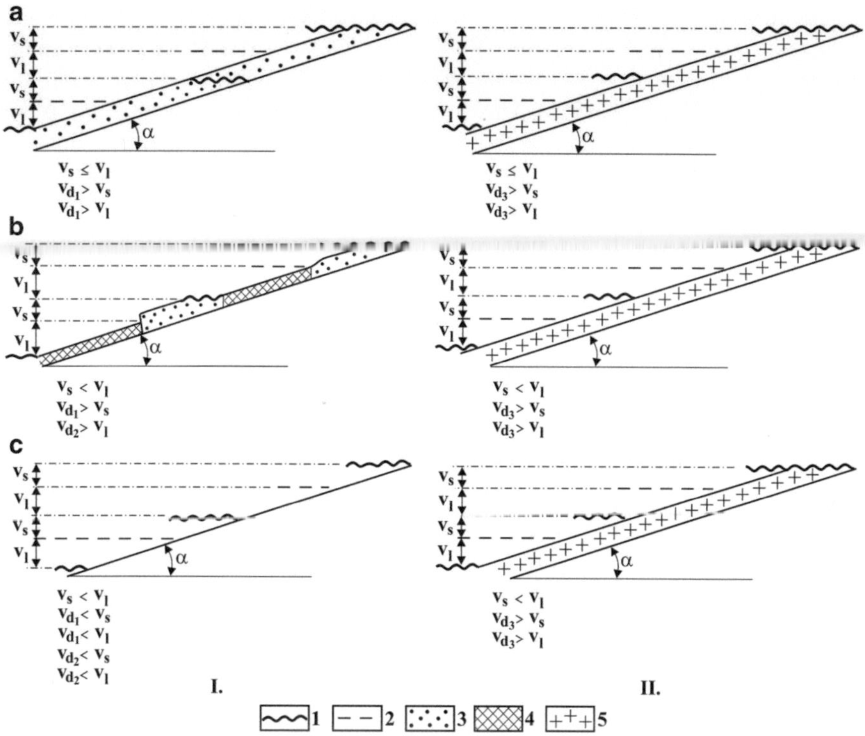

Fig. 6.20 Model of sedimentation in the experimental basin in case of water level sinkings interrupted by stagnant water levels. (**a**) Low rate of water level sinking, (**b**) higher rate of water level sinking, (**c**) high rate of water level sinking, *I* the sediment of the suspension is not coagulated, *II* the sediment of the suspension is coagulated, *1* stagnant water level of the lake of the experimental basin, *2* shifting water level, when the rate of water level sinking changes, *3* sediment with a grain size of 1–63 μm, *4* coarser fraction of the sediment with a grain size of 1–63 μm, *5* sediment with a grain size coarser than 63 μm, V_s initial rate of water level sinking, V_l later rate of water level sinking, V_{d_1} : sedimentation rate of the sediment with a grain size of 1–63 μm (thus that of fine-grained sediment too), V_{d_2} : sedimentation rate of the coarser fraction of the sediment with a grain size of 1–63 μm, V_{d_3} : sedimentation rate of the sediment with a grain size coarser than 63 μm, α. dipping of the floor of the experimental basin

6.3.2 *Sedimentation Under Natural Conditions*

The water inflow in the depressions carries various substances, which derive from the cover sediments of the hinterland, from the bedrock, soil or natural or cultivated vegetation (plant waste). Material transport either takes place on the floor (rolled or saltated) or suspended and dissolved. The suspended load of the depression may be rapidly settling fine sand with diameters of 63–125 μm, slower settling silt with 3.9–63 μm grain size, clay (colloidal like material) which diameter is 1–3.9 μm, not settling real colloid which diameter is less than 1 μm and plant waste of various sizes. The size of the plant waste may be from some millimetres to even several tens of centimetres, and its shape is either fibrillary or lamellar.

The particles transported on the floor deposit from flowing water even before the formation of the flood lake (Fig. 6.6b). The coarser but still suspended (non-colloidal) substance deposits from the water of the developing flood lake according to Stokes' law (Hiemenz 1986). This fraction includes particles above 1 μm. If the settling rate of the sediment in the lake is lower than the rate of water level sinking, as it was demonstrated in laboratory experiments, the sediment is transported into the karst with water. The exception are only those substances which are precipitated or adhere to various surfaces or onto trees in the depression (colloids) or get stuck on other obstacles (plant waste). With their large surfaces and charges, colloids are bound to surfaces through both adsorption (Pais 1981) and adhesion (Stefanovits 1981).

At wastewater treatment, in primary settling tanks, the anions of the solvent adhere to the surface of suspended particles through physical adsorption (Benedek and Valló 1982). Because of the repellent forces caused by their charges, colloids do not settle. At wastewater treatment, deposition is induced in two ways: with salt formation, the weight of particles is increased or their charge is eliminated, and as a result, they flocculate to form larger flakes (Barótfi 2003).

In natural systems the following processes contribute to the settling of the suspended material:

- The weight of suspended but non-colloidal substance increases through adsorption.
- Dissolved carbonate adheres to the particles.
- Colloids coagulate.
- The plant waste is loaded through adsorption, and furthermore, non-colloidal substance deposits on the waste.
- Plant waste adheres on the various surfaces or gets stuck on uneven surfaces (of trees, bushes, human structures).
- Plant waste also contributes to the settling of suspended material and colloids (see the results of the above-described experiment).

Sediment from flood lakes deposits in the following manner:

- The coarser suspended material sinks to the floor of the depression which bears the lake.
- Plant waste is caught by the trees of the depression.
- Plant waste is loaded with suspended load, colloids and dissolved load and sinks to the floor. The sediment input into the lake can increase the sinking rate of plant waste. The salts transported into the lake increase water density and, thus, reduce the sinking rate.
- Plant waste is bound by the floor through adhesion.
- Colloids adhere to the various surfaces.
- Colloids are deposited on the floor in two ways: they absorb dissolved material (chemical adsorption), and they coagulate because they lose their charges.

Deposition from flood lake takes place either uniformly (sediments of higher than 1 μm grain size, colloid, plant waste) or in zones (colloid, plant waste). Uniform deposition happens if the rate of water level lowering is continuously below that for suspended sediment. From the suspension, however, only material deposits the rate of deposition of which is higher than the dropping of lake water level (for colloids and additional condition is coagulation). Zonal deposition occurs if lake water lowering is not continuous but interrupted with stagnant water levels or if the colloid and the plant waste are positioned close to some surface (e.g. tree trunks in the depression, lake floor at the margins where water depth is small). At stagnant water level, both deposition and adherence to the floor (colloids, plant waste) take place. Subsequently, when the lake water level resumes to drop at a higher rate, it is not only adherence that could stop but less and less material deposits. Thus, adherence and deposition stop and the zone wedges out. A new zone develops if another stagnant water level forms in the lake.

6.3.2.1 Deposition in Ephemeral Lakes (of Rapidly Sinking Level)

In such lakes there is no stagnant water level and only sediment of rapid settling can deposit. Grains larger than 63 μm belong into this class. Plant waste cannot settle, but it can only adhere to the floor or it can accumulate in non-karstic pipes or in the small dissolution features of the walls of the shafts (Veress 1982) (Fig. 6.21).

6.3.2.2 Deposition in Lakes of Longer Duration (Fig. 6.22, Table 6.4)

The sediment finer than 63 μm deposited from flood lake continuously covers the doline floor, while plant waste is either continuous (Figs. 6.22b and 6.23) or zonal (Figs. 6.22c, d and 6.24). The plant waste and sediment cover are continuous if there is abundant plant waste (or sediment) in the lake or the settling rate of the waste is higher than the sinking rate of lake water level (Fig. 6.23). For the development of continuous plant waste cover, not only much plant waste is necessary but also large amounts of suspended and/or dissolved material. (This is a condition to the sinking of the waste to the bottom.) The zonally developed plant waste, as it was mentioned, accumulates at the margin of the lake with stagnant water level. Therefore, a further condition to the formation of the plant waste zone is shallow water depth. The width of the zone is marked by shallow water. The plant waste has a coating of colloids and other deposits. In case of an enduringly stagnant water level, the plant waste, in lack of loading, adheres to the lake floor too (Fig. 6.8d).

Zonal plant waste is either homogeneous or heterogeneous. In the former case, the zone shows no subzones, and the zones are separated by areas without plant waste. The heterogeneous plant waste zone can be divided into two subzones of plant waste. The outer (and higher) subzone is composed of continuous plant waste, while in the inner (and lower) subzone, the occurrence of plant waste is not continuous on the floor of the depression and becomes more and more rare towards the interior of the depression (Fig. 6.24). This pattern can also repeat itself.

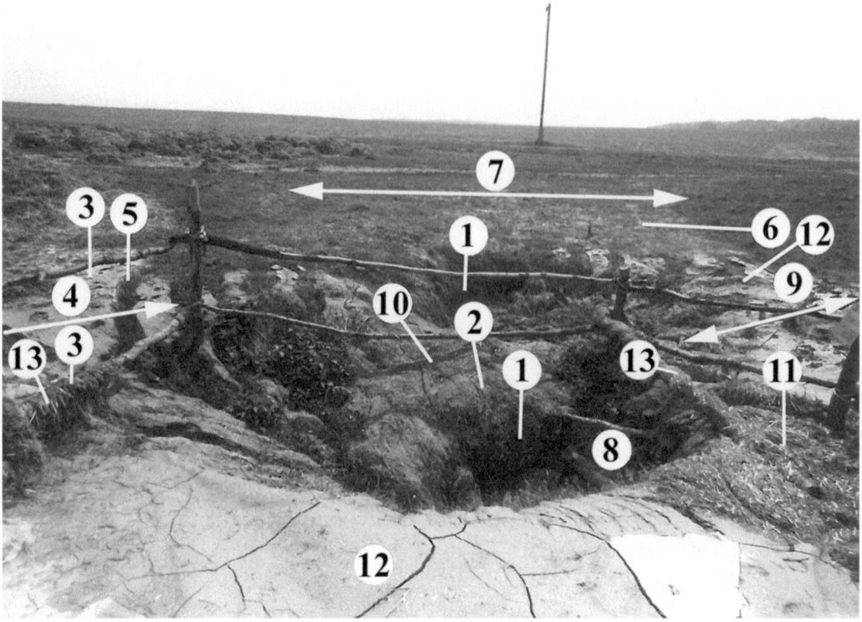

Fig. 6.21 Sediments of an ephemeral flood lake in dropout doline Gy-12 (following the activity of 9 May 1980). *1* Partial depressions, *2* Sill, dividing wall, *3* Coarser sediment or plant waste created by intensive water inflow, *4* Zone of intensive surface water inflow, *5* Rills created in sediment, *6* Discolouring of colloidal origin on vegetation, attesting to slow surface water inflow, *7* Zone of surface water inflow, *8* Ephemeral section of the former lake (intensive water flow and conduction), *9* Longer-duration section of the former lake (limited water flow and conduction), *10* Plant waste deposited on the sill because of shallow water depth, *11* Plant waste accumulation attesting to overflow, *12* Sediment created by overflow, *13* Plant waste stuck on an object

The outer part of the heterogeneous waste zone develops at the lake margin at stagnant water level. In this case, at low water depths, the accumulation of plant waste is continuous at the same spot. An inner zone forms when water level sinking begins, but at slow rate, only some parts of waste fragments have sufficient time to reach the floor. The heterogeneous plant waste zone corresponds to the zone of thickening continuous deposition created in the laboratory experiment (Fig. 6.19).

The colloid can adhere to the mentioned plant waste, tree trunks and foliage. The colloid coating on tree trunks is either uniform or becoming gradually paler downwards, divided into rings (Fig. 6.25). A whole series of such rings are occasionally created on the trunks of trees in the depression if the sinking stages of water level are interrupted by a series of stagnation spells (Fig. 6.25).

According to the relative rates of water level sinking and plant waste, the following forms of deposition occur in the depressions:

- If the dropping of water level takes place at a slower rate than the sinking of plant waste but it has a uniform rate, a continuous plant waste series forms on the depression floor and a continuous colloid coating on the tree trunks.

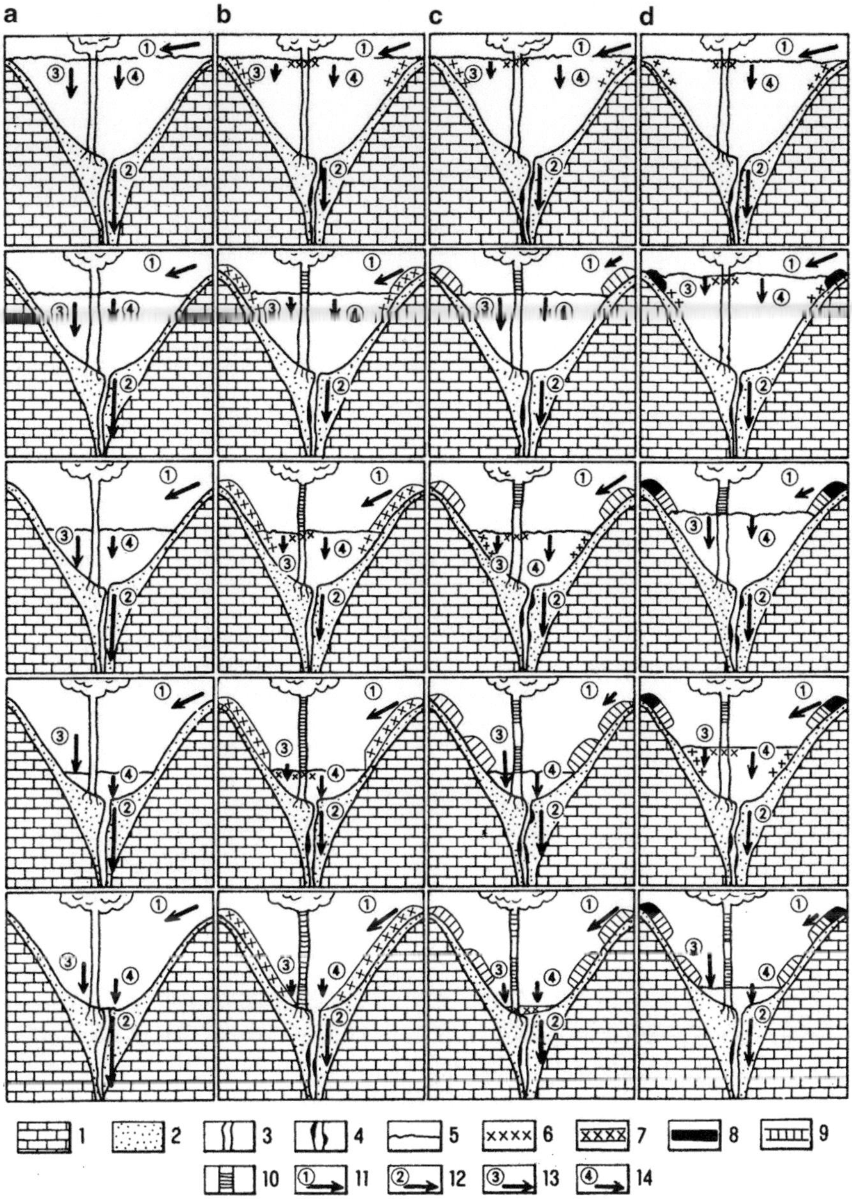

Fig. 6.22 Sedimentation in longer-duration lakes (Veress 1987). *1* Limestone, *2* Sediment fill, *3* Non-karstic pipe, *4* Partially clogged non-karstic pipe, *5* Water level in intermittent lake, *6* Sedimentation, *7* Contiguous sequence of plant waste or silt and clay with colloid coating, *8* Outer part of the heterogeneous plant waste zone, *9* Homogeneous zone of plant waste or inner part of the heterogeneous plant waste zone, *10* Colloid coating, occasionally coating of plant waste on tree trunk, *11* Water inflow into karst depression, *12* Water conduction, *13* Sinking water level, *14* Sinking rate of plant waste; (**a**) The sinking rate of water level always exceeds that of the sinking rate of plant waste, silt and clay; (**b**) The sinking rate of lake level is always lower than that of the sinking rate of plant waste, silt and clay; (**c**) The sinking rate of water level sometimes exceeds and sometimes falls below that of the sinking rate of plant waste, or the sinking of the water level is interrupted by stagnant water levels; (**d**) The sinking of water level is 0, then sinking is slow and is approximately the same as the sinking rate of plant waste

Table 6.4 Sedimentation of longer duration lakes

Rates and places of sedimentation	Water level subsidence takes place during water inflow				Water level subsidence takes place without water inflow
Relation between sedimentation and rate of water level subsidence	$V_1 > V_2$	$V_1 < V_2$	$V_1 < V_2$ interrupted by spells (with “n” frequency), when $V_1 > V_2$	$V_1 = 0$ $V_1 < V_2$ interrupted by spells when $V_1 > V_2$	$V_1 < V_2$
On the floor of the depression	There is no plant waste development and sedimentation	Continuous settling form and continuous plant waste series with settling veneer	‘n’ number plant waste zone with settling veneer	‘n’ number plant waste zone with settling veneer separating into two parts	Plant waste series with settling veneer
On the tree trunks of the depression	There is no colloidal veneer formation	Colloidal veneer	‘n’ number ring with colloidal veneer	‘n’ number ring with colloidal veneer	Colloidal and plant waste veneer

Veress (1987)
V_1 rate of water level subsidence of the lake
V_2 subsidence rate of the sediment or the adherence rate of the colloid onto the surface

- If the sinking of water level is interrupted by stagnation spells (but the rate of sinking exceeds the rate of sinking of waste or that of the adherence of colloids), both the plant waste and the colloid will show zonal patterns.
- If the rate of sinking of the water table invariably exceeds that of plant waste sinking or of adherence of colloid onto surfaces, neither plant waste nor colloid rings develop.

6.3.2.3 Deposition in Enduring Lakes

In enduring lakes not only the sediment of 1–63 μm grain size (silt and clay) are deposited after a long period of existence but also the colloids retained in the depression during evaporation.

According to observations around Hárskút, following intensive activity, the sediments of lakes formed in completely clogged, largely filled karst depressions in arable field environment consist of a lower, coarser, silty lamina of light colour and an upper (colloidal), finer lamina of darker colour (of jellylike consistence in fresh state) (Fig. 6.26).

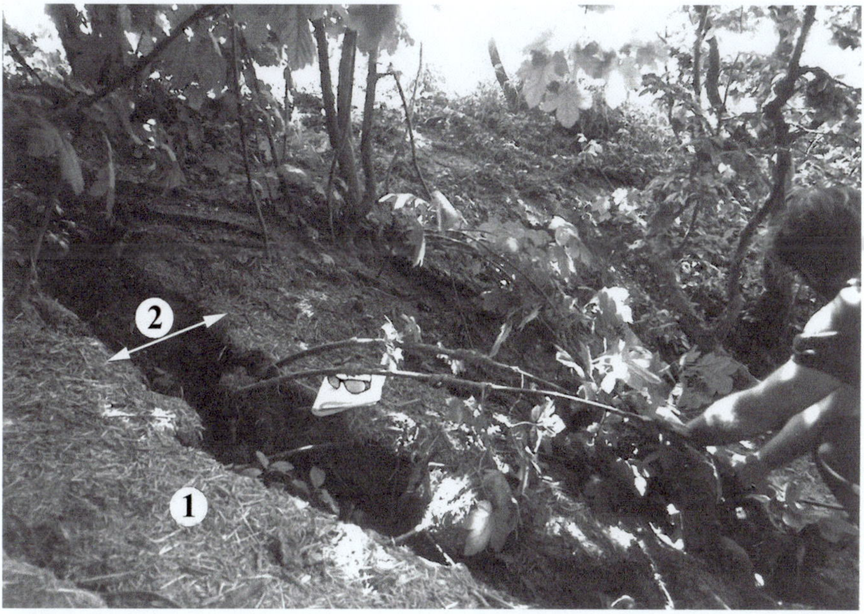

Fig. 6.23 Sediments of longer-duration lake in the suffosion doline Gy-11 (Hárskút Basin, Bakony Mountains), generated during the activity of 9 May 1980. *1* Continuous plant waste, *2* Rainwater rill, its depth equals the thickness of the sediments deposited when the lake existed

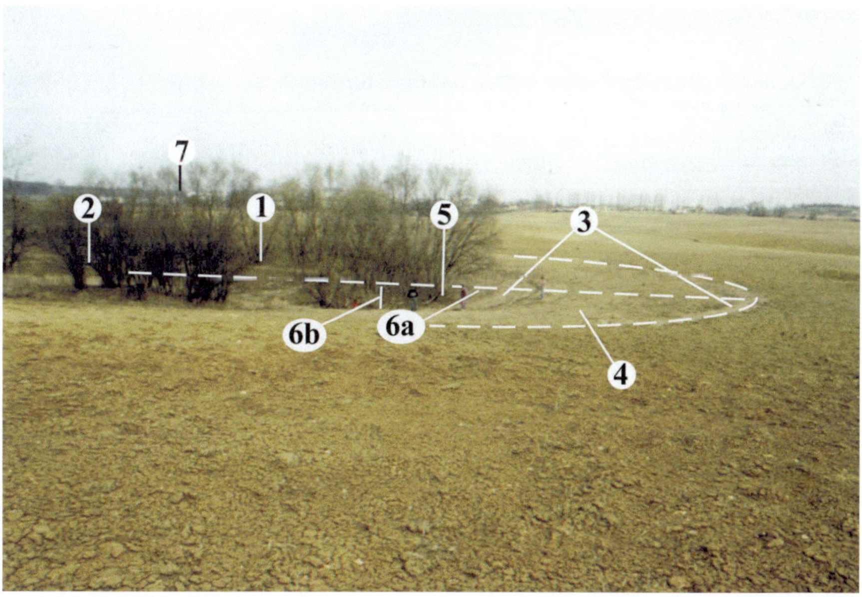

Fig. 6.24 Deposition in longer-duration lake fed by meltwaters in a doline near the village Dudar (Bakony Mountains) (shooting date in 1982). *1* Karst depression, *2* Water replenishment, *3* Outer parts of plant waste zones with abundant waste and sharp outward edge, *4* Inner part of the plant waste zones, *5* Terrain without plant waste, *6* Maximum extension of the lake horizontally (**a**) or vertically (**b**), *7* Dudar village

Fig. 6.25 Annular colloid coating on trees of the doline presented in Fig. 6.24. *1* Maximum height of lake level, *2* Colloid rings indicating slowly sinking or stagnant water levels (a_1–a_7), *3* Colloid-free tree trunk detail (the water conduit at the tree base generates local acceleration of water flow), *4* Rainwater rills in the lake sediment

Concluding from the observations, laminite formation takes place in the following stages:

- The coarser suspended sediment (silt and clay) partly settles from lake water.
- Lake level is reduced by evaporation to the extent that the non-deposited colloids are transformed into colloid gel, which is mixed with the remaining silt and clay. The colloid content of a sample taken from gel coating in a roadcut was determined and found to be 2.3 % in the upper part of the measuring cylinder and 2.8 % in the lower part (the substance did not settle from the water in the measuring cylinder even after several weeks).

It derives from the characteristics of laminite development that laminite series are of relatively moderate thickness and small lateral extension and the laminae are horizontal. Similarly to varves, they form in lakes within very short time intervals. The reason for the rapid development of laminae lies in the abrupt changes of the conditions of sedimentation. A summarising description of varve formation is provided by Zolitschka (2007). As a matter of course, varves occur in different (glacial) environment.

Distinct pairs of laminae only develop in karst depressions where there is no considerable loss by percolation but evaporation is remarkable. Under other

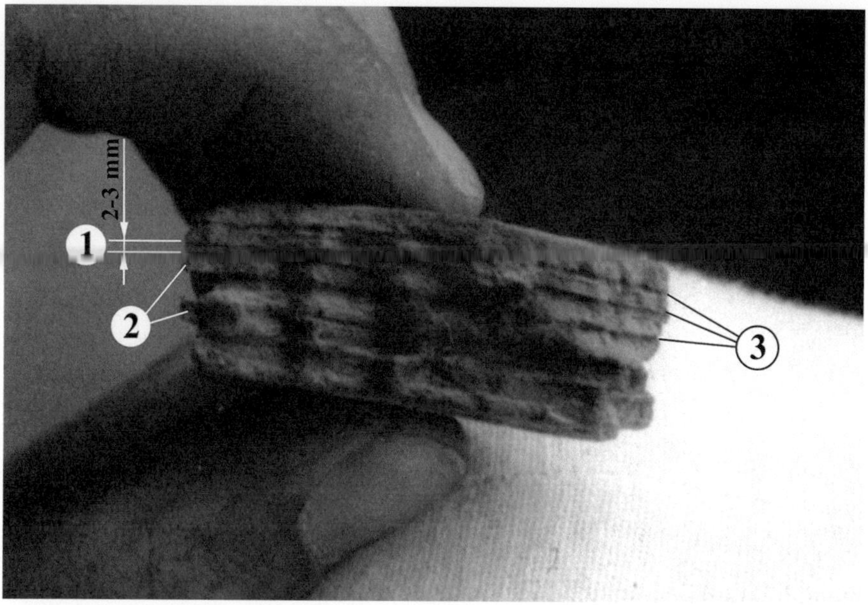

Fig. 6.26 Fragment of laminite series from the subsidence doline Ho-8 (Homód Valley) *1* Lamina pair, *2* Lamina of coarser deposit, *3* Lamina of finer deposit

conditions, instead of the laminae pair, an upward fining sequence without sharp boundaries comes about. A purely colloidal coating can result not only through water loss by evaporation but also through loss by very slow percolation (Fig. 6.27). Then colloids from the percolating water adhere to the floor of the karst depression, while silt is washed out from the depression.

A laminite series can be found in the fill of the karst depression Hu-1 (Hárskút, Fig. 6.28). There the series at 0.25–0.80 m depth in the fill consists of two kinds of laminae: a finer one of several millimetres' thickness and a coarser (silty) lamina of similar thickness. From this laminite sequence of almost a hundred pairs of laminae at 0.7 m depth below ground, a 'Globus' tin was recovered. In the sequence yellowish-grey rounded limestone fragments of 15–20 mm diameter as well as thin stripes of plant detritus or litter also occur. The coarser laminae are deposited from old lakes, while the finer laminae are of colloids and silts left over after the evaporation of (the remnants) of the same lakes. According to remembrances of locals, in the 1930–1940s the depression was still a 'ponor' of rocky walls. The available topographical maps of 1:10,000 scale also prove that the accumulated doline Hu-1 could have filled ca 55–80 years ago – also attested by the date of manufacturing on the tin found in the laminite series.

Eventually, in the enduring lakes of karst depressions, pairs of laminae form which build a laminite sequence over several decades. Since the formation of lakes is highly dependent on weather conditions, the time interval of complete clogging can only be estimated from the number of the lamina pairs or, in lack of overlying

Fig. 6.27 Colloidal deposit from an enduring lake (subsidence doline of the ponor shown in Figs. 37a and 6.6)

layer, the beginning of clogging. The observations made near Hárskút indicate that in the same filled doline at least twice and at most 10–15 times lakes develop. It seems probable, however, that there are years with no lake in the dolines.

The laminite sequence is typical first of all of karst depressions in vegetation-free environments. The lack of vegetation is not caused by climatic conditions but human impact (arable cultivation). The laminite sequence can also develop in forested environment, e.g. the doline Ho-8 (Homód Valley, Bakony) (Fig. 6.29) and the dolines Mb-50 and Mb-41 on Mester-Hajag (Fig. 6.30).

In the environments of recent dolines, laminite sequences theoretically occur in two forms:

- The laminite series next to each other do not constitute levels (Fig. 6.31i), e.g. in the case of the environs of the doline Mb-50 (Fig. 6.30). Such occurrence of laminite may indicate the previous existence of a doline row filled up by now. The present-day dolines developed independent from the members of the old doline row.
- The laminite sequences occasionally constitute levels in the environs of the present-day doline (Fig. 6.31II). In this case the present doline formed in the fill of a larger infilled doline.

Doline evolution may reach an active or an inactive phase (Veress 1986, 2000). The doline is active if no enduring lake has as yet formed in its interior and inactive in the opposite case. The transition from active to inactive phase is marked by the clogging of the non-karstic pipe. When becoming inactive, it accumulates silt and clay and then plant waste, which probably deposit in the temporary (longer-duration)

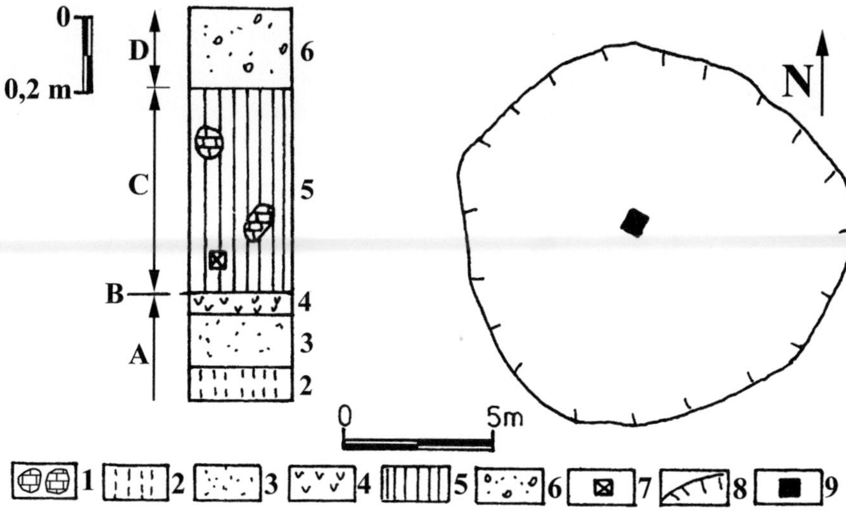

Fig. 6.28 Fill of inactive karst depression (doline Hu-1, Hárskút Basin) (Veress 2000). *1* Limestone debris, *2* Bluish-greyish silty clay with quartz gravels of 5–10 cm diameter, *3* Light grey silt, *4* Plant waste, *5* Laminite, *6* Unstratified clay and silt, *7* Tin, *8* Karst depression, *9* Exploration pit, (**A**) Inactivisation, (**B**) Closing of water conduit, (**C**) Inactive phase, (**D**) Rapid infilling, burial

lakes of the doline. If the chimney or non-karstic pipe of the doline is clogged or buried, the doline becomes inactive (clogged doline). Subsequently, as it has been mentioned, laminite accumulates (in an enduring lake) (Fig. 6.28). The inactive phase can become final and fossil doline results with a permanent lake. In this final stage, in lack of water conduction, groundwater accumulates below the doline floor. The inactive phase, however, can be followed by another active phase, indicated by the appearance of silt (or possibly fine sand) on the floor. Then the non-karstic pipe of the doline is cleared to a degree (or a new non-karstic pipe comes about) that an ephemeral flood lake emerges in the doline again (Figs. 6.28 and 6.29).

The characteristics of a fossil doline and its environment are the following:

- In the environs of the doline, the slope of the ground is gentle towards the doline, or because of erosion, even counterslope conditions are observed. Therefore, hardly any sediment reaches the doline. However, even this small amount is sufficient for the filling or burial of the doline.
- In the doline a permanent lake (or a lake existing for a long period, for several months) emerges. If its environs are eroded, water flows out of the doline. The water may derive from precipitation or from groundwater.

In the exploration pits of fossilised dolines, deposits bearing charcoal (or more properly twigs turning to charcoal), iron oxides, limonite and carbonates are found in mostly clay sequences (Fig. 6.32). Table 6.5 summarizes the formations associated with the various phases of fossilisation. Charcoal may occur as nodules and as a charcoal series, while limonite and carbonates form concretions of 1–2 cm diameter.

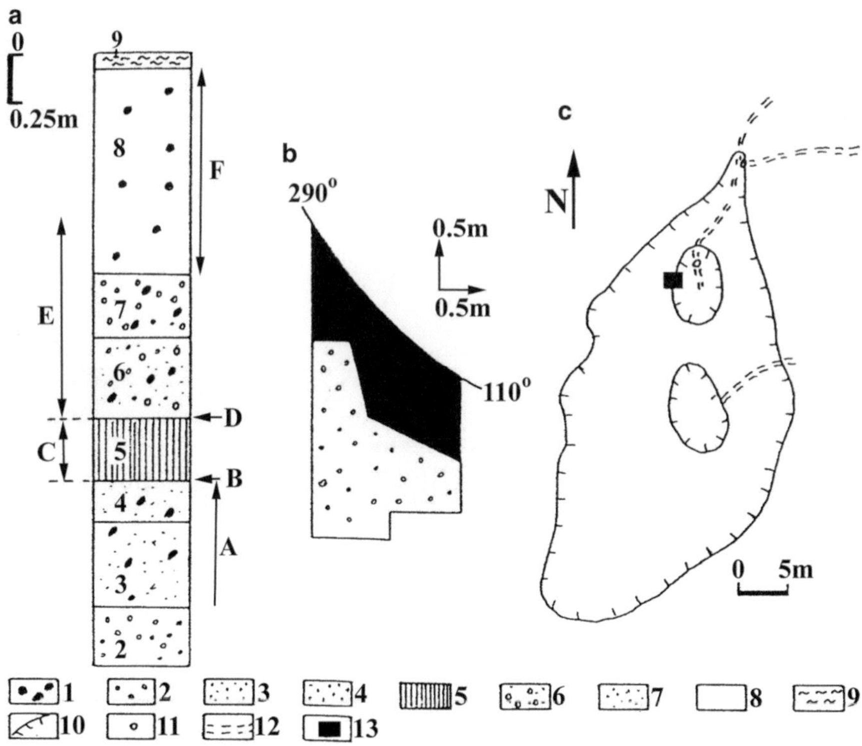

Fig. 6.29 Profile of sediment fill of a karst depression of complex evolution (doline with ponor Ho-8, Homód Valley) (Veress 2000). (**a**) Cross-section of the exploration pit of the karst depression, (**b**) longitudinal profile, (**c**) groundplan of the karst depression, *1* Sporadic gravels in the fill, *2* Yellow silt, *3* Dark grey clay, *4* Greenish grey clay, *5* Laminite, *6* Slightly clayey silt, *7* Yellow silt, *8* Brownish-yellow soil, *9* Humus, *10* Edge of karst depression, *11* Water conduit, *12* Gully, *13* Exploration pit, (**A**) Inactivisation, (**B**) Closing of water conduit, (**C**) Inactive phase, (**D**) Opening up of water conduit, (**E**) Activisation, (**F**) Active phase

Iron oxide builds a series and limonite develops from iron oxide. Limonite also co-occurs with iron oxide, charcoal and nodular charcoal. The joint occurrence of iron oxide and charcoal series, however, is not typical.

Iron oxide forms if the pH of groundwater increases as charcoal adsorbs acids (Fekete 1988). Limonite formation is associated with changes in groundwater level (Fekete 1988). With rising groundwater table, in reductive environment, ferrous oxide (Fe^{++}) emerges, while with sinking groundwater, ferric oxide (Fe^{+++}) is generated. Carbonate concretions (mostly above charcoal) develop if the acidity of infiltrating solutions decreases since charcoal binds acids. Acidity is reduced and carbonates precipitate.

Charcoal forms under anaerobic conditions and nodular charcoal if the tree fragments transported into the doline are buried. Anaerobic conditions are also generated under enduring water inundation.

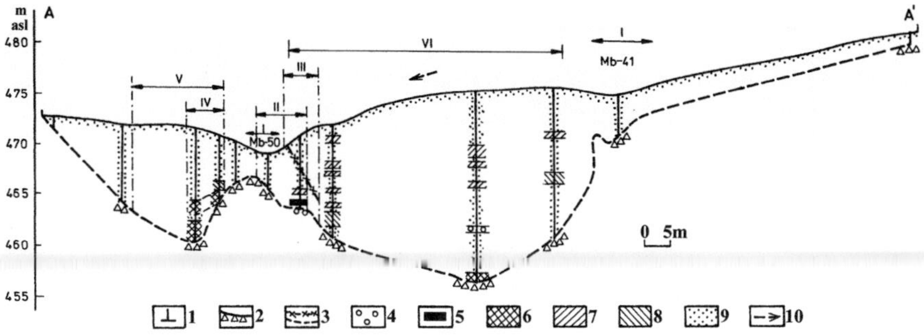

Fig. 6.30 Sediment geological profiles of the environs of the karst depression Mb-50 on Mester-Hajag (Veress 2000). *1* Borehole site, *2* Limestone bedrock reached by borehole, *3* assumed former and present-day limestone bedrock, *4* Gravel, *5* Brown clay, *6* Reddish-brown clay, *7* Laminite, *8* Buried soil, *9* Silt, clayey silt, *10* Former material transport, *I* Recent depressions formed above hidden rock boundary (Mb-50: on top of buried elevation truncated by karstification; Mb-41: on the side of elevation), *II* Elevation destroyed by karstification (the lack of reddish-brown clay indicates that it has been an eroding surface rising above its surroundings), *III* Side of elevation once affected by karstification, *IV* Reddish-brown clay of thickness increased through redeposition, *V* Presumably fossil depression (filled with reddish-brown clay, silt and then deep soil), *VI* Various laminite series: height differences indicate origin in several (two or three) former dolines, from the enduring lakes in them

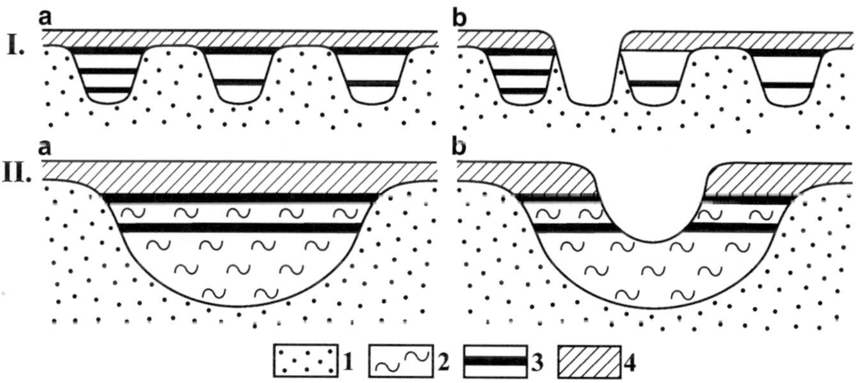

Fig. 6.31 The position of laminite series relative to recent dolines. *1* Cover deposit on the bottom of depression, *2* Sediment fill of depression, *3* Laminite series, *4* Sediment covering the dolines, (**a**) Infilling depressions, (**b**) Formation of new depression among (*I*) or in filled dolines (*II*)

In summary, charcoal forms during enduring lake conditions and burial. Iron oxide and carbonate concretions from infiltrating water, while limonite is the outcome of fluctuations in groundwater table. The condition of the fossil doline indicates the gradual appearance of groundwater in the doline. At this time, there is no water conduction from the doline into the karst anymore.

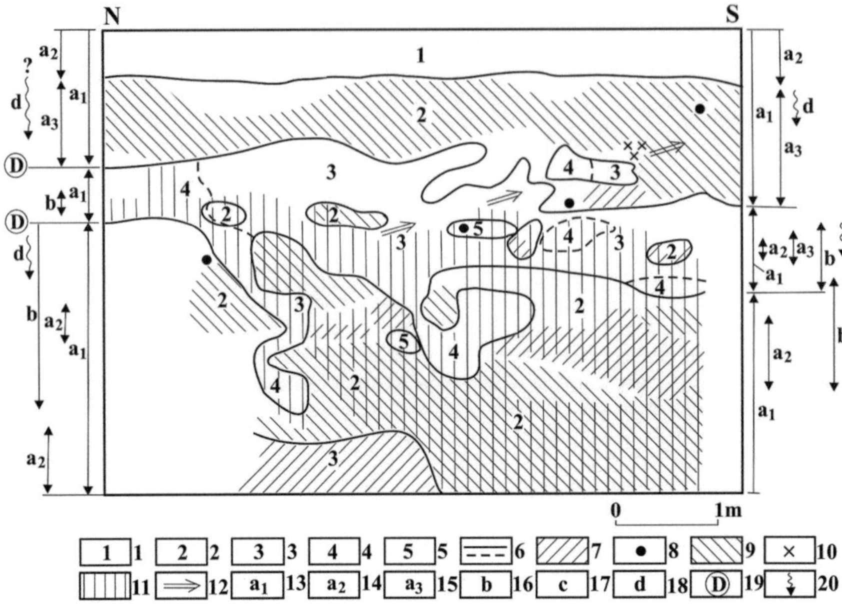

Fig. 6.32 Sediments in a fossil doline (Homód Valley area). *1* Recent sediment fill and soil, *2* Grey clay, *3* Grey-yellow mixed clay, *4* Slightly yellowish clay, *5* Variegated mixed clay, *6* Material boundary, *7* Series with charcoal, *8* Nodular charcoal, *9* Iron oxide, *10* Limonite, *11* Carbonate series, *12* Probable shift of site of water conduction during infilling, *13* Undifferentiated fill, *14* Slow infilling, *15* Rapid infilling (burial), *16* Redeposition within the depression, infilling in partial depression, *17* Carbonate precipitation, *18* Limonite formation, appearance of groundwater, *19* Erosion (discontinuity surface), *20* Infiltration

The characteristic properties of the different subsidence dolines (active but clogging doline, clogged doline, fossil doline) with regard to doline morphology, the manner of water conduction, the resulting sequence as well as the environment of the origin of sediments and the conditions in the hinterland are summarized in Table 6.5.

6.4 Development of Partial Features

6.4.1 Examples of the Intensity of Geomorphic Evolution of Covered Karst

Covered karst dolines or their partial features develop rapidly. As an example of rapid doline development, the Flint River Valley can be cited where in 1994, following intensive rainfalls during a tropical storm, at least 312 dolines emerged in the alluvium within a time span of 48 h (Hyatt and Jacobs 1996). During the storm the

Table 6.5 Relation among the condition, environs and sediments of the subsidence dolines

Doline condition	Doline morphology	The developed series	Water conduction from the doline	Environs of the development of the sediment	Condition of the hinterland
Active, but clogging doline	Its depth is several metres, it has a conduit	Buried soil, plant waste, colloid and sediment of grain size smaller than 63 μm	Water conduction through passage	From lakes of longer duration (ca. 1–2 h)	Mainly arable terrain or forested terrain with larger slope angle
Clogged doline	Its depth is several metres, but it has no conduit	Laminite	Seepage and evaporation	From enduring lakes (several days)	Gentle slope, fine-grained sediment is denuded and transported
Fossil doline	Its depth is 1–2 dm	Series of calcareous concretion	Evaporation, water seepage	There is no environs with lack of oxygen (no charcoal develops, little sediment arrives or there is no lake situation)	Transport of limestone debris into the depression
Fossil doline	Its depth is 1–2 dm	Iron oxide	Insignificant water seepage, evaporation	Environs with lack of oxygen, lake condition	Relatively large amounts of sediment transported into the depression
Fossil doline	Its depth is 1–2 dm	Series with charcoal	Evaporation, water outflow	Environs with lack of oxygen, enduring lake (several months)	Very gentle slope (it does not lean towards the doline partly), denudation of a small amount of the sediment is possible from the area of the doline
Fossil doline	There is no depression or there is a depression with a depth of some centimetres only	Limonite concretions, charcoal	Evaporation, water outflow	Acid groundwater, reductive environment	There is no sediment denudation, the surface is planar or does not lean towards the doline

old dolines too (predating the storm) broadened considerably (Hyatt et al. 1999). Another case can be mentioned from the state of Florida, where, according to Beck and Sinclair (1986), new dolines emerge week by week with the frequency of 0.05 km^{-2} $year^{-1}$ (Waltham et al. 2005). Magdalene and Alexander (1995) observed in a loess and till landscape of Minnesota state that 34 new dolines came about in 10 years' time. Kemmerly and Towe (1978) recorded 18 new dolines in Montgomery County, Tennessee, between 1937 and 1972.

The duration of doline development and its frequency also depend on the type of cover deposit. The age of dolines in silt of Montgomery County was estimated at 25 years, in clay for 38 years and in loess for 65 years. Doline evolution is even quicker on evaporites. On gypsum 0.01–3 dolines develop in 1 km^2 area (Waltham et al. 2005). Northeast of Ripon, on Ure Bank Terrace (North Yorkshire, UK), a doline 10 m across and 5.5 m deep emerged on cover sediment above gypsum in April 1997 (Cooper and Waltham 1999). Also near Ripon, at Sharon the above authors mention a subsidence doline formed on 1 February 1982.

Doline deepening is also rapid on cover sediment above gypsum. On alluvial deposits covering gypsum, doline subsidence rates of 39 mm $year^{-1}$ or even 64.5 mm $year^{-1}$ have been recorded (Soriano and Simón 2001). Landform evolution is even more intensive on the cover of rock salt. In Winkler County (Texas) two depressions, called sinks, developed: Wink Sink 1 on 3 June 1980 and reached a diameter of 360 ft within 24 h and 110 ft maximum depth in 3 days (Baumgardner et al. 1982), while the depression Wink Sink 2 formed on 21 May 2002 and grew to a size of 450 times 300 ft by March 2003 (Johnson et al. 2003). In the surroundings of these depressions, also rapid changes took place. Between 1970 and 1999 subsidence in one of the areas amounted to 23 ft while in the other 28 ft (Johnson et al. 2003). In the subsiding area, fractures (atectonic cracks) emerged in 1 ft width and 6.5 ft depth.

Particularly on evaporite karst, human activity (mineral oil extraction, drilling, etc.) can induce rapid landform development and growth. Thus, below the above-mentioned Wink Sink 1, water conduction along a borehole caused salt solution (Johnson 1987; Johnson et al. 2003). The Berezniki caprock doline (Ural), formed above a potash, also belongs here (Andrejchuk 2002). Klimchouk and Andrejchuk (2002) pointed out the activation of collapses of pipes and the development of new collapses induced by water extraction related to mining on the gypsum karst of Ukraine. Doline development is not induced by human activity on the cover sediment of evaporites only. At Interlacken, Florida, a doline came about on the cover of limestone during drilling a well on 16 August 1985 (Beck and Sinclair 1986). A multitude of dolines can result from water extraction. Near the Chinese town of Liupanshui (Guizhou province), water extraction was carried out for 8 years from 1 to 2 m depth, which led to the generation of 1023 new subsidence dolines (Waltham and Smart 1988).

It is not only observed that individual dolines form rapidly or numerous dolines emerge simultaneously, but where the area is particularly prone to doline development, new dolines come about almost continuously over a longer period of time. Such rapid doline formation was recorded at Ripon, as mentioned above (Cooper and Waltham 1999).

The shape and size of the resulting landforms also changes rapidly and to a great extent. Thus, in Ukraine a doline (chimney) that originated from the opening of a breccia pipe was transformed through the erosion of its walls into a bowl-shaped feature over 3 months (Klimchouk and Andrejchuk 2002). The dimensions of the mentioned Berezniki doline reached an extension of 40 times 80 m at the margins, while its depth, measured to the water level of the lake formed on the bottom, was 150 m. From its emergence on 25–26 July 1986 to 1992–1993, its dimensions had changed extremely. With the erosion of its slopes, its diameter grew to 200 m, while its depth was reduced to 80–100 m (Andrejchuk 2002).

Large-scale collapses, mostly occurring during the development of caprock dolines, can be accompanied by numerous phenomena. At the above-mentioned doline of Berezniki, simultaneous to collapse, flash light, sound effect, material outburst and vibration (weak seismicity) were recorded. The ejected material was deposited in a circle of several hundreds of metres around the doline (Andrejchuk 2002). The explosion was due to the compression of gases accumulated in the hollow by the collapsed material. Collapses can be followed by lake formation or, in an opposite case, the disappearance of an existing lake.

The water from karst lakes can be rapidly conducted into the karst during collapses or the formation of non-karstic pipes. This is also an indirect indication of rapid changes on the covered karst. The lake the water of which is conducted away may form in clogged depressions or in their vicinity. If a non-karstic pipe forms outside the lake, the lake survives (Fig. 6.33). In this case, at most lake overflow can reach the passages during floods. If the non-karstic pipe forms under the lake, the lake is rapidly drained (Fig. 6.34).

It may occur that next to the lake another doline forms and the lake is drained by that. Lake Jackson in Texas (area: 10 mile^2) was drained in spring 1982 when a doline formed in its vicinity (Beck and Sinclair 1986). The swampy plain of Paynes Prairie south of Gainesville, Texas, can also be mentioned. Its water has also been drained by a doline, but when water conduction ceased in 1873, a lake formed (with even ferry traffic), and when the activity of the doline resumed in 1891, a swamp developed through the drainage of lake water into the karst (Beck and Sinclair 1986). Lake Grady, formed through the impoundment of a river in 1972, can also be cited. Its water flowed into a dropout doline formed in May 1974 but its drainage began in April 1974 through sand (Beck and Sinclair 1986).

6.4.2 *Characteristics of Partial Feature Development*

As mentioned in Chap. 5, the partial features of subsidence dolines are either of karstic or non-karstic origin. Non-karstic partial features are the scars of mass movements (collapses, landslides) in the area of dolines, their accumulational features, rainwater rills, gullies and ravines and their accumulational features. Among the partial features of karstic origin, minor collapsed inner depressions of the cover sediment (doline varieties D_8 and S_3) and non-karstic pipes are distinguished.

Fig. 6.33 Collapse in the environs of a fossil doline on the Pádis Plateau. (**a**) Photograph taken in July 2011, (**b**) in July 2012

Fig. 6.34 Dropout doline rejuvenated by collapse near Sanica (Bosnia) (its lake was 20 m across and 8 m deep) (**a**) fossil doline, (**b**) dropout doline formed in the place of the fossil doline (50 m across, 30 m deep) developed in November 2013

The collapses of the cover, i.e. the development of karstic partial or interior features, are not related with a direct causal relationship to intensive rainfalls. The appearance of such features, however, can be an indication of intensive activity associated with intensive rainfall and the ensuing intensive processes. Collapse features develop if the increased weight of the cover due to material deficit exceeds the cohesion of the cover sediment. This probably results from a single rainfall event (single activity) but more often due to several consecutive events (multiple activities). Collapses occur during rainfall events or following them. The time elapsed between two collapses can be different for the individual dolines or for the individual collapses of the same doline. The development of collapse features, however, can be rapid.

Destructive erosional features (rills or gullies) very seldom result from an activity related to a single intensive rainfall event. The development of accumulation features, however, can happen during a single activity phase (within some hours or days). They can also evolve over longer periods, during several activities (over months or years). The following processes lead to the formation of such features (Table 6.6):

- A single collapse of short duration (some seconds)
- A series of collapses: collapses are interrupted by collapse-free intervals of various durations
- Subsidence (slow deepening of the bottom at different rates lasting long and interrupted by stagnation periods of various duration)
- Removal or redeposition of cover sediments; the cover sediment is accumulated in the passages of the doline/depression
- A combined effect of the above

The date of emergence and the duration of development are to be distinguished. The date of emergence is identified first of all by observation. It is a rather rare occasion, however, that features form right at the time of observation. The date of emergence can be established in the following ways:

- The date of emergence can always be given with high precision if the doline forms in a man-made structure (road, railway) (Fig. 5.21).
- If there are grassed (Figs. 5.36 and 6.35) or cereal patches on the floor of the collapsed form (most often small dropout doline). In both cases the collapse leading to doline formation took place in the vegetation period of the year of observation.
- Through repeated observation, the age of young changes can be established. If the change occurred between two observations, the time interval of development can also be identified. The closer the dates of the two observations are, the more precisely the date (Fig. 6.33) and time interval of change (form development, see below) can be given. If the observation was supplemented with photo documentation, the changes not remarkable in aftermath could also be identified.
- Upturned trees in dolines may indicate mass movements (primarily collapses on the doline floor and slopes). Their maximum age is younger than that of the tree (identified from the tree rings).
- Curved trunks (Fig. 6.36a) and modified root systems (Fig. 6.36b) show increased slopes in the doline or the sliding of the cover sediment towards the doline interior. The ages of such movements can also be established from the study of tree rings. Growing slope angles and mass movement intensity may point to doline formation or indicate changes in the bedrock (e.g. opening of a chimney or more sediment output than input).
- Recent change (of some years' duration related to the observation) is indicated by a sharp doline edge, steep side slopes truncating soil profiles and different cover sediments (failure front) and the collapse heaps on the doline floor. The roots outcropping on the failure front (mostly fine roots of herbaceous plants) also make it probable that recent collapses (younger than 1 year) happened.

The time interval of development and the rate of landform evolution can be given in the following ways:

Table 6.6 Duration and start of development of some partial features of dolines

Feature	Formation process	Duration of development	Determination method	Start of development related to observation	Determination method
Plant 'beard'	Fall	1–2 s	Estimation	Less than 1 year	In the growing season
Vertical surface in sediment cover	Fall	1–2 s	Estimation	Less than 1 year	No rainwater impact
Collapse mound	Fall	1–2 s	Estimation	Less than 1 year	No rainwater impact
Plant residue or soil residue on doline floor	Fall	1–2 s	Estimation	Less than 1 year	In the growing season
Tilted or uprooted (dead) tree	Series of falls/slides	From 1 to 2 days to several years	Estimation	Maximum the age of the tree	Tree ring counting
Partially buried tree trunk	Infilling	From 1 to 2 years to several decades	Tree ring counting	Maximum the age of the tree	Tree ring counting
Buried object	Infilling	From 1 to 2 years to several decades	At earliest the date of manufacture	From 1 to 2 years to several decades	At earliest the date of manufacture

Veress (2013)

- Using data from measurements or photo documentation at two dates. Thus, the re-measurement of depth allows the calculation of the rate of deepening or infilling if the time elapsed between the two dates is known (Johnson et al. 2003; Veress 2013). Repeated photo documentation allows the establishment of growing doline density in the study area (Currens et al. 2012).
- Artefacts buried in the doline (Fig. 6.37) point to accumulation in the doline. If it is known when the object is incorporated in the sediment (or its approximate date can be estimated), from the thickness of the cover above the object, the rate of accumulation can be estimated.
- In knowledge of the age of buried trees (Fig. 6.38) and the thickness of fill, the rate of accumulation can also be estimated.

Fig. 6.35 Living grassed patches on the floor of a dropout doline (Pádis, area P-2, taken in 2011)

Fig. 6.36 Deformed tree trunks in dolines in 2012. (**a**) Curved tree in one of the suffosion dolines of Kab Mountain, (**b**) curved tree indicating mass movement in the doline with ponor Gy-9 (Adjusted to the movement, roots on the slope below the tree differ from those above it, Hárskút Basin 1979)

6.5 Change in Size

The formation of features in dolines was investigated in some dolines of the Tés Plateau in 2010 (Veress 2013; Veress et al. 2013, Table 6.7). This year had been selected since in 2010 the frequency, intensity and amount of precipitation were extremely high. Changes compared to the period before 2010 can be identified as the studied dolines were visited at least once a year between 2004 and 2009. The table shows that landform generation was considerable and unique to the effect of rainfall in 2010 in the dolines. In dolines lying close to one another, destructive (non-karstic pipes, internal suffosion dolines, gullies) and accumulational features (infilled floors or floor details, Fig. 6.39) could equally come about. Among the dolines under study, the most remarkable changes were observed in the depression of cover deposit D-2 (doline I-31), where maximum deepening amounted to 120 cm in 7 years (Fig. 6.40, Table 6.8). This means a 171 mm year^{-1} average deepening rate, which is regarded considerable given that the rate of deepening for subsidence dolines above a gypsum bedrock was found to be 64.5 mm year^{-1} (Soriano and Simon 2001). The partial features of the above-mentioned DSD also underwent remarkable transformation. The inner of marginal doline developed on the eastern margin of the depression was infilled by 2010. Between 2004 and 2010 (not known more precisely) another doline formed on its western edge and merged with the main doline by 2010.

Doline depth either grows or reduces. The change can be described with a single value (pointlike depth change) if measurements were made at the same site at different dates. Alternatively, it can be described with an isopleth map if measurements were made at several sites (see Chap. 3), and thus data are obtained on both the areal distribution of depth change and on the change of doline shape.

6.5.1 Pointlike Depth Change

Measuring doline infilling in the karst regions of Hungary between 2004 (2006) and 2010, it is claimed that the greatest extent of infilling was found for the solution dolines of the Aggtelek Karst (Veress 2013). The average rate of infilling of subsidence dolines, compared to solution dolines, was 33 %. Subsidence dolines filled in at a rate of 1.3 cm year^{-1}, while solution dolines at 3.9 cm year^{-1}, in spite of the fact that cover sediments were absent in the environs of the latter or was restricted to solution residue at most. (It is assumed that no transported sediment left the solution dolines, which have no conduit, into the karst, while this was the case for subsidence dolines.) Therefore, it is estimated that at least 67 % of the sediment input into subsidence dolines was further transported into the karst in the period indicated (Table 6.8).

Comparisons among the infilling of subsidence dolines investigated can be made according to the vegetation cover of their catchments (Table 6.8). For depressions with grassed catchments, the average rate of infilling was 0.8 cm year^{-1} between 2004 and 2010 (between 2006 and 2010 for Nagymező, Bükk Mountains), for those on forested catchments 1.2 cm year^{-1} (between 2003 and 2010, Homód Valley, Hárskút Basin, Bakony Mountains) and for those in arable environments 2.0 cm year^{-1} (between 2004 and 2010, Tés Plateau, Bakony Mountains). A similar rate was experienced for the infilling of dolines with arable environment when the infilling rate was calculated from the cover thickness above the tin in the fill of the doline Hu-1 (see Chap. 3). Its value was 2.8 cm year^{-1}. (The tin emerged in 1980 and it has been estimated that the tin got into the doline at the beginning of the 1950s.)

Fig. 6.37 Buried tyre near Žabljak (Durmitor) in 2012

Fig. 6.38 Buried tree trunk in the covered karst ponor K-1 (Hárskút Basin, Bakony Mountains). *1* Internal depressions, *2* Buried trunk, *3* Filled eastern side of the depression, *4* Depression floor, almost plane due to infilling

Table 6.7 Partial features of dolines developed in 2010

Doline code	Feature	How many?	Process	Morphological situation
I-15	No	0	–	Flat terrain
I-17	No	0	–	In depression D-3
I-18	Non-karstic pipe	1	Opening by fall	
I-19	Non-karstic pipe	2	Opening by fall	
I-22	No	0	–	In depression D-4 on valley floor
I-23	Inner suffosion doline	1	Fall	
I-24	Flat floor	1	Filling of cleaned shaft	
I-25	1. Flat floor in inner suffosion doline	2	1. Infilling	
	2. Non-karstic pipe	1	2. Opening by fall	
I-104 E	Flat floor	0	Infilling	
I-104 W	1. Non-karstic pipe	1	1. Opening by fall	
	2. Inner suffosion doline	1	2. Fall	
	3. Channels	5	3. Rill/gully erosion	
I-26	No	0	–	Flat terrain
I-27	Inner suffosion doline	1	Fall	Flat terrain
I-28	Non-karstic pipe	ca 10–15	To the effect of water percolation	Flat terrain
I-29	1. Flat floor in inner subsidence doline	1	1. Infilling	On the floor of Tábla Valley
	2. 3–5 non-karstic pipe in margins	3–5	2. To the effect of water percolation	
I-31	1. Channel	1	1. Rill/gully erosion	In the depression D-2 in the side of Tábla Valley
	2. Broadening of inner subsidence doline	1	2. Induced by fall	
	3. Non-karstic pipe in inner subsidence doline	1	3. Opening by fall	
I-32	1. Non-karstic pipe	1	1. Opening by fall	In the depression D-1 on the floor of Tábla Valley
	2. Flat floor	1	2. Infilling	
I-32 L	No	0	–	
I-33	No	0	–	

Tési Plateau, Veress (2013)
Notice: Observation date of the data presented in the table is 08.10.2010, former observations between 2004 and 2009 once a year and in January 2010

Thus, the rate of infilling for covered karst landforms in arable environment is double of those on grassed or forested catchments. This is explained by the more intense removal of soil and cover sediments on arable land due to the lack of vegetation than on surfaces covered by grass or trees. Since in the period 2004–2010 precipitation was extremely high in 2010, infilling was probably the highest in this

Fig. 6.39 Subsidence doline filled up in 2010 (eastern partial depression of subsidence doline I-104, Tés Plateau, 8 October 2010). *1* Margin of fill

Fig. 6.40 Changes in depression D-2 (doline I-31) by 2010 (Tés Plateau, 2010). *1* Well rings exposed by the formation of the internal doline, *2* Internal doline

Table 6.8 Infilling or deepening of karst depressions in the 2000s

Tés Plateau (Bakony Mountains), measured in April 2004 and in June 2010					Homód-árok (Bakony Mountains), measured in June 2003 and in July 2010				
Code	Change in depth (cm)	Rate (cm year^{-1})	Catchment vegetation	Genetic type	Code	Change in depth (cm)	Rate (cm year^{-1})	Catchment vegetation	Genetic type
I-33	+16	2.3	Arable	F_1	Ho-1	−20	2.5	Forest	F_1
I-32	+10?	1.4	Arable	F_1	Ho-2/a	0	0	Forest	F_1
I-32_L	+13	1.6	Arable	F_1	Ho-2/b	−22	2.7	Forest	F_1
I-31	−120	17.1	Arable	F_2	Ho-3	−30	3.7	Forest	F_1
I-22	0	0	Forest	F_1	Ho-4	0	0	Forest	F_1
I-23	−25	3.6	Forest	F_1	Ho-6	+19	2.4	Forest	F_1
I-24	−14	2.0	Forest	F_1	Ho-7	+22	2.7	Forest	F_1
I-25	−10	1.4	Forest	F_1	Ho-8	+23	2.9	Forest	F_1
I-26	−17	2.4	Arable	F_1	Ho-9	0	0	Forest	F_1
I-28	+21	3.0	Arable	F_1	Ho-10	+29	3.6	Forest	F_1
I-15	+16	2.3	Arable	F_1	Ho-11	+12	1.5	Forest	F_1
I-17	+21	3.0	Arable	F_1	Ho-12	0	0	Forest	F_1
I-18	−32	4.6	Arable	F_2	Ho-13	+16	2.0	Forest	F_1
I-19	+16	2.3	Arable	F_1	Ho-14	0	0	Forest	F_1
I-27	+13	1.9	Arable	F_3	Ho-15	+10	1.2	Forest	F_1
Nagymező (Bükk Mountains), measured in July 2006 and in July 2010					Ho-16/A	−39	4.9	Forest	F_1
N-1	−19	3.8	Grassland	F_1	Ho-16/B	−12	1.5	Forest	F_1
N-3	+10	2.0	Grassland	F_1	Ho-17	0	0	Forest	F_1
N-8	0	0	Grassland	F_1	Ho-20	0	0	Forest	F_1
N-6	+6?	1.2	Grassland	F_1	Ho-21	0	0	Forest	F_1
N-4	0	0	Grassland	F_1	Ho-22	+10	1.2	Forest	F_1
N-9	+13	2.6	Grassland	F_1					
N-5	0	0	Grassland	F_1					
N-2	0	0	Grassland	F_1					
Aggtelek Karst, measured in July 2006 and in July 2010									
A-1	+32	6.4	Forest	O					
A-2	+21	4.2	Forest	O					
A-3	+25	5.0	Forest	O					
A-4	0	0	Forest	O					

Veress (2013)

+ infilling, − deepening, *E* forest on its catchment, *Gy* grassland on its catchment, *Sz* arable on its catchment

O solution doline, *F* subsidence doline (F_1 suffosion doline, F_2 dropout doline, F_3 fossil suffosion doline) year. This makes it probable that in this year the actual rate of infilling was higher than calculated (as an annual average).

6.5.2 *Areal Distribution of Depth Change*

The morphological changes of the above-mentioned DSD D-2 (or doline I-31) were studied through computing the differences between contour lines on maps prepared at two dates (2004 and 2010) (see Chap. 3). In the depression subsiding partial areas are identified (Fig. 6.41), which are divided by less intensively deepening or infilling surfaces (Veress 2013). The intensively subsiding details of the floor and the features indicating sediment loss (e.g. the development of inner and marginal dolines) probably point to explored caves or shafts of considerable depth below the depression (Táblavölgy dripstone cave). Not only the removal of sediment contributed to the deepening of the doline (an exploration was carried out in the cave) but also the fact that well rings were installed above the cave entrance. Because of the removal of sediment, the sediment inwash into the cave increased. Well rings ensured intensive sediment removal into the cave as they did not clog as the non-karstic pipes did. Sediment transport into the cave took place at the upper edge of the well rings or at the contact of well rings. This and the deepening are indicated by the position of the edge of the uppermost well ring above the floor in 2010 (Fig. 6.40). The uppermost well ring rising above the floor shows destruction of the floor. The inner doline could have formed due to material transport from the cover into the cave by piping through the slots between well rings. Subsequently, the evolution of the inner doline further intensified the destruction of the floor of the depression.

The surface around the depression, which is on the western slope of the Tábla Valley of N-S strike, has an eastern slope and, consequently, receives its fill material from the west. At the same time, on the western margin of the depression, which – as mentioned – is deepened in the west, a new partial depression emerged. Therefore, in the depression changes in depth are less controlled by sediment input. If this were the case, the western part could have deepened less intensively than the eastern part, but rather the karstic processes and generated sediment transport in its area and its environs were of great significance (Fig. 6.41).

6.5.3 *Material Budget of Dolines*

Conclusions on the material budget of doline can be drawn from the joint analyses of landforms and depth change within a time period. Regarding landform evolution in dolines showing change, we observe that in deepening and infilling dolines, features indicating sediment input and output equally develop (Tables 6.7 and 6.8).

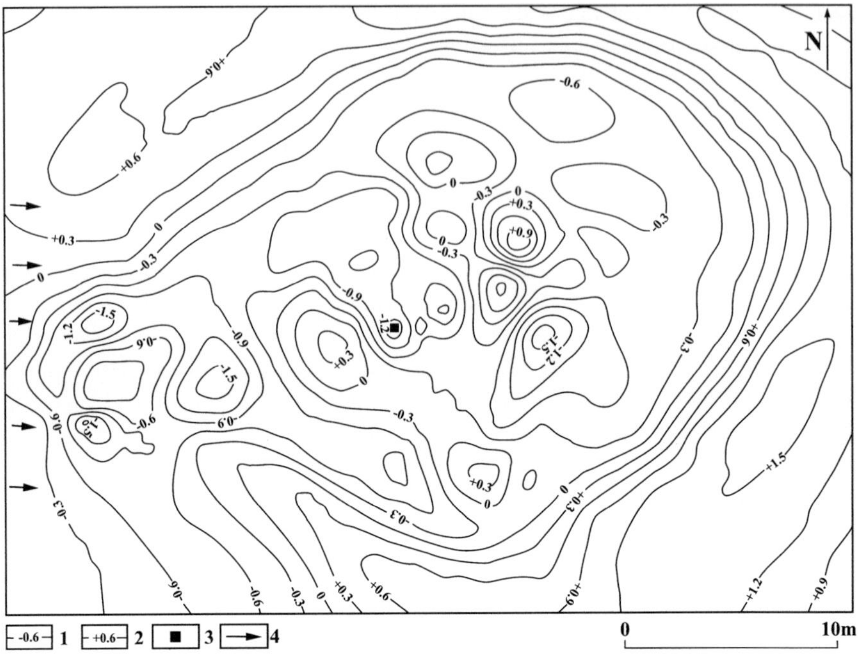

Fig. 6.41 Map of depth changes in depression D-2 (Veress 2013). *1* Deepening (dm), *2* Infilling (dm), *3* Shaft (cave) of depression with well rings, *4* Direction of sediment input (depths measured in 2004 and 2010)

Based on the above, the following varieties of subsidence dolines are distinguished according to depth change (Veress 2013, Table 6.9).

If no non-karstic (accumulational) landform is generated on the doline floor but doline depth has increased, there was material transport out of the doline but probably no sediment input into the doline took place, or if there was a sediment input, it did not accumulate in the doline (type A). If there are destructive features in the doline and its depth has increased, sediment input happened but at a lower rate than sediment output and the input preceded output in time (variety B_1). If an accumulational feature occurs in the doline and doline depth has increased, outward transport also surpasses inward transport but in this case sediment output preceded input (variety B_2).

If accumulational features are observed in the doline and doline depth has been reduced, it indicates previous material input into the doline without considerable outward transport (type C). The doline belongs to this class even if the growth of doline depth has not been accompanied by the formation of characteristic features. If destructional features are generated in the doline and simultaneously doline depth has been reduced, outward material transport from the karst was of lower degree than that of inward transport (type D). Naturally, if there has been no change either in depth or in morphology, no inward or outward transport is indicated (type E).

Table 6.9 Relation among the sediment budgets, depth changes and morphology of dolines

Doline code	Depth change (cm)	Features	Type based on sediment budget
I-15	+16	x	C
I-17	+21	x	C
I-18	−32	m	B_1
I-19	+16	m	D
I-22	0	x	E
I-23	−25	m	B_1
I-24	−14	a	B_2
I-25	−10	m, a	B_1
I-26	−17	x	A
I-27	+13	m	D
I-28	+21	m	D
I-29	?	m, a	?
I-31	−120	m	B_1
I-32	+10	m, a	D
I-32 L	+13	x	C
I-33	+16	x	C
I-104 eastern partial doline	?	a	?
I-104 western partial doline	?	m	?

Veress (2013)
+: degree of infilling
– degree of deepening
m: feature referring to sediment transport from the doline into the karst
X: the features of the doline did not change
a: feature referring to infilling
Esz: sediment transport from the doline into the karst
Bsz: sediment transport into the doline
A: no feature development in the doline, but there is increase in depth; there is Esz, but there is no Bsz
B_1: feature development by sediment transport from the doline into the karst and there is increase in depth, but sediment transport into the karst precedes deepening, Esz > Bsz
B_2: there is infilling feature and increase in depth, but sediment transport into the karst happened after deepening, Esz > Bsz
C: there is infilling feature development and decrease in depth, there is no Esz, but there is Bsz
D: there is feature development by sediment transportation from the doline into the karst and there is decrease in depth; Esz < Bsz
E: there is no feature development and change in depth in the doline; there is neither Esz nor Bsz

6.6 Conclusion

The conditions of activity of karst depressions (water inflow) were analysed. The activity was characterised by its nature (winter and summer activities). Types of water inflow (surface, rivulet, linear and valley floor inflow), activity phenomena (flood lake, overflow, intermittent spring, onfilling) and types of activity (surface inflow, outward seepage, latent activity, complex activity) were identified and described. The characteristic properties of winter and summer activities and their role in the evolution of the passage system of karst were demonstrated. The flood lakes were typified according to their mode and place of development and lifetime. Ephemeral, longer-duration and enduring flood lakes were distinguished and the conditions of their origin and sedimentation were described.

In laboratory the origin of sedimentation in flood lakes were investigated. Sedimentation can take the form of precipitation (due to intensive evaporation and dissolved material), coating and deposition. The coating form is annular, linear, continuous or thickening continuous (on basin wall or floor). The sedimentation features depend on the properties of the suspension, the position of the surface where settling happens, the rate of water level sinking and the presence of stagnant water intervals. In laboratory we pointed out that continuous water level sinking involves the development of continuous sedimentation forms, while if it is interrupted by stagnant water spells, either continuous sedimentation form or thickening continuous sedimentation form results. During increasing the rates of water level sinking, in suspensions without coagulation or with little coagulation, the sediment thickens at stagnant water level, since in case of initial (or stagnant) rate of water level sinking, the fine-grained sediment is deposited too. At later water level sinking (farther from the stagnant water level), a sediment of smaller thickness develops as here only the coarser sediment can deposit. From suspensions where coagulation happens, the coarser grains are able to deposit at various rates of water level sinking (thus also at high rate). The developing settling form will be continuous and of similar thickness. The sediment finer than 63 μm in the fill of the depression indicates the development of flood lakes at the time of the development of the fill. The presence of the sediment coarser than this in the fill refers either to the existing conditions for the inward transport of the material of such grain size or to the presence of a material favouring coagulation in the water of the lake. The settling form created in the laboratory is similar to the properties of plant waste in karst depressions. Thus, the thickening continuous form of deposition corresponds to the plant waste sequence of heterogeneous zones. The upper, thickened part of the former corresponds to the outer part of the plant waste zone, while the lower part corresponds to the inner subzone of the plant waste series.

The conditions of sedimentation from flood lakes are intensive activity (water inflow), retarded water conduction and the presence of clay (colloid) on the catchment. Continuous appearance is typical if the lake water level sinks continuously and at a lower rate than the rate of sinking for sediment and plant waste. They are arranged in zonal (annular) form if the lake has stagnant water level(s). An additional condition of the development of a plant waste zone is the gentle slope of the doline and the presence of plant waste in the hinterland.

From flood lakes, colloid (particles smaller than 1 μm), silt, clay (particles of 1–61 μm) and plant waste settle. Plant waste and colloids adhere to surfaces and plant waste, silt and clay deposit. At stagnant water level or if the rate of sinking of water level is below that of the mentioned materials, silt, clay and the loaded plant waste deposit continuously. If the rate of water level sinking is above that of these materials, they do not deposit or only that part deposits which has a higher rate of settling (the sinking rate of loaded plant waste) than the rate of water level sinking.

At the bottom of the longer-duration lakes, plant waste accumulates (if water level sinking is continuous) or series of plant waste zones and colloid ring coatings form (if water level sinking is interrupted by stagnant water spells). Zonal accumulation is either homogeneous or divided into subzones. The water of enduring lakes is not drained because the depressions are clogged but affected by evaporation, which results in laminite formation. The increasing clogging (or occasionally its initial opening) of the depression is indicated by the plant waste series, while the presence of laminite points to complete clogging.

In doline evolution, accounting for laminite and plant waste as well as the grain size of the filling sediments, active (non-karstic pipe emerged in the doline) and inactive phases (clogging of non-karstic pipe or shaft in the doline) are distinguished.

The inactive doline may be transformed into a fossil doline. During the fossil stage, the lake becomes permanent and water conduction from the doline into the karst gradually ceases. In the cover below the doline, groundwater accumulates. Sediment and water input into the doline also cease gradually as the surrounding terrain is eroded, and its elevation and slope are reduced. Input is gradually replaced by water flow and sediment transport outward the doline.

On some Hungarian karst areas, our depth measurements show that between 2004 and 2010 the infilling of covered karst dolines was slower than that of the solution dolines. This slower sedimentation is explained by the greater extent of sediment transport into the passage system of the karst. The extent to which dolines are filled in depends on the vegetation in the hinterlands of dolines (on the type and density of vegetation during humid periods) and the development stage of the doline conduits.

The material budget of suffosion dolines was analysed. According to material turnover, five types were identified. Into type A dolines, there is no sediment input. The sediment input into type B dolines is at a lower rate than sediment transport from the doline into the karst (the difference between dolines of B_1 and B_2 varieties lies in the temporal sequence of the two processes). Outward transport from type C dolines is not considerable, while from type D outward transport is considerable but does not exceed the rate of sediment input. Finally, into type E dolines, neither input nor output is observed. Out of the 15 investigated dolines of the Tés Plateau, in six cases outward sediment transport surpassed that of inward transport, while in eight cases the opposite was found. For one doline neither inward nor outward transport could be identified.

References

Andrejchuk V (2002) Collapse above the World's largest potash mine (Ural, Russia). Int J Speleol 31:137–158

Barótfi I (2003) Környezettechnika (Environmental technology). Mezőgazdasági Kiadó, Budapest, 1049 p. (in Hungarian)

Baumgardner RW, Hoadley AD, Goldstein AG (1982) Formation of the Wink Sink, a salt dissolution and collapse feature, Winkler Country, Texas. University of Texas at Austin, Bureau of Economic Geology, Report of Investigations 114:38 p
Beck BF, Sinclair WC (1986) Sinkholes in Florida: an introduction. Florida Sinkhole Research Institute Report 85-86-4, 16 p
Benedek P, Valló S (1982) Víztisztítás-szennyvíztisztítási zsebkönyv (Handbook of sewage treatment). Műszaki Kiadó, Budapest, 697 p. (in Hungarian)
Böcker T (1972) A karsztvizek mozgásviszonyai természetes körülmények között (Karstwater movements under natural conditions). In: Szádeczky-Kardoss E (ed) II. Anyag- és energiaáramlási ankét. Akadémia Kiadó, Budapest, pp 107–121 (in Hungarian)
Brook GA (2004) Africa, Sub-Sahara. In: Gunn J (ed) Encyclopedia of caves and karst science. Fitzroy Dearborn, New York, pp 20–23
Cooper AH, Waltham AC (1999) Subsidence caused by gypsum dissolution at Ripon, North Yorkshire. Q J Eng Geol 32:305–310
Currens JC, Paylor RL, Beck EG, Davidson B (2012) A method to determine cover-collapse frequency in the Western Pennyroyal karst of Kentucky. J Cave Karst Stud 74(3):292–299
Deák G, Samu S, Veress M (2013) Szuszpenziós rendszerek ülepedésének és kiválásának vizsgálata modell kísérletekkel (An investigation of veneer development of suspension systems with model experiments). Karsztfejlődés XVIII:49–64 (in Hungarian)
Fekete J (1988) Trópusi talajok (Tropical soils). Akadémia Kiadó, Budapest, 502 p. (in Hungarian)
Ford DC (2004) Bear rock karst, Canada. In: Gunn J (ed) Encyclopedia of Caves and Karst Science. Fitzroy Dearborn, New York, pp 137–138
Ford DC, Williams PW (2007) Karst hydrogeology and geomorphology. Wiley, Chichester, 561 p
Hiemenz P (1986) Principles of colloid and surface chemistry. Marcel Dekker, New York, 815 p
Hyatt JA, Jacobs PM (1996) Distribution and morphology of sinkholes triggered by flooding Tropical Storm Alberto at Albany, Georgia, USA. Geomorphology 17:305–316
Hyatt JA, Wilkes HP, Jacobs MP (1999) Spatial relationship between new and old sinkholes in covered karst, Albeny, Georgia, USA. In: Beck BF, Petit AJ, Herring JG (eds) Hydrogeology and engineering geology of sinkholes and karst. Balkema, Rotterdam, pp 37–44
Jakucs L (1956) A barlangi árvizekről (Die Höhlen-Überschwemmungen). Földrajzi Közlemények 80(4):381–402 (in Hungarian)
Johnson KS (1987) Development of the Wink Sink in West Texas due to salt dissolution and collapse. In: Beck BF, Wilson WL (eds) Karst hydrogeology: engineering and environmental implication. Balkema, Brookfield, pp 127–136
Johnson KS, Collins EW, Seni SJ (2003) Sinkholes and land subsidence due to salt dissolution near Wink, West Texas, and other sites in western Texas and New Mexico. Oklahoma Geol Surv Circ 109:183–195
Kemmerly RR, Towe SK (1978) Karst depressions in a time context. Earth Surf Process Landf 35:355–361
Klimchouk A, Andrejchuk V (2002) Karst breakdown mechanisms from observations in the gypsum caves of the western Ukraine: implications for subsidence hazard assessment. Speleogenesis and Evolution of Karst Aquifers (www.speleogenesis.info) 1(1):20 p
Komac B, Zorn M (2013) Extreme floods Slovenia in September 2010. In: Lóczy D (ed) Geomorphological impacts of extreme weather. Case studies from Central and Eastern Europe. Springer, Dordrecht, pp 121–139
Korzhuev SS (1961) Merzlotnyi karst Srednego Prilen'ya i nekotorye osobennosti yego proyavleniya. (The Middle-Lena frozen karst and its characteristics). In: Sokolov NI, Gvozdetskiy NA, Balashov LS (eds) Regionalnoe karstovedenie. Izdatelstvo AN SSSR, Moscow, pp 207–220
Magdalene S, Alexander E (1995) Sinkhole distribution in Winona Country, Minnesota revisited. In: Becks BF (ed) Karst Geohazards. Balkema, Rotterdam, pp 43–51
Pais I (1981) Általános szervetlen és analitikai kémia (General inorganic and analytical chemistry). Mezőgazda Kiadó, Budapest, 317 p. (in Hungarian)

Pulina M (2005) Le karst et les phenomenes karstiques similaires des regions froides In: Salomon, JN, Pulina M (eds) Les karsts des regions climatiques extremes. Karstologia Mémoires, 14, Presses Universitaires de Bordeaux, Bordeaux, pp 11–100

Rohrsetzer S (1991) Diszpergált részecskék ülepedése (Settling of dispersed particles). In: Rohrsetzer S (ed) Kolloidika (Colloidics). Tankönyvkiadó, Budapest, pp 295–297 (in Hungarian)

Soriano MA, Simón JL (2001) Subsidence rates of alluvial dolines in the central Ebro basin, Northeastern Spain. In: Beck BF, Herring JG (eds) Geotechnical and environmental applications of karst geology and hydrology. Balkema, Lisse, pp 47–52

Stefanovits P (1981) Talajtan (Soil science). Mezőgazda Kiadó, Budapest, 382 p. (in Hungarian)

Sweeting MM (1973) Karst landforms. Columbia University Press, New York, 362 p

Veress M (1982) Adatok a Hárskúti-fennsík karsztmorfogenetikájához (Data on the karst morphogenetics of the Hárskút Plateau). Karszt és Barlang II:71–82 (in Hungarian)

Veress M (1986) Feltárás előrejelzése a karsztos üledékek vizsgálatával (Prediction of exposure by the study of karst sediments). Karszt és Barlang II:95–104 (in Hungarian)

Veress M (1987) Karsztos mélyedések működése bakonyi fedett karsztokon (Activity of karst depressions on the covered karst in the Bakony Mountains). Földrajzi Értesítő 36(1–2):91–114 (in Hungarian)

Veress M (2000) Covered karst evolution in the Northern Bakony Mountains, W-Hungary. A Bakony Természettud. Kut. Eredményei, Bakonyi Természettudományi Múzeum, Zirc 23:167 p

Veress M (2013) Intense rainfall on subsidence karst doline. In: Lóczy D (ed) Geomorphological impacts of extreme weather. Case studies from Central and Eastern Europe. Springer, Dordrecht, pp 327–345

Veress M, Futó J (1990) Fedett, paleokarsztos térszíneken végbement lepusztulás és felhalmozódás kimutatása a Bakony-hegységben (Erosion and accumulation on covered paleokarst surfaces in the Bakony Mountains). Foldtani Kozlony 120(1):55–67 (in Hungarian)

Veress M, Puskás J (2007) Adalékok az Eleven-Förtési töbörcsoport (Bakony-hegység karsztosodásához (Contributions to karstification on the Eleven-Förtes doline group (Bakony Mountains). Karsztfejlődés XII:171–192 (in Hungarian)

Veress M, Péntek K, Schläffer R (2013) Az intenzív csapadékhullások hatása a karsztos formákra (Impact of intensive rainfalls on karst landforms). Karszt és Barlang 2011(I–II):41–50 (in Hungarian)

Veress M, Unger Z, Mitre Z, Gárdonyi I, Deák G (2015) The relationship between suspended sediment settling velocity and water table sinking rates in intermittent lakes in karstic depressions. Proc Geol Assoc 126:417–425

Waltham AC, Smart PL (1988) Civil engineering difficulties in the karst of China. Q J Eng Geol 21:2–6

Waltham T, Bell F, Culshaw M (2005) Sinkholes and subsidence. Springer, Berlin, 382 p

Zhang Z (1980) Karst types in China. GeoJournal 4(6):541–570

Zolitschka B (2007) Varved lake sediments. In: Elias SA (ed) Encyclopedia of quaternary science. Elsevier, Amsterdam, pp 3105–3114

Chapter 7
Landform Evolution and Development

Abstract In this chapter, the development and further evolution of subsoil karren, caprock dolines, subsidence dolines, ponors, DSDs and remnant caves are represented. For the above landforms and their varieties, the influencing geological, morphological and hydrological conditions are presented, and the origin, evolution and transformation of landforms are demonstrated. In the presentation of the development of karren, the role of the cover (mainly its grain size) and water movement in the cover is detailed. The classification of caprock dolines relies on the depth of stoping pipes. When studying subsidence dolines, the influence of terrain slope, elevation, karst water and the properties of the cover sediment (grain size, $CaCO_3$ and clay contents) are investigated. The contribution of landform development on the bedrock and processes in the cover (piping, compaction) to the formation of the depression is analysed. In the genetic classification of ponors, the distribution of non-karstic rocks and the position of karst water are considered. In the description of the evolution of DSDs, the karstic and non-karstic influences are overviewed, and it is investigated from what landforms (subsidence doline, ponor, katavothron) the studied features developed.

Keywords Capillary porosity • Aggregate porosity • Breccia pipe • Stoping up • Collapse • Piping • Conditions of origin • Cover thickness • Indirect inheritance • Direct inheritance

7.1 Karren Formation

7.1.1 General Characteristics of Karren Development

Karren on bare surfaces originate from water flow and water seepage (White 1988; Veress 2010). In covered karsts, water motion is retarded. Therefore, covered karren develop by seepage or weak flow. If water penetrates through the cover, water motion is seepage, and if it exists between the bedrock and the cover, it can be weak flow, too.

The soil and the cover influence karren formation through water circulation and CO_2 production. The soil (and also the cover) dissipate water on the rock surface

M. Veress, *Covered Karsts*, Springer Geology,
DOI 10.1007/978-94-017-7518-2_7

(Trudgill 1985), but as it is stored in the cover, the duration of the solution period increases. The direction and pattern of water motion and the location and extension of contact with the bedrock depend on the occurrence of groundwater in the cover, of impermeable intercalation and of the grain size of the cover.

Total porosity is composed of the pores between particles (aggregate porosity) and the pores within particles (capillary porosity) (Stefanovits 1981). In our investigations, lowest porosity was found in cover sediment of 0.250–0.500 mm particle size, and it grows with both increasing and decreasing particle size. With decreasing particle size, total porosity can grow (Table 7.1) because capillary porosity grows. In the pores of aggregate porosity, water moves gravitationally. Capillary water is situated in the pores of capillary porosity and on the surface of small particles. The lifting effect of capillarity affects capillary water. Therefore, in cover sediment where aggregate porosity is higher (coarse-grained cover), water percolates in vertical direction towards the bedrock and, when it reaches the bedrock, if it does not infiltrate, flows along the bedrock surface. Where capillary porosity is higher (fine-grained cover), water percolates upwards and laterally – even if there is no impermeable intercalation in the cover. This statement is also proved by both our water lifting and water overlifting experiments (Table 7.1). The water lifting experiment was infinite because the experiment was carried out during the increase of water level. However, in the case of the 24-h water lifting experiment, the same was experienced. Thus, the water rise was 480 mm in the sediment with a grain size of 0.001–0.063. This value decreased by the increase of grain size (the following 24-h water lifting values belong to sediments with grain sizes of 0.063–0.125 mm, 0.125–0.250 mm, 0.250–0.500 mm, 0.500–0.1000 mm, 1.000–2.000 mm and 2.000–2.500 mm: 440 mm, 390 mm, 150 mm, 50 mm, 30 mm and 25 mm). In the case of the sediment with a grain size of 2.000–2.500 mm, there was no water lifting. In the case of the water overlifting experiment with a 10 cm high glass tube, the total amount of water (30 cm^3) got into the sediment when the grain size of the sediment was 0.125–0.250 mm. From this amount, 24 cm^3 remained in the sediment, and the other part got into the water container from where the water is originated. However, in the case of a sediment with a grain size of 0.500–1.00 mm, during the experiment, no water got into the water container from where the water is originated. Thus, no water overlifting happened in the latter case.

Solution is not only enhanced by CO_2 originating from the soil but also that from the cover. Thus, CO_2 is produced in terra rossa (Zámbó 1970) or (weathered) basalt (Song and Liang 2009). Solution can be periodical. The water between the cover and the bedrock is saturated when solution is interrupted and renewed when the water percolating through the cover takes up CO_2 and unsaturated water presses out saturated water. Karren formation is affected by the insoluble residue of bedrock and weakness of rock. Solutional residue reduces water motion or prevents water from contacting the rock. Zones of weakness are represented by joints and bedding planes, which are sites of water conduction, where karren formation begins.

The composition of the cover also influences karren formation on the bedrock. With higher or increasing carbonate content of the cover sediment, infiltrating waters are partly or fully saturated before reaching the bedrock. Thus, there is hardly any solution on the bedrock (see Sect. 7.3.3.3).

Table 7.1 Relationship between particle diameter and various properties of the cover (density, pore volume, capillary rise)

Particle size [mm]	Clay	0.063>	0.063–0.125	0.125–0.250	0.250–0.500	0.500–1.000	1.000–2.000	2.000–2.500	2.500–5.000
Density [g/cm^3]	2.3671	2.276064	2.288475612	2.330081438	2.343629635	2.540943382	2.60042679	2.576621005	2.595140576
Porosity V/V%	50.7456	48.538	44.2504	38.2104	36.4432	39.2186	39.7362	40.87	42.38
Capillary rise (cm)[a]	–	99	64	42	25	7.5	2.5	–	–

[a]The period of capillary rise

During karren formation, the dismembering of the rock mass leads to interaction between the cover and the (surface of) bedrock. The cover sediment intrudes into the resulting depressions and solution continues along joints (Curtis et al. 1976). The contact surface between the cover and the bedrock increases and this enhances solution. The cover expands, and its porosity increases which favours water seepage. With the dismembering of the bedrock through solution, its fragments are incorporated into the cover (Veress and Péntek 1996). This modifies the solution capacity of waters percolating in the cover.

Covered karst karren can form under soil only or under cover sediment. The latter are regarded typical covered karren. Covered karren can develop through the burial of karren on bare surfaces (Figs. 7.1a–c) under solutional residue (Fig. 7.1e) or under non-karstic cover.

Water can arrive between the cover and the bedrock by way of seepage through the cover or at the bedrock outcrops. If the bedrock is impermeable, water can reach the surface of the bedrock where it outcrops or, in the case of consolidated rock, along the joints. Solution intensifies on the bedrock, where percolating water adds to the water flowing between the cover and the bedrock (Song 1986).

It is common that karren formation under cover sediment is not coupled with the development of surficial features. The reason for this is that karren are mostly of too small dimensions to cause subsidence in the cover. Subsidence can also be inhibited by the thickening of the cover from below, from the insoluble bedrock. In certain cases, the cover can subside, particularly if the cover is thin and non-cohesive or if the karren features are associated with the cavities of the epikarst (Figs. 7.1d and 7.2, Trudgill 1985).

Slabe (2009) studied the origin of covered karst features in an experiment on plaster. He identified solution zones under cover on the gypsum columns. Thus, he recognised a micro recess zone on top of the columns. In the middle of the columns, the size of the features is described as scallops grew. At the fractures, channels of semicircular cross-section developed. At the bottom, where inundation was permanent, channels came about, while at the level of inundation, notches emerged. At the lower levels of columns, channels of Ω cross-section could be observed on the ceiling. In the course of another experiment (with clay cover), vertical channels and shafts developed on the columns.

In our experiment (see Chap. 3 and below), karren-like features have developed on the surface of the plaster table through the so-called continuous dissolution, under coarse cover sediment.

Slabe and Liu (2009) identified covered karren with respect to the water flow and inundation conditions of the cover, relying on the results of experiments by Slabe (2005, 2009). They distinguished between karren below cover and those at soil level. Karren below cover can develop through water percolation, flow or complete inundation (the latter in the case of periodical water level fluctuation). Percolating water produces channels, flutes, scallops and minor depressions in the rock. Under inundated cover, channels and notches develop. At the soil level, half-bells and notches come about. Here we should mention that half-bells are typical where

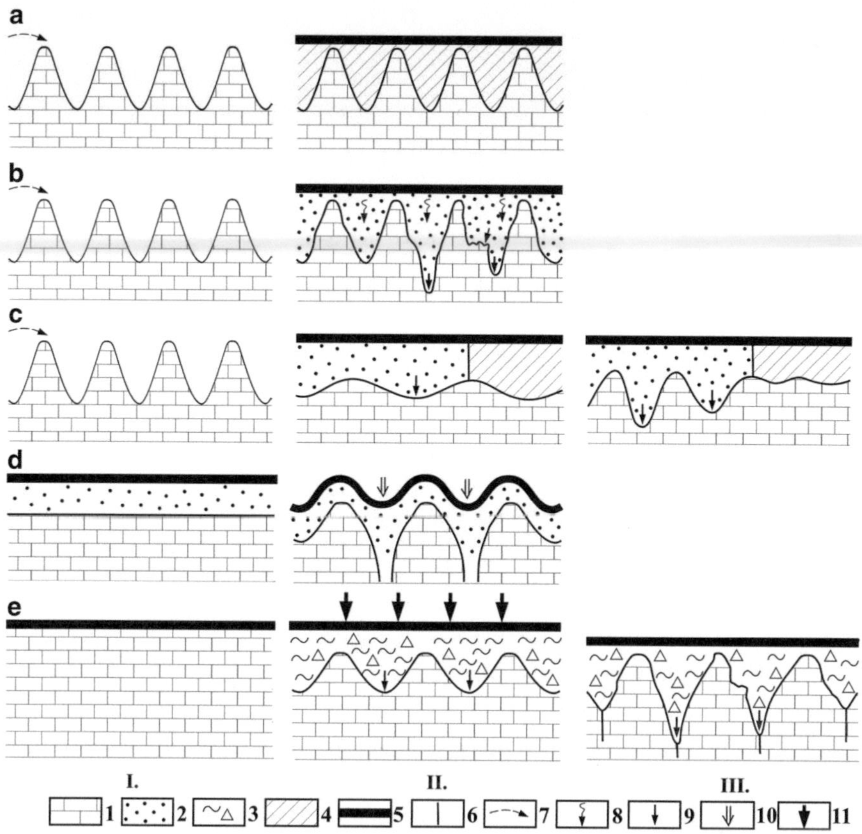

Fig. 7.1 Types of subsoil karren. *1* Limestone, *2* permeable cover, *3* weathering residue, *4* impermeable cover deposit, *5* soil, *6* fracture, *7* accumulation of cover, *8* infiltration, *9* solution, *10* subsidence of the cover above karren, *11* surface lowering due to material loss by solution, (**a**) transformation of bare karren into cryptokarren by accumulation, (**b**) transformation of bare karren into concealed karren by accumulation, (**c**) truncation of bare karren and rejuvenation of karren formation under permeable cover and conservation of the karren surface under impermeable cover, (**d**) inheritance of karren features from the bedrock over to the cover by cover subsidence, (**e**) karren formation under the weathering residue, *I* state before karren formation, *II–III* formation of karren features

channels conduct water from the bordering cliff onto the cover. Channels under inundation probably only form on plaster since such features are not associated with katavothra (Fig. 5.12).

The development of covered karren is presented following the classification in Sect. 5.2.2. Irregular features include subsoil cavernous karren and subcutaneous tubes. In the formation of such forms, the role of rock structure is emphasised (Slabe 1999; Zseni 2009). Solution does not take place on the surface of the karstifying surface but in the interior, along fractures and bedding planes.

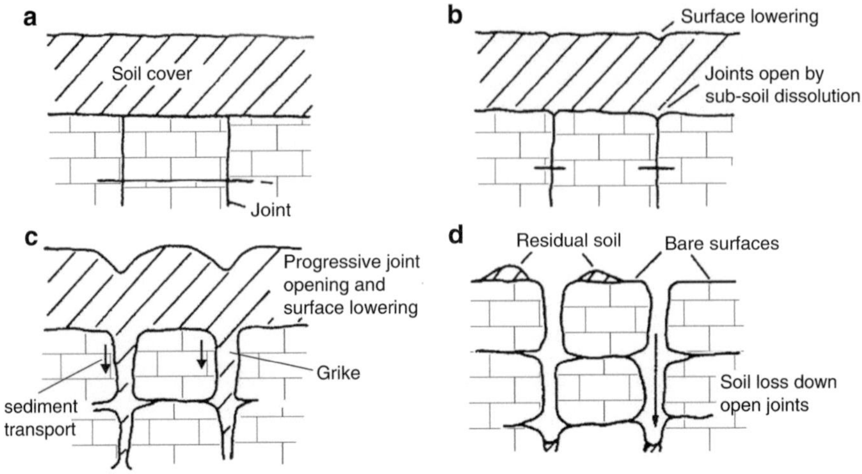

Fig. 7.2 Impact of cover on the formation of grikes and the denudation of the cover by grikes (Trudgill 1985) – modified

7.1.2 Formation of Vertical Karren

Vertical features include solution pipes and subsoil pinnacles. Solution pipes form at the fractures of the bedrock. The clay cover probably also contributes to their formation. On the one hand, on gypsum such features develop under clay (Penck 1924), and on the other, Slabe (2009) reported on the formation of shafts in his model experiment if the cover sediment was clay.

On the coast of Madagascar (see Chap. 5), large-scale chimneys occur on bare rock, but have no fill. Such features may derive from the root zones of one-time trees on the coast. Following their formation, when the pipes crossed the rock, the fill was removed and the trees were destroyed (Veress et al. 2008).

Pinnacles (pinnacle karst) are residual landforms. The tropical karst varieties under cover are the stone forest, arête and pinnacle karst (Ford and Williams 2007). The development of such forms begins with the emergence and widening of chimneys, shafts and grikes. Adjacent chimneys and shafts merge. This merging and the widening of giant grikes of different direction leave behind gradually narrowing elevation (pinnacles). Afterwards, as it was also described for the Lunan stone forest (Song and Liang 2009), along the fractures below the cover, during solution pinnacles are becoming narrower and lower, resulting in pillars and then stone teeth. A similar process happens at the arête and pinnacle karst varieties according to Osmaston (1980) and Osmaston and Sweeting (1982). Here pinnacles develop between perpendicular fractures. The origin of pinnacle karst is presented in detail below related to grike formation. The development of pinnacles is demonstrated by the experiment of Slabe (2009).

7.1.3 *Formation of Linear Karren*

Linear features include channels and grikes. A variety of channels is rundkarren (rounded solution channel), which, as already presented in Chap. 5, developed from the rinnenkarren (channels) of bare surfaces through subcutaneous solution (Bögli 1960; Veress 2009a).

The channels under cover form during water flow between the cover and the bedrock. To their origin rock fractures also contribute, while in the development of the channels of bare terrain fractures have no role to play (Veress 2010). Their development is also influenced by the water percolating through the cover. Broadening sections come about where the water percolating through the cover mixes with water flowing between the cover and the bedrock (Song 1986).

Similar to grikes on bare terrain, those under cover mostly (but not exclusively) develop along fractures. The water percolating through the cover sediment is moving along fractures and dissolves the rock. The rock walls of fractures retreat by lateral corrosion (Zámbó 1970).

The depth of grikes corresponds to the thickness of the epikarst and their floor coincides with the level of saturation. (Exceptions are presented by recent, later developed grikes which are shallower, and their floor has not reached down to the lower surface of the epikarst.) If the epikarst is deep and the karst water table is close to the surface, the floors of grikes reach down to the karst water. If the karst water table sinks, the grikes keep on deepening until they extend down to the karst water table or the level of saturation. The grikes cannot deepen under the level of the permanent water cover because the water levels act a kind of impermeable series. However, they become broader (see later). In the absence of an impermeable series, they can deepen, and this has been proved by our laboratory experiments too: the grikes were formed in plaster deepened until the impermeable series. The deepening was larger on the cover with a grain size of 2.5–5.0 mm and took place until the clay (Table 7.2). The grike floor reached the clay only partially in the case of a fill of finer

Table 7.2 The change of the grike size in the case of a sediment fill of 2.5–5.0 mm if there is no impermeable series on the floor

	Original state		At the end of the experiment		Change			
	Width (mm)	Depth (mm)	Width (mm)	Depth (mm)	Width		Depth	
					mm	%	mm	%
At the place of water supply	–	–	26	39	–	–	–	–
At 4 cm from the place of water supply	23	16	25	34	2	8.7	18	112.5
At 15.5 cm from the place of water supply	23.2	17.8	25	33	1.8	7.8	15,2	85.4
At 27 cm from the place of water supply	26.6	14.2	26	27	–0.6	–2.3	12.8	90.1
At the end of the grike	–	–	26	22.5	–	–	–	–

grain size (0.063–0.125 mm) (Table 7.3). In this case the reason for smaller deepening is that the capillary water seeped partly laterally in the cover which is also proved by the fact the grike filled with fine grain broadened to a larger extent (Tables 7.2 and 7.3). At the level of fill, the broadening depended on the distance from the place of water supply (its extent decreased by the increase of the distance). As we have already mentioned, it depended on the grain size of the cover too. The subsidence of the surface of the fill depends on the grain size as compared to the subsidence of the bedrock. On a fine-grained fill, the subsidence is the same or similar to that of the coarse-grained fill (Tables 7.4 and 7.5). In spite of the fact that the subsidence of the grike floor under fine-grained sediment is significantly smaller than under coarse-grained sediment. This also proves a lateral water motion in the fine-grained sediment. In the case of this kind of fill, the subsidence of the fill is large as compared to the dissolution of the floor, which is caused by the lateral transportation of the fill. This may happen by lateral seepage.

On tropical karsts, the depressions of the pinnacle surface (probably grikes or their modified varieties) may be 20–30 m deep (Bergado and Selvanayagam 1987, Figs. 5.5 and 7.3). Grike depth is typical of the individual climates since the position of the grike floor is controlled by the saturation depth of infiltrating water, which depends on CO_2 production (i.e. climate) and the denudation rate by solution for the limestone terrain between grikes.

Solution on the rock walls bordering the fracture can take three forms.

– Solution rate reduces downwards and with reduced solution grikes wedge out.

Table 7.3 The change of the grike size in the case of a sediment fill of 0.063–0.125 mm if there is no impermeable series on the floor

	Original state		At the end of the experiment		Change			
	Width (mm)	Depth (mm)	Width (mm)	Depth (mm)	Width		Depth	
					mm	%	mm	%
At the place of water supply	–	–	54	28	–	–	–	–
At 4 cm from the place of water supply	28.1	23.2	43	34	14.9	53	10.8	46.6
At 15.5 cm from the place of water supply	26.6	24.8	35	26	8.4	31.6	1.2	4.8
At 27 cm from the place of water supply	29.3	23.3	32	24	2.7	9.2	0.7	3
At the end of the grike	–	–	33	21	–	–	–	–

Table 7.4 The subsidence of the surface of the grike fill in the case of a grain size of 2.5–5.0 mm if there is no impermeable series on the floor

At the place of water supply	15 mm
At 4 cm from the place of water supply	13 mm
At 15.5 cm from the place of water supply	10 mm
At 27 cm from the place of water supply	9 mm
At the end of the grike	6.5 mm

Table 7.5 The subsidence of the surface of the grike fill in the case of a grain size of 0.063–0.125 mm if there is no impermeable series on the floor

At the place of water supply	20 mm
At 4 cm from the place of water supply	12 mm
At 15.5 cm from the place of water supply	5 mm
At 27 cm from the place of water supply	4 mm
At the end of the grike	4 mm

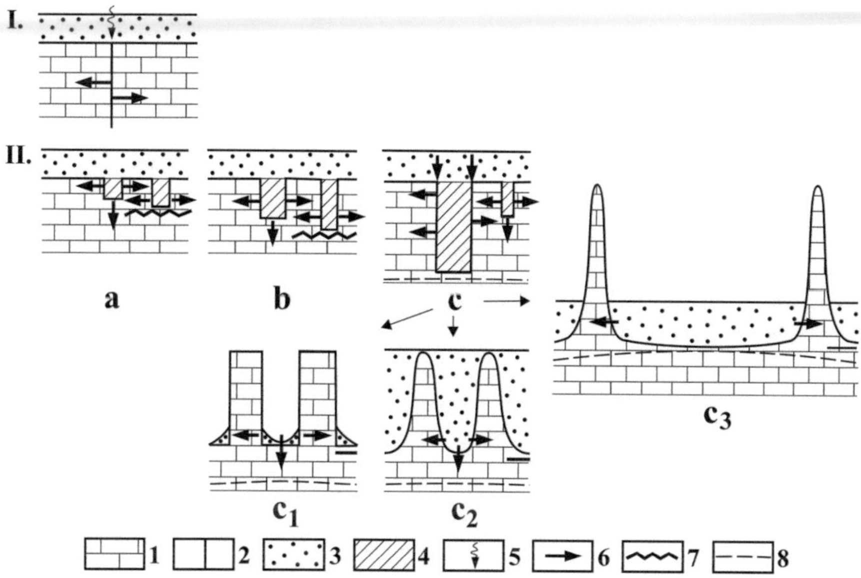

Fig. 7.3 Grike evolution on karsts under different climates. *1* Limestone, *2* fracture, *3* cover, *4* grike fill, *5* infiltration, *6* solution, *7* level of saturation, *8* karst water table. *I* Initial stage. *II* Mature stage: on temperate karst (**a**), mediterranean karst (**b**), tropical karst (**c**). Varieties of tropical karst by development stage: stone forest karst (C_1), pinnacle karst (C_2) and inselberg karst (C_3)

– Solution rate remains similar to the level of saturation (which may also be the lower surface of the epikarst). In this case grike walls show parallel retreat. As a consequence, grikes with parallel walls come about.
– Solution rate increases downwards and with more intensive solution at greater depths grikes broaden. The intensity of solution may increase downwards if the grain size of the cover changes and, thus, becomes finer downwards. In this case, on the deeper part of the grike, water flow will be lateral in the cover. There in the fine-grained cover, the water dissolves the wall to a larger extent as compared to coarse-grained cover situated above. If the grike is only partly filled with cover or there is an impermeable series (karst water) on the grike floor, grike broadening downwards comes about too. During our laboratory experiments, the walls between the grikes with clayey floors (their width was 5–6 cm) completely dissolved, so the grikes widened downwards independent of the grain size of the filling material. However, the widening of the grikes at the level of the filling was

Table 7.6 The change of the grike size in the case of a sediment fill of 2.5–5.0 mm if there is an impermeable series on the floor

	Original state		At the end of the experiment		Change			
					Width		Depth	
	Width (mm)	Depth (mm)	Width (mm)	Depth (mm)	mm	%	mm	%
At the place of water supply	–	–	25	24	–	–	–	–
At 4 cm from the place of water supply	21.2	19.4	22	22	0.8	3.8	2.6	13.4
At 15.5 cm from the place of water supply	23.4	20.6	24	22	0.6	2.6	1.4	6.8
At 27 cm from the place of water supply	22.9	16.8	25	16	2.1	9.2	–0.8	–4.7
At the end of the grike	–	–	28	17	–	–	–	–

Table 7.7 The change of the grike size in the case of a sediment fill of 0.063–0.125 mm if there is an impermeable series on the floor

	Original state		At the end of the experiment		Change			
					Width		Depth	
	Width (mm)	Depth (mm)	Width (mm)	Depth (mm)	mm	%	mm	%
At the place of water supply	–	–	30	25	–	–	–	–
At 4 cm from the place of water supply	23.5	22.8	27	22	3.5	14.9	–0.8	–3.5
At 15.5 cm from the place of water supply	21.4	21.8	25	21	3.6	16.8	–0.8	3.7
At 27 cm from the place of water supply	22.3	20.8	25	19	2.7	12.1	–1.8	–8.7
At the end of the grike	–	–	23	18	–	–	–	–

significantly smaller, and it depended on the grain size (Tables 7.6 and 7.7). All these things prove that if the grike floor cannot subside, but the water has the ability to solve, the grike will widen.

The Lunan stone forest partly formed under the karst water table (Ford et al. 1997; Yuan 1991; Zhang et al. 1997). In this region, epikarst thickness amounts to more than 100 m (Mangin 1997). The origin of the Lunan stone forest or the experiment carried out by Slabe (2009) on plaster reconstructs the initial conditions. The pillars and teeth of the Lunan stone forest developed through the broadening of the rock fractures (Song and Liang 2009), i.e. by way of grike formation.

Peng et al. (2007) described the evolution of stone forests by a triplex erosion model, which, according to those authors, can also be applied to the pinnacle karst. In the model the above authors identify subaerial solution, subsoil solution and soil erosion. Pinnacles form along grikes in three manners. If subaerial solution takes place at a slow rate and subsoil solution is rapid, pinnacles tend to grow. If soil erosion is great, the pinnacles become exhumed. If soil erosion is slow and subaerial and subsoil solution is rapid, pinnacles become buried.

If the terrain is uplifted, the floors of grikes emerge above the karst water table. With a given rate of CO_2 production (even if it is very high), solution declines at some point if the (laterite) cover is too thick. Cover sediment removal is a precondition to the deepening of the floors of the grikes. The geomorphic evolution through karren formation takes a course similar to the savanna-type peniplanation described by Büdel (1957): the resulting laterite is thinning out through pluvial erosion. With the erosional thickening out of the cover (which may happen not only by surficial removal but by the transportation of the cover into the karst), the level of saturation acquires a deeper position and grikes and shafts begin to deepen. The deepening of grikes and shafts, with their floors reaching down to the level of saturation, only continues if the surface of the bedrock between them is affected by solution, i.e. the bedrock surface is lowering. This, however, results in the recharge of laterite from below. Too thick laterite inhibits solution. The solution of the bedrock can only succeed the thinning out of the laterite. Eventually, the solutional lowering of the bedrock surface is controlled by the removal of the laterite. With deepening and then widening grikes, a pinnacle karst forms as the blocks between grikes (giant grikes) transform into crests (Fig. 7.3c_2).

The pinnacle karst can develop into a stone forest if the walls of grikes perform parallel retreat and the pinnacles are exhumed (Fig. 7.3c_1). The formation of steep walls of grikes is favoured by the high rates of CO_2 production, observed from the Lunan stone forest. To high CO_2 production, a suitable cover is an important condition (Song and Liang 2001). Through the dismembering and exhumation of crests along fractures (Waltham 1997; Day and Waltham 2009; Williams 2009), an arête and pinnacle karst emerge (Fig. 7.4). If the karst water table remains at the bedrock surface for a longer period, widening grikes produce an inselberg karst (Paton 1964), Fig. 7.3c_3).

The karst types developing from grikes presented above cannot be strictly distinguished from one another. On the stone forest karst gentler elevations (often denuded slopes of exhumed landforms) or narrow ridges also occur and elevations with vertical sides also appear on the pinnacle karst.

7.1.4 Formation of Horizontal Karren

Notches are horizontal features formed in two types of environments: at the level of the soil or of temporary water inundation (Slabe 2009; Slabe and Liu 2009). The development of notches at the level of soil cover is favoured by water conduction through the soil to the rock. The growth of such features is caused by the downward shift of the level of solution following the denudation of the surface (Slabe and Liu 2009; Zseni 2004, 2009). Large-scale varieties of notches of several metres height develop under tropical climate by continuous solution under enduring inundation

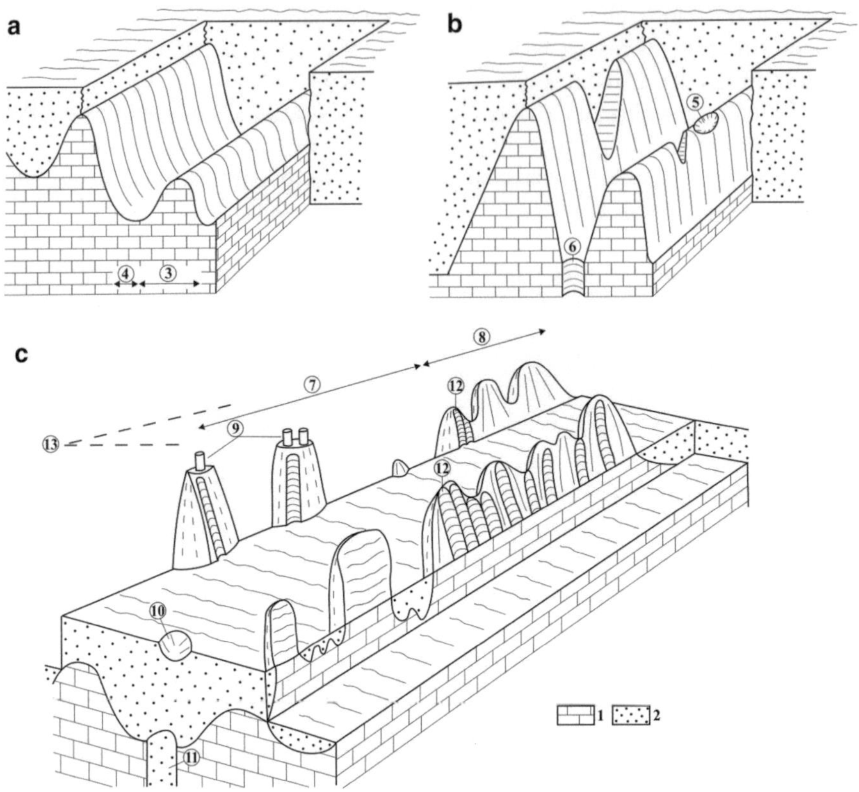

Fig. 7.4 Development of pinnacle surface. *1* Limestone, *2* superficial deposit (solution residue), *3* karren ridge, *4* grike, *5* chimney on the ridge, *6* chimney on grike karren, *7* stone forest karren, *8* arête and pinnacle karst, *9* pinnacle, *10* suffosion doline, *11* chimney filled with superficial deposit, *12* chimney ruin which lost its cover deposit, *13* level of former inundation, (**a**) grike formation under the cover, (**b**) further development of grike karren (grikes deepen, ridges are dissected), (**c**) narrowing ridges exhumed with the denudation of the cover, development of pinnacles with vertical walls (stone forest) and gentler slopes (arête and pinnacle karst)

and intensive biological activity. The swamp slots, which are formed in wetland environment (Wilford and Wall 1965), are transitional to notches at water table.

7.1.5 *Formation of Karren with Circular Platform*

Karren of circular platform include subsoil cups, subsoil scallops and tafoni-like features. According to Slabe and Liu (2009), cups develop where the waters percolating through the cover sediment and arriving vertically at the bedrock perform local solution, and where surface slope is gentle, the cover is thin and porous. Rock jointing and vegetation significantly contribute to their development. According to

Slabe and Liu (2009), subsoil scallops form to the effect of water flow at the contact between the rock and the permeable cover. Jennings (1985) explains their formation with periodical fluctuation of solution capacity.

7.1.6 Karren Formed Under Temporary Inundation

According to Slabe and Liu (2009), karren features formed at or under the level of temporary inundation include notches, subsoil spongy surfaces, subsoil tubes and subsoil channels.

Karren formed by groundwater develop continuously, mainly at the contact between the cover and the karstic rock. The karren which are due to karst water are less bound to the contact between the rock and the cover, but primarily develop along fractures and bedding planes. They grow from bottom to top. If, however, the cover sediment is impermeable, the probability increases that the karst water, reaching the surface of the bedrock, flows laterally between the cover and the bedrock and dissolves the surface of the bedrock too. If the cover is permeable, karst water flow is not directed laterally but upwards. Thus, dissolution will be more local.

Because of the upward motion of karst water, landforms terminating blindly or formed in the bedrock are common, like the karren features of Cerkniško polje (see Chap. 5).

7.1.7 Formation of Rock Salt Karren

On rock salt karren may develop on gently or steeply sloping surfaces. They form under cover and along the cover or through selective denudation. The observations at Parajd show that on gently sloping terrain grikes dominate.

On sloping surfaces, rinnenkarren and rillenkarren are typical. The formation of rinnenkarren can take several courses as presented in Fig. 7.5.

If under desert climate, there are insoluble rock fragments on the surface, and karren tables develop (Fig. 5.18b). Little precipitation is sufficient to generate selective solution, but is not enough to induce considerable surface runoff which, flowing laterally, could remove the part of these formations built up of rock salt.

7.1.8 Pseudokarren Formation

Pseudokarren can develop in the fills of karren (Fig. 7.6I). First karren are filled with soil, which can be removed by pluvial erosion (Fig. 7.6IIIa, b). If fill material is transported by suffosion (piping), porosity increases (Fig. 7.6IIc). Subsidence after compaction results in the formation of suffosion pseudokarren grikes

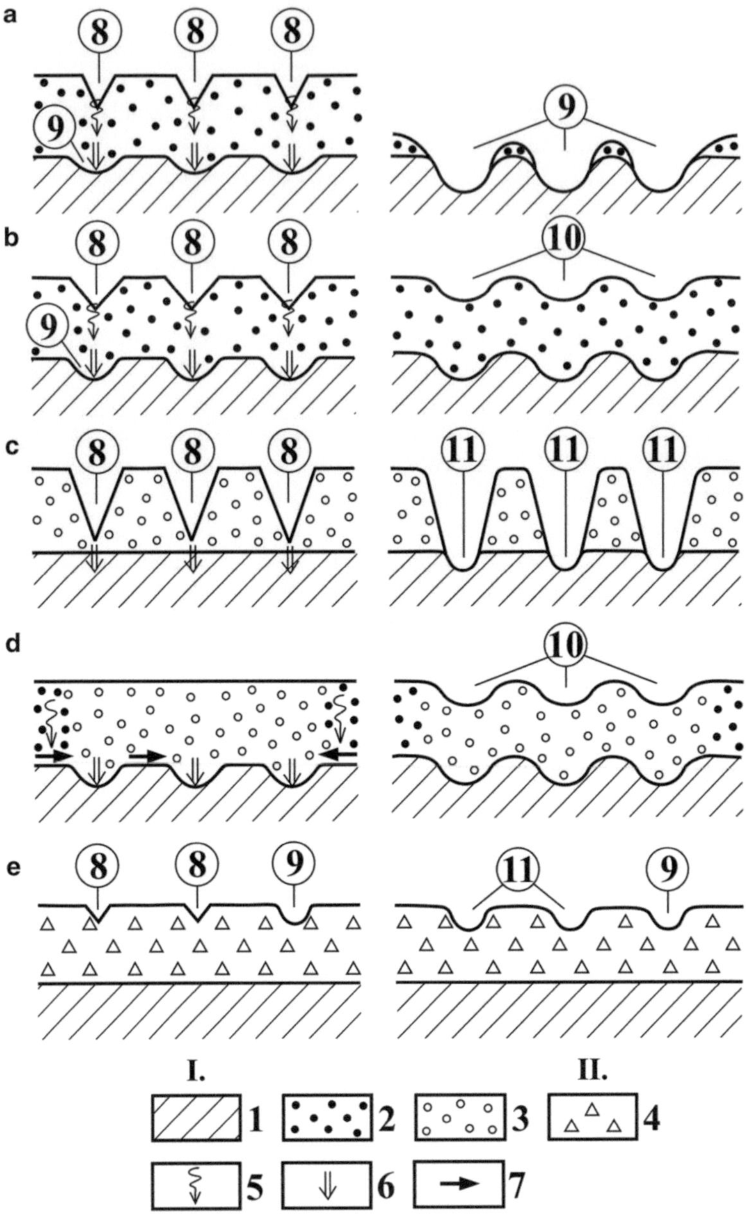

Fig. 7.5 Formation of rinnenkarren on rock salt. *1* Rock salt; *2* permeable cover deposit; *3* partly or completely impermeable cover; *4* salt breccia; *5* water percolation; *6* solution; *7* water percolation on rock salt; *8* feature of erosional origin (rainwater rill); *9* solutional landform (channel); *10* landform originating from cover subsidence (channel); *11* rainwater rill transformed by solution; *I* initial stage; *II* mature stage; (**a**) channels develop on the bedrock under the rainwater rills of the permeable cover and the cover is denuded; (**b**) channels develop under the rainwater rills of the permeable cover by solution and the cover sinks into them; (**c**) the rainwater rills of the impermeable cover are inherited over the rock salt, where they further develop by solution; (**d**) water flow (percolation) under the impermeable cover forms channels on rock salt by solution and the cover sinks into them; (**e**) on the salt breccia rainwater rills are created by erosion and transformed into channels by solution or channels are formed by solution

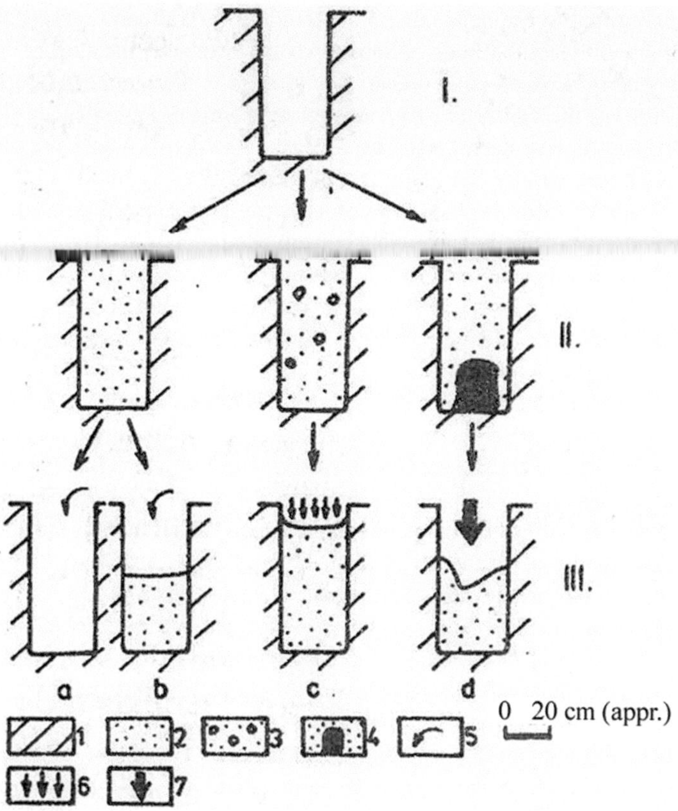

Fig. 7.6 Formation of subsidence pseudogrikes. *1* Karstic rock, *2* grike fill, *3* small cavity partly formed by removal of material, *4* pseudokarren cave, *5* sheet wash, *6* subsidence, *7* collapse, *I* grike formation, *II* grike infilling, *III* partial or complete removal of material from grikes and pseudokarren formation, (**a**) totally exhumed grike, (**b**) partially exhumed grike, (**c**) suffosion pseudokarren grike, (**d**) dropout pseudokarren grike

(Fig. 7.6IIIc). Pseudokarren caves also occasionally emerge during the removal of fill (Fig. 7.6IId), and through the collapse of their roofs, dropout pseudokarren grikes develop (Fig. 7.6IIId).

Pseudokarren pipes either form if the roof of pseudokarren caves locally caves in or if the fill of solution pipes in the bedrock collapses. In the first case, pipes are divided by earth bridges, which are remnants of the roofs of pseudokarren caves.

7.2 Formation of Caprock Dolines

In previous publications (Jennings 1985; Trudgill 1985), caprock dolines were exclusively associated with consolidated rocks, but recently this opinion is refuted. Today their origin is not always explained by the collapse of near-surface cavities, but by stoping up of chimneys (shafts). According to Waltham et al. (2005), they originate wherein insoluble rock collapsing extends towards the surface, i.e. in the cover pipe emerges. A precondition to this process is the reduced stability of the rock constituting the cavity (cave) roof, but appropriate geological build-up and sufficient time are also necessary. The process is sometimes accelerated or even induced by human activity (Waltham et al. 2005).

Typical caprock dolines occur where the entire cover is composed of consolidated rocks. It is possible, however, that consolidated rocks are overlain by nonconsolidated rocks. In this case, the process induced by material deficit spreads over the nonconsolidated rock. The whole doline can be deepened into nonconsolidated rock (and, consequently, secondary processes rapidly make its slopes gentle). In the cover the proportion of consolidated rocks can be so subordinate that chimney formation also basically takes place in nonconsolidated rock, for instance, on the gypsum karst of Ukraine. Klimchouk and Andrejchuk (2002) distinguish between five stages of doline formation, the second, third and fourth of which involve the stoping of the chimney upwards in nonconsolidated rock (merely the first stage occurs in consolidated rock, while the fifth stage is collapse on the surface). If the entire cover is nonconsolidated rock, transitional features from caprock dolines to subsidence doline result. In this case no chimney formation is observed. The actual type of the doline depends on the dimensions of the cavity formed in the bedrock and on cover thickness. The larger is the cavity, the more the character of the doline will approach a caprock doline, the smaller is its size and the thinner is the cover, and the more probable is the development of a subsidence (or even suffosion) doline. Caprock dolines form through direct or indirect inheritance. For direct inheritance, the process on the bedrock (collapse) and the form generation in the cover are simultaneous, but for indirect inheritance, they are not since the development of features on the cover takes a longer time.

7.2.1 Formation of C_1 Caprock Dolines

C_1 dolines come about either with or without chimney or shaft formation. If formation takes place without chimney formation, indirect inheritance happens and the bedrock and the cover cave in simultaneously. The collapse of the bedrock results from the broadening of the cavity through solution and instabilisation of its roof. It is common that roof collapse is preceded by the subsidence of karst water table and ceasing of support for the roof.

This kind of evolution is similar to the origin of subsidence dolines of direct inheritance since the latter also involves the collapse of a cavity (see below). But in that case, the size of the cavities is an order of magnitude smaller than that of the cavities causing caprock doline formation. Therefore, subsidence dolines only develop on unconsolidated rocks. Small-size cavities cannot inherit over consolidated rocks.

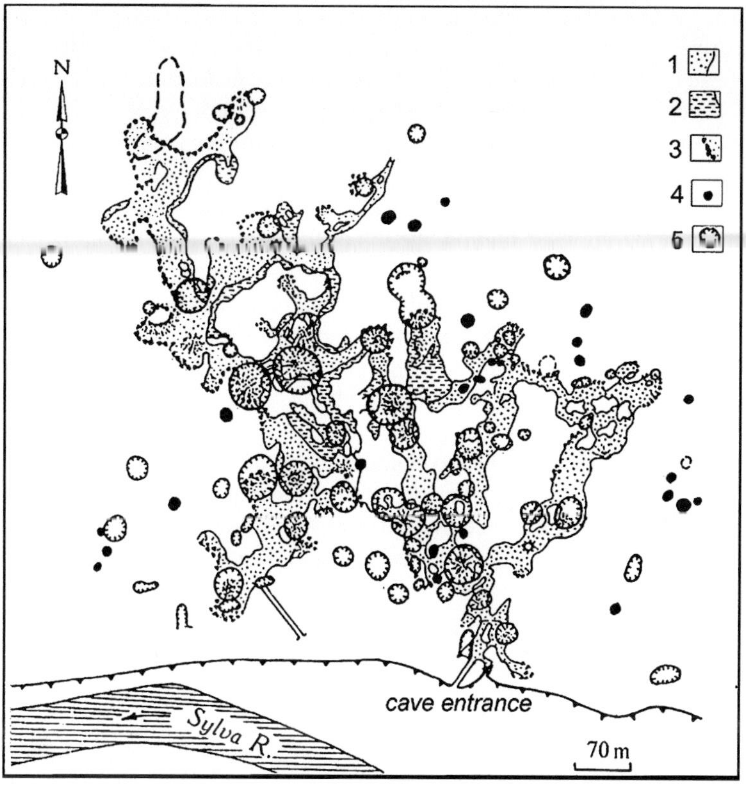

Fig. 7.7 Correlation between caprock sinkholes and cave passages in the buried gypsum at the Kungur Caves in Russia (Andrejchuk and Klimchouk 2002). *1* Dry cave passages, *2* cave lakes, *3* contours by the breakdown material, *4* suffosion dolines, *5* caprock dolines

Chimney (shaft) formation is associated with indirect inheritance. From the roof of a near-surface, horizontal allogenic cave – similarly to C_2 dolines – upward stoping takes place. If the roof is thinning out above blind chimneys, caprock dolines form through collapse and align in a row above the cave (Fig. 7.7). Upward stoping is caused by collapse(s) from the ceiling of the blind shaft. In addition to or instead of collapses, rockfalls can also occur. Collapse(s) happens if chimney diameter is large and the rock is very compact. In both cases, the process is favoured by the dense jointing of the rock. Collapses and rockfalls are not sharply distinct, and sometimes there are transitions between them. Collapses and rockfalls can alternate during chimney evolution. Chimney formation is promoted by infiltration from the surface (particularly water inflow is pointlike), for instance, in karst areas covered by sandstone (as in South Wales) and solution from below, from the direction of the horizontal cave. This may be caused by the solution effect of flooding in the horizontal cave system as well as ascending artesian water (Klimchouk and Andrejchuk 2002). Doline formation occasionally promotes chimney development (Thomas

1974; Bull 1980). Smaller dolines in sandstone are able to develop since the limestone surface is dissolved because of concentrated water flow and minor caprock dolines emerge on the sandstone. This favours water conduction into the horizontal cave of deeper position and the collapsing of the cave roof.

To the effect of pointlike water inflow, on gypsum both solution and collapse contribute to chimney formation (Klimchouk and Andrejchuk 2002).

According to Klimchouk and Andrejchuk (2002), a condition to chimney development is that during its evolution, the top of the debris fan in the horizontal cave should not reach the ceiling of the blind chimney. (If this happens, the evolution of the chimney stops.) This is less probable if the cover is thin or the growth of porosity is also small. The size of the bearing (horizontal) cave is another important factor. If it is larger, a broader and less high debris fan accumulates on its floor. The cave stream and its floods can also promote chimney formation through the removal of debris from the fan.

The process terminates when the chimney opens up to the surface by collapsing. The diameter of the resulting doline depends on the diameter of the chimney, while its depth depends on the distance from the top of the chimney fill to the surface, the thickness of the roof above the blind chimney and the pore volume of the collapsed material.

7.2.2 *Formation of C_2 Caprock Dolines*

The formation of C_2 caprock dolines is associated with the development of pipes and shafts (breccia pipes), which derive from non-allogenic caves. The cavities from where the stoping starts are deep below the surface (even at several hundreds of metres depth). The development is usually a longer process, during which pipes gradually approach the surface, and when they reach it, caprock dolines originate (Walters 1977; Wassmann 1979). The formation of breccia pipes is independent from water recharge (with the exception of artificial water inflow). It may happen that caprock dolines do not develop above breccia pipes by collapse, but by shifts along ringlike fractures (atectonic faults). Crater Lake in south-western Sakatchewan in Canada developed in this way (Christiansen 1971).

In the solution of cavities, water of different origin play a part: karst water for limestone (present or former), ascending thermal water for limestone (Waltham et al. 2005), circulating groundwater for gypsum and rock salt (Waltham et al. 2005) or water inflow of artificial origin from the surface (Walters 1977).

In the case of postgenetic development, pipes form on the ceiling of an old paleokarst cavity. Postgenetic development equally occurs in limestone and evaporites. In the case of syngenetic development, the cavity and the pipes come about almost simultaneously. Syngenetic development is typical of evaporites. In limestone, pipes mostly form above paleokarst cavities (Waltham et al. 2005).

In limestone the formation of pipes is favoured by the thin stratification and jointing of the rock (Waltham et al. 2005). The conditions for pipe formation (either syngenetic or postgenetic) are more favourable on evaporites. Since the dissolution of evaporites is rapid, not only large-scale cavities form, but the collapsed material also gets into the dissolution (see Chap. 5). Since the evaporites are less solid rocks, their roof can collapse more easily.

Theoretically, there are three ways of pipe formation:

- Since rock stability is reduced, collapses affect the ceiling of the cavity. Reduced stability is due to earthquake, increased pressure (e.g. water saturation, loss of support) or solution along fractures. The process can take place in any karstic rock if the previously formed or presently forming cavities are of sufficiently large dimensions. Pipe formation follows cavity formation.
- The existing cavity is renewed since it receives water from the surface; in the soluble rock, collapse is replaced by solution or in non-karstic rock collapse will be more intensive. This process mostly induces pipe formation in the cavities of evaporites. Pipes form after cavity formation.
- The resulting cavity is vertically aligned. Here cavity evolution continuously develops into pipes without a sharp transition. Cavity and pipe formation takes place simultaneously. This evolution is typical of evaporites. Collapses take place when the process spreads over the non-karstic rock.

7.2.3 Formation of C_3 Caprock Dolines

A typical location of C_3 dolines is Kab Mountain. On Kab Mountain groups of pipes (shafts) evolved under basalt (occasionally under a single depression) (Fig. 7.8). The individual members of the shaft groups approach the basalt or the surface to various degrees. Secondary shafts often branch out from the main shafts. The origin of C_3 dolines is due to their pipe groupings. The following ways of evolution are reconstructed.

Where the basalt cap is thinner (Móga and Németh 2005) (see the karst windows in Sect. 4.6.2.3.2) and densely jointed, blind shafts emerge (Fig. 7.9a). The origin of dead-ended blind shafts is presented in Sect. 7.3.3.4.1. Above blind shafts, the limestone is caving in and this process also extends over the basalt (Fig. 7.9bI). The unsupported basalt mass is divided into columns, which fall into the shafts (Fig. 7.9cI). The secondary shafts also develop until they reach the basalt. Over the secondary shafts of small diameter, the displacement of collapsed material is minimal, and thus, on the margins of the doline, deepening is also more moderate. Above the collapsed basalt columns, a caprock doline of large diameter and gentle slopes emerge (Fig. 7.9dI). If in the limestone main shafts occur next to each other, coalescing dolines develop (Fig. 7.9dII). On the margin of caprock dolines, younger caprock dolines and subsidence dolines come about (Fig. 5.24).

The development of caprock dolines or their further development to ponor is particularly intensive on the undrained or almost undrained terrains enclosed by a

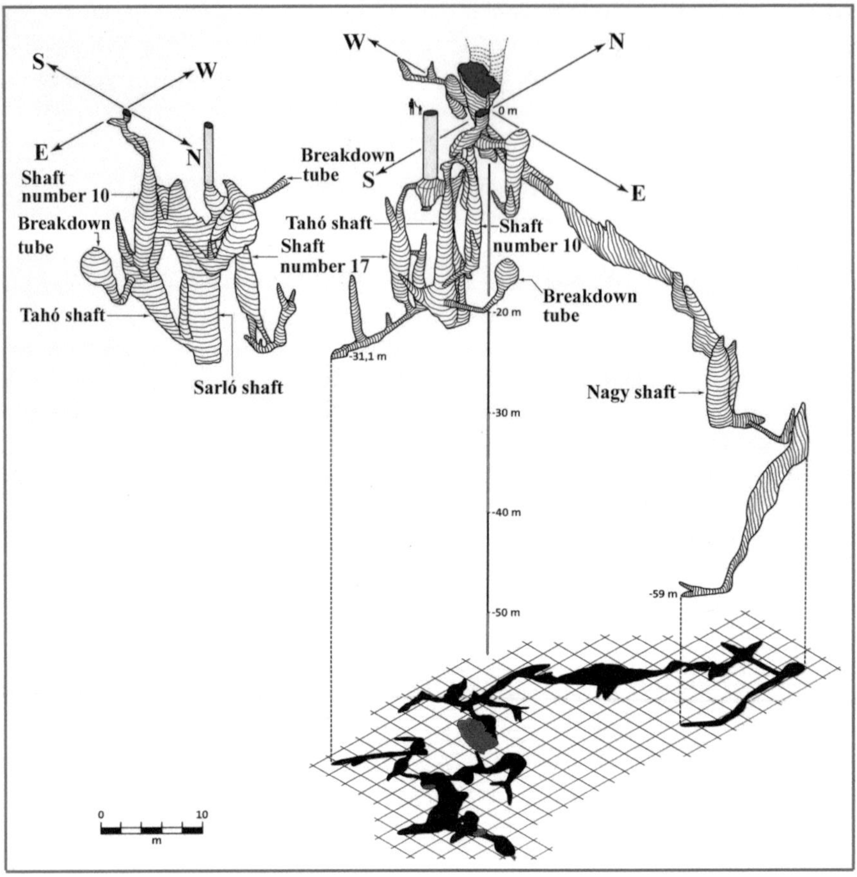

Fig. 7.8 Shaft system of the Öreg-köves ponor (Kab Mountain, Veress 2015)

basalt rampart (Figs. 4.59c and 4.60). The reason for this is that the surface runoff from these terrains is restricted and the probability of pointlike infiltration and water inflow and, thus, shaft formation increases.

Fig. 7.9 (continued) blocks; *7* soil; loess; *8* former bedrock surface and chimney wall before the collapses; *9* main blind chimney and subsidiary blind chimney; *10* opened-up main and subsidiary chimney (reaching the bottom surface of the basalt); *11* doline detail above the main chimney; *12* doline detail above the subsidiary chimney; *13* partial doline; *14* threshold between partial dolines; *I* development of single doline; *II* development of composite doline; (**a**) blind chimneys develop in limestone; (**b**) subsidiary blind chimneys form, the limestone above the main chimney caves in, and blocks of columnar basalt are detached; (**c**) detached basalt columns subside above the main chimneys, and the limestone above the subsidiary chimneys also caves in; (**d**) the subsided basalt columns are separated into blocks and subside even above subsidiary chimneys, and dolines develop above the subsided basalt columns

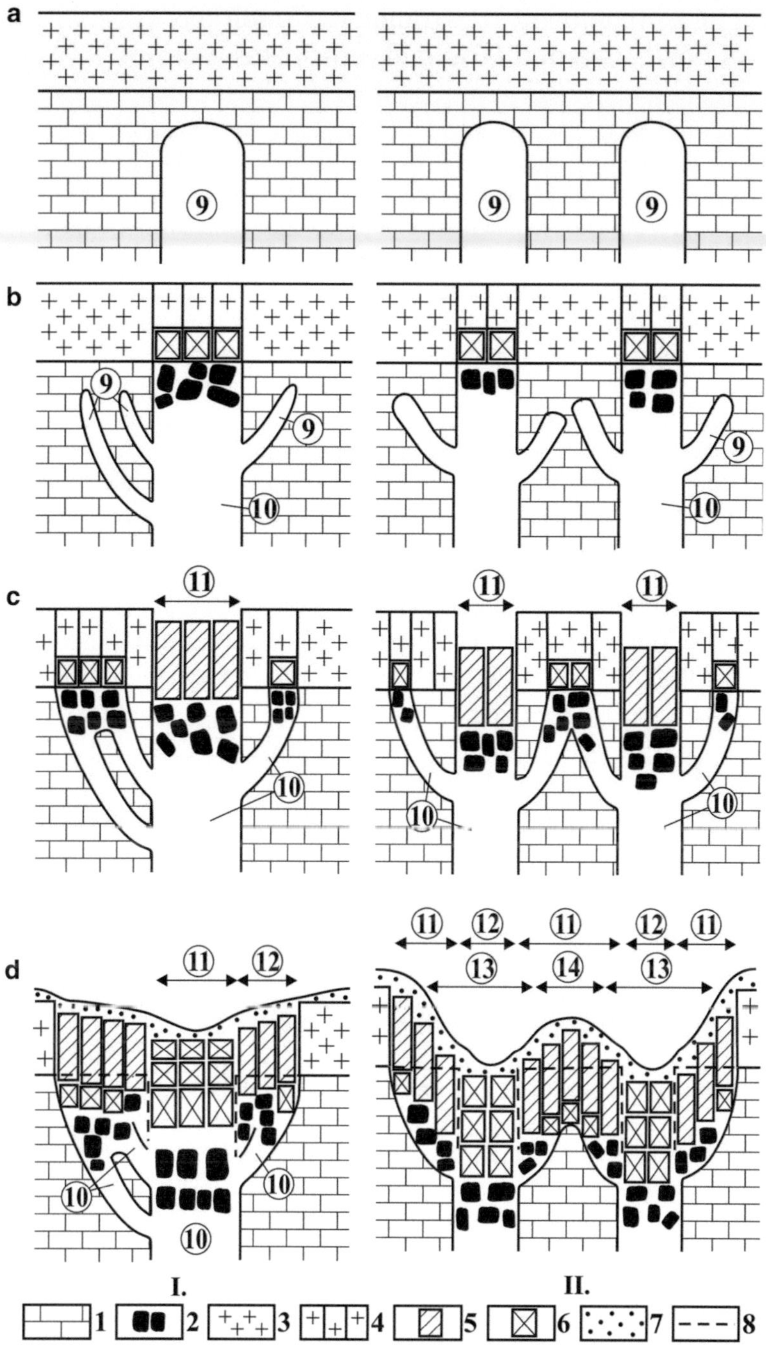

Fig. 7.9 Origin of inherited doline on Kab Mountain. *1* Limestone; *2* limestone breakdown blocks; *3* basalt; *4* jointed basalt; *5* subsided (collapsed) basalt columns; *6* basalt breakdown

7.3 The Development Environment, Conditions of Formation and Origin of Subsidence Dolines

Subsidence dolines develop on hidden rock boundary and half rock boundary. Since subsidence dolines often occur where the cover thins out, the hidden rock boundary is the location where the cover thins out. Here water with solution capacity reaches the bedrock. Piping and the inheritance of pore volume increase over the cover are also dependent on cover thickness. The thicker is the cover, the less efficient is material loss through piping and the less probable is the extension of pore volume growth to the surface. At the same time, both transport by piping and pore volume growth also depend on other properties of the cover (see below).

Subsidence dolines emerge if material deficit develops in the bedrock and features develop and the cover material is transported into these features. The solution in the bedrock governs the amount of material the bedrock can incorporate. The cover material is transported by water flowing on or included in the cover and by means of mass movements. The extent, nature and rate of transport from the cover depend on the properties of the cover and their changes. The changes are not only induced by natural phenomena (infiltration from precipitation, volume change, earthquake, water level alteration) but also human impact such as water conduction or extraction, vibration, which can also be caused by construction works or vehicle traffic (Waltham et al. 2005), or pointlike loading of the surface (Chen and Beck 1989).

7.3.1 Modelling the Development of Subsidence Dolines

Doline formation was also investigated on plaster in the laboratory. (For the description of the experiment, see Chap. 3.)

7.3.1.1 Features Resulting from the Experiment

The features formed on the cover are basins of gentle slopes, basins of vertical walls, rainwater furrows, small basins and irregular and elongated basins, longitudinal and a transversal grikes, radial grikes and arcuate grikes. The basin of gentle slopes and the basin of vertical walls (Fig. 7.10) are found at the site of water supply. The former is more typical of coarser cover over the basin of gentle slopes (kamenitza) in the bedrock, while the latter occurs where a feature (shaft) crossing the plaster bed developed. The small basin has 1–2 cm diameter and gentle slopes formed in greater distance from the site of water, at the junction of the channels of the bedrock (Fig. 7.11). The rainwater furrow is elongated downslope, usually created by erosion from the overflow water of ponds in the basins of gentle slopes. Elongated basins are particularly typical of cover with 5 cm thickness. They are closed features with gentle slopes and without basins on the plaster beds below. The

Fig. 7.10 Basin with vertical sides on the cover of the plaster tableland. *1* Basin with vertical sides, *2* shaft, *3* slope direction of the surface. Remark: particle size of the cover, <0.063 mm; slope of gypsum tableland, 5°; mode of water recharge, on the cover from above

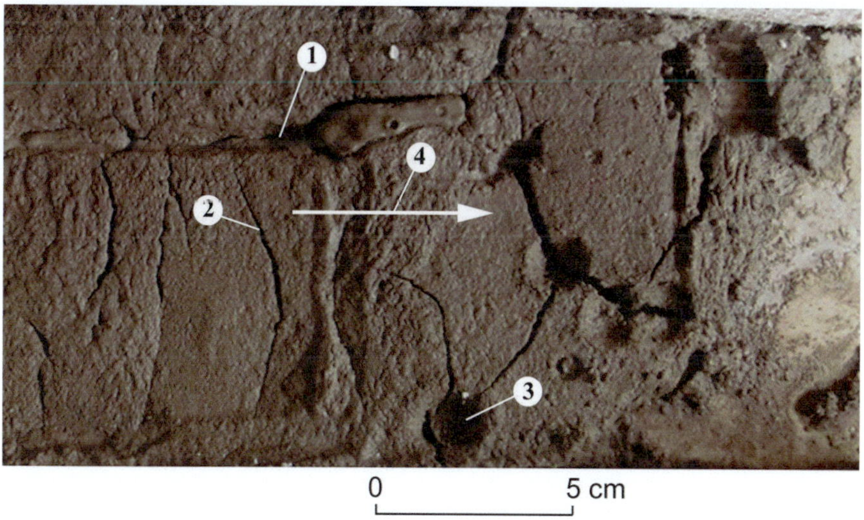

Fig. 7.11 Grikes and small basins on the cover of the plaster tableland. *1* Longitudinal grike, *2* transverse grike, *3* small basin, *4*, slope direction of the surface. Remark: particle size of the cover, <0.063 mm; slope of plaster tableland, 5°; mode of water recharge, laterally on the cover

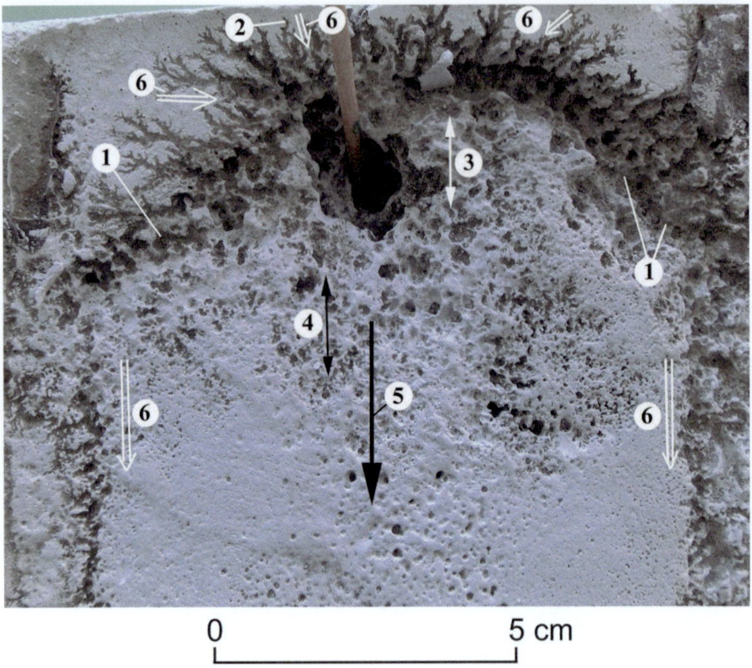

Fig. 7.12 Decantation channel on the bedrock. *1* Decantation channel, *2* inflow channel, *3* shaft, *4* chimneys, *5* slope direction of the surface, *6* water seepage on the surface of the plaster. Remark: particle size of the cover, 0.250–0.5 mm; slope of plaster tableland, 5°; mode of water recharge, on the cover from above.

radial and arcuate grikes are associated with the basins or parallel with their margins. The latter are created if the dip of the plaster bed is 0°. Similar to the irregular basin, which is not a closed form on the cover. The grikes parallel with the basin margin point to the subsidence of blocks of the cover. Such structures of the cover developed around subsidence dolines are held by Waltham et al. (2005) to be of suffosional origin.

The features of the bedrock are kamenitzas, shafts, fingertips, longitudinal and transversal channels, inflow channels, chimneys, decantation channels, channel next to cover, grike next to cover and ceiling channel. The decantation channel is a linear feature branching from the shaft or kamenitza (Fig. 7.12), while inflow channels are associated with them (Fig. 7.12). The depth of chimneys (maximum some millimetres) is larger than their diameter (Fig. 7.12). A kamenitza is a closed shallow depression with gentle slopes (Fig. 7.13b). A shaft is a closed depression with vertical walls crossing the plaster bed (Fig. 7.10). The fingertips are semispherical features larger than the chimneys (maximum several tens of millimetres) with diameter larger than depth (Fig. 7.13b). The longitudinal channels are aligned in dip direction, while the transversal channels are aligned along the strike on the bedrock (Fig. 7.13a), formed below the grikes of the cover. The channels and grikes along

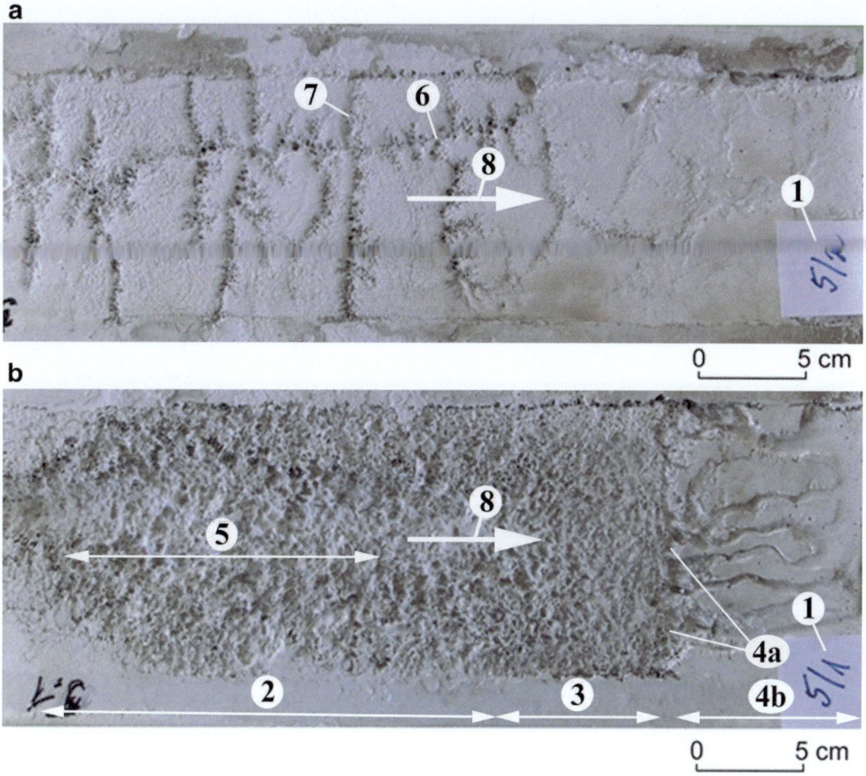

Fig. 7.13 Channels and fingertips on the bedrock. *1* Identification code of plaster tables, *2* fingertip zone, *3* chimney zone, *4a* grike along the cover, *4b* channels along cover, *5* kamenitza, *6* longitudinal channel, *7* transverse channel, *8* slope direction of the surface, (**a**) plaster Table 5.2, (**b**) plaster Table 5.1. Remark: for plaster Table 5.1, particle size of the cover, 2.5–5 mm; dip of plaster table, 5°; water recharge, laterally on the cover. For the plaster Table 5.2: particle size of the cover, <0.063 mm; dip of the plaster table, 5°; water recharge, laterally on the cover

the cover are features created on the bare bedrock at the termination of the cover (Fig. 7.13b). The ceiling channels are features developed on the bottom surface of the plaster bed.

7.3.1.2 Solution on the Bedrock

The development of features on the bedrock (through solution) is either local or continuous or transitional between these types (partly local, partly continuous). The various types of form development result from local, continuous or partly continuous solution. Local features on the bedrock are kamenitzas and shafts. Fingertips and chimneys are products of continuous solution and channels are created by partly continuous solution.

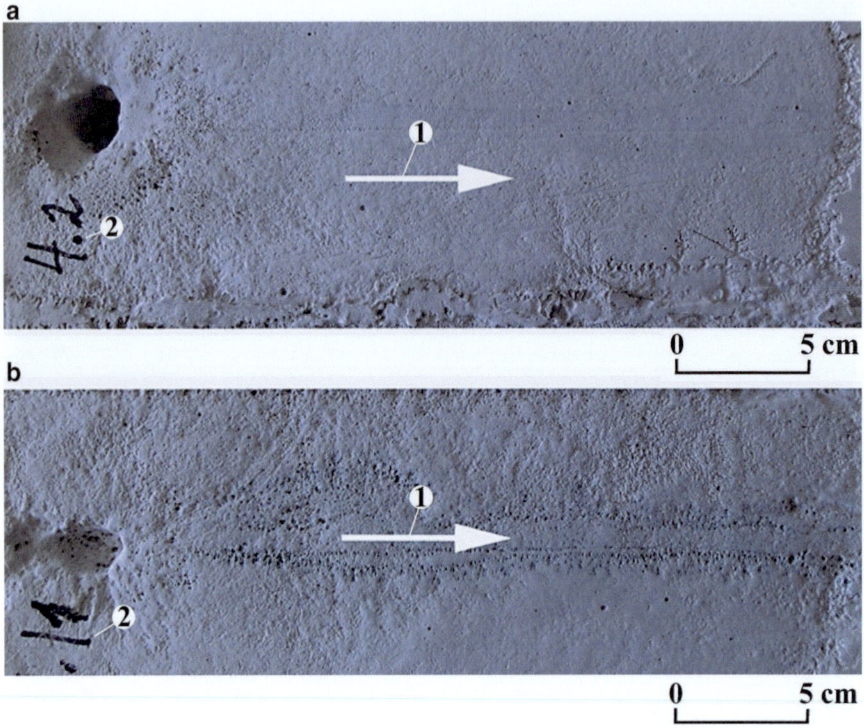

Fig. 7.14 Solution of plaster tables with water recharge from above to the cover (**a**) and with water recharge laterally on the bedrock (**b**). *1* Slope direction of the surface, *2* code of plaster table. Remark: dip of the plaster table, 5°; particle size of the cover, <0.063 mm

On coarse-grained cover (e.g. 2.500–5.000 mm particle diameter), solution is continuous (Fig. 7.13b). With fine-grained cover (with less than 0.063 mm diameter), solution on the bedrock is minimal (Fig. 1.14b) or local (Fig. 7.14a). If the cover is 5 cm thick, the character of solution also depends on particle size. Below a cover of 0.125–0.250 mm particle size, water supply onto the bedrock creates only kamenitzas, and consequently, solution was local (Fig. 7.15a). With the same cover thickness, increasing particle size leads to more and more continuous solution (Figs. 7.15b, c). However, in the case of a coarser-grained cover, a shaft developed on the bedrock.

The relationship between the type of solution and particle diameter is explained by the change in the proportion of the two kinds of pore volume (capillary and aggregate). On fine-grained cover, where capillary pore volume is higher, the cover stores capillary water. This water leaves the system through percolation sidewards or infiltrates downwards (in the interruptions of recharge) and reaches the top surface of the gypsum bed below the site of recharge (Fig. 7.15a). With 5 cm thick cover, capillary water motion in the cover can reach such dimensions that solution can partly or entirely cut across the plaster bed along the marginal grikes of the cover.

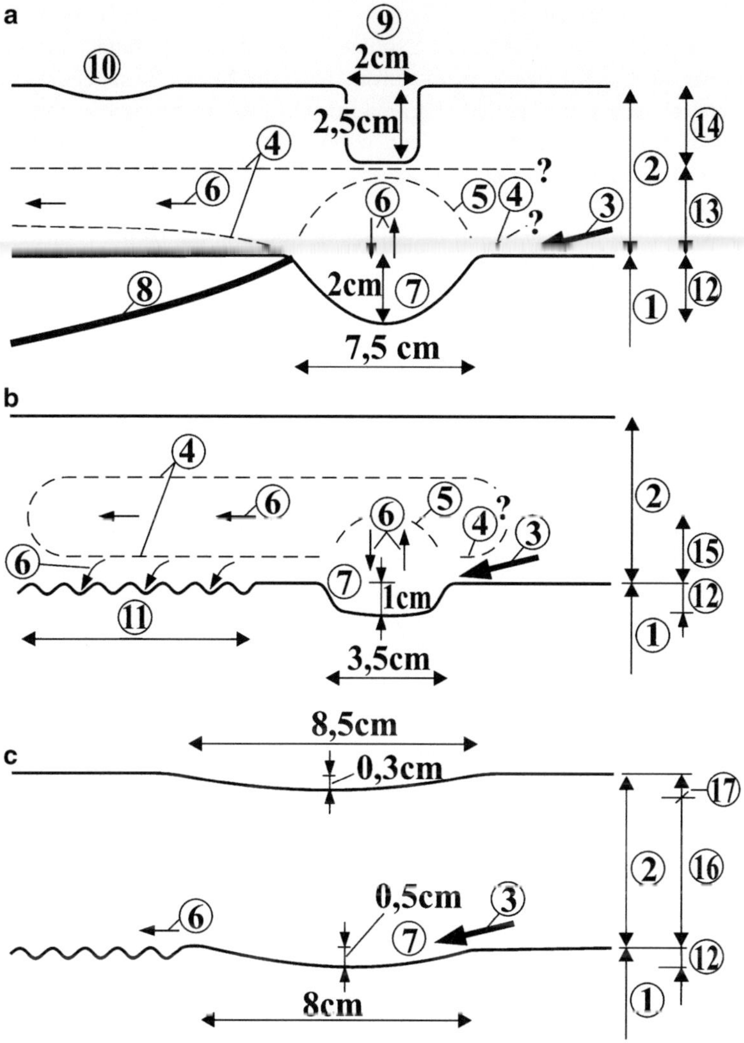

Fig. 7.15 Relationship between the particle size of the cover and capillary water position in the laboratory with 5 cm cover thickness. (**a**) Particle size of cover, 0.125–0.250 mm; (**b**) particle size of cover, 0.5–1 mm; (c) particle size of cover, 1.00–2.00 mm; *1* plaster table; *2* cover; *3* water recharge; *4* assumed boundary of the extension of capillary water with water recharge; *5* assumed boundary of the extension of capillary water in the breaks of water recharge; *6* water motion and material transport; *7* kamenitza; *8* floor of decantation channel; *9* basin with vertical sides; *10* elongated basin; *11* continuous solution (fingertips and chimneys); *12* grain fall and suffosion; *13* suffosion, growth of pore volume, cavity formation; *14* collapse; *15* suffosion, growth of pore volume; *16* growth of pore volume, compaction; *17* subsidence

This happens if the intensity of water recharge increases. Occasionally water flow onto the bedrock is secondary, and capillary water does not contact the bedrock directly but through the cracks of the cover (resulting from volume changes of the cover) and channels are created (Fig. 7.13a). On a coarser-grained cover, where aggregate porosity is higher, if capillary water gets among the grains, it can reach the bedrock through the fissures of the grains (Fig. 7.15b). If the bedrock is even coarser, since capillary porosity is low, the water recharged on the bedrock spreads in the cover in a limited way and percolates or flows on the bedrock surface (Fig. 7.15c). In the case of water recharge onto the cover, the water also reaches the bedrock through the fissures of the aggregate pore volume and, percolating on its surface, induces continuous solution. Finally, with further increase of particle size (above 0.250 mm), solution will be local again. Because of the high aggregate pore volume, the water percolates through the cover rapidly, entirely and vertically, and it reaches the bedrock at one point during water supply. Since abundant water reaches the bedrock, solution cuts through it and cannot percolate on its surface. Consequently, no continuous solution is possible.

A major control of the types of features is particle size. With fine-grained cover, irrespective of the way of water recharge, steep-walled formations emerge on the bedrock. It is possible if the water reaching the bedrock, although risen in the cover, in the interruptions of water recharge downward percolation takes place with local solution. The evolution of features was also modified by the way of water recharge if the cover was 1 cm thick with lateral water recharge onto the bedrock, then a kamenitza formed instead of a shaft (Fig. 7.14b), and solution was continuous on other parts of the plaster bed. On cover with the same particle size (less than 0.063 mm) but with direct water recharge onto the cover, a shaft formed on the plaster bed and no continuous solution happened (no fingertips or chimneys came about) (Fig. 7.14a). Similar observations could be made for covers of 5 cm thickness. With water recharge onto the bedrock, the type and dimensions of features developed on the bedrock (and also on the cover) depended on particle size (Fig. 7.15).

Landform evolution also depends on the angle of the bearing slope. With 0° slope solution is local, but the shape of the resulting shaft is controlled by the way of water recharge. If water arrives laterally onto the bedrock, an elongated shaft results (Fig. 7.16a), and if from above, it will have a circular groundplan (Fig. 7.16b).

7.3.1.3 Landform Development on the Cover

Inheritance from the bedrock over the cover mostly takes place in the case of local and partly continuous solution and only rarely with continuous solution. During local solution, inheritance takes the following forms. If water recharge onto the cover occurs, kamenitzas or shafts may develop on the bedrock. In the former case, part of the particles of the cover is redeposited into the depressions of the bedrock through piping, followed by compaction and the subsidence of the cover above the

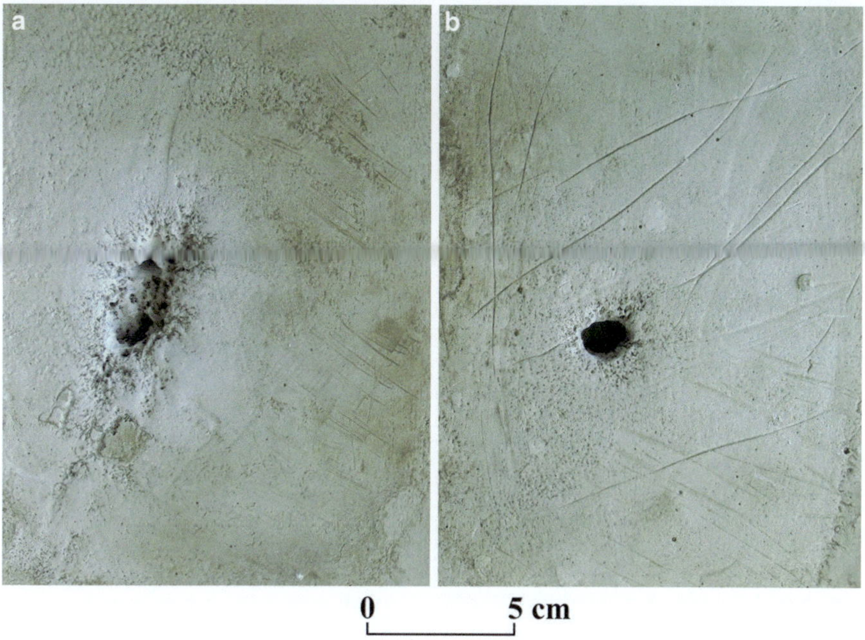

Fig. 7.16 Dissolution of plaster tables if water recharge onto the bedrock takes place laterally (**a**) and from above (**b**). Remark: particle size of the cover, <0.063 mm; dip of the plaster table, 0°; (**a**) plaster table G/2; (**b**) plaster table G/3

compacted part. Subsidence results in the formation of a depression on the cover (Fig. 7.17a). If solution cuts through the bedrock, the particles of bottom part of the cover are transported by piping through the shaft onto the bearing surface of the plaster bed (Fig. 7.17b). Then larger-scale water deficit comes about in the cover, cavities may also form, and collapse ensues.

If water recharge affects the bedrock surface, kamenitzas or shafts can also develop. In this case no piping occurs. In the former case, particles from the bottom part of the cover also fall into the kamenitza of the bedrock resulting in increasing pore volume and subsidence (Fig. 7.17c_1). A depression in the cover also forms if the material is transported laterally as well as downwards and accumulated in the fingertips and chimneys of the bedrock. This way primarily small basins develop (Fig. 7.11). Occasionally no depression comes about on the bedrock below the depression in the cover. In this case the material is removed from the portion under cover through piping.

If water recharge takes place onto the bedrock, but a shaft forms on the bedrock, particles from the bottom part of the cover fall onto the bearing surface of the plaster bed through the shaft. At the bottom part of the cover, the pore volume grows (occasionally with cavity formation), followed by the collapse of the top part of the cover (Fig. 7.17d).

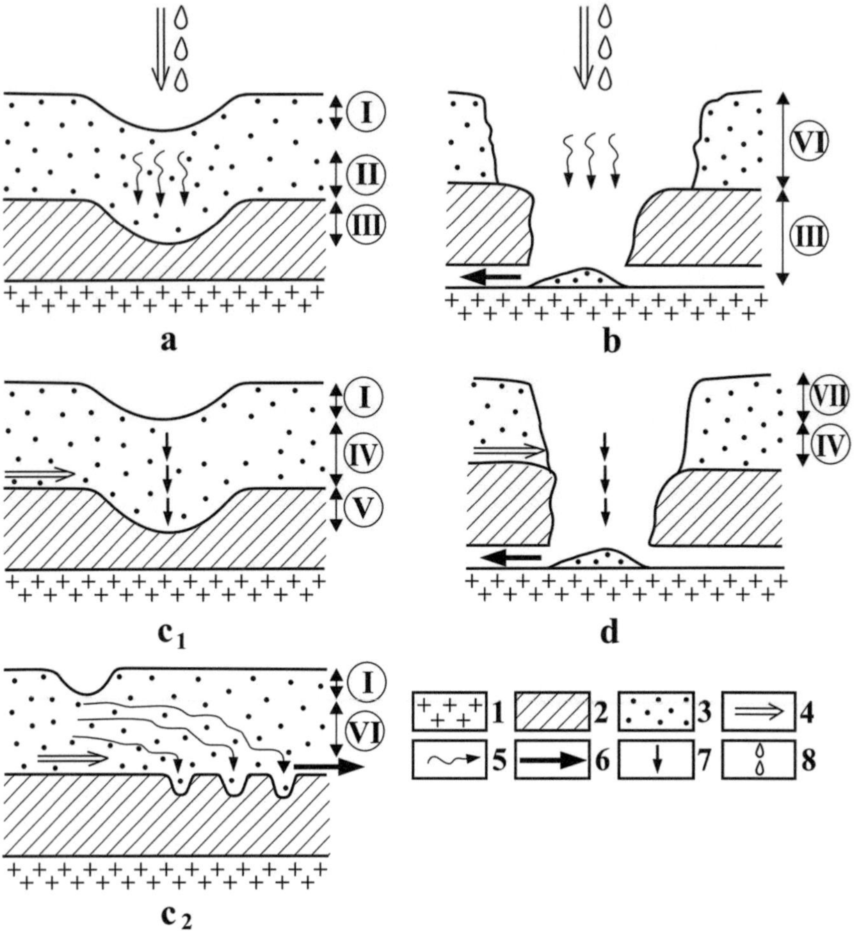

Fig. 7.17 Landform development on the cover of the plaster table with 1 cm cover thickness. *1* Metal plate bearing the plaster table, *2* plaster table, *3* cover, *4* water recharge, *5* suffosion, *6* material transport on the bearing metal plate, *7* fall of grains, *8* dripping, *I* subsidence zone, *II* suffosion–compaction, *III* solution–suffosion–compaction, *IV* fall of grains (increasing pore volume, followed by compaction), *V* solution-falling grains, *VI* suffosion, *VII* collapse, (**a**) suffosion–compaction–subsidence, (**b**) suffosion, (**c_1**) falling grains–growing pore volume–compaction–subsidence, (**c_2**) material transport by suffosion–growing pore volume–compaction–subsidence, (**d**) falling grains–growing pore volume–collapse

The experiments prove that the dimensions of the features on the bedrock depend on the dip of the plaster bed, the way and rate of recharge and the grain size and thickness of the cover. The dimensions and shape of the features developed on the cover depend on the dimensions and size of the features of the bedrock, on the thickness and grain size of the cover and the way and rate of water recharge. Depressions can form on the cover even if there are no solutional features underneath.

7.3.2 *Conditions of Doline Formation*

Dolines form if the angle of the bearing slope, the elevation of the bearing surface and the properties of the bedrock and the cover are suitable for this process. Further conditions are the surface morphology of the bedrock (mounds of the paleokarst terrain or cuestas created by glacial erosion), ground frost and water in the cover. Slope inclination, elevation and bedrock properties have been statistically investigated in the following karst regions: in the Hochschwab and Mecsek mountains (Plan and Decker 2006; Lippmann et al. 2008) and in the Miroč Mountains (Telbisz et al. 2007). In the Bakony Mountains, the distribution of subsidence dolines by elevation has been studied (Veress 1983), but during data processing, subsidence dolines were not distinguished from solution dolines. It is claimed, however, that most of the dolines in the studied areas (and all of them in the Bakony Mountains) are subsidence dolines.

7.3.2.1 Slope Angle

The relationship between doline density and slope angle was investigated in the Mecsek Mountains by Lippmann et al. (2008). The data of Lippmann et al. (2008) show that dolines are most abundant on terrains of 2°–7° slope (but their absolute number is the highest on 2°–4° slopes), while on slopes steeper than 13°, no doline occurs. Similar results are achieved by Telbisz et al. (2007) for the Miroč Mountains, where 85 % of dolines are found on slopes gentler than 12° and only 23 % on slopes above 12° inclination. On slopes steeper than 19°, only 1 % of dolines occur. Unfortunately, the genetic distribution of dolines (solution or subsidence dolines) is not known on this karst either.

The predominance of dolines on gentle slopes indicates that their origin is controlled by the ratio between surface runoff and infiltration. Large-scale infiltration, possible at low slope angles, favours doline formation on either covered or uncovered karst. Since on uncovered karst percolation loss is more probable, on this karst type sufficient amounts of water probably reach the rock even at higher slope angles to induce doline formation. On covered karst, however, more water has to percolate away to reach the bedrock since part of the water is stored in the cover. Therefore, the change of slope inclination has a greater influence on the formation of subsidence dolines than on that of solution dolines.

7.3.2.2 Elevation Above Sea Level

In the Hochschwab Mountains, dolines occur at elevations of 1300–2300 m. Between 1700 and 1800 m, doline density is the highest. Plan and Decker (2006) investigated the altitudinal distribution of dolines distinguished according to volume size. The small-scale dolines (probably subsidence dolines) are found at 1700–1900 m and contribute to the great density of dolines at this level. Medium-size dolines are of uniform distribution, while the large-size dolines (only 4 % of all dolines) occur above 1900 m, on the highest sections of the plateau. These are probably solutional paleodolines.

On the Mecsek karst dolines are concentrated at two levels: at 230–310 m and 370–400 m (Lippmann et al. 2008). The dolines of the lower level are located on the abrasional terrace of Lake Pannon (Lippmann et al. 2008), and their altitudinal distribution coincides with that of this surface.

In the Bakony Mountains, subsidence dolines occur between 321 and 620 m elevations (the mountains are 200–700 m high). The 97.57 % of the 577 dolines are located between 361 and 520 m elevation. A significant level of doline formation is at 441–460 m elevation with 31.37 % of dolines (Veress 1983). The lower limit of doline distribution is probably controlled by the low karst water level of the mountains, which ranges from 160 m (on the margins) to 260 m (in the central parts). The lower limit of the distribution is probably also influenced by the lower occurrence, higher thickness and broader distribution of impermeable cover sediments. In higher-lying blocks where a karst water storey developed (with water tables up to 400 m elevation), the lower limit of doline distribution is obviously marked by the level of karst water storey.

If karstification is differentiated by block types, dolines are most common on exhumed horsts in summit position (at 350–520 m elevation). (This type includes the Hárskút Basin, the Mester-Hajag, the Tés Plateau and the area between Kőris and Som mountains.) On the blocks of this type, at the mentioned level (441–460 m), doline density is 4.33 dolines/km^2. The karstification of these blocks is promoted by the low dissection of their surfaces. For this reason, gentle slopes are common and the loess mantle is less eroded here. At the same time, they are at relatively high elevations and the one-time impermeable cover (Csatka Gravel Formation) has been completely removed from the surface or only retained in patches.

On the more elevated horsts in summit position (at 600–700 m), doline density is only 0.47 dolines/km^2, while on blocks in lower position (cryptopeneplain or threshold surfaces below 350 m elevation), doline density is 0.07 and 0.17 dolines/km^2, resp. On the blocks of the former block type, low density is explained by the large-scale removal of cover sediment (patches of the permeable cover, i.e. loess, are only found in patches). In the case of the latter block type, low doline density is due to the great thickness of the cover (cryptopeneplain) or by the total absence of cover sediment and the position of their surface close to the karst water table.

The altitudinal distribution of exhumed horsts in summit position (Hárskút Basin, Tés Plateau, Mester-Hajag, area between Kőris and Som mountains) is variable, manifested in:

- Elevations belonging to the highest frequency of occurrence.
- The pattern of distribution (symmetrical or asymmetrical): with symmetrical distribution, the number of dolines decreases with distance from the dominant level of doline formation, at the same rate both downwards and upwards (Hárskút

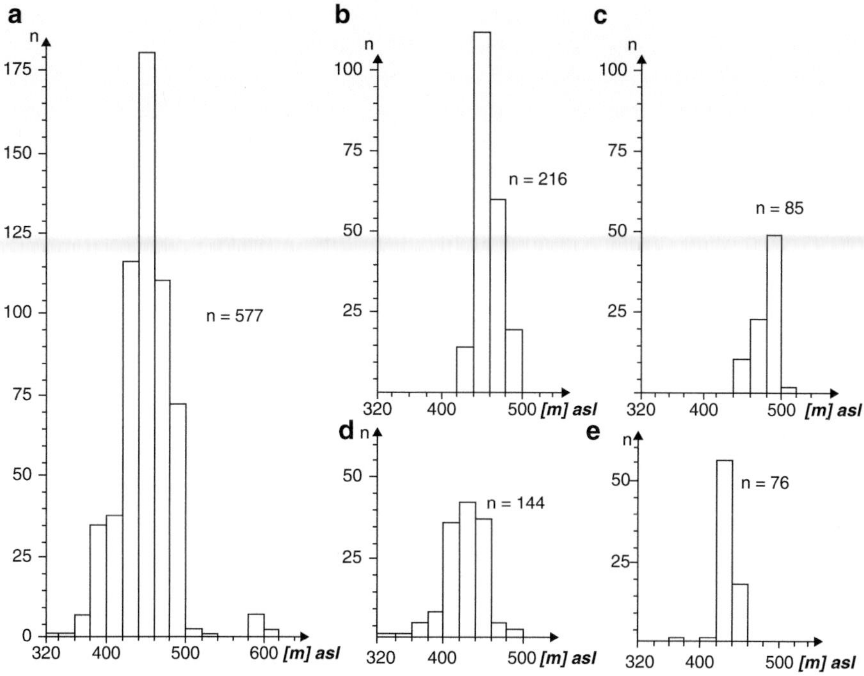

Fig. 7.18 Altitudinal distribution of subsidence dolines on the concealed karst of the Northern Bakony Mountains (Veress 1983). (**a**) For the whole Northern Bakony, (**b**) for the Hárskút Basin, (**c**) for the Mester-Hajag, (**d**) for the Tés Plateau, (**e**) for the area between Kőris and Som mountains

Basin and Tés Plateau); this is not valid for the areas of asymmetrical distribution (Mester-Hajag, area between Kőris and Som mountains) (Fig. 7.18).

– In the variable amplitude of distribution (difference in elevation between the highest and lowest dolines), which is the highest on the Tés Plateau.

Doline formation is assumed to be the earliest and most rapid at the main level of doline formation. At ever lower or higher elevations, less and less dolines can be counted since:

– The rate of doline formation reduces moving away from this level.
– With increasing distance from this level, the process began ever later.
– Dolines were destroyed (buried or eliminated) through the accumulation or removal of the cover with distance from this level.

Any of the above reasons may play a part in the above presented distribution of dolines. One of them, however, can be decisive. Thus, on the Mester-Hajag, dolines formed at various dates since the exhumation of the block is of greatest extent on the summit level. The denudation of the lower parts is of limited scale, and in these sites, even accumulation can take place as sediment is redeposited from higher services. Therefore, the development of dolines extends from the higher portions of the block over to the lower parts which are later exhumed.

Among the four areas presented, cavity formation is of the largest scale on the Tés Plateau and the amplitude of distribution is also the largest. Therefore, there could be a relationship between the extent of cavity formation and the distribution of doline development. The shafts in the bedrock could have formed below surfaces of variable elevation, and this favours doline formation at variable elevations. The larger dimensions of shafts promote doline formation, while their increased density involves greater doline density.

7.3.2.3 The Bedrock

Bedrock properties (lithology, structure) are probably only manifested indirectly in the formation of subsidence dolines, through the modifying impacts of the cover. (Among them, the significance of structure will be detailed below.) For instance, the alignment of dolines is rather controlled by valley directions than by faultlines (Lippmann et al. 2008). Even on the Hochschwab Mountains, there was a difference between the direction of faultlines and doline alignment (Plan and Decker 2006). The indirect control of faultlines is underlined by the investigations by Waltham and Hatherley (1983). Our maps prove that the occurrence of subsidence dolines is not related closely to the fractures in the bedrock. Similarly, according to Magdalene and Alexander (1995), the locations of dolines are not dependent on rock structure. Dolines, however, are widespread where the dip of limestone strata is high (Waltham et al. 2005; Hyatt et al. 1999).

7.3.2.4 Karst Water

The character and evolution of the covered karst is fundamentally controlled by the position of the karst water table, which is of three types: at least temporarily higher than the surface (1), reaching to the surface (2) or lower than the surface (3). Particularly in the two latter cases, covered karst processes can be influenced by the position of the karst water relative to the bedrock surface. The covered karst processes are also influenced by the frequency of water level fluctuation, the intensity and range of fluctuation.

To the impact of ascending water level – as it has been mentioned at the demonstration of karren formation – the chimneys of the bedrock are stoping upwards until they reach the cover. If the water level rise extends over the cover, cavities or passages also develop there. With water level rise, the porosity of the cover is reduced and its weight grows. The ascending water carries particles upwards and they deposit in the environs of the resulting depression. If the karst water level is lowering, the particles of the cover are moving downwards and accumulate partly in the bottom part of the cover and partly in the chimneys and cracks of the bedrock. With lowering water level, the sucking effect increases and the support of the cover reduces. Water loss is followed by compaction. With dropping water level, material from the upper part of the cover collapses and is washed in into the cavities and passages not filled in by water.

7.3.2.5 Cover Thickness

Since the size (thickness) of the doline depends on cover thickness only in some cases and to a limited extent, the dimensions of the doline cannot be predicted from cover thickness (Hyatt et al. 2001). Thus, if the cover is thin, only shallow dolines develop. This, however, does not mean that a thick cover involves exclusively deep dolines. Cover thickness, however, shows a close relationship with the origin and density of dolines.

In China 60 % of the several thousand subsidence dolines formed on cover sediment thinner than 5 m and more than 85 % on cover less than 10 m thick (Yuan 1987; Chen 1988). According to Beggs and Ruth (1984), in Florida most of the dolines came about on cover sediments with less than 20 m thickness, while according to Upchurch and Littlefield (1987) on covers less than 25 m thick. Currin and Barfus (1989) claim that the majority of dolines form on covers thinner than 30 m. In Florida, however, although on rare occasion, dolines occur on covers thicker than 60 m (Wilson 1995) and in China even with more than 70 m cover thickness (Xiang et al. 1988). Sites of subsidence dolines on the Pádis are known where cover thickness amounts to 85 m (Silvestru 1997). In the case of extreme cover thickness, only individual dolines emerge: in Florida at 150 m cover thickness (Arrington and Lindquist 1987) or in Italy at 200 m thickness (Waltham et al. 2005). Here ascending waters due to geothermal effect contribute to the development of the doline (Beck 2000).

The relationship between cover thickness and doline density was studied by Klimchouk and Andrejchuk (2002). They found that doline density decreases with increasing cover thickness, but, depending on the properties of the cover, to various extent even in the case of the same (gypsum) bedrock (Fig. 7.19).

Within the same karst area, a relatively close relationship is found between cover thickness and the site of doline occurrence (Veress 2008, 2009b). On the covered karst of the Northern Bakony, inner (doline floor) cover thickness was less than 4 m for 78.23 % of the 37 dolines studied. The outer (doline margin) cover thickness was less than 4 m for 32.43 % and less than 6 m for 75.68 % of the dolines. Regarding only the 21 syngenetic dolines out of the total 37 landforms, the outer cover thickness was below 6 m in 80.95 % of them. Even more strikingly, 78.38 % of the 37 dolines were located above mounds in the bedrock (85.71 % of the syngenetic dolines). For syngenetic dolines, outer cover thickness was invariable less than 6 m. The dissolution on the mounds of the bedrock can be increased by the increase of lateral seepage in the cover. Since the amount of lateral seepage depends on the grain size of the cover, the degree of dissolution may be greater on the mounds of the bedrock covered with finer grains than in the case of the cover with coarser grains. The investigated depressions were not exclusively located above the buried

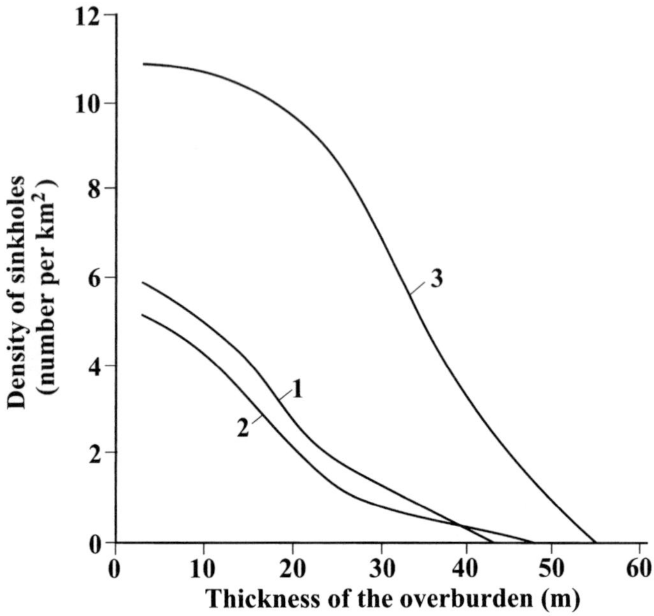

Fig. 7.19 The density of sinkholes versus thickness of the overburden (Klimchouk and Andrejchuk 2002). *1* In the Seret-Nichlava interfluve (entrenched karst), *2* in the Cerny Potok area (subjacent karst), *3* in the Zoloushka cave area (subjacent karst)

mounds of the bedrock, but those in valley floor position were also common (34 out of the total 37 dolines). Consequently, the dolines studied are not merely associated with thin cover but also with its local thinning out. At the same time, the subsidence dolines of the Pádis, where cover thickness is great (mostly above 10 or even 20 m and it is coarse grained), are less clearly bound to locally thinned-out cover deposit. Neither on the karst of the Mecsek Mountains does the location of dolines depend on cover thickness as dolines occur where clayey series terminate (Fig. 7.20, Veress 2011). Consequently, dolines occur with high probability where the cover thins out (hidden rock boundary) or where its composition changes laterally (half rock boundary).

Lei et al. (2001) claim that subsidence doline origin is more dependent on cover type and structure than on its thickness. In our opinion, cover thickness exerts four kinds of influence on doline origin and density:

- The thicker is the cover, the more probable it is that the material deficit is compensated merely through increasing pore volume instead of ground subsidence.
- The thicker is the cover, the more probable it is that the cover include carbonate series (consuming the solution capacity of water) or impermeable intercalations.
- The thicker is the cover, the more substance must be removed to induce ground subsidence (collapse). The removal of more substance from the cover is increasingly difficult partly because of the limitation of transport and partly because of

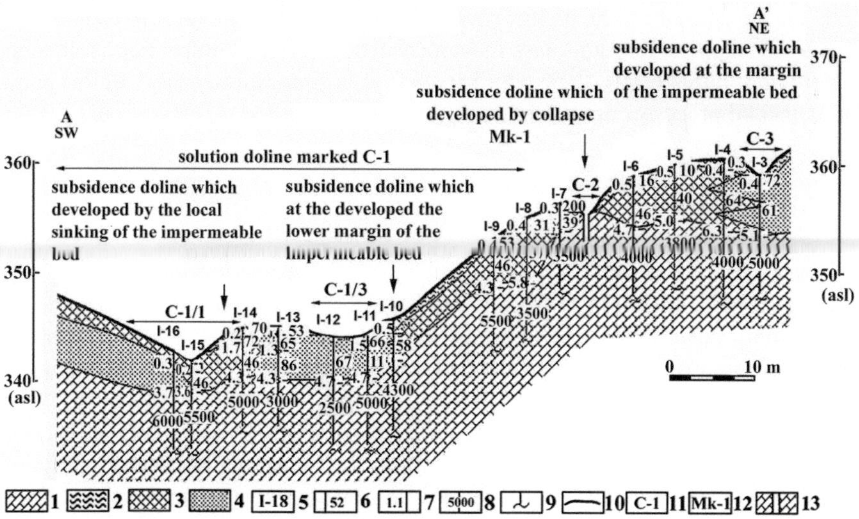

Fig. 7.20 Subsidence dolines formed at the termination of clay series from the Mecsek Mountains (Veress 2011). *1* Limestone; *2* soil, sand, silt; *3* clay (with limestone debris, sand); *4* sandy loess (with limestone debris); *5* site of VES measurement with identification code; *6* geoelectric resistance of the series (Ohmm); *7* basal depth of the geoelectric series (m); *8* geoelectric resistance of the bedrock (Ohmm); *9* approximate penetration of VES measurement; *10* boundary of geoelectric series; *11* code of karst depression; *12* site of bedrock outcrop; *13* shaft

the limited capacity of the cavity in the bedrock. For the removal of more material, a larger cavity is necessary, but the probability of its formation is lower than for a smaller cavity.

– On a thicker cover, the non-karstic pipe is longer than on a thinner cover, and it is more likely to clog even before the formation of a depression on the ground above it.

7.3.3 *Formation of Subsidence Dolines*

7.3.3.1 The Origin of Depressions in the Cover

The material deficit in the bedrock is compensated for in the cover in various ways, depending on the properties of the cover (clay content, $CaCO_3$ content, grain size) as well as on the presence of groundwater or ground ice in the cover. Until the cover remains stable, there is no material redeposition or landform development on it. Stability depends on cover thickness, the size of the cavity in the cover, velocity of groundwater flow, porewater pressure and cave air pressure (He et al. 2003). During subsidence doline formation, cover material can be transported into the bedrock, into the non-karstic pipe of the cover or into the pores of the cover particles. Transport happens through mass movement (solifluction), compaction, suffosion, dissolution, pluvial erosion or with karst water or groundwater flow. Below some fundamental processes are presented.

Suffosion (Piping)

The finer-grained material of the cover deposit is transported to depth or into the karst. Although suffosion often includes transport in solution too (Butzer 1976), here it only covers sediment transport since it also occurs in sediments without soluble components. Others also emphasise the redeposition of fine sediment in suffosion (Williams 2004).

Material transport through suffosion takes place from the surface or its vicinity or from the bottom part of the cover deposit. In the former case, the infiltrating water either transports the particles of the cover into depth or the material is washed in into the non-karstic pipes of the cover. The non-karstic pipe itself can originate from material transport by suffosion.

Particles are transported by infiltrating rainwater, groundwater or karst water table drop. In both cases, material transport in the cover is only possible if groundwater is present there only temporarily. In the presence of groundwater, transport by suffosion takes a rather lateral direction, and therefore no dolines are able to develop. This is proved that there are no dolines on the catchment of covered karst ponor G-6/b (see Chap. 6), where groundwater was detected in boreholes (Veress and Futó 1990).

During suffosional material transport, not only suffosion dolines but also compaction dolines are able to develop (Wilson 1995).

Pore Volume Growth

Cover porosity is characterised by the actual pore volume typical of the given rock and the loosening pore volume. Actual pore volume is typical of the rock and, as mentioned above, is composed of capillary and aggregate porosity.

In the case of large pore volume, water saturation of the cover will be high. Water conduction to the karst may be interrupted even in the absence of impermeable intercalations, and the rate of suffosion may also decline. Because of the compaction of particles due to soil moisture, suffosion may be replaced by falls and collapses in the cover.

In a coarse-grained cover, high aggregate porosity induces gravitational water motion in vertical direction and promotes sediment transport by suffosion. This kind of cover is capable of taking considerable amounts of fine material transported by suffosion. The increase of pore volume may happen because of suffosion (the fine-grained material is washed out), of the decrease of water content and of loosening. The cause of loosening is the expansion of the cover because the material is transported into the chimney that developed on the bedrock.

There can be either primary or secondary pore volume (Axaixiang and Yezhy 2008), it is primary if it derives from rock expansion on the first occasion and secondary if it comes about subsequent to that.

In the cover of covered karst primary loosening pore volume is generated when the material deficit in the bedrock extends over the cover, but the cover remains in place. In this case the space between the grains increases. Secondary pore volume results from further growth of pore volume because material deficit increases in the bedrock again. Final pore volume growth is calculated as the product of primary and secondary pore volumes (Axaixiang and Yezhy 2008).

Subsidence

The growth of pore volume causes compaction which is followed by subsidence. If the growth of pore volume extends to the surface, compaction also happens to the surface; thus, the surface subsides. If the growth of pore volume does not reach the surface, only the part of deeper position in the cover is compacted until the pore volume increased. In this case, the cover part above the compacted part involutes into the compacting part.

Szalai et al. (2006) and Veress (2009b) pointed out horizontal and vertical increase of pore volume from the cover in the environs of a doline (in the Homód Valley) (Fig. 7.21a). The 'units' of different pore volume could develop because the loosening of the cover deposit extended or shifted. The material loss at the bottom generated pore volume growth first locally at the bottom of the cover (Fig. 7.21b). Then it extended upwards (Fig. 7.21b3) and reached the ground surface (Fig. 7.21b4). The upward shifting of pore volume growth, in our opinion, accompanies the formation of non-karstic pipes (Fig. 7.21b5) or ground subsidence. Other geophysical measurements indicate that pore volume growth alters the structure of the cover (Kruse et al. 2006; Benson and Yuhr 1987), i.e. involutions form in the cover series (Fig. 7.22).

Mass Movements

The adjustment of the cover to material deficit is also possible through mass movements. The mass movements leading to depression formation are collapses, rockfalls and solifluctional processes.

Collapses occur in the top part of the cover if larger cavities develop in its bottom part. The cavities probably form as a consequence of collapse in the bedrock or rapid development of a large-scale shaft on the bedrock.

If the cover is coarse grained (rock debris), large fragments can fall from its bottom part into the shafts of the bedrock. This rockfall increases the pore volume of the cover, inducing ground subsidence or suffosion, or results in thinning out of the cover and collapse.

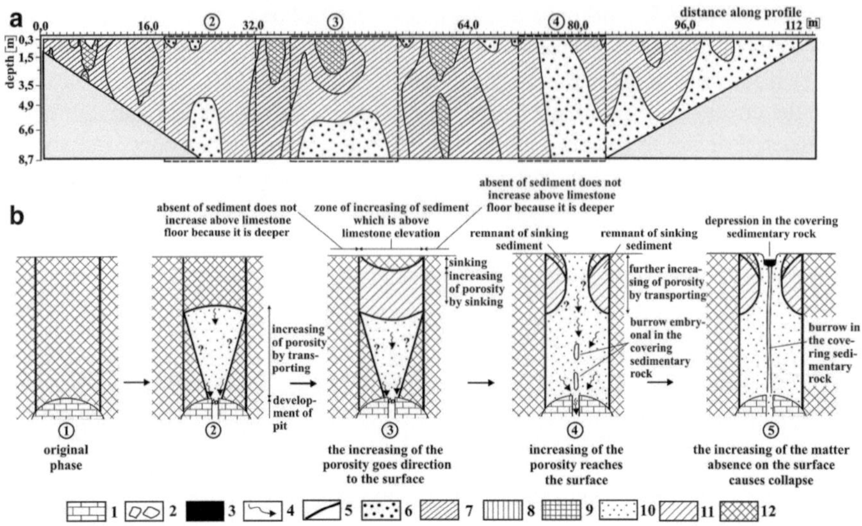

Fig. 7.21 Development phases of non-karstic pipe in the sedimentary cover (Veress 2008). *1* Limestone, *2* breakdown which consists of limestone, *3* breakdown which consists of superficial deposit, *4* water infiltration and suffusion in the superficial deposit, *5* boundary of sedimentary rock with various porosity. (**a**) Measured specific resistance: *6* 1–5 Ωm, *7* 5–50 Ωm, *8* 50–120 Ωm, *9* 120–400 Ωm. (**b**) Theoretical porosity by using specific conduction: *10* porosity of the superficial deposit is high. *11* porosity of superficial deposit is medium, *12* porosity of superficial deposit is low. The subfigures *2*, *3* and *4* of the figure were constructed with the parts of the measured figures: (**a**) marked with *broken lines*

Rock fragments fall if the shaft is too wide, the cover is loaded from above or the ice cement of the cover melts. This process is typical of glaciokarst or tundra karsts. Fine cover particles may fall into the cavity of the cover if porewater pressure is increased there (Waltham et al. 2005; Currens et al. 2012) or if water level drops in the cover deposit (Xu and Zhao 1988).

The bottom part of the cover deposit can be transported into the shafts of the bedrock by solifluction too (Drumm et al. 1990; Tharp 1999; Waltham and Fookes 2003; Williams 2004).

7.3.3.2 The Impact of Processes on Depression Formation

To present the impacts, we employed geoelectric–geological profiles, primarily for the Pádis plateau, where cover structure is simple and easily detectable. Unfortunately, the VES measurements are not able to identify shafts in the bedrock and non-karstic pipes in the cover deposit. Thus, their existence can only be concluded from the sediment structure.

Suffosion is either local or areal. Suffosion is pointed out from the involutions of surfaces of the cover series or of the ground surface. With local suffosion, the

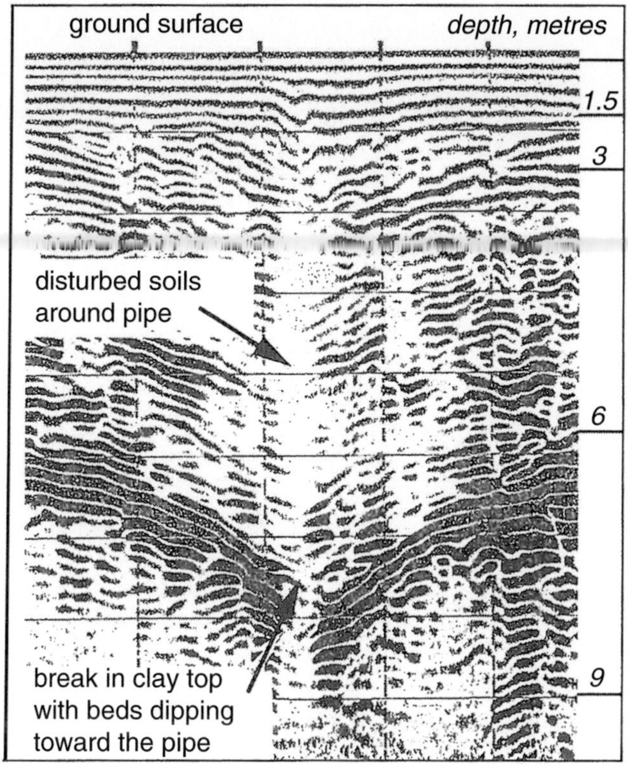

Fig. 7.22 An image from ground probing radar that could not reach bellow the clay (Benson and Yuhr 1987) – modified

process is limited to the cover under the subsidence doline, while in the case of areal suffosion, the upper surface of the bottom cover series subsides over an area which may embrace several dolines (Fig. 7.23, between VES measurement sites R21 and R27). It is assumed that with local suffosion, material transport is bound to a non-karstic pipe, while this is not the case with areal suffosion.

If the series of higher position thinned out and bent into the cover of lower position, a non-karstic pipe took shape, which is fed from the material of the upper series through local suffosion. If both the cover of lower position and the one of higher position bended inwards, the non-karstic pipe receives material from both the lower and the upper series. This is probably possible if the non-karstic pipe terminates on the boundary of the two series, as it is seen at the VES measurement site R24 (Fig. 7.23). If only the upper series of the cover bended inwards, but the underlying series did not, the non-karstic pipe extends into the upper series. The material loss of the upper series through suffosion results in bending in and thinning out. This situation is visible at the observation site R22 (Fig. 7.23). It may occur that the surface of the lower series is planar and the upper series thins out towards the centre of the depression (Fig. 7.24, observation site R54). In this case the upper

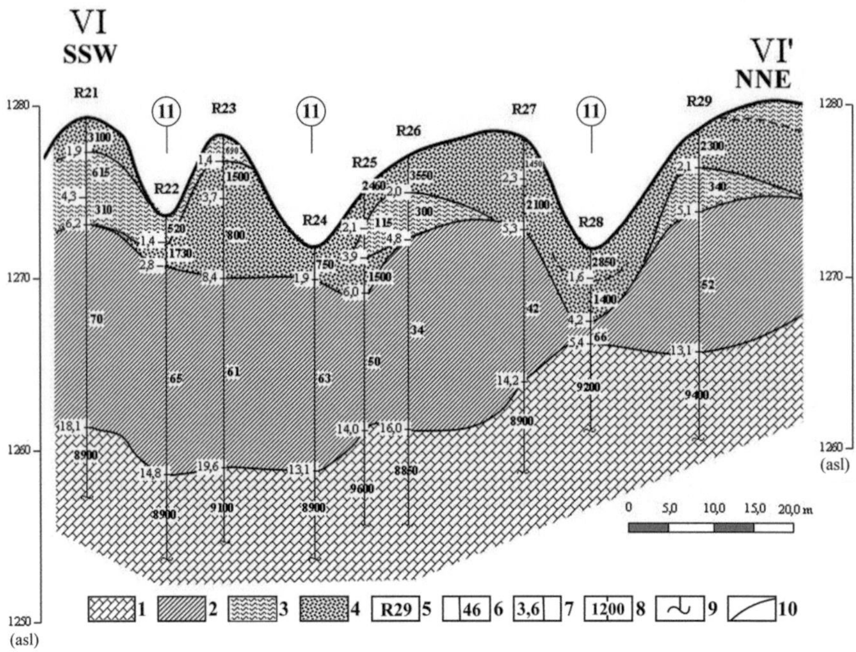

Fig. 7.23 The geoelectric–geological profile VI–VI′ from the Padis-1 (Uvala Račhite) area. *1* Limestone, *2* clayey silt, *3* mixed rock debris (with sandstone and limestone debris), *4* mixed rock debris (sand, sandstone and limestone debris), *5* VES observation site, *6* geoelectric resistance of the series (Ohmm), *7* basal depth of geoelectric series (m), *8* geoelectric resistance of the bedrock (Ohmm), *9* approximate penetration of the VES measurement, *10* boundary of geoelectric series, *11* subsidence doline

series thinned out by means of pluvial erosion if the non-karstic pipe opened up to the one-time surface, or if it lost its material through suffosion, the non-karstic pipe also extends into the upper series and only received material from there.

If the lower series bended inwards to a great extent (Fig. 7.23, at VES observation site R28), the non-karstic pipe receives much material from this series. Then, because of material loss, the pore volume of the lower series increases, followed by compaction and subsidence of the surface of this series. This is made possible by the termination of the non-karstic pipe in the lower series. The upper series is adjusted to the bending (subsidence) of the surface of the series in two ways: either bends into the subsided part or its pore volume increases, followed by compaction and subsidence of the surface.

Depending on the thickness of the upper series, the following situations occur. If the thickness of the upper series is uniform all over the doline, only the bearing series of the doline bent and subsided, but no sediment input through suffosion took place from the upper series (Fig. 7.23, VES measurement site R28, phase 2 in Fig. 7.25b). The transport from the lower series can reach a degree that the upper

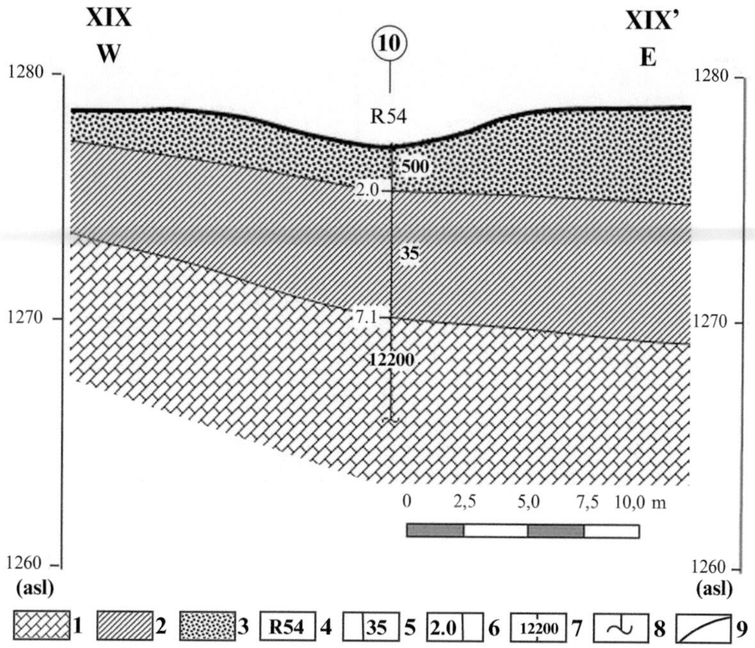

Fig. 7.24 The geoelectric–geological profile XIX-XIX′ from the Padis-1 (Uvala Račhite) area. *1* Limestone, *2* clayey silt, *3* mixed rock debris (with sandstone and limestone debris), *4* VES observation site, *5* geoelectric resistance of the series (Ohmm), *6* basal depth of geoelectric series (m), *7* geoelectric resistance of the bedrock (Ohmm), *8* approximate penetration of the VES measurement, *9* boundary of geoelectric series, *10* subsidence doline

series reach(es) the bedrock during bending (subsidence) (phase 3 in Fig. 7.25b, doline E-6 in Fig. 7.26). If the thickness of the upper series is reducing towards the centre of the doline (Fig. 7.25d), suffosion from the upper series was at higher scale or the series was also affected by pluvial erosion. If the thinning out of the upper series was caused by suffosion since more material was transported from there than from the lower series, the upper surface subsided to a greater degree than the lower (Fig. 7.27, VES measurement site R19). Then the non-karstic pipe probably also developed in the upper series. If the upper series is getting thicker towards the centre of the depression, either accumulation or the subsidence of the lower series happened and suffosional transport from there exceeded that from the upper (Fig. 7.25c, VES measurement site R105 in Fig. 7.28).

Dolines can develop from the suffosion of uniform cover series and the ensuing subsidence. In this case the material from the overlying series is redeposited into the non-karstic pipe formed in the series and subsidence follows. During the subsidence of the cover, the non-karstic pipe is consumed and the doline takes shape through the inward bending accompanying the subsidence (Figs. 7.25a and 7.29).

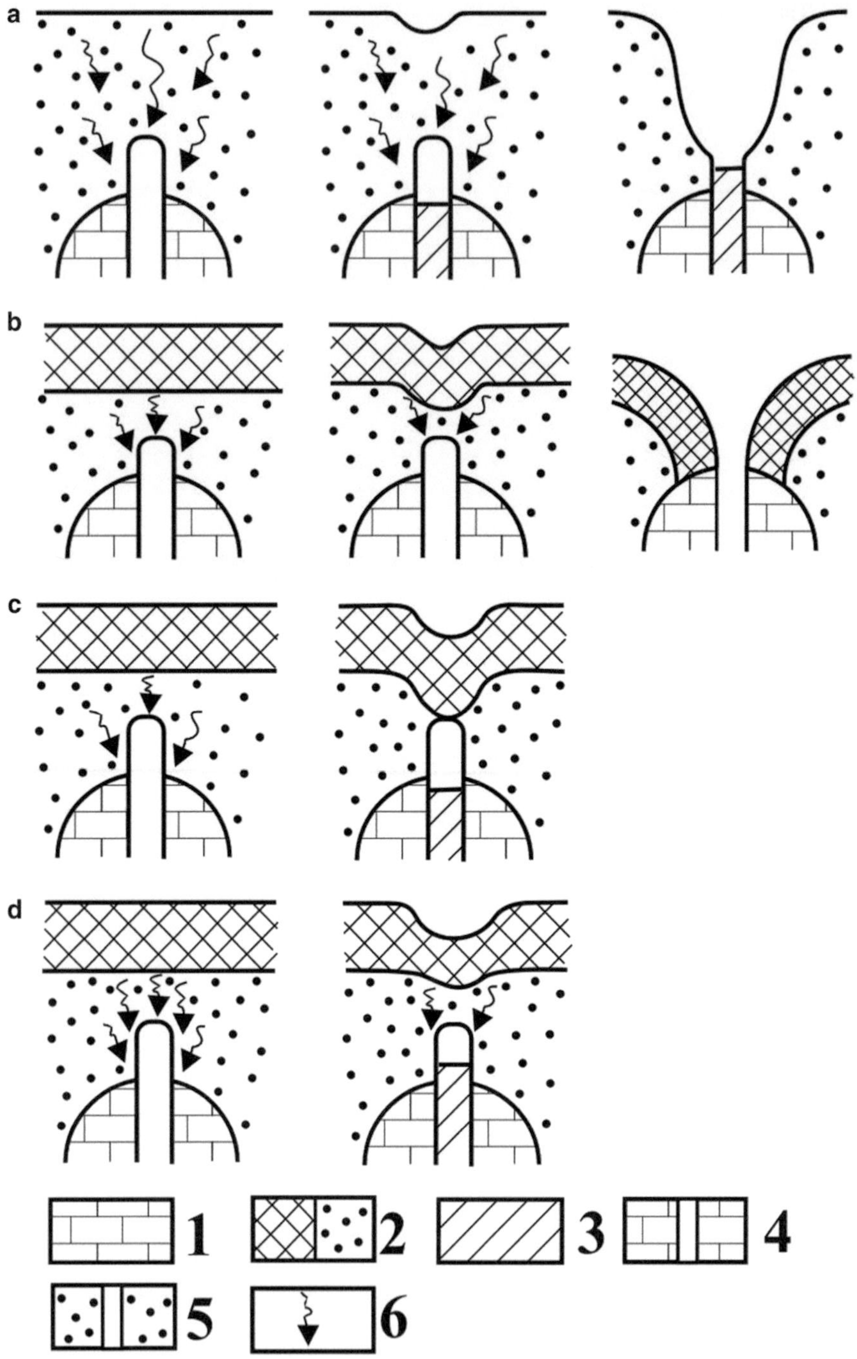

Fig. 7.25 Changes in cover bound to suffosion. *1* Limestone, *2*, cover deposit, *3*, redeposited cover, *4*, shaft, *5*, non-karstic pipe, *6* water percolation and suffosion. (**a**) Water percolation and suffosion in the cover, the surface bends in; (**b**) the top series bends in; (**c**) during bending the top series is thickening: as a result of material transport from the bedrock series by filling or suffosion (in the latter case, the subsidence of the upper surface of the lower series is followed by the subsidence of the lower surface of the upper series); (**d**) during bending, the top series is thinning (as a result of pluvial erosion or of material transport by suffosion from the top series)

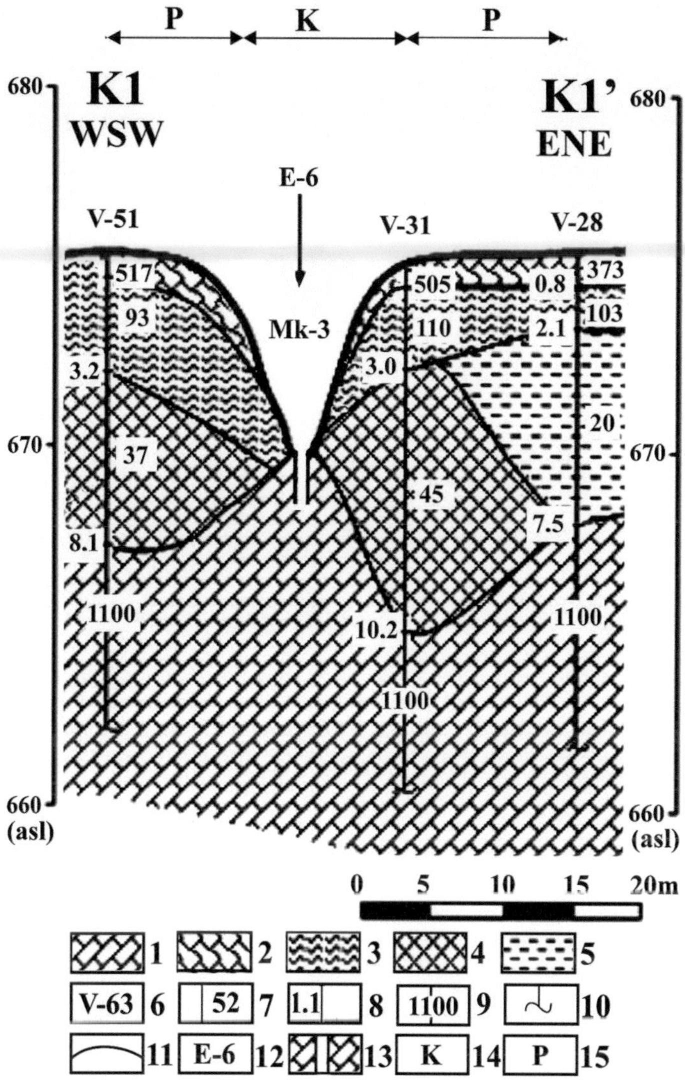

Fig. 7.26 The K1–K1′ geoelectric–geological profile from the Eleven-Förtés area (Kőris Mountain, Bakony Mountains, Veress and Puskás 2007). *1* Limestone; *2* limestone debris; *3* clayey limestone debris; *4* clay (with loess, limestone debris); *5* clay; *6* VES measurement site; *7* geoelectric resistance of the series (Ohmm); *8* basal depth of geoelectric series (m); *9* geoelectric resistance of the bedrock (Ohmm); *10* approximate penetration of VES measurement; *11* boundary of geoelectric series; *12* code of subsidence doline; *13* chimney, shaft; *14* paleokarst mound; *15* paleokarst doline

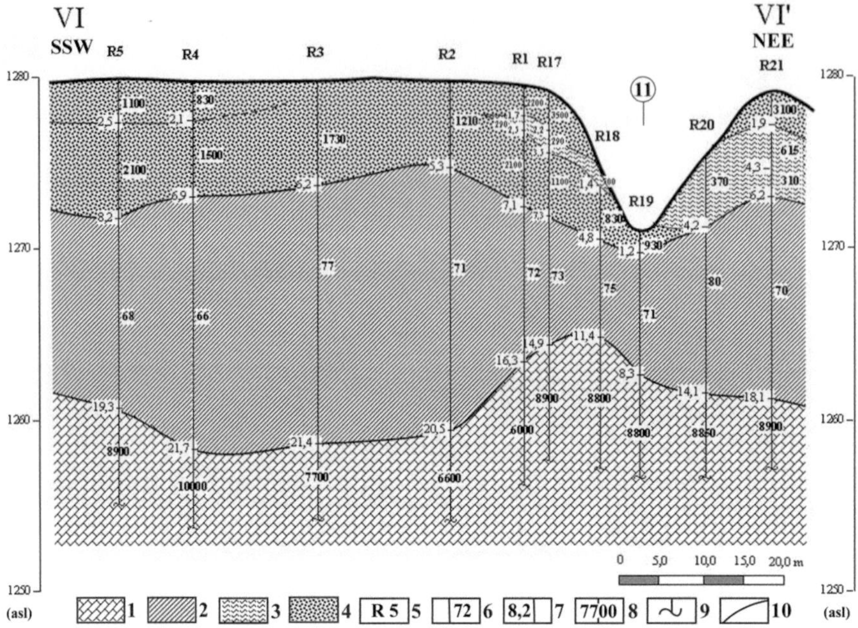

Fig. 7.27 The geoelectric–geological profile VI–VI′ from the Padis-1 (Uvala Račhite) area. *1* Limestone, *2* clayey silt, *3* mixed rock debris (with sandstone and limestone debris), *4* mixed rock debris (sand, sandstone and limestone debris), *5* VES observation site, *6* geoelectric resistance of the series (Ohmm), *7* basal depth of geoelectric series (m), *8* geoelectric resistance of the bedrock (Ohmm), *9* approximate penetration of the VES measurement, *10* boundary of geoelectric series, *11* subsidence doline

7.3.3.3 Impact of the Composition of the Cover on Doline Formation

The bedrock under the cover suffers dissolution if water reaches the bedrock, and this water is capable of solution. Solution capacity depends on the CO_2 content of water and on the amount of carbonate in the cover. The calcareous or carbonate-bearing cover sediments of the concealed karst are loess and morainic deposits. In the cover, vertical or horizontal water percolation could be observed. The latter happens if there is groundwater (karst water), ground ice, and impermeable bed or if the cover has a high capillary pore volume. The laterally percolating water is influential in karstification if the limestone bedrock has an uneven surface dissected with mounds. In this case, for instance, the water percolating over the impermeable series reaches the bedrock at a mound (Figs. 4.1c and 4.3g). In the cover, water percolates either permanently or temporarily. Temporary percolation takes place if in the cover conditions with and without ground ice or if the cover has a clay content. In the latter case, the percolation stops when the clay minerals swell.

Jones (1965) observed that the bedrock does not dissolve under carbonate-bearing morainic deposits. Williams (1966) explained this with the saturation of

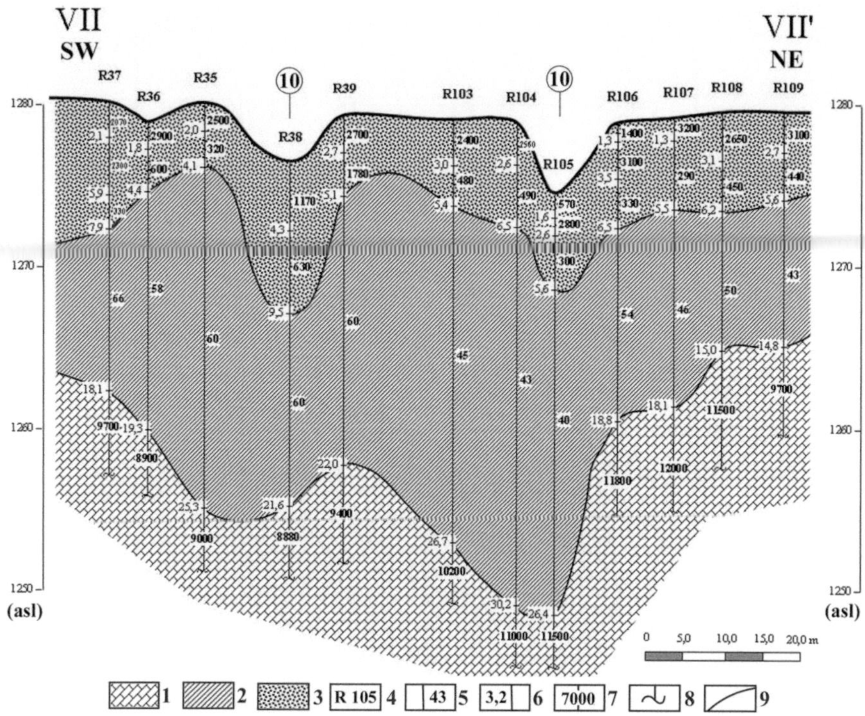

Fig. 7.28 The geoelectric–geological profile VII–VII′ from the Padis-1 (Uvala Račhite) area. *1* Limestone, *2* clayey silt, *3* mixed rock debris (sand, sandstone and limestone debris), *4* VES observation site, *5* geoelectric resistance of the series (Ohmm), *6* basal depth of geoelectric series (m), *7* geoelectric resistance of the bedrock (Ohmm), *8* approximate penetration of the VES measurement, *9* boundary of geoelectric series, *10* subsidence doline

water percolating through the calcareous cover (morainic material with carbonate content) of the karst region (glaciokarst in Ireland) before it reaches the bedrock. The solution capacity of water on the bedrock was measured by Jennings (1977) and Trudgill (1976a). According to Trudgill (1972, 1975, 1976b, 1985), the dissolution of the limestone bedrock takes place under a cover, which has pH 4–7 and $CaCO_3$ content below 1 %, but no solution is possible if pH is 7–9 and carbonate content is above 10 %.

The saturation level of the water percolating into the bedrock marks the lower boundary of the epikarst zone, i.e. to where solution occurs. Its extension depends on the amount of CO_2 production. In temperate karsts, the epikarst zone reaches to 10 m (Al-fares et al. 2002), while in tropical karsts even 100 m (Mangin 1997). On glaciokarsts, it is some metres thick (Ford and Williams 2007) – partly because of the low CO_2 content and partly because of glacial erosion. This is confirmed by our observations. The depth of grikes is 1–2 m, while that of the solutional pipes is maximum 3 m (Veress 2010). On glaciokarst there was neither sufficient time for its thickening nor sufficient amounts of CO_2 available in the infiltrating water.

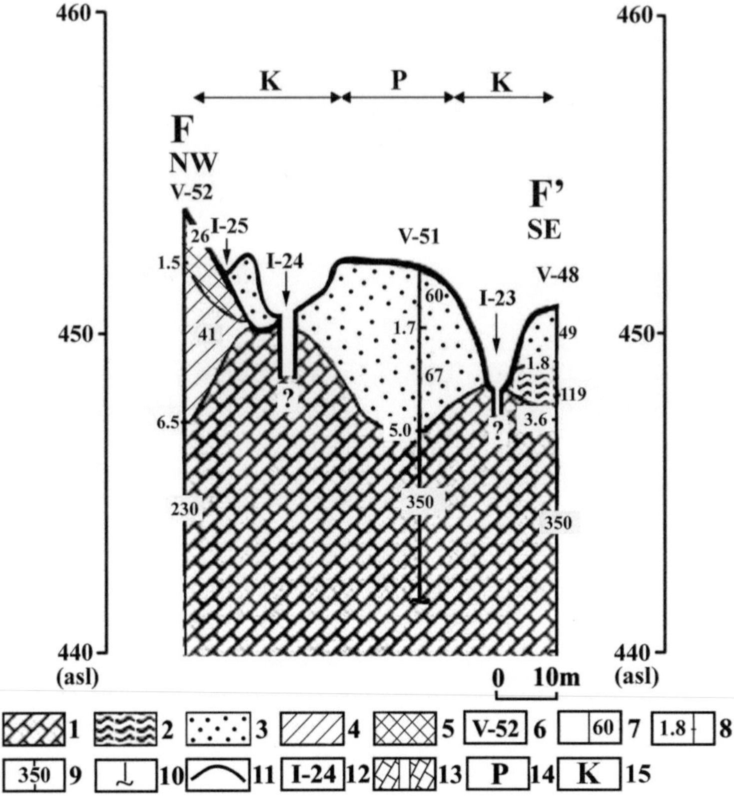

Fig. 7.29 The geoelectric–geological profile F–F′ from the Tés Plateau (Bakony Mountains). *1* Limestone; *2* limestone debris (with clay); *3* loess (with sand and limestone debris); *4* loess (with clay and silt) or clay with limestone debris; *5* clay (with loess and limestone debris); *6* VES measurement site; *7* geoelectric resistance of the series (Ohmm); *8* basal depth of geoelectric series (m); *9* geoelectric resistance of the bedrock (Ohmm); *10* approximate penetration of the VES measurement; *11* boundary of geoelectric series; *12* subsidence doline with code; *13* chimney, shaft; *14* paleokarst depression; *15* paleokarst mound

If the carbonate content is lower, for thicker cover, the bedrock can also be dissolved or solution will be more intensive. Thus, in some samples from the Tés Plateau, according to Kiss et al. (2007), carbonate content is 21–31 %, while Füzesi (2007) estimates it up to 34 %. At the same time, in the Mecsek Mountains (Hevesi 2001) on the Pádis plateau, the $CaCO_3$ content was 0 % (Silvestru 1997). On both karst, doline density is high; in the Mecsek it is 73.4 dolines in km^2 (Hevesi 2001), but in some terrains (in an area of almost 2 km^2) may be up to 164 dolines in km^2 (Lippmann et al. 2008). Since the above authors provide doline density together with solution dolines on the map available for us, we only calculated the density of subsidence dolines for a smaller area (40,179 m^2) and found an even higher density than the above data (237 dolines in km^2). On the Pádis it is 29.2 dolines per 1000 m^2 as opposed to the Tés Plateau, where it is 0.035 dolines in km^2. This means that low carbonate content is coupled with high doline density. Doline density, however, is not controlled solely by the carbonate content of the cover (or the thickness of the limestone cover), but a range of other factors. From the aspect of the saturation of the infiltrating water, the thickness of the cover is important. With the same (low) amount of carbonate, the increasing thickness of the cover raises the chance of saturation before the infiltrating water reaches the bedrock. CO_2 production, the clay content of rainwater and the density of shafts and grikes on the bedrock, which also depend on further factors, are also of significance. Therefore, on glaciokarsts, in spite of a cover with high carbonate content, doline density can be high (on the alluvial fan of a Dachstein paleouvala: 0.031 dolines in m^2.) This high doline density of glaciokarsts is explained by the high density of grikes (see below).

The clay content of the cover influences surface runoff. Informative are the groundwater conditions in the covered karst of the Hochschwab paleouvala, where three zones are distinguished. The first zone covers the area encircled by subsidence dolines, where the cover is permeable (only in this case dolines can develop) in spite of clay content above 30 % ($CaCO_3$ content is 1.7 %). In this zone clay content increases surface impermeability and surface runoff to the extent that erosion gullies and ravines become associated with some dolines. East of this, a strip of several 10 m width (i.e. the second zone) percolating rainwater accumulates as groundwater in the cover, evidenced by the intermittent springs on the margin, where gullies begin (Fig. 5.91). The impermeable cover interbeddings in the area of the second zone are confirmed by the presence of groundwater and, thus, water cannot reach the karst from this area. In the third zone, which is cryptokarst of the depression, the cover surface is completely impermeable. Surface waters in this zone generate erosional landforms (gullies and ravines) and then, coalescing, flow in a concentrated manner into the karst through the ponor.

Both in Hochschwab and at the Tés Plateau, the clay content is high, and it ranges between 11.9 and 24.8 % (Füzesi 2007). Many subsidence dolines of the Tés Plateau also have gullies or ravines. In contrast, on the Pádis plateau, where clay content is maximum 8 % (Silvestru 1997), particles often reach diameters of some tens of centimetres, and none of the several hundreds of dolines show an associated gully or a ravine. Consequently, the ratio of surface runoff to infiltration depends on the clay content (and particle size) of the cover. Where clay content is low and particle size is large, water percolates away at many sites, and therefore, many dolines develop without gully or ravine, while there where clay content is high and particle size is small, water percolation is restricted to few sites and dolines develop in smaller numbers but with ravines.

7.3.3.4 Impact of Shaft Formation on Doline Generation

Shaft Formation

The structure and features of the epikarst are influential in the development of depressions in the cover. Therefore, we present the formation of epikarst features and their effect on the cover.

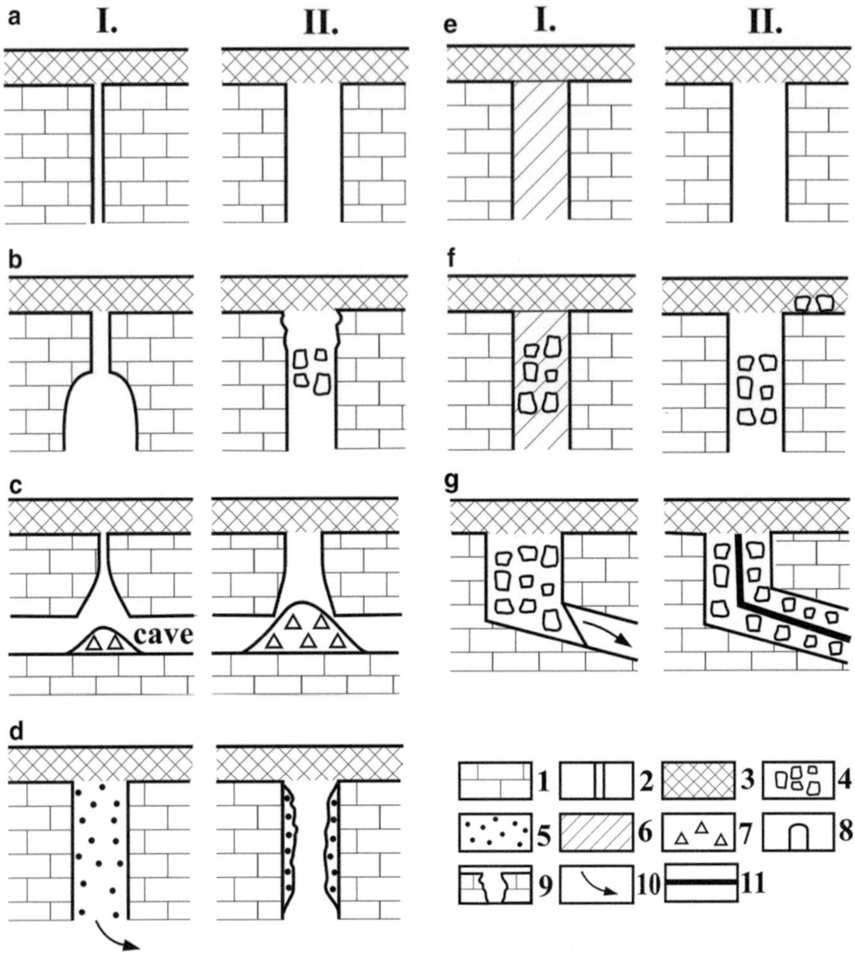

Fig. 7.30 Theoretical cases of the formation of material deficit in the bedrock. For syngenetic karstification: (**a**) broadening grike, (**b**) ceiling of blind chimney caves in, (**c**) cave ceiling caves in. For postgenetic karstification: (**d**) sediment fill transported from the chimney, (**e**) melting of the ice fill of the chimney, (**f**) melting of the ice cementing the sediment fill of the chimney, (**g**) the morainic mound of the shaft is transported into the deeper part of the shaft by mass movements and a conduit forms in the morainic deposit. *1* Limestone; *2* fracture; *3* cover deposit, sediment fill; *4* morainic deposit; *5* fine-grained deposit; *6* ice; *7* collapsed blocks; *8* blind chimney, blind shaft; *9* surface created by collapse; *10* sediment removal; *11* conduit in morainic material; *I* initial stage; *II* chimney development

Solution features on the bedrock are karren (grike), pinnacle surface, bowl-shaped depressions, chimneys (shafts) and cavities. Thus, on the bedrock with pinnacle surface, depressions may occur into which the cover sinks during the deepening of the karst terrain (Brink and Partridge 1965). Shafts or grikes either form after the deposition of the cover (syngenetic covered karst formation, Fig. 7.30a–c) or after the loss of the fill of the shaft which originated before the deposition of the cover (postgenetic covered karst formation, Fig. 7.30d–g). Syngenetic karst formation either happens after the broadening of the grike (Fig. 7.30a), through the caving-in of the blind chimney (Fig. 7.30b) and through the collapse of the cave ceiling (Fig. 7.30c). Similar phenomenon was described from Indiana state, where a subsidence doline developed above Blue Spring Cave (Waltham et al. 2005). Postgenetic karstification can happen through the transport of sediment fill by water (infiltrating water, karst water, Fig. 7.30d), melting of the cementing ice (Fig. 7.30e) and removal of the fill through mass movement (Fig. 7.30g).

Below we present the role the structure of the cover and the bedrock plays in the process of shaft (chimney) and depression generation. In the shafts of the depressions in the Bakony Mountains, investigated in detail (such as the shaft of doline Gy-12), there is a collapse zone above (Fig. 7.31), and below it there is a dissolution chimney, on the walls of which dissolution features developed (Fig. 7.32).

The shafts do not always open to the surface but end in the rock (as blind shafts or chimneys, Fig. 7.33), closed on the top. The grikes could grow so wide that only the dividing walls of the shaft walls are preserved between them (Fig. 7.34). The presence of the blind chimney or the collapse zone indicates that solution was not the most intensive at the bedrock surface, as on bare karst (Jakucs 1977), but we have to reckon with an inversion of solution intensity.

The depth figures of numerous shafts of the Bakony Mountains show that shafts reaching down to various depths occur. Therefore, the epikarst locally broadens at the shafts. With the downward shifting of the level of saturation, the lower front of shaft formation reaches or approaches the karst water table – even if it lies at relatively greater depth. The local broadening of the epikarst is allowed by the presence of the cover sediment (see below) and the fact that at the dolines of the shafts concentrated and increased water conduction into the karst takes place.

The presence of collapsed material at the top part of the shafts of the Bakony proves that they developed from blind chimneys through the collapse of their ceiling. The process of shaft formation takes place through the stages of grike and blind chimney development in the following way (Veress 2000, Fig. 7.35).

Grike karren form along fractures. If the cover fully fills them, grikes develop further. On the walls of the filled grike parts, solution ceases or is reduced, particularly if the clay content of the cover is high. The clay forms a coating on the walls of the filled sections of grikes, while the walls of unfilled grike sections continue to be dissolved and, thus, widen in lack of protective coating. Moreover, the intensity of solution increases since CO_2 production increases in the grike fills. The unfilled grike sections widened to coalesce into a blind chimney or blind shaft (Fig. 7.33). Because of this coalescence on the walls of the blind chimneys, only ruined remnants of the former grikes survived (Figs. 6.12, 7.33 and 7.34). The thinned-out ceiling of the blind chimney caves in. The collapse may extend over the cover if that is thin and a depression emerges on the cover. From this point, shaft evolution is governed by pointlike water inflow through the depression. To the effect of inflowing water, the shaft is widening and deepening. Beck (1986) presented a similar course of development. In his opinion, the depression of the cover formed after the collapse of the cavity in the bedrock.

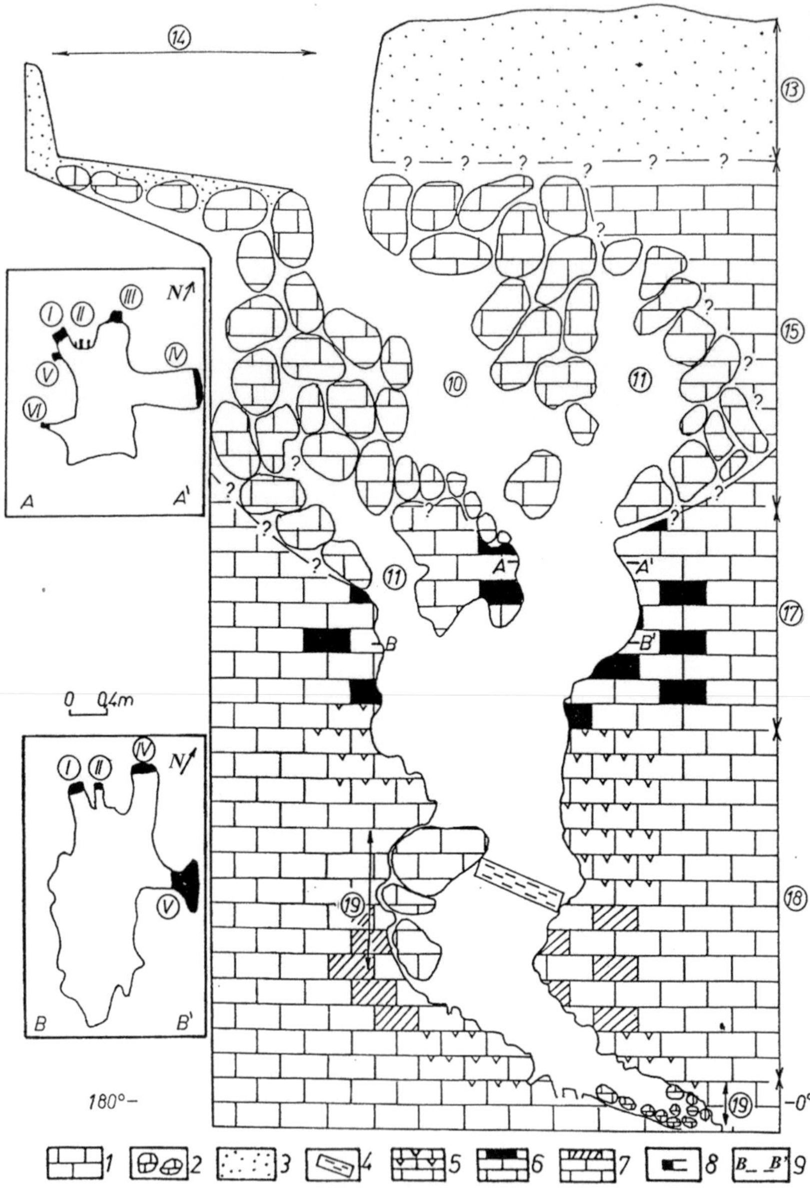

Fig. 7.31 Morphogenetic map of the shaft of the subsidence doline Gy-12 (Hárskút Basin). *1* Limestone; *2* collapsed debris; *3* cover deposit (loess); *4* wooden scaffolding; *5* zone of unfilled primary ruined grikes; *6* zone of primary ruined grikes filled with soil; *7* zone of primary ruined grikes partially filled with clay, redeposited gravel and rock debris; *8* primary ruined grike with soil fill (with identification code on cross-section); *9* site of profile; *10* main shaft; *11* subsidiary blind shaft; *12* zone of opening up; *13* collapse in the cover deposit; *14* surface depression formed by collapse and sheet wash; *15* collapse in the bearing rock; *16* zone of blind shaft formation; *17* zone of solutional blind shaft formation by upward expanding solution; *18* blind shaft detail formed by lateral coalescence due to solution (collapsed material have been partly extracted); *19* collapsed material redeposited to deeper position

Fig. 7.32 Solutional features on the wall of the shaft of the doline Gy-12

The morphology and pattern of the resulting shafts as well as that of the dolines above the shafts are controlled by the structure of the bedrock rock, detailed below.

Shafts form solely along fractures, solely along bedding planes or alternately along fractures and bedding planes (Veress 1982, Fig. 7.36). If solution takes place along fractures (with rock strata horizontal or dipping at less than ca 10°), solitary vertical shafts result (Figs. 7.36Ia and 7.37d). If the density of fractures is high, there is an increased probability that several adjacent shafts or subsidiary shafts branching out from the main shaft develop (Fig. 7.36Ib).

The shaft formed along a bedding plane will be oblique (Figs. 7.36IIa and 7.37a). Shafts (caves) along oblique bedding planes emerge, as our measurements show, if the dip of strata is 11°–42°. Shaft formation can happen along strata with high dip or vertical position (Fig. 7.36IIIa). In this case also vertical or subvertical shafts develop through solution along the bedding plane. The shaft will be elongated in the direction of the strike of rock strata.

If the dip is higher (ca 11°–42°), but the rock is jointed, shafts develop at several levels above one another along bedding planes since water flow is conducted upon the bedding planes of lower position along the shaft sections following fractures. Composite shafts may result (Figs. 7.36IIb, 7.37b, 7.38, 7.39 and 7.40). These shaft systems are composed of oblique shaft sections formed along bedding planes and

Fig. 7.33 Blind shaft and grike remnants of the doline Gy-3 (Hárskút Basin). *1* Blind shaft, *2* grike remnant

subvertical shaft details originated along fractures. The topmost section of the shaft system is joined with the doline. The shaft section formed along fractures terminates above the shaft section of higher position and creates blind shafts or blind chimneys connecting the various parts of the shaft systems of different position (Figs. 7.37b, 7.38, 7.39 and 7.40).

If there is no doline above the blind shafts of the shaft system and, thus, there is no pointlike water conduction, they can only develop from bottom to top by pressing the water filling the shaft against the ceiling of the blind shaft. The solutional effect of the water pressed against cave ceilings is well known in literature (Bretz 1942, 1956; Renault 1968; Ford 1971; Slabe 1995). This phenomenon is caused by the infilling of the passages of the phreatic zone. Passages grow upwards until the karst water table is risen by impoundment. For the shaft systems, it happens if the bottom part of the shaft is filled in with sediments or the section along the bedding plane below the shaft is clogged by sediment. The blind shaft of the shaft system can be filled with sediment if the composite shaft is filled with the inflowing water (as the doline of the shaft system receives much water during intensive activity see Chap. 6 and Fig. 6.10). The blind shaft is able to develop until the impounded water table reaches the ceiling. If the blind shaft is properly close to the surface, a new doline may take shape (Fig. 7.40I).

Fig. 7.34 Grike remnants on the wall of the shaft of the doline Gy-3. *1* Remnant of dividing wall, *2* collapsed material, *3* soil fill in ruined grike

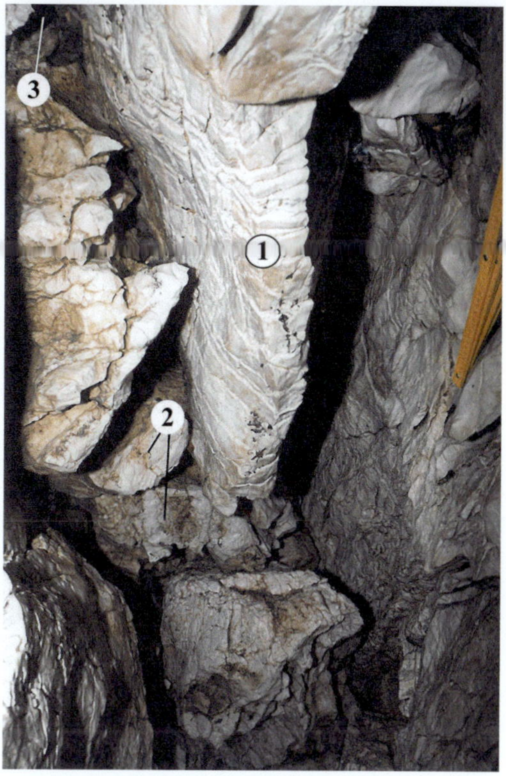

As a result, a doline row develops. The oldest of the dolines is at the starting point of the shaft system, while the younger dolines are located above the former blind chimneys of the shaft system. The resulting doline rows mainly form on valley floors. The blind shafts of the system may also connect with shafts above blind shafts (Fig. 7.40II), exemplified by Alba Regia Cave (Fig. 7.39). In this case too, doline rows form on the valley floor.

On covered karst, composite shaft systems often come about. Examples can be mentioned from the former or present-day covered karst of the Murge karst region (Apulia, Southern Italy) (Parise 2011).

Shaft system also develops in the case of vertical strata where the shaft sections along different bedding planes are connected by shaft sections formed along oblique fractures (Fig. 7.36IIIb).

Relationship Between Shaft and Doline Morphology

The thinner is the cover, the better feature formed in it reflects the shape and size of the feature in the bedrock. This is particularly so if the compensation for the material deficit happens through collapse or fall of particles. The shafts of different

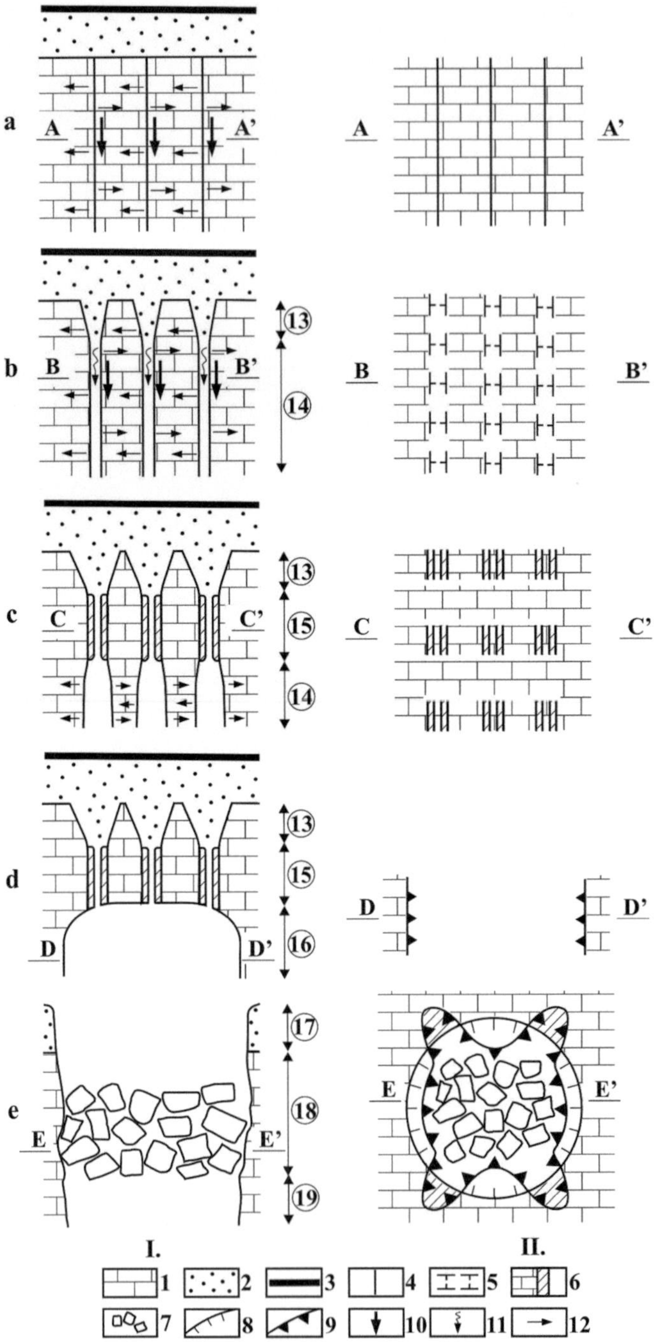

Fig. 7.35 Shaft and doline formation in cryptokarst. *1* Limestone; *2* cover deposit; *3* soil; *4* fracture; *5* primary grike; *6* cover and soil coating on rock wall or cover and soil fill in the grike; *7* collapsed material; *8* doline in the cover; *9* doline, shaft in the bedrock; *10* infiltration; *11* redeposition of the

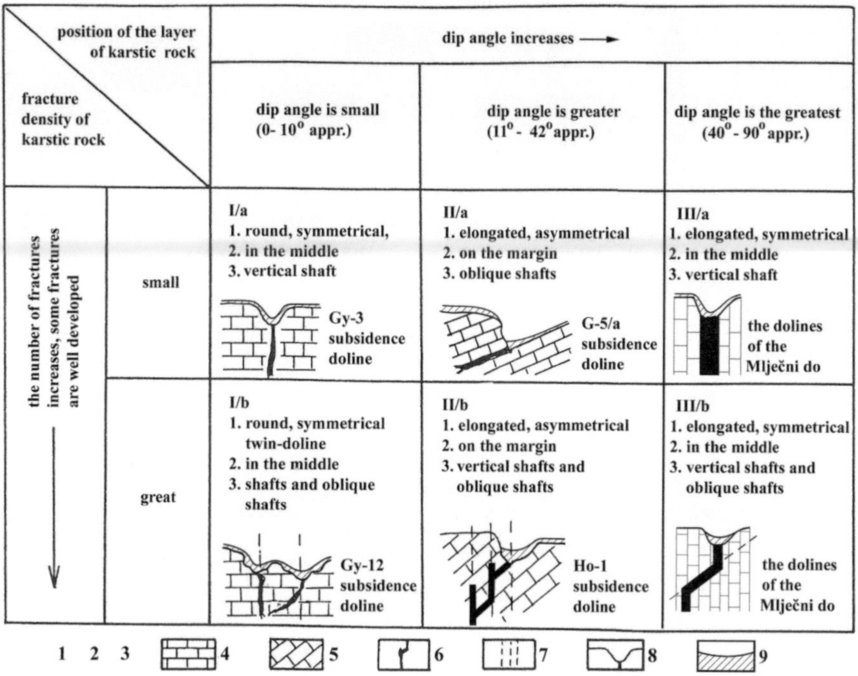

Fig. 7.36 Relationships between rock structure, shafts and karst landforms (the directions of theoretical profiles equals the dip of strata). *1* Shape of karst depression in plan and lateral view, *2* location of the conduit in the depression, *3* nature of the cave, *4* horizontal limestone strata, *5* tilted limestone strata, *6* cave entrance (shaft), *7* fracture planes, *8* karst depression, *9* superficial deposit

positions and the different patterns of shaft groups result in the formation of the following varieties in doline morphology:

- If the doline developed above a shaft formed along a solitary fracture, the doline will be circular and of symmetrical cross-section (Fig. 7.36Ia).
- Depending on the surface distance between the subsidiary shaft and the main shaft, the dolines developed above the shafts are distinct from each other (Fig. 7.41a), in marginal position (Fig. 7.41b), or coalescing (Fig. 7.41c).
- If the strata are (sub)vertical, the dolines are elongated in strike direction and arranged in rows in strike direction (Figs. 7.42 and 7.43).

Fig. 7.35 (continued) cover; *12* solution; *13* zone of grikes; *14* zone of broadening primary grikes; *15* zone of non-broadening primary grikes and shafts; *16* zone of secondary grikes and chimney (blind chimney); *17* doline formed in the cover; *18* collapse zone of the doline; *19* shaft of the doline. (**a**) Solution along fractures; (**b**) grike formation with primary grikes in their continuation; (**c**) on the upper part of the walls of primary grikes, coatings of cover material emerge, during which solution ceases and the grikes without coating are further broadening; (**d**) coalescence of broadening primary grikes by solution resulting in a secondary grike or a blind shaft; (**e**) the ceiling of the blind shaft caves in, collapse is inherited over the cover, and a dropout doline originates. *I* cross-section, *II* plan view

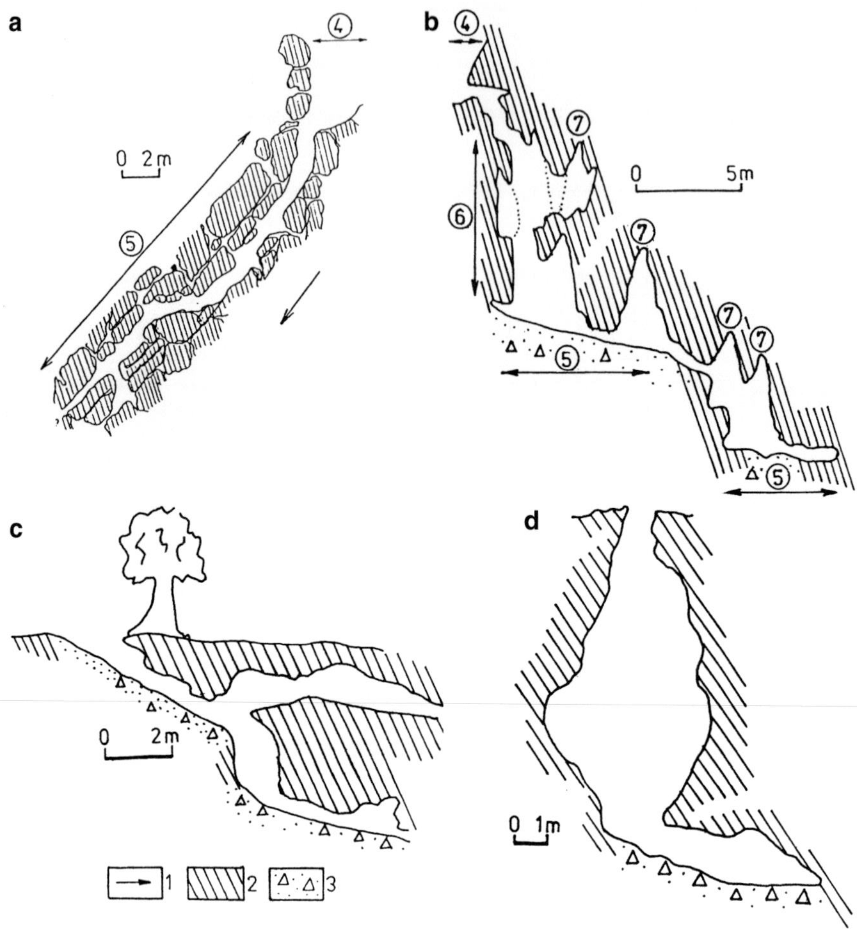

Fig. 7.37 Cave profiles from the Bakony Mountains: G-5/a (**a**), Ho-1 or Homód Valley cave (**b**), Gyenespuszta Cave, Hárskút Basin (**c**) and the Cseres Aven (**d**). *1* Dip direction of bearing rock, *2* bearing rock, *3* soil, debris, other redeposited deposit, *4* subsidence doline, *5* shaft or shaft detail formed along bedding plane, *6* shaft detail formed along fracture, *7* blind shaft

– If the dolines formed above a grike instead of a shaft, a row of elongated dolines (Fig. 7.44c), grike-like dolines (Fig. 7.44d) or rows of grike-like dolines results (Fig. 7.44h).

7.3.3.5 Inheritance over the Cover

The Process of Inheritance in General

The shape, size and density of features in the bedrock influence the formation of features on the cover. According to Cramer (1941), the formation of single doline is caused by the transport of cover deposit into several minor karren developed on the

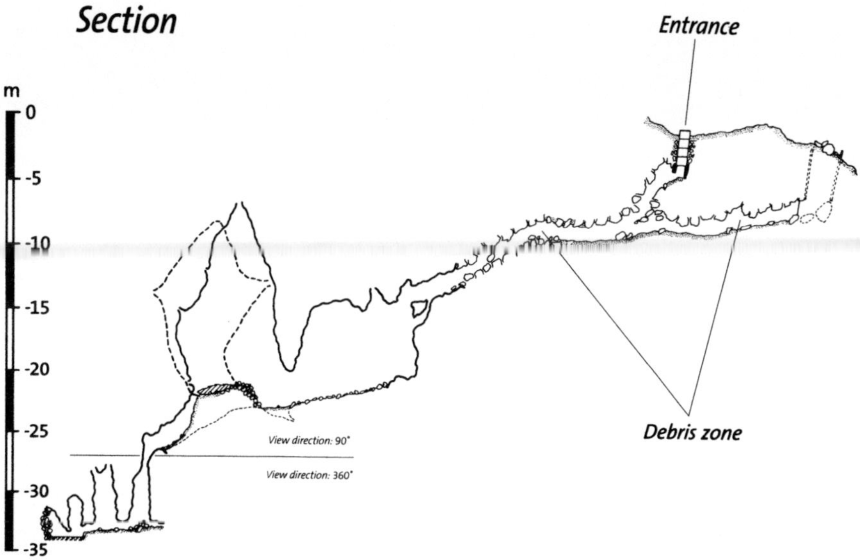

Fig. 7.38 Composite shaft under the covered karst ponor of Zsófiapuszta (Kab Mountain)

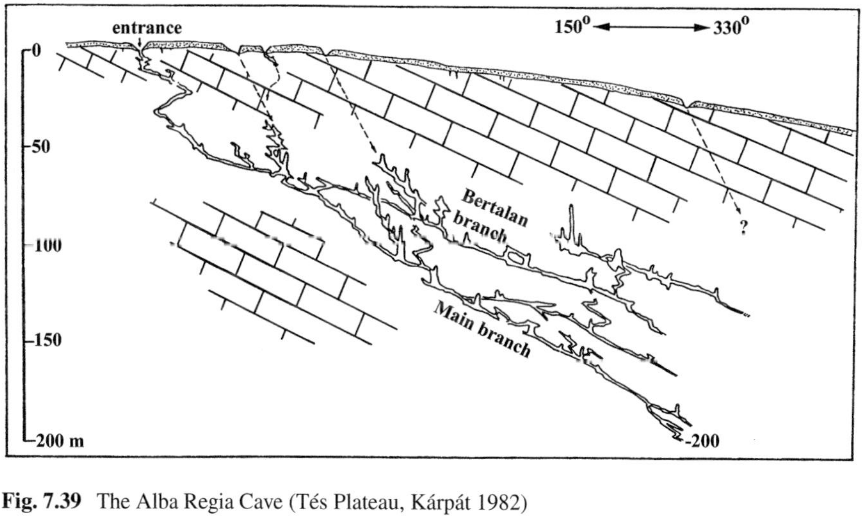

Fig. 7.39 The Alba Regia Cave (Tés Plateau, Kárpát 1982)

bedrock (Fig. 7.45). According to Clayton (1966), a depression develops on the cover because a depression develops on the surface of the bedrock by solution too (Fig. 7.46). On the cover a doline can even form above a single chimney or shaft of the bedrock (Jammal 1984) or above a network of grikes (Crawford 2001). It is common that the feature formed on the bedrock surface only conveys the cover material to the cavities of the deeper karst. The bearing form of the sediment can be composed of partial forms of different shape in a given site. According to Spooner

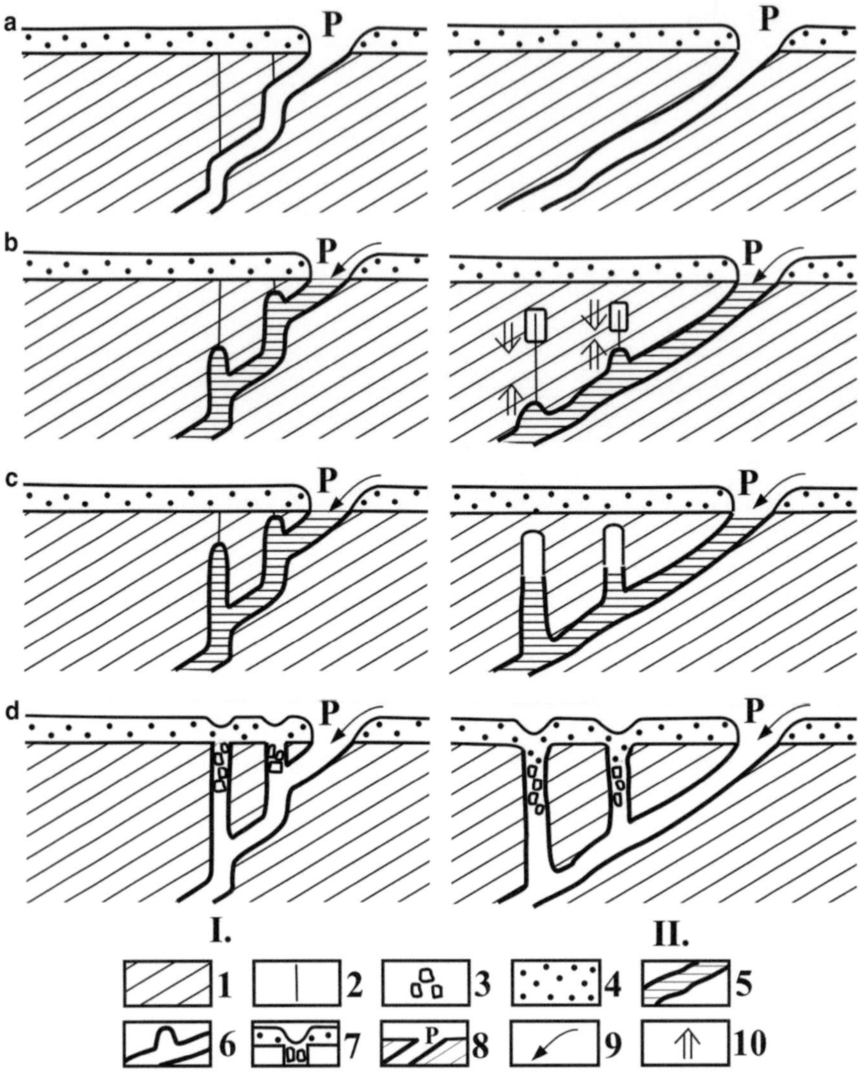

Fig. 7.40 Doline formation through extension of blind shafts of the composite shaft system towards the surface (*I*) and through coalescence with blind shafts developed close to the surface (*II*). (**a**) Formation of the composite shaft system of the doline (*I*) or its oblique shaft (*II*); (**b**, **c**) development of upward extending blind shafts of the composite shaft system (*I*); for the oblique shaft, the blind shafts starting from the oblique shaft extend in opposite directions from *bottom* to *top*, while the blind shafts close to the surface extend from top to bottom (*II*); (**d**) where the blind shafts reach the surface, dolines come about (*I*); dolines form where the shafts of two types coalesce (*II*). *1* Bedding plane, *2* fracture, *3* collapsed material, *4* cover, *5* shaft temporarily filled with water, *6* blind shaft, *7* subsidence doline, *8* subsidence doline functioning as ponor, *9* water recharge, *10* shaft growth

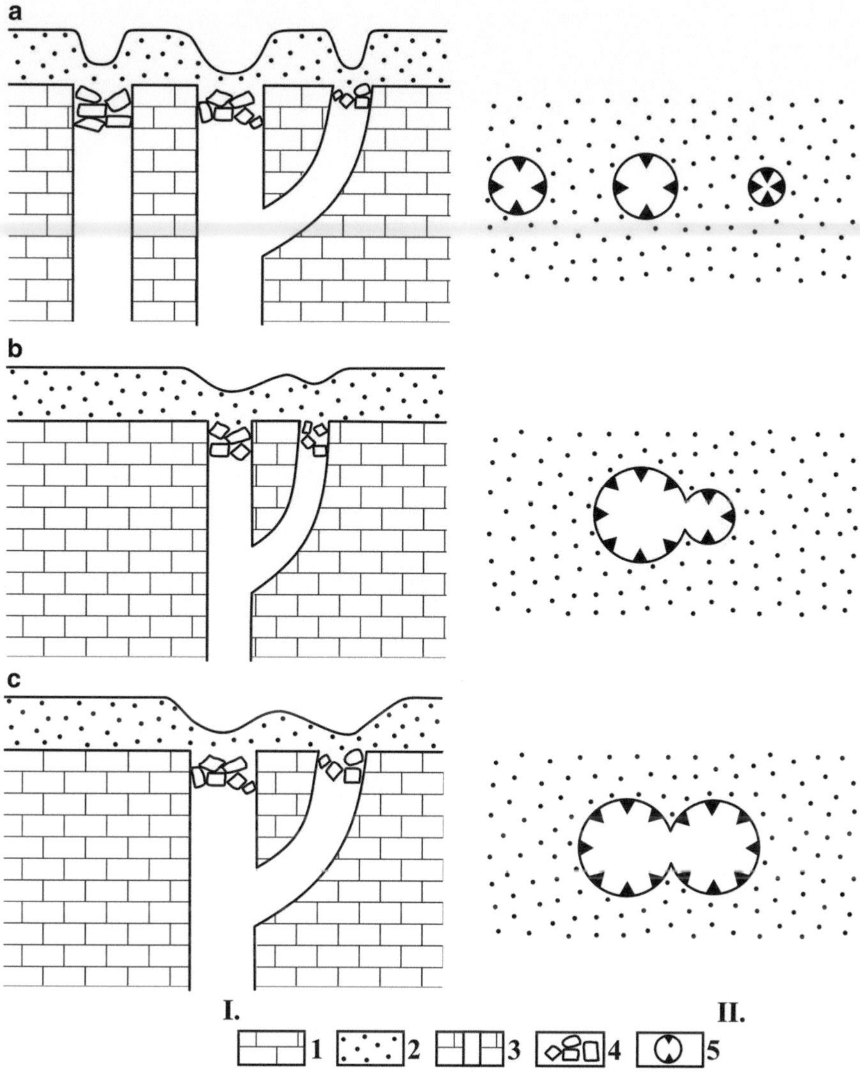

Fig. 7.41 Theoretical relationship between doline types and the chimneys of dolines. *1* Limestone; *2* cover; *3* chimney, shaft; *4* collapsed material; *5* doline in plan view. *I* lateral view, *II* cross-section. (**a**) With the relatively long distances between chimneys, dolines isolated from each other develop; (**b**, **c**) with higher density of chimneys or because of the subsidiary chimneys, marginal dolines (**b**) and uvalas (uvala-like features) (**c**) develop

Fig. 7.42 Solution doline formed along bedding plane (Mlječni do)

(1971), in Zambia for the emergence of a giant subsidence doline, 700,000 m^3 of cover deposit had to be transported. Out of this amount, 300,000 m^3 accumulated in a mine and 400,000 m^3 arrived into the joint network of the bedrock.

According to Waltham et al. (2005), the nature of the resulting landform on the cover and, thus, the nature of inheritance depend on the composition of the cover deposit. On non-cohesive rock, the material loss of the cover makes the surface sink (suffosion doline), while on cohesive rock, a cavity forms on the cover (Fig. 7.47). The cavity expands towards the surface since fragments of different size are detached from the ceiling (Beck 1991; White and White 1992; Currens et al. 2012). The collapse of the cavity leads to the formation of a dropout doline (Fig. 7.47). The cohesive cover is not a necessary condition to collapse and dropout doline formation. According to Beck (1991), White and White (1992) and Currens et al. (2012), cavity formation also takes place in non-cohesive cover. According to White (1988), suffosion dolines originate above shafts, while dropout dolines above grikes if there is a cavity in the cover. Beck and Sinclair (1986) claim that on the bedrock a cavity forms instead of a shaft, and the collapse of this cavity is followed by the cave-in of the cover. The collapse is induced by dropping karst water level.

Inheritance over the cover may take two ways: through direct or indirect inheritance. Direct inheritance takes place during the collapse of the bedrock (Fig. 7.48). A precondition to that is the large dimensions of the cavity of the bedrock and its

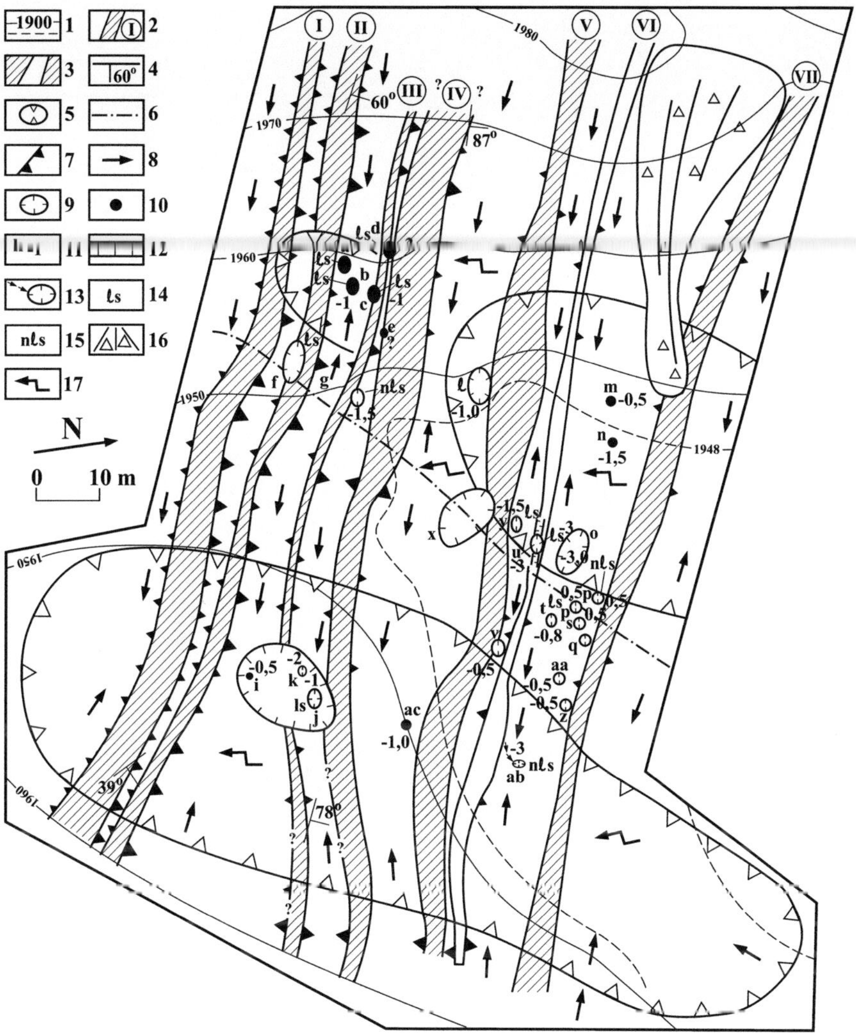

Fig. 7.43 Structural and morphological map of a part of Mlječni do. *1* Contour line, *2* thick bed, *3* series of thin siliceous beds, *4* strike and dip of bed, *5* rock basin, *6* top of roche moutonnée, *7* hogback (rock blade), *8* dip of surface, *9*, solution doline, *10* suffosion doline, *11* code and depth of doline, *12* grike, *13* solution doline functioning as ponor, *14* doline elongated in strike direction, *15* doline not aligned in strike direction, *16* debris fan, *17* direction of ice motion

proximity to the ground surface. With direct inheritance, the depression forms through collapse or partially through collapse even if in non-cohesive rock. The collapse of the blind chimney or cavity and of the cover happens simultaneously (Fig. 7.48). This takes place if the cover is thin.

With indirect inheritance, the material deficit of the bedrock gradually extends over the cover and on the bedrock collapse does not necessarily happens. This kind

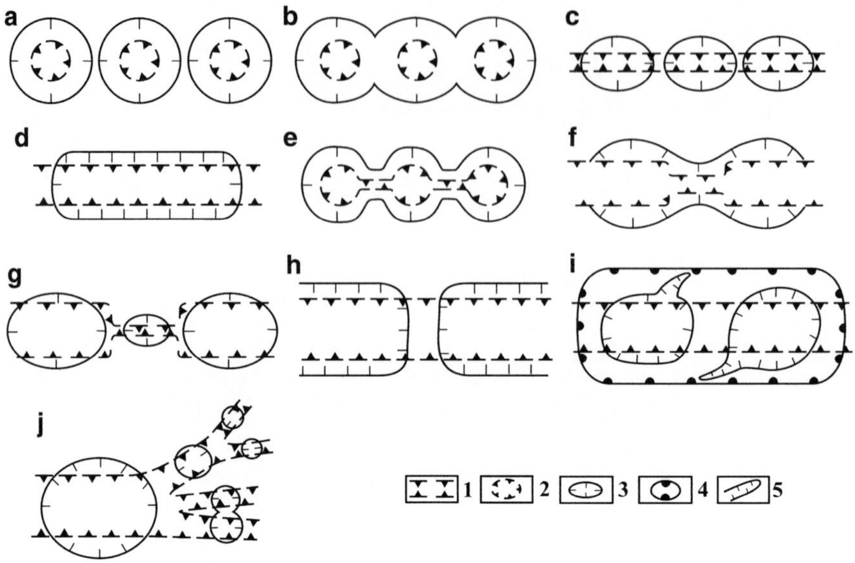

Fig. 7.44 Doline shapes and patterns controlled by solution features on the bedrock. *1* Covered grike; *2* covered chimney; *3* doline; *4* DSD; *5* gully, ravine. (**a**) Dolines above chimneys and shafts; (**b**) uvalas above chimneys close to one another; (**c**) row of elongated dolines formed above grikes; (**d**) elongated doline formed above grike; (**e**) uvalas formed above shafts at greater distance from each other; (**f**) elongated uvala formed above grikes of variable width; (**g**) row of elongated dolines formed above grikes of variable width; (**h**) row of elongated dolines formed above grike; (**i**) doline row and DSD formed above grike; (**j**) doline group formed above the branches of giant grikes (bogaz)

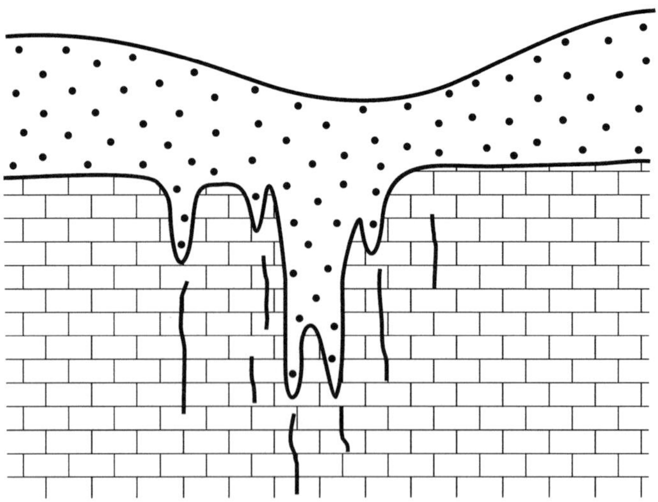

Fig. 7.45 Formation of subsidence doline (in previous terminology: alluvial doline) according to Cramer (1941) – modified

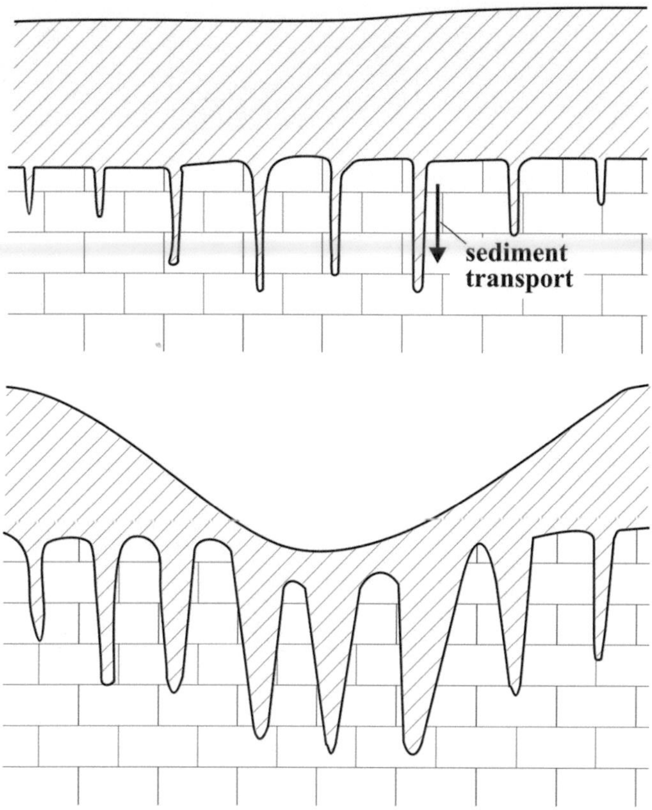

Fig. 7.46 Formation of subsidence dolines (shake holes) according to Clayton (1966) – modified

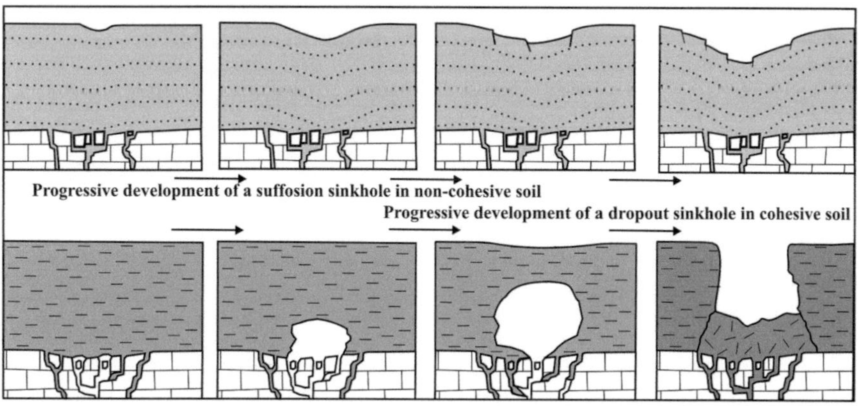

Fig. 7.47 Suffosion doline developed on non-cohesive cover (*top figures*) and dropout doline formed on cohesive cover (*bottom figures*), which was generated by the collapse of the cavity in the cover (Waltham et al. 2005)

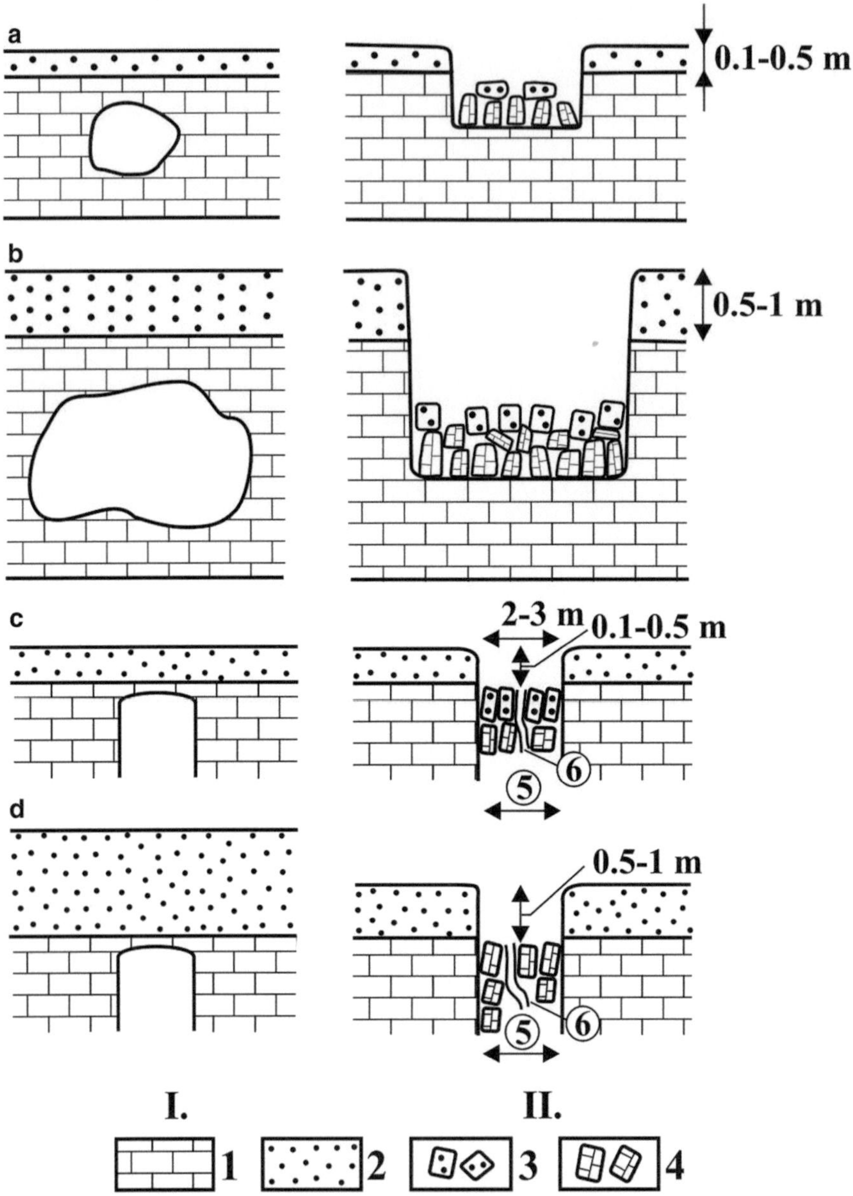

Fig. 7.48 Doline formation with direct inheritance. *1* Limestone; *2* cover; *3* collapsed cover material; *4* collapsed blocks of the bedrock; *5* opened-up chimney, shaft; *6* conduit in collapsed material. (**a**) With thin cover, small cavity in the bedrock collapses; shallow and steep-sided doline emerges with ruins of the cover and the collapsed material of the bedrock on its floor. (**b**) With thick cover, large cavity of the bedrock collapses: a doline of great depth and with steep sides develops with cover on its floor. (**c**) Opening up of blind shaft: doline with shallow depth and non-karstic pipe forms. (**d**) Opening up of the blind shaft with thick cover: steep-sided doline with great depth and non-karstic pipe forms. *I* initial state, *II* developed state

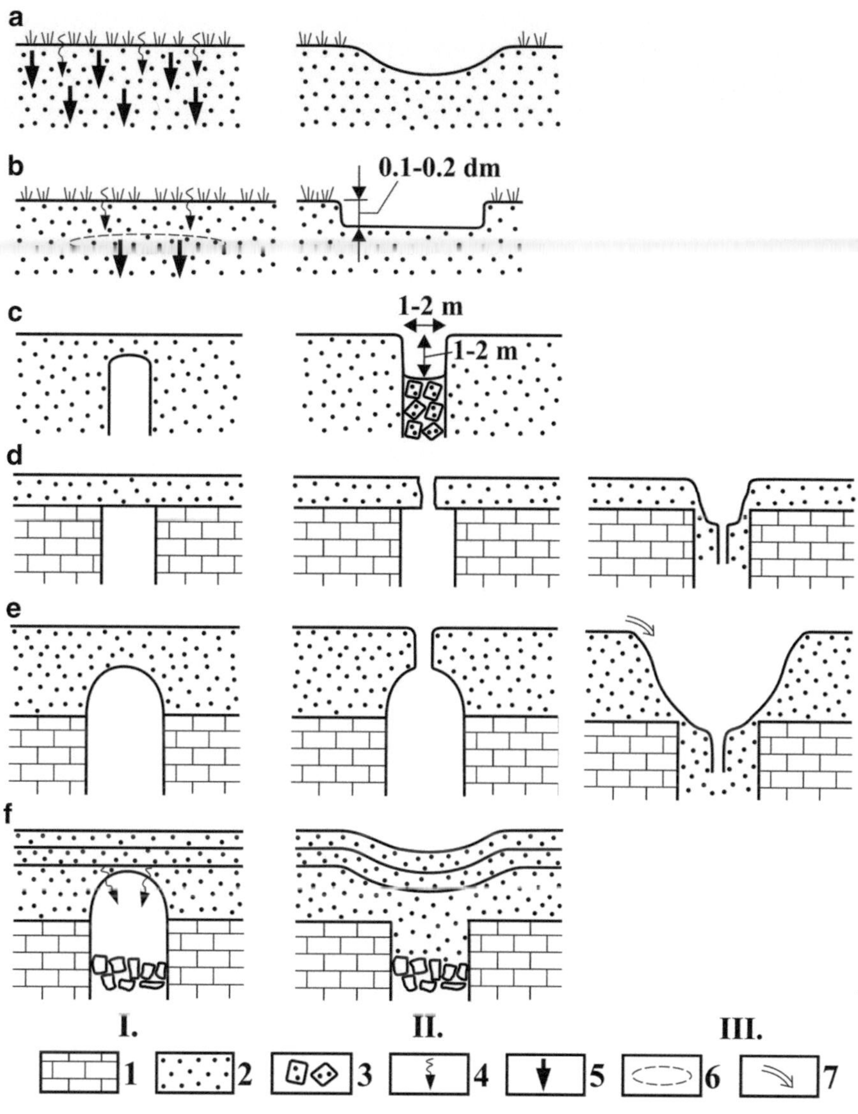

Fig. 7.49 Doline formation with indirect inheritance. *1* Limestone; *2* soil and cover deposit; *3* collapsed blocks of the cover; *4* infiltration; *5* suffosion; *6* part with high pore volume in the cover; *7* pluvial erosion. (**a**) Infiltrating water redeposits the cover from the surface into depth; (**b**) in the cover, an area with high pore volume is generated by suffosion into which the material above it drops (collapses), and a doline of small depth and steep sides forms; (**c**) steep-sided depression forms by the collapse of the non-karstic pipe; (**d**) in the thin cover, non-karstic pipe develops above the shaft, and the conduit is transformed into a doline of gentle walls; (**e**) the blind non-karstic pipe opens to the surface and the wall of the non-karstic pipe becomes more gentle; (**f**). the series above the ceiling of the non-karstic pipe sinks into the area of material deficit. *I* initial stage, *II* mature stage, *III* further evolved stage

of inheritance is mainly typical of thick cover or if a small-scale feature (shaft, cavity) forms on the bedrock. The direct inheritance is always associated with syngenetic karstification, while indirect inheritance can take place with both syngenetic and postgenetic karstification. Indirect inheritance can be accompanied by several types of movements in the cover: cover collapse, subsidence and suffosion (Fig. 7.49). With indirect inheritance, groundwater or karst water is often present.

Inheritance over the cover takes place either by way of subsidence or collapse. Subsidence creates dolines in three different manners: (1) Pore volume increases in the bottom part of the cover from where material removal can only happen. Then this part of the cover is compacting and compaction results in the bending of the top part of the cover. The extent of bending is becoming reduced downwards with beds located below one another (Fig. 7.50a). (2) The sediment filling the shaft is transported to depth and the whole thickness of the cover sinks into the unfilled shaft space. In this case the superimposed series are of similar bending (Fig. 7.50b). (3) Finally, if pore volume increases in the whole thickness of the cover, the cover is compacting in its whole thickness. The series are not bending, the cover is getting thinner, and the ground surface is subsiding during bending (Fig. 7.50c).

The doline may also develop from a non-karstic pipe if that opens up to the surface. Then the wall of the pipe is affected by pluvial erosion, the side slopes become gentler, and its diameter grows (Fig. 7.50d).

Collapse may be induced by the opening up of the blind chimney (Veress 2000) or the caving-in of the cavity in the cover (White 1988; Waltham et al. 2005). The material from the cavity is either directly transported into the bedrock (Waltham et al. 2005) or the cover under the series (in which the cavity developed) is removed. Such processes were described, for instance, from Michigan state, where the sand below the clay was transported into the grikes of the gypsum bedrock and material deficit resulted leading to cavity and doline formation (Benson and Kaufmann 2001).

The cover mostly collapses if it contains a cavity or cavities. For the generation of a collapse, the emergence of a cavity is not a necessary condition. At the same time, if there exists a cavity in the cover, its formation must have been preceded by material transport (by suffosion, fall of grains). An important precondition to collapse is rapid material transport, large-scale material deficit and a cover of fine deposit, which is not necessarily cohesive. The cave-in of the cavity is promoted if the cavity is broad, and the cover is thin and loosened by water saturation or melting of ground ice. According to Beck (1991) and White and White (1992), in the

Fig. 7.50 (continued) *17* shaft fill. (**a**) Material transport from the bottom of the cover, where compaction takes place and the area above sinks; (**b**) subsidence of the cover into the chimney of the bedrock; (**c**) part of the cover is transported into the shaft, pore volume increases in the entire thickness of the cover; during compaction, the surface of the cover subsides; (**d**) denudation of the cover by surface waters; (**e**) the shaft of the bedrock caves in, and the collapse is inherited over the cover; (**f**) cavity formation in the cover and collapse of the cover above it; (**g**) ground ice in the cover melts above the chimney and the melted part collapses; (**h**) sinking karst water table, collapse of the cover (or subsidence or redeposition) into the shaft. *I* initial stage, *II–III* mature stage

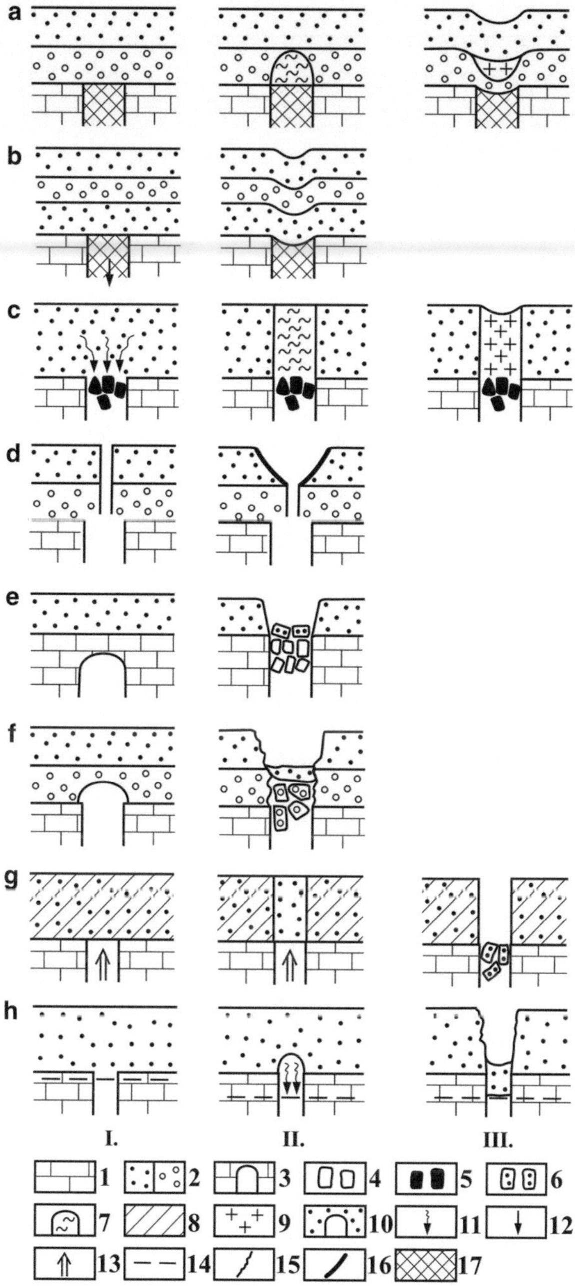

Fig. 7.50 Modes of doline formation. *1* Limestone, *2* cover deposit, *3* blind shaft, *4* collapsed material of the bedrock, *5* morainic deposit, *6* collapsed material of the cover, *7* loosened cover (with higher pore volume) due to material removal, *8* ground ice, *9* compacted cover, *10* non-karstic pipe, *11* water infiltration, *12* redeposition of cover deposit, *13* air current, *14* karst water table, *15* doline slope developed by collapse, *16* doline slope developed by the denudation of cover,

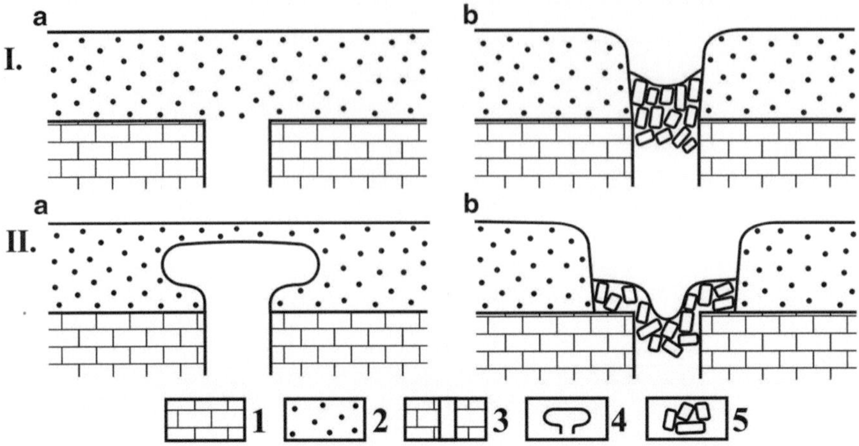

Fig. 7.51 Theoretical relationship between the size and mode of inheritance of dropout dolines. *I* With direct inheritance, the size of the doline depends on the size of the shaft of the doline; *II* with indirect inheritance, the size of the doline depends on the size of the cavity in the cover. **a** Initial stage, **b** mature stage. *1* Limestone; *2* cover; *3* chimney, shaft in the bedrock; *4* cavity in the cover; *5* collapsed blocks

beginning small cavities emerge in the cover (which can also be non-cohesive) and grow with increasing loading and reducing cohesion. The growth of cavities through the coalescence of small cavities is enhanced upwards (Drumm et al. 1990). The ceiling of the resulting large-scale cavity is wetted by percolating waters. The ceiling becomes heavier and in the sediment pore water pressure increases. As a consequence, fragments are detached from the ceiling (Currens et al. 2012; Waltham et al. 2005). The collapses of the cover are promoted by dropping water levels in the cavity. The emerging vacuum enhances suction (Xu and Zhao 1988) and also reduces the support of the ceiling. In our experiment, we have found a similar phenomenon when hot air flow was generated from below onto the cover with ground ice (Veress et al. 2012). From the bottom surface of the cover, melting detached fragments fell into the cavity under the cover. The collapse can be preceded by subsidence on the surface (Crawford 2001). In this case shear stress emerges in the cover above the cavities of the cover which causes subsidence of small extent.

There are four ways of doline formation by collapse: (1) In the case of a thin cover, the collapse of the blind shaft extends over the cover too. The cover deposit, which can also be non-cohesive, caves in into the shaft (Fig. 7.50e). (2) With thicker cover, a cavity forms in the cover (generated from a blind chimney or following the transport of the cover into existing shaft(s)) and collapses subsequently (Fig. 7.50f). The width of the resulting doline depends on the width of the collapsing cavity in the cover (Fig. 7.51). (3) Collapse or series of collapses occur if ground ice in the cover melts (Fig. 7.50g). The melting and the collapse of the cover were presented by a model experiment (see Chap. 3). (4) With the subsidence of karst water table (or groundwater), cavity may also originate in the cover and then collapse

(Fig. 7.50h). Naturally, dropping water level may also generate subsidence in the cover. At the bottom, the cover material is only partially transported away. With direct inheritance, the size of the doline is similar to that of the shaft in the bedrock, while with indirect inheritance, the doline can be larger than that (Fig. 7.51).

To the above presented modes of inheritance, however, supplements are needed for glaciokarst, permafrost and in the case of groundwater or karst water oscillates in the cover.

Special Cases of Inheritance

Inheritance on Glaciokarst

On glaciokarsts, subsidence dolines develop by the transportation of the cover (moraine) into grikes (Erhartič 2012) and into shafts (Ford and Williams 2007). If the cover consists of limestone debris, the solution of the bedrock depends on the saturation of the water percolating through the debris. The bedrock is only able to dissolve if the water percolating through the debris is not saturated at the surface of the bedrock. Therefore, material deficit on the bedrock emerges in the following ways:

- The cover is transported into grikes or shafts which formed previously before its accumulation (Ford and Williams 2007). The grikes and shafts were mostly created by meltwater from glaciers (Ford 1984). In this case dolines developed during postgenetic karstification. With postgenetic karstification, the morainic material or the debris is transported into shafts generated by glacial meltwater. This happens through the transport of finer substance into pores of the coarse material (Fig. 7.52a), or through the melting of ice (snow) cementing the debris of the shaft or filling its pores (Fig. 7.52b), and the material is transported into the place of the ice (snow). If there is ice under the debris in the shaft, its partial melting causes the displacement of debris to greater depths (Fig. 7.52c). In the case of a composite shaft, the debris fill reaches a deeper position through collapse or redeposition, and these processes lead to doline formation (Fig. 7.52d). Finally, if the top part of the shaft is broadening, the doline develops by the falling of fragments of the cover into the shaft (Fig. 7.52e).
- Dolines develop if the cover deposit is transported into presently evolving grikes or shafts. The solution of the bedrock has two reasons: the cover is either thicker than several metres but with low carbonate content or even if it is with high carbonate content but small thickness (see Sect. 7.3.3.3). In both cases an embryonic epikarst zone develops in the bedrock. Because this zone is underdeveloped, there is no swallow hole on the floor of the subsidence dolines. Subsidence dolines form if the debris is transported into the poorly developed joints, cracks and karren. Since the joints, cracks and karren occur in high density, doline density is also high, but because of the underdevelopment of the epikarst, their dimensions are small. The size and morphology of the dolines, and, therefore, also their degree of dissection, is probably a function of the number and extent

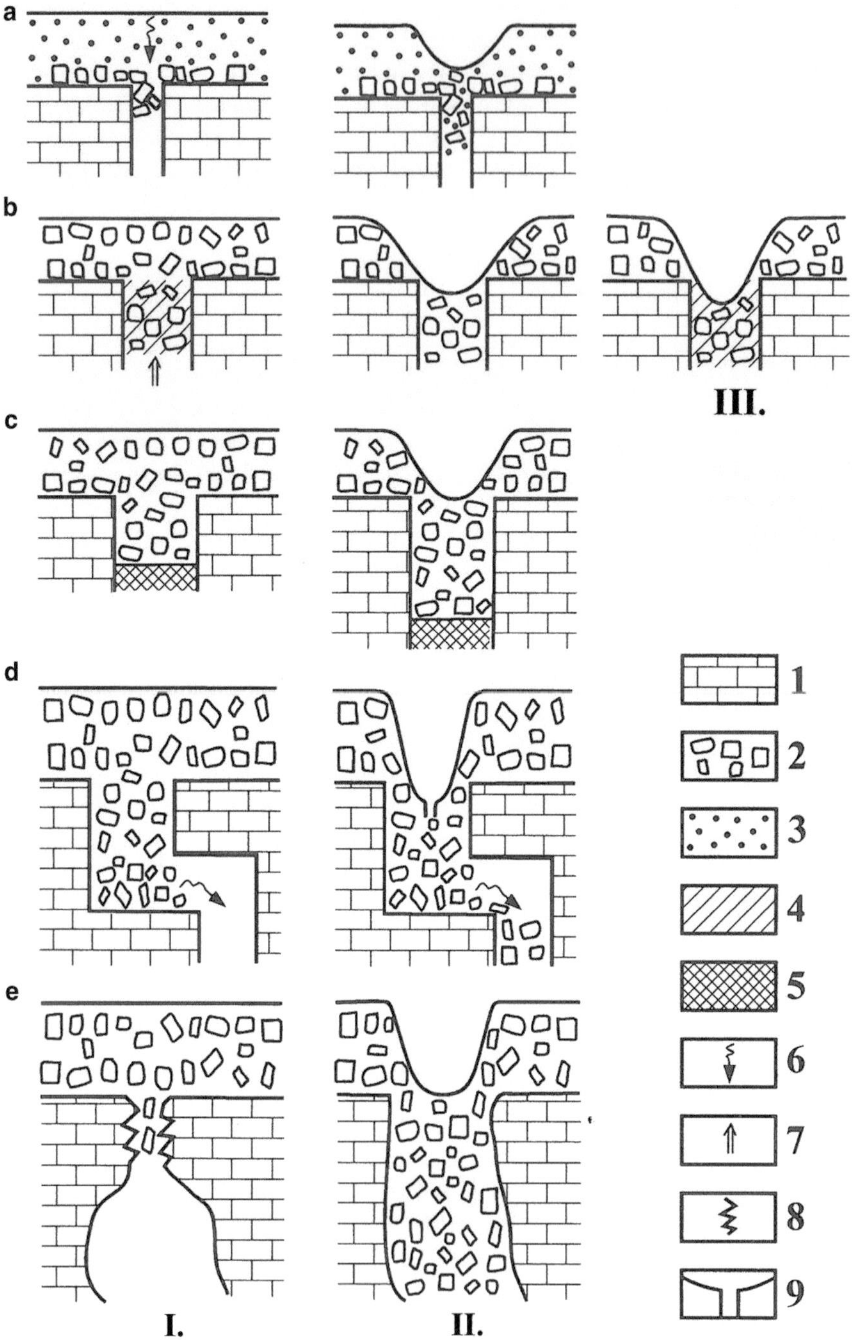

Fig. 7.52 Modes of postgenetic doline formation above grikes and shafts of the mountain karst. (**a**) The frost-shattered debris is transported into depth among rock blocks; (**b**) the ground ice between rock blocks melts and the blocks are displaced downwards in the grike; (**c**) under the

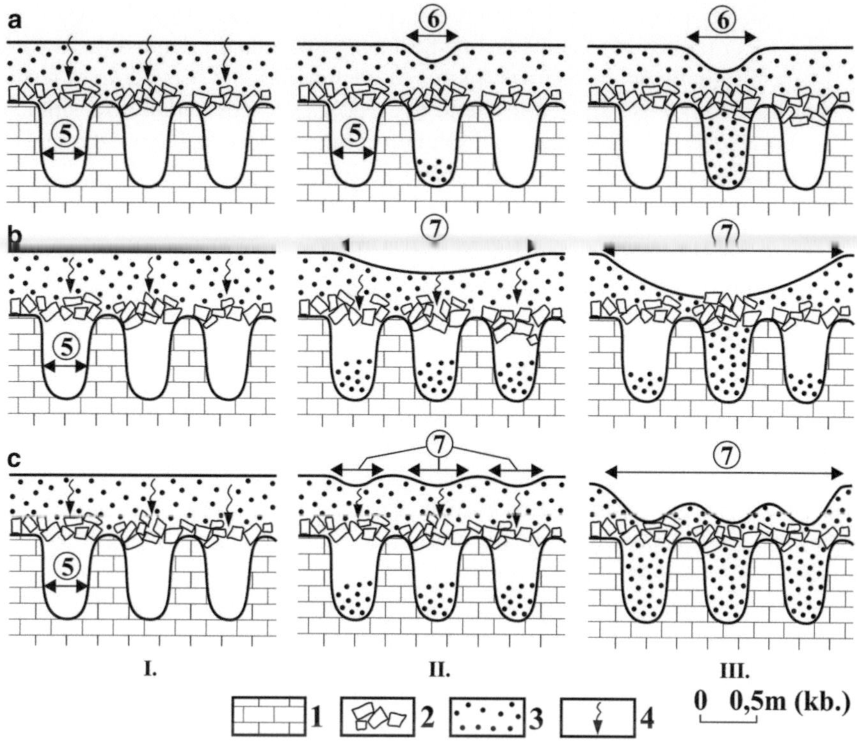

Fig. 7.53 Suffosion dolines formed on cover above grikes. *1* Limestone, *2* collapsed blocks or morainic deposits, *3* frost-shattered debris, *4* water infiltration and debris redeposition, *5* grike, *6* small doline with some metres diameter, *7* larger doline with some metres diameter, (**a**) small-scale doline above grike, (**b**) larger doline formed above several grikes, (**c**) development of larger doline with different rates of redeposition of debris into the grike, the floor of which will be uneven. *I* Initial stage, *II–III* doline formation

of infilling of the grikes (Fig. 7.53). If, during development, the bedrock features (grikes) under the dolines are filled with the cover deposit, their further growth is only possible if solution continues in the epikarst zone.

– Dolines may also form if limestone debris is generated from the dissolution of the uncovered limestone (covered with soil patches at most). Because of the short distance of percolation (solution), the water percolating through the limestone debris of low thickness does not become saturated even then. The dolines

Fig. 7.52 (continued) blocks, the top part of the snow and ice fill melt and the blocks above the fill are displaced downwards; (**d**) the collapsed material is redeposited from the vertical shaft section into the bottom of the shaft; (**e**) the top part of the shaft is broadened by solution and the collapsed material falls into the shaft. *1* Limestone, *2* collapsed and morainic material, *3* frost-shattered debris, *4* ground ice, *5* snow, *6* water infiltration, *7* upward air current, *8* surface formed by solution, *9* doline, *I* initial stage, *II–III* mature stage of the cover do not originate through transport of the debris into the shafts of the bedrock or falling into the shafts, but the debris sinks into the depression formed by local solution on the bedrock. In this case numerous low-depth dolines result and the bedrock is exposed in them. Such dolines are essentially transitional between covered karst dolines and solution dolines. Under such dolines, there is no capacity of the bedrock to store sediment (Figs. 4.6 and 5.65b).

On glaciokarsts dolines are often aligned in rows (see Chap. 4). The rows are sometimes perpendicular to the glacial trough (Fig. 7.54). Then in the foreland of the retreating glacier, grikes are often generated subparallel with the ice front. Suffosion dolines either form postgenetically – if the karren terrain is later buried – or syngenetically – if jointing happened after the burial with morainic deposits. Suffosion dolines are occasionally aligned in rows parallel with the glacial trough (Fig. 7.55). In this case the morainic deposits are reworked and accumulated in striations (widened by solution) (Fig. 7.55).

On glaciokarsts, doline morphology is controlled by the stratification of the bedrock or the terrains developed by glacial erosion. On schichttreppenkarst (Bögli 1964), the cover deposit accumulates along the cuestas and is transported into the grikes along the cuestas. Doline rows result with circular or elongated dolines (Fig. 7.56a). With broad grikes, the dolines are irregularly placed on the cover (Fig. 7.56b). In the case of vertical beds, glacial erosion created hogbacks (e.g. at Mlječni do). Below the cover deposit filling the depressions between hogbacks, solution along bedding planes takes place. The cover deposit is transported into the depressions between bedding planes. The ground surface subsides in the area of the patch of morainic deposit and a row of larger dolines emerges (Fig. 7.56c). If broader passages (grikes) come about along the bedding planes, rows of smaller dolines originate on the cover (Fig. 7.56d). Rows of elongated dolines appear on the cover of rock basins (Fig. 7.56e). If dolines develop on cover of roches moutonnées, they are of circular shape, smaller size and irregular arrangement (Fig. 7.56f).

Inheritance on Permafrost

It is a contradiction that on permafrost dolines occur (particularly on the taiga karst of Siberia) without the conditions of water flow being present. On taiga karst, ground ice is often exposed on the surface or lies close to the surface even in summer. According to Korzhuev (1961), karstification is mainly concentrated along fractures and faults, where water flow is possible. Low rates of water conduction are indicated by the presence of intermittent ponds in dolines in wet periods (Korzhuev 1961). Below the subsidence dolines along the Lena River, the cracks of the bedrock are filled in with cover material (Korzhuev 1961). The above does not only prove the dissolution of the bedrock material but also point to the suffosion of the cover. The mentioned author also claims that the permafrost is not uniform, but its horizons are separated by taliks. Water with high dissolved material content circulates indicating that the water of the taliks is capable of solution (consequently, karstification is active) or the taliks receive such water. The water of the karst springs along

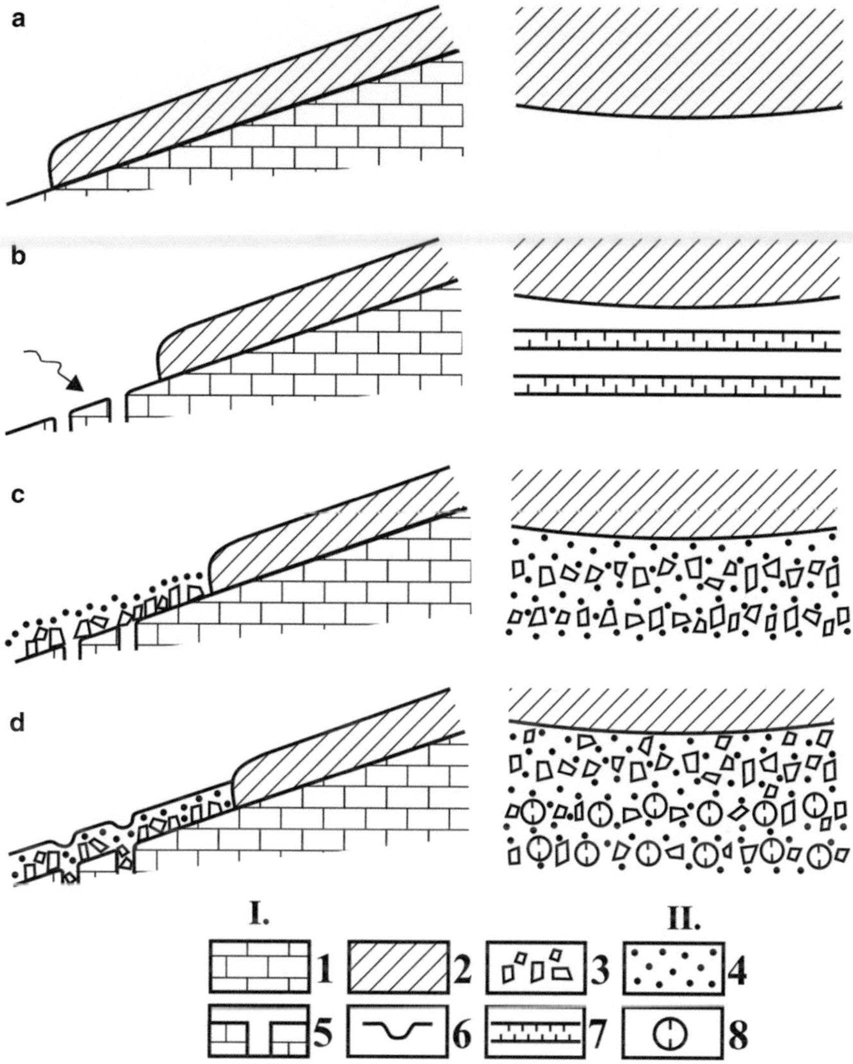

Fig. 7.54 Suffosion dolines formed above grike in the foreland of retreating glacier. *I* Lateral view: *1* limestone, *2* glacier, *3* blocks of mountain collapse or morainic deposits, *4* frost-shattered debris, *5* grike, *6* doline. *II* cross-section: *7* grike, *8* doline. (**a**, **b**) Grike formation in the foreland of retreating glacier; (**c**) the terrain of grikes is covered with the material of mountain collapse and by frost-shattered debris; (**d**) blocks fall into the grikes and dolines emerge on the surface

the Lena River indicates water flow and their dissolved material content points to solution. Numerous authors (Lauriol et al. 1997; Lungersgauzen 1966; Korzhuev 1972; Pulina 2005) are on the opinion that paleokarst occurs both on the tundra karst and taiga karst and paleokarst passages may also occur. According to Ford and

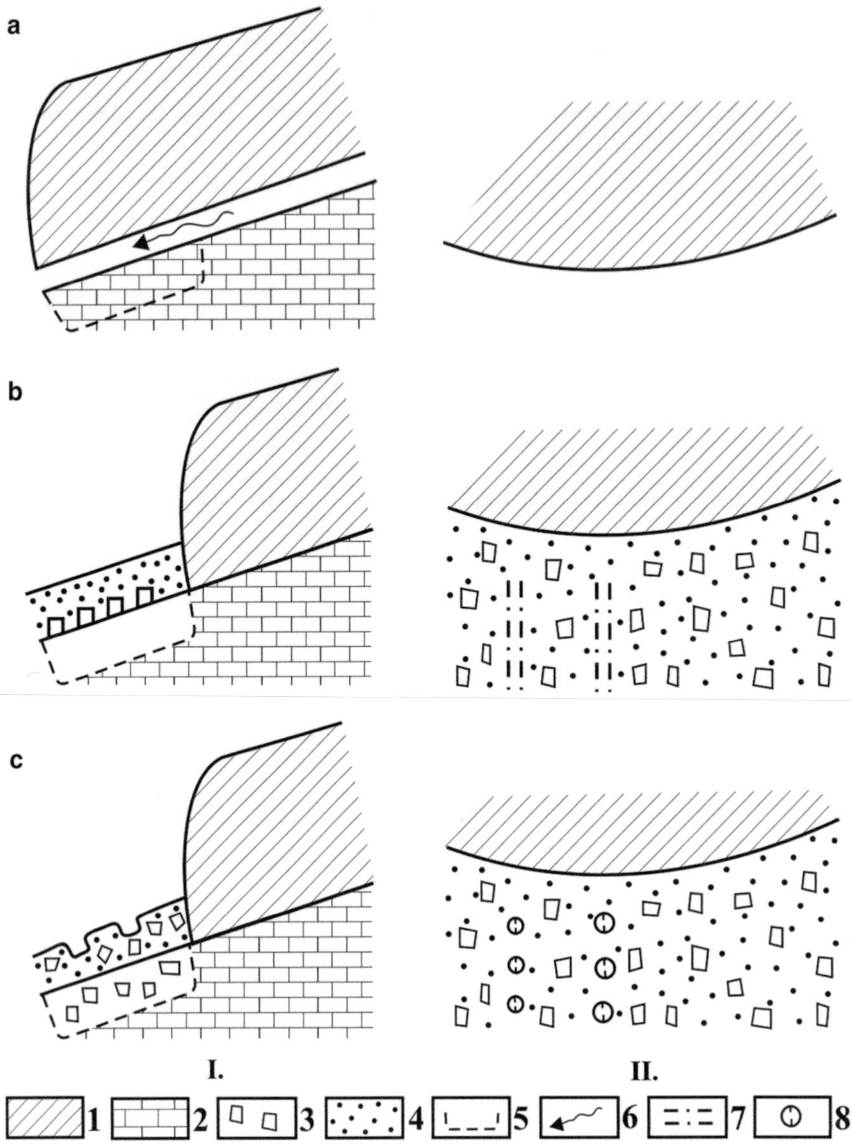

Fig. 7.55 Suffosion dolines formed above grike in glacial environment. *1* Glacier, *I* lateral view: *2* limestone, *3* morainic material, *4* frost-shattered debris, *5* grike in lateral view, *6* meltwater. *II* Plan view: *7* grike in plan view, *8* subsidence doline. (**a**) Grike formation under glacier; (**b**) with repeating glacier, grikes emerge and become buried under morainic material; (**c**) blocks falling into grikes and doline formation on the surface

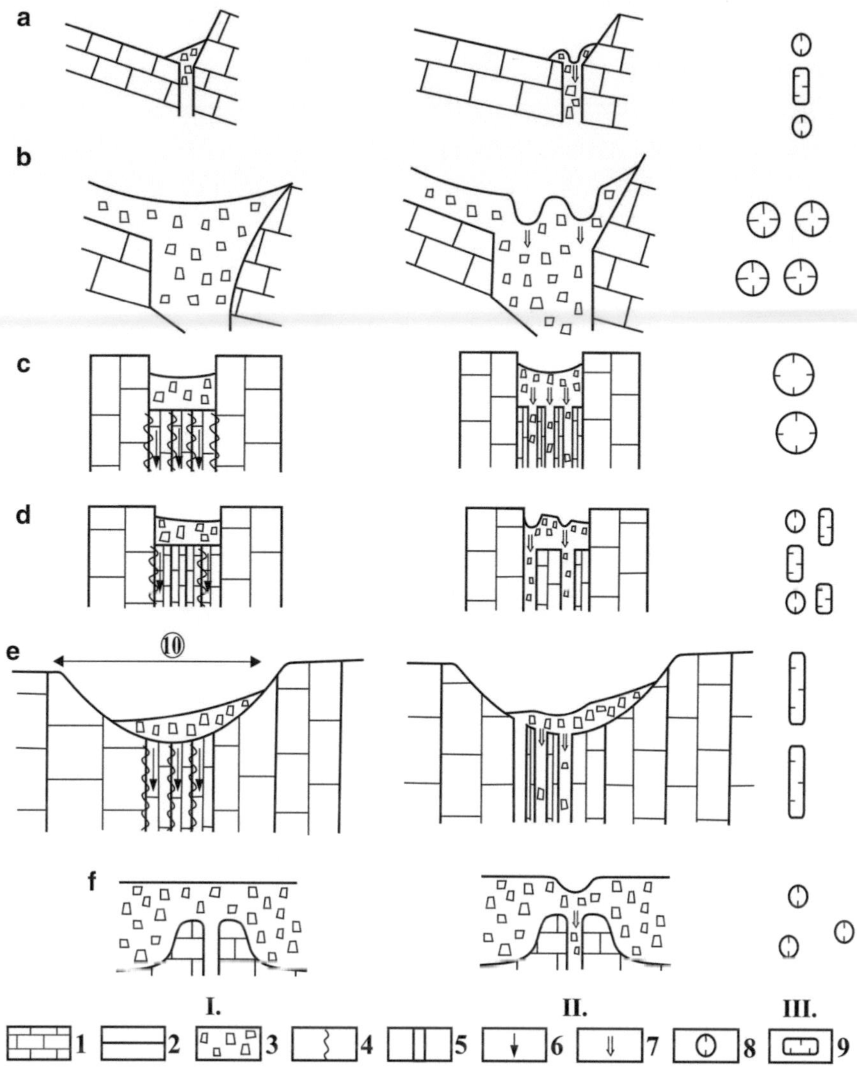

Fig. 7.56 Formation of subsidence dolines on terrains affected by glacial erosion. In cross-section: *I* initial stage, *II* mature stage, *III* mature stage in plan view, *1* limestone, *2* bedding plane, *3* morainic deposit, *4* solution along bedding plane, *5* solution grike along bedding plane, *6* water infiltration, *7* debris fall, *8* subsidence doline of circular groundplan, *9* subsidence doline of elongated groundplan, *10* rock basin. (**a**) Doline row of dolines with circular and elongated groundplan develops in the debris accumulating above the solution grike at the front of the cuesta; (**b**) doline row of large dolines with circular groundplan develops in the debris above giant grike at the front of the cuesta; (**c**) doline row of large dolines with circular groundplan develops in the debris covering the more denuded terrain on thin beds between hogbacks; (**d**) doline row of dolines with circular and elongated groundplan develops in the debris above thin beds between hogbacks; (**e**) doline row of dolines with elongated groundplan develops in the debris of a rock basin formed above the thin beds of terrain with hogbacks; (**f**) dolines with circular groundplan develop in the debris above chimneys formed on roches moutonnées. Remark: In the parts **a**, **b** of the figure, glacial cuestas are shown on terrain with strata of 10°–20° dip, while in the **c**–**e** parts, it is shown that on the terrain built of subvertical strata (the profile in c and d is perpendicular to the strikes of strata, while in e it is in the direction of strike), the ice created hogbacks with depressions between them. In the parts **c**–**e**, actual solution takes place along the bedding planes

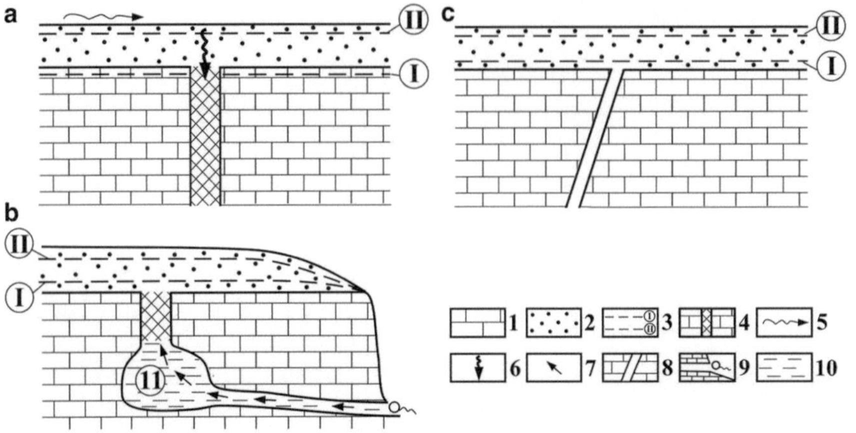

Fig. 7.57 Possible ways of conduit formation on permafrost. *1* Limestone, *2* permeable cover, *3* upper limit of permafrost in summer (*I*) and in winter (*II*), *4* ice-filled paleokarst conduit, *5* surface watercourse, *6* water infiltration, *7* air current, *8* joint, *9* spring with paleokarst conduit leading to it, *10* water, *11* talik. (**a**) Water percolating on the surface infiltrates and melts ice in the paleokarst conduit; (**b**) air current from the karst towards the surface melts ice in the paleokarst conduit; (**c**) the joint is not filled with ice to the effect of infiltrating water and ascending air

Williams (2007), the ice plug of karst passages (paleokarst passages) can be breached by surface waters and dolines result from this process. The possible origins of the passage are shown on Fig. 7.57. The three possible ways of doline formation are the following:

1. The ice fill of the paleokarst passage melts to the effect of water percolation from the surface. The permafrost thaws around the passage leading to ground subsidence and material transport by suffosion (Fig. 7.58I).
2. Doline formation on valley floor happens if patches of cover deposit and exposed limestone alternate (Fig. 7.58II). Water percolating at the contact melts ground ice and the surface of the permafrost subsides. Percolating from the resulting depression into the limestone, the water creates a passage. Through the erosion of side slopes, the depression is broadening (Fig. 7.58IIc).
3. Doline formation is also mainly concentrated on the valley floor if the cover is of variable thickness there. Water flows all along the valley floor, infiltrates and melts the top part of the ground ice (Fig. 7.59b), and the cover subsides. The subsidence is more intensive where the cover is thicker. The water percolating from the cover into the bedrock generates a passage in the limestone into which the cover material is redeposited through suffosion (Fig. 7.59c).

Inheritance with Intermittent Inundation of the Cover

The decisive role of change (lowering) of water level is proven by the effect of water pumping on the process (Sinclair 1982). Subsidence dolines form by water extraction in large numbers within a short time, for instance, more than a hundred emerged

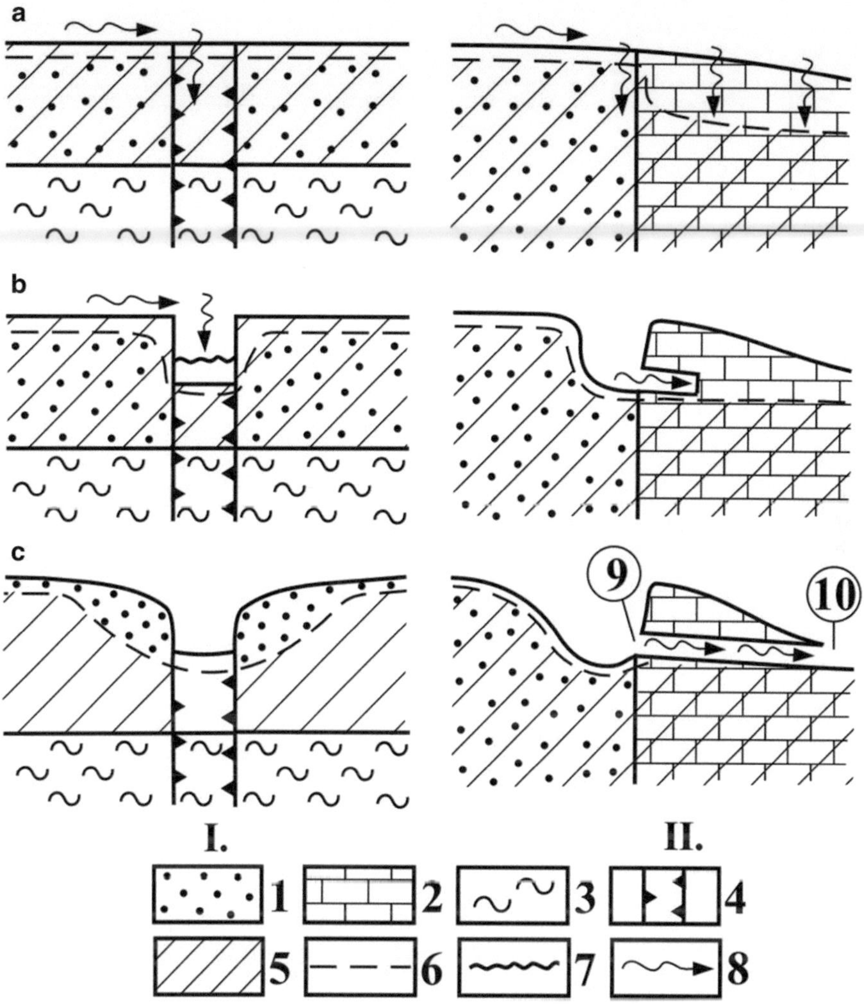

Fig. 7.58 Doline formation on permafrost. *I* Postgenetic karstification, *II* syngenetic karstification, *1* cover deposit, *2* limestone, *3* talik, *4* paleokarst conduit, *5* permafrost, *6* permafrost surface, *7* lake, *8* water flow and percolation, *9* conduit, *10* spring, *I* doline formed at paleokarst conduit: (**a**, **b**) ice in the conduit melts at the top and melting extends over the bordering cover; (**c**) subsidence of the cover, exposure of the paleokarst conduit in the interior of the resulting depression. *II* doline formed on rock boundary: (**a**) water infiltrating at limestone outcrop locally melts ground ice; (**b**) ground ice melting extends laterally over to the cover deposit and because of thawing the cover surface sinks; (**c**) formation of embryonic conduit in the limestone by solution

in Pennsylvania state (Foose 1953), more than 2000 in Alabama state (LaMoreaux and Newton 1986) and more than 10,000 in China (Enkou mine area, Hunan province) (Li et al. 1983) due to water extraction. The chance of their development is particularly high in the wet season (Waltham et al. 2005). At the same time,

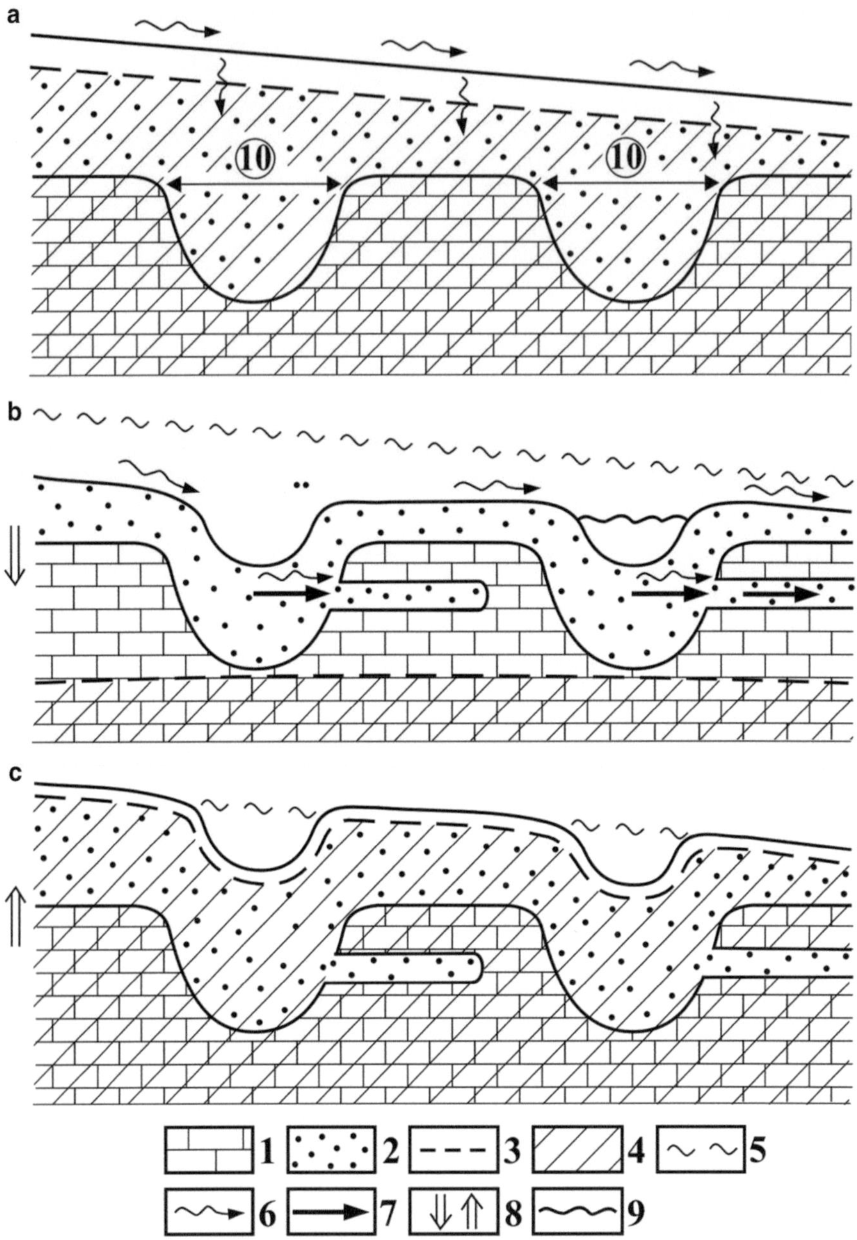

Fig. 7.59 Doline formation on permafrost at the local thickening of the cover. *1* Limestone, *2* cover, *3* surface of permafrost, *4* permafrost, *5* initial surface, *6* water flow and percolation, *7* material transport (through redeposition and solution), *8* surface subsidence and then to repeated rise due to the thawing of permafrost, *9* lake, *10* paleokarst doline; (**a**) in early summer, to the effect of warming and infiltrating waters, the permafrost melts and the surface sinks; (**b**) by late summer, the cover thaws in its whole thickness, and water percolates from the developing doline into the limestone, where it dissolves the rock and transports the deposit under the doline into the rock; subsidence will be more intensive where the cover is thicker (at paleodoline); (**c**) in winter permafrost re-emerges in the cover, which is rising; in the dolines of the cover, this rise is more limited than elsewhere since in summer material is removed from the cover below the doline

inundation reduces this chance, for instance, if a lake is established on the surface (Beese and Creed 1995). Doline also forms if the water loss of the storing system is not due to water extraction but to other human activity. Thus, the water is conducted into a tunnel (Guo 1991) or a mine (Yuan 1987; Xu and Zhao 1988). In the latter instance, the process can only be stopped by abandoning the mining activity (Chen 1988). Experience shows that subsidence dolines developed on the very first day of pumping in the eastern United States (Newton 1987) or within some days in Florida (Bengtsson 1987), Alabama (Newton and Hyde 1971) or China (Waltham 1989). Occasionally dolines take shape several months or even years after water pumping (Foose 1953, 1968; Sowers 1996). This type of origin is especially typical on inselberg karsts and particularly on its intermountain plains (Chen 1988; Guo 1991; Waltham and Smart 1988). Dolines mostly develop if the water table sinks below the bedrock surface, and the rate of infiltration from the surface is considerable (Foose 1968; LaMoreaux and Newton 1986; Lei et al. 2001). The drop of water table is an important factor too. Thus, on the Guizhou karst, 6.8 m drop of water table is not yet suitable to generate doline formation, but at a drop of 24 m, dolines emerge (He et al. 2003). The origin of dropout dolines with water table dropping is explained by the water loss of the cavity of the cover (Waltham et al. 2005). Following the water table decline, compaction dolines can also develop with the compaction of the cover (Jennings 1966).

There is intensive doline formation if water loss occurs both in the cover and the bedrock – either for natural or artificial reasons. This happens if the karst water (groundwater) table is near the surface (e.g. in Florida) and its drop is of large scale. Pore volume change is induced by the water loss and gain accompanying the fluctuation of water level, particularly on fine-grained cover because of water outflow from the capillaries. This generates cracking in the cover. Along the cracks, water flows from the cover into the bedrock – as confirmed by our laboratory experiments (see Sect. 7.3.1). Under natural conditions, the desiccation of the cover also induces cracking, particularly in sand, which promotes that water should reach the bedrock (Tharp 1999).

Dolines of similar origin particularly occur on the intermountain plains of inselberg karsts and at the katavothra of poljes (Figs. 5.12a, b, 5.92a, b and 5.93). At these sites, karst water table is not only close to the surface, but its level may fluctuate considerably. The stages of landform evolution above katavothra are presented in Fig. 7.60.

7.3.3.6 Development of Doline Varieties

The origin of the different dropout doline varieties depends on cover thickness and the way of inheritance. In the case of direct inheritance, the cover caves in through the opening up of the shaft in the bedrock (Figs. 7.48c, d and 7.50e). This is how D_4 dropout dolines originate (Figs. 5.44 and 5.60). The D_3 variety dolines formed on evaporites have a similar origin, but the doline is deeper since a larger shaft develops in the bedrock (Fig. 5.59).

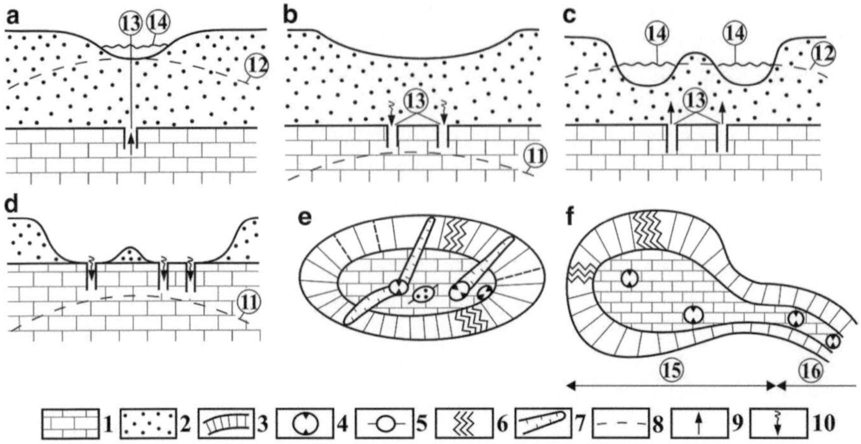

Fig. 7.60 Types of depression formed above katavothra (based on examples from the Cerkniško polje). *1* Limestone, *2* cover deposit, *3* side slope of depression, *4* katavothron, *5* mound, *6* rainwater rill, *7* gully, *8* karst water table, *9* loosening of the cover during water flow above the karst water, *10* descending karst water transporting the cover into the conduit, *11* low karst water table, *12* high karst water table, *13* shaft, karst passage, *14* intermittent lake, *15* basin-like section of DSD, *16* valley section of DSD, (**a**, **b**) doline formation above the katavothron, (**c**) formation of doline group above the katavothron, (**d**) DSD formation above katavothra, (**e**) denudation of the cover by rainwater rills and gullies around the DSD, (**f**) valley formation by retreat from the DSD (**a**–**d** lateral view; **e**, **f** plan view)

The D_1 variety dolines emerge in thick cover, also through direct inheritance, attested by the bedrock exposed in dolines (Fig. 5.57). Direct inheritance is allowed by the large width of the blind shaft of the bedrock. With the shaft opening up, the cover immediately loses its stability. Thus, collapse is not preceded by cavity formation in the cover. If during inheritance from the bedrock over the cover a non-karstic pipe develops, the collapse of its ceiling leads to D_5 variety dolines (Figs. 5.29a and 5.61). The origin of D_6 variety dolines is explained from their morphology. Relative to their depth, their diameter is larger. On their floors, the bedrock or its debris is not exposed. Therefore, they exclusively result from the collapse of a near-surface cavity in the cover with large width and low height.

In thicker cover, larger dropout dolines (D_2 variety) come about. The cover is cohesive (Waltham et al. 2005). The cavity in the cover either forms through the broadening of the shaft or through the loss of shaft fill (postgenetic karstification and postgenetic doline). On non-cohesive rock, the doline can form by opening up a blind shaft (Fig. 5.33) and a cavity develops, which rapidly extends towards the surface.

S_1 variety dolines occur if there are large-scale cavities on the bedrock. If the cover is not too thick, a large doline results (Beck and Sinclair 1986). The mentioned authors present the formation of a subsidence doline associated with a large cavity of the bedrock from Florida state. The dolines are broadening through the denudation of their side slopes.

The S_2 variety dolines are of young age. Their small size also results from the material deficit in the cover. They occur where the cover above the non-karstic pipe sinks.

S_3 variety dolines emerge on the floors of previously formed S_1 variety dolines closely associated with the material turnover of the bearing (S_1 variety) doline. The part of the cover above the non-karstic pipe which took shape in the fill of the bearing doline subsides.

S_4 and S_5 variety dolines develop on morainic deposits. Their origin is due to small-scale solution on the bedrock under the cover of limestone debris. They are most typical on glaciokarsts (see Sect. 7.3.3.5.2).

The side slopes of S_6 variety dolines with different inclination either result from variation in denudation or from the fact that heads of bed form their steeper slopes. The latter are most typical on the cuestas of glaciokarsts, but also occur on mediterranean karsts.

S_7 variety dolines are elongated landforms. They either develop through the linear, instead of pointlike, removal of the cover or if at the end of the existing gully, S_2 or S_4 variety dolines emerge and the gully will have no outlet.

Variety S_8 is a fossil doline. The clogging and subsequent infilling and inactivation of the doline (with ceasing water conduction from the doline into the karst) and the accompanying sedimentation were presented in Chap. 6.

7.3.3.7 Doline Evolution

General Characteristics of Doline Evolution

Doline evolution can be described by the analyses of changes in doline type and dimensions, of renewed doline formation and of the relationships between the doline and its environs. Doline width is limited by doline depth (Waltham et al. 2005).

A close relationship was presented between depth and width on two karst areas too (Fig. 3.4). It can be seen that the increase of width relative to the depth is different on the two karst areas. The different denudation of the doline slopes probably plays a significant role in the fact that their slopes are different: the denudation of dolines of the Bakony Mountains is greater since they developed in loess than the dolines of Pádis plateau which developed in debris with sandstone. It can also be caused by the fact that the deepening of the dolines of the Pádis plateau is larger. It can be explained by the greater material transport from the floors of the dolines because of more cavities in the bedrock.

It can also be experienced that on both areas the width of some dolines which are about 2 m deep exceeds the width of dolines which is maximum 8 m in case of such depth according to literature (Waltham et al. 2005). Therefore, the cause of these such great doline widths is not the denudation of doline slopes to a large extent, but that the surface subsides on a great area even at the beginning of doline development.

It is also remarkable that the width of dolines with a depth of 2 m in the Bakony Mountains is similar, while on Pádis plateau, the width of dolines increases by the growth of their depth. Also these differences probably have genetic causes.

As far as dimensions are concerned, small-scale (suffosion and dropout) dolines form and fossilise or develop into those with medium size. The medium-size infilled dolines can be fossilised or S_3 variety dolines emerge in their interior. The medium-size suffosion and dropout dolines expand into large suffosion dolines. In the interior and on the margin of large dolines also S_3 variety dolines come about (Fig. 7.61). Doline growth ceases sooner or later and infilling and then burial begin. The existence of former dolines is attested by the lenticular interbeddings of the cover deposit. The lenticular fill could be covered (Fig. 7.62, VES measurement site R45). On the infilling surface, renewed doline formation can take place. In Fig. 7.63 at the VES measurement site R142, there are two lenticular interbeddings above each other. The top surface of the lower lenticular interbedding is less bent than the lower surface. This indicates more intensive suffosion in the series under the lenticular interbedding than from the bearing series of the lenticular interbedding or the infilling of the doline and the development of a new doline in this fill. The upper lenticular interbedding (bearing the present-day doline) shows that on the more and more intensively infilling surface, a new (second or third) doline emerged. The fill of this doline is ever thicker towards the centre. Therefore, suffosion also had a lower rate here (the upper lenticular interbedding) than below, or the doline was filled in or then reactivated. Consequently, at the VES measurement site 142, as many as three or four dolines could have been originated (including the present doline).

Another possible way of doline evolution is demonstrated in Fig. 7.64. The suffosion doline is expanding and then inner dolines appear in its interior. The doline system fills up and then a new doline develops in the main doline which can be filled up later too. The evolution of the doline and its environs happens in three possible ways:

- The doline exists for a longer period, and the bounding surface is not filled up or not eroded below its level. In this case, the terrain between neighbouring dolines is lowered by pluvial erosion on the joint catchment (if no watercourse is associated with the dolines). The slopes are directed towards the dolines (Fig. 7.65a). During pluvial erosion, the terrains sloping towards the dolines are broadening (Fig. 7.65b) until they merge (Fig. 7.65c), a divide emerges between them, and the joint catchment is divided into two.
- In the area of the infilled fossil doline and its environs, denudation takes place. The sediment fill of the doline is thinning out and a new doline may form in the fill (Fig. 7.66I).
- There is accumulation in the environs of the doline. The doline is filled up and then buries (Fig. 7.66II). During the infilling and burial, lenticular interbeddings come about one above the other in the cover deposit.

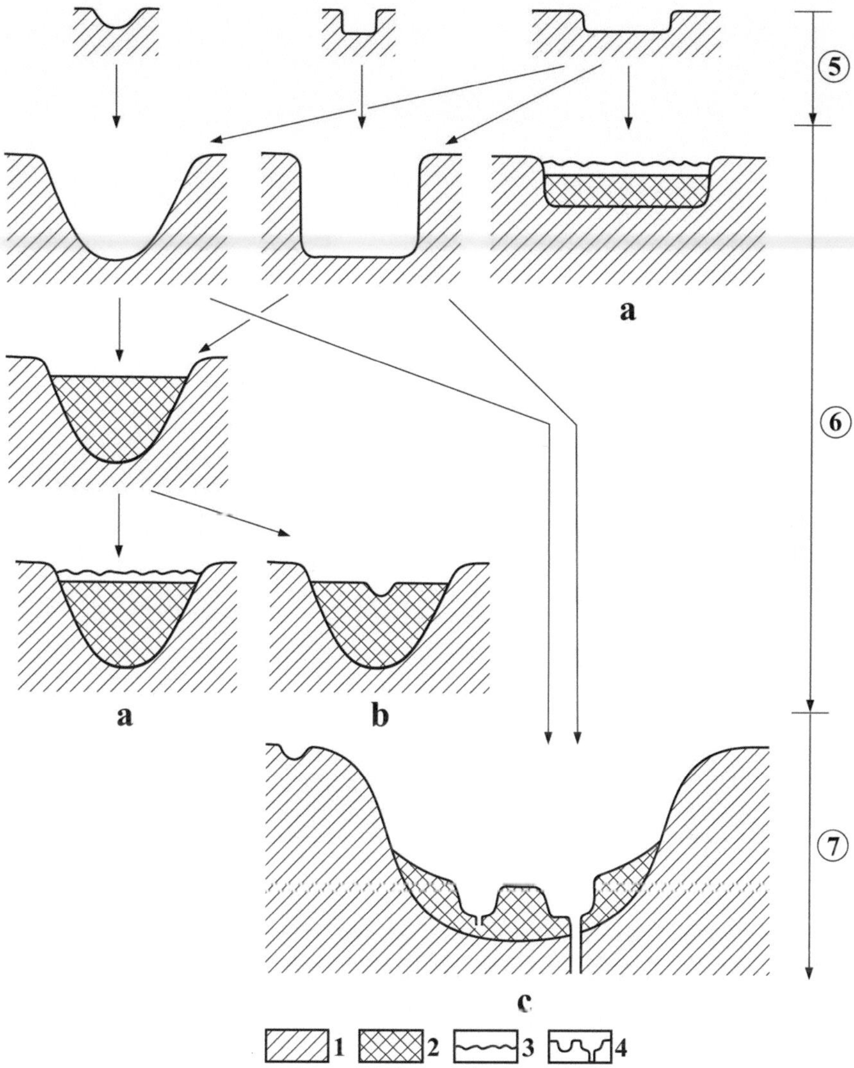

Fig. 7.61 Development and transformation of subsidence dolines. *1* Cover deposit, *2* doline fill, *3* lake, *4* S_3 variety suffosion doline, *5* small-scale doline (less than 1 m diameter and depth), *6* development of medium-size dolines (diameter 5–10 m, depth 1–5 m), *7* development of large-scale dolines; certain dolines are transformed into fossil dolines (**a**); others are filled up but then reactivate (**b**), while still others expand, are partially filled up and then reactivated (**c**)

Doline Evolution in Some Study Areas

The (L, M and K) parameters of subsidence dolines on glaciokarsts (Júlian Alps, Asiago Plateau, Totes Gebirge) and middle-mountain karsts (Pádis, Bakony, Mecsek) were analysed in the parameter space (Figs. 7.67 and 7.68). L shows doline

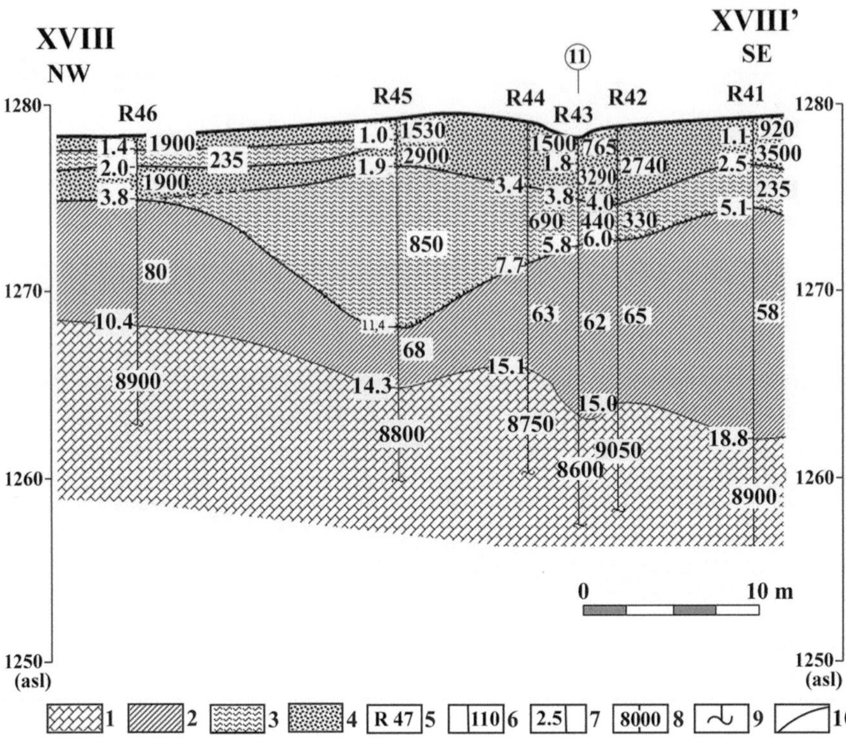

Fig. 7.62 The geoelectric profile XVIII–XVIII′ from the Padis-1 area. *1* Limestone, *2* clayey silt, *3* mixed rock debris (sandstone and limestone debris), *4* mixed rock debris (sand, sandstone and limestone debris), *5* VES measurement site, *6* geoelectric resistance of the series (Ohmm), *7* basal depth of the geoelectric series (m), *8* geoelectric resistance of the bedrock (Ohmm), *9* approximate penetration of VES measurement, *10* boundary of geoelectric series, *11* doline

depth, and M is a negative value characteristic of doline diameter, while K represents slope shape. The higher is K ($K \geq 1$), the more concave is the slope (or the shorter is the convex slope segment), the lower is its value, and the more convex is the slope (the longer is the convex slope segment).

It was found that the diameters of dolines on glaciokarsts show a wider range (from 0 to −1.25) than that for middle-mountain karsts (from 0 to –0.75), while their doline depths present a narrower range (from 1 to 4.8 m) than that for middle-mountain karsts (1–8 m). In harmony with our previous statement, it means that on glaciokarsts dolines with smaller diameter than on middle-mountain karsts also occur, while dolines in middle mountains can be deeper than those on glaciokarsts too.

Regarding the diameter/depth ratio, it is claimed – in accordance with literature data (Waltham et al. 2005) – that the dolines becoming ever deeper also get ever broader. The dolines on glaciokarsts, however, are deepening at a lower rate during broadening that those in middle mountains. This indicates that the cover on the floor

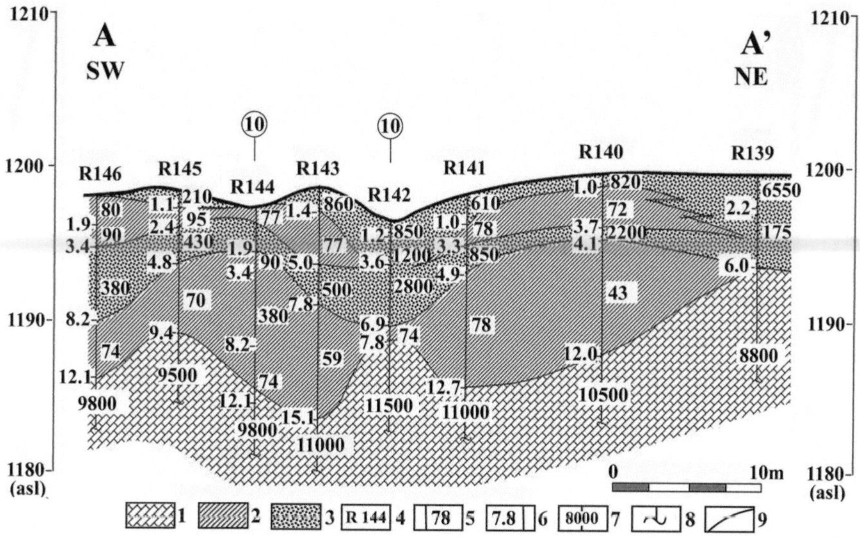

Fig. 7.63 The geoelectric profile A–A′ from the Padis-3 area (see also Fig. 4.45). *1* Limestone, *2* clayey silt, *3* mixed rock debris (sand, sandstone and limestone debris), *4* VES measurement site, *5* geoelectric resistance of the series (Ohmm), *6* basal depth of the geoelectric series (m), *7* geoelectric resistance of the bedrock (Ohmm), *8* approximate penetration of VES measurement, *9* boundary of geoelectric series, *10* subsidence doline

of the doline under them is either thinner (and thus hinders their deepening) or removed at a lower rate.

The average values of the K parameter so dolines on glaciokarsts (K = 1.54) are above those for dolines in middle mountains (K = 1.47). This indicates more concave side slopes in the case of the former. Most of the K values of dolines on glaciokarsts fall between 1 and 1.6, while for dolines in middle mountains are mostly between 1 and 1.9. Consequently, the slope shapes are more diverse for the latter, probably explained by rock properties. (The doline slopes of the latter are formed in various rocks and this favours variable shapes.) The K values of dolines on glaciokarsts show a remarkable narrow range, particularly if compared to the K values of the solution dolines in middle mountains with a range of 1–4.0.

Plotting the K parameter against the M or L parameters, K shows hardly any change. (Although on glaciokarsts, extreme K values also occur.) This indicates that slope shape is largely independent from both diameter and depth variation. Thus, the data in literature which point to doline slopes becoming gentler with doline growth are only valid in certain cases and for certain rocks (e.g. for the dolines in Florida state, the slopes of which are formed in sand).

At the same time, it is also observed that different K values belong to the same doline diameter or depth. The reason behind that could be that local denudation or lithology can modify slope shape.

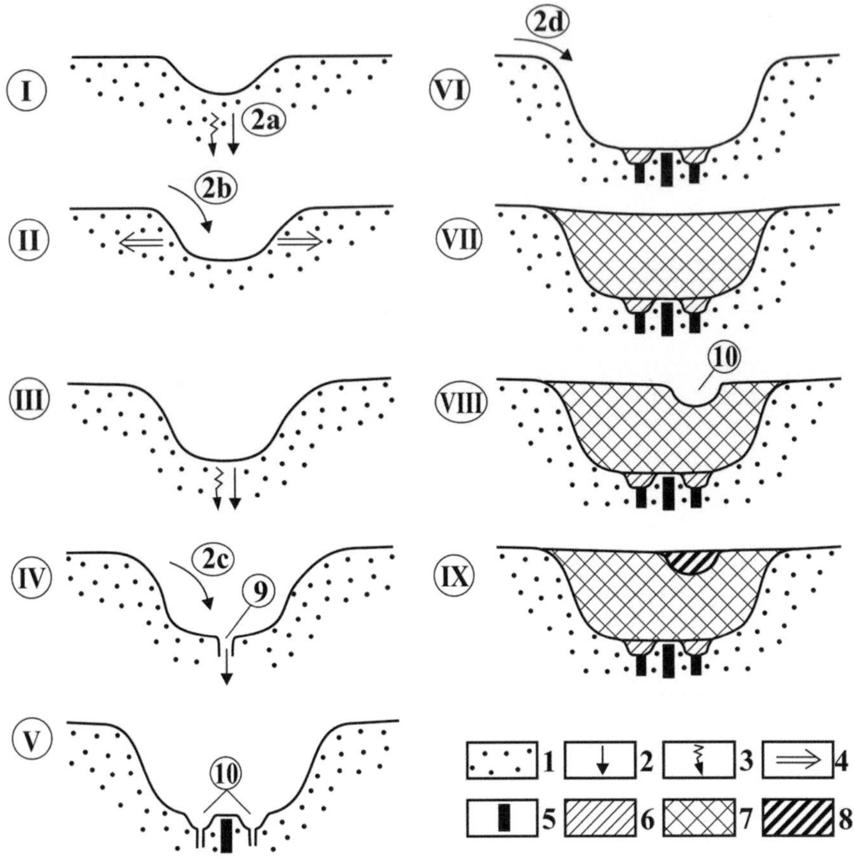

Fig. 7.64 Stages of suffosion doline formation. *1* Cover deposit, *2* material transport (*2.a* suffosion, *2.b* denudation of doline slope by pluvial erosion, *2c.* material removal from doline slope into the karst, *2.d* material input from the environs of the doline), *3* deepening of the doline, *4* broadening of the doline, *5* fill of non-karstic pipe, *6* fill of old doline, *7* fill of older doline, *8* fill of doline formed in the fill of the older doline, *9* non-karstic pipe, *10* dropout doline, *I* doline formation by subsidence, *II* doline broadening, *III* doline deepening, *IV* development of the non-karstic pipes of the doline with material removal from the doline through the pipe, *V* conduit infilling (dropout doline formation in the doline), *VI* infilling of dropout dolines, *VII* infilling of the suffosion doline, *VIII* formation of new doline in doline fill, *IX* infilling of the new doline

7.3.4 *Origin of Subsidence Pseudokarst Depressions*

The material deficit generating a depression derives from a non-karstic process. The material deficit is caused by collapse of the ceiling of non-karstic pipes (Fig. 7.69a) or suffosion if the top material of finer particles fills the pores of the bottom coarser fragments (Fig. 7.69b). Material transport can also take place in solution. The shallow loess dolines (Fig. 7.69c) are partly of this kind of origin – although in their material, loss suffosion plays a significant part.

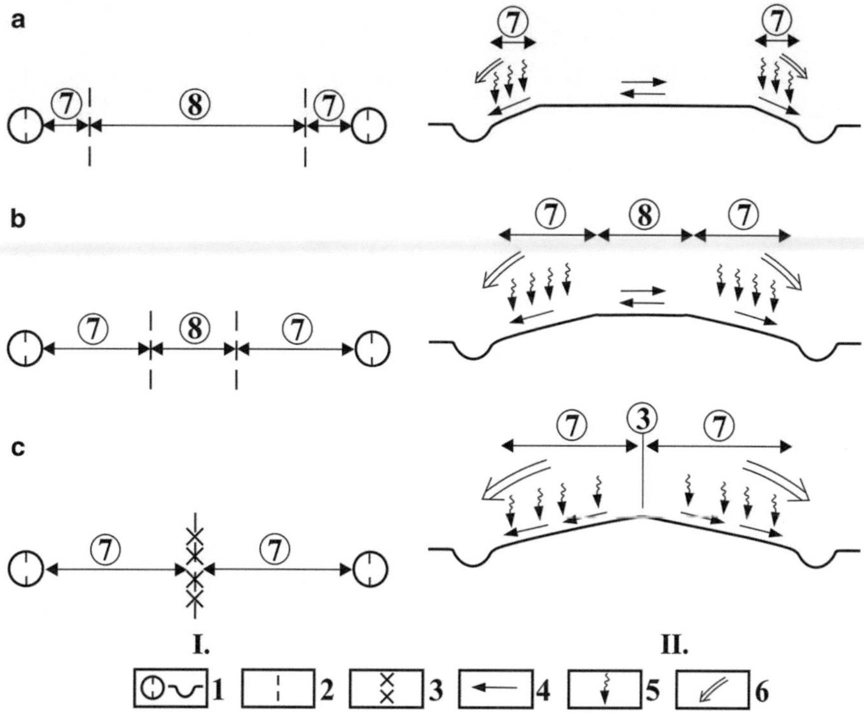

Fig. 7.65 Development of divide between dolines on covered karst. *1* Doline, *2* boundary of catchment of doline, *3* divide between dolines, *4* water flow, *5* pluvial erosion, *6* transport of cover deposit, *7* catchment of the doline, *8* joint catchment, *I* plan view, *II* cross-section. (**a**) Due to pluvial erosion, surfaces sloping towards the dolines come about (doline catchment); (**b**) doline catchments are broadening; the initial area (joint catchment) is shrinking; (**c**) the doline catchments consume the joint catchment

In volcanic regions where enduring snow (firn, ice) cover is present, the formation of pseudokarst depressions is associated with melting. The snow, firn and ice covered by pyroclasts melt, and as a consequence, the cover caves in (Fig. 7.69d). As an example, the caldera of Askja (Iceland) can be cited where numerous dropout pseudokarst depressions developed after the 1875 eruption of Askja. The pumice ejected covered the snow which was affected by firnification. Where the pumice mantle was thinner, the firn melted. Meltwater flows downslope and created melt channels in the firn. In the pumice above the channels, dropout pseudokarst depressions emerged by collapse. Pseudokarst dropout and suffosion dolines which had developed above the melting dead-ice were described from the Ojos del Salado volcano (Mari et al. 2014).

Subsidence pseudokarst depressions may form through direct inheritance, too, when the material deficit in the bedrock under the cover is inherited over the cover (Fig. 7.69e) by means of the collapse of cavities in the bedrock. The size, shape and density of the resulting depressions depend on the size, shape and density of the cavity in the bedrock.

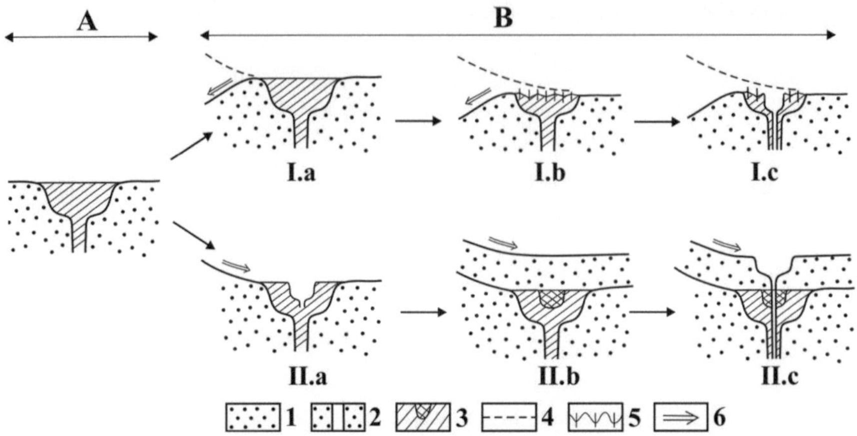

Fig. 7.66 Surface conditions influencing the evolution of covered karst depressions (Veress 2008, modified). (**a**) Fossil depression, (**b**) evolution of fossil depression. *I. a* the environs of the depression are denuded; the depression acquires a relatively high position; *I. b* the depression loses its deposits; *I. c* new depression comes about as the fill is thinning out; *II. a* the environs of the depression are filling up; the surface slopes towards the depression and thus receives sufficient amounts of water for the conduit to lose its deposit and for a new doline to form; *II. b* the depression and its environs are buried; *II. c* although sediment thickness increases, the terrain above the buried depression receives again sufficient amounts of water for the conduit to lose its deposit and for a new depression to form. *1* Cover deposit, *2* non-karstic pipe, *3* fill of depression, *4* initial surface, *5* truncated depression fill, *6* redeposition of deposit

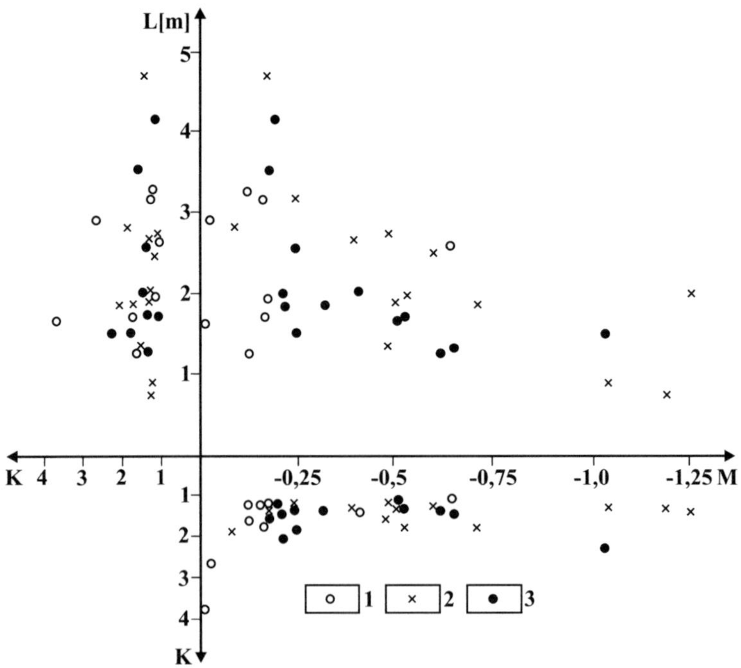

Fig. 7.67 Parameter space of high-mountain subsidence dolines. *1* Asiago Plateau, *2* Totes Gebirge, *3* Seven Lakes Valley (Julian Alps)

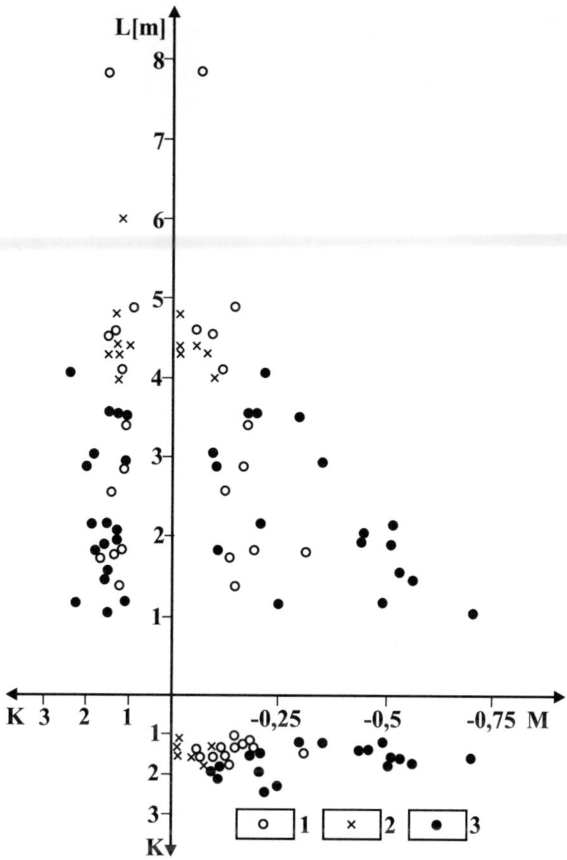

Fig. 7.68 Parameter space of middle-mountain subsidence dolines. *1* Padis, *2* Bakony, *3* Mecsek

Indirect inheritance leads to pseudokarst depressions of lava flows, as the dolines of a lava flow on Hekla (see Chap. 5). The small size of dolines shows that they developed above the small cavities (gas bubbles) of the basalt instead of lava caves. Their gentle slopes point to material loss of the cover by suffosion. It is probable that suffosion dolines are generated above gas bubble cavities from which chimneys opening up to the surface of the lava flow during outgassing (Fig. 7.69f). Between the dolines of the lava flow, steep-sided dropout pseudokarst depressions also occur (Fig. 5.71). Their origin is probably related to the collapse of gas bubble cavities (consequently, they are of direct origin; Fig. 7.69e).

Pseudokarst depressions may be generated above artificial cavities at great depths as exemplified by Dohányos Hill in the Bakony Mountains, where coal mining in Cretaceous limestone has been performed since 1865. Horizontally extended cavities were created. At chambers coal was extracted from neighbouring galleries, always filling in the abandoned gallery with spoil. Mechanised front production created cavities of 2.6 m height and in width corresponding to the distance between galleries. The caved-in material fills in these cavities. With collapsing pore, volume

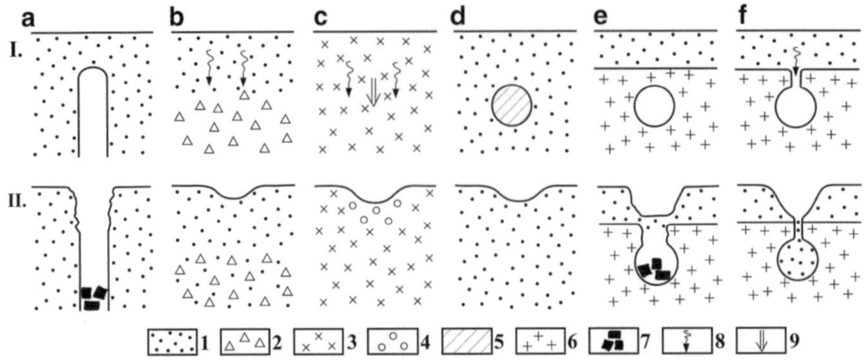

Fig. 7.69 Formation of a pseudokarst subsidence doline. *1* Fine-grained cover deposit, *2* coarse-grained cover deposit, *3* loess, *4* decalcified loess, *5* ice, *6* basalt, *7* collapsed material, *8* suffosion, *9* solution. *I* Initial stage, *II* mature stage. (**a**) collapse of the cover of non-karstic pipe; (**b**) transport of fine-grained deposit into the coarse-grained cover, followed by subsidence; (**c**) subsidence following suffosion and decalcification of loess; (**d**) subsidence following the melting of ice; (**e**) collapse of cavity in lava rock; (**f**) suffosion of the cover deposit of lava rock

increases. Our calculations show that a hollow of 2 m height and the above-lying cavity created by collapse are filled by the collapsed material of 7.7 m thickness. The collapse cannot reach up to the surface since above the galleries and chambers of Dohányos Hill, the thickness of limestone is much greater than 7.7 m, more than 100 m.

Pseudokarst depressions form above galleries and particularly above chambers by the subsidence of rock masses in wide extension as a single block, without collapses, into the cavities of galleries and chambers. The subsided rock block is separated from its environment by atectonic faults. Depressions on the ground surface come about above the entire subsided block if it sinks at a uniform rate (broad, shallow and composite graben form) or where the block did not subside in a uniform way but it was tilted (asymmetrical graben). This latter situation occurs where the chamber wedges out. Features of low width develop in the interior of uniformly subsided blocks where extensional cracks appeared. At these sites narrow grabens develop. The cracks wedging out upwards only reach the surface along short sections, where shaft-like features come about. In the bedrock, circular depressions may also emerge at the junctions of cracks.

The evolution of the feature may extend over the cover, too. On the cover, landforms either develop by direct or by indirect inheritance. Direct inheritance happens through the displacement of the cover following subsidence of the bedrock. Between the broad grabens, steep-sided features occur with bedrock outcrops in their side. In this case the atectonic fault was also inherited over the cover. Gentle-sided broad grabens also occur. In this case atectonic faults were not inherited, but the subsidence of the cover followed that of the bedrock. If there is a landform (shaft) of limited extension but great depth and wedging out downwards, the inheritance generates landforms like dropout dolines on the cover by collapse.

With indirect inheritance, there is material deficit in the cover. Suffosion (or the collapse of the cavity in the cover) creates suffosion or dropout dolines (Fig. 5.73). Dolines originated from indirect inheritance probably develop where the cracks wedging out upwards reach the surface of the bedrock.

7.4 Origin of Ponors

7.4.1 Main Characteristics of Ponor Formation

According to Ford and Williams (2007), the following conditions are necessary for allogenic inputs: input discharge, hydraulic conductivity, hydraulic gradient, location (lateral or vertical), geological context, process environment and time. The conduits are adjusted to the direction of the hydraulic gradient (Ford and Williams 1989). Water conduction is becoming more and more localised and ponors emerge where the hydraulic gradient is more steeper and the current velocity in allogenic watercourses reduces (Ford and Williams 1989). Although in the zone of karst water fluctuation, passages operating as ponors (katavothra) may develop, a major condition of ponor formation is that the karst water table should be located at the rock boundary below the surface. Water conduction into the karst is only possible if surface waters are drained through conduits. Not only such conduits should be present but they should not be filled with water.

Ponor formation has the following conditions: rock boundary, favourable position of the karst water table, missing or limited surface runoff, the presence of an underground cavity (conduit) and its opening up to the surface as well as the link between the surface watercourse and the cavity opened up to the surface.

It is a general observation that there is no water conduction from the surface and, consequently, no ponor formation, in spite of the presence of a surface watercourse if the karst water table is located at the surface. Ponors can form if the karst water table is deeper relative to the valley floor of the watercourse. This happens if the karst is uplifted (or the base level sinks).

Surface runoff is mostly lacking in closed depressions (paleodolines, poljes), and this situation is also promoted by karstic elevations of different size. An example is the Pádis plateau, where ponors are aligned at the feet of karstic elevations. On tropical karsts, karstic elevations impede the drainage of the surface, and thus they promote ponor formation. Balázs (1984) reports about ponors developed at the base of exhuming inselbergs, and Sweeting (1973) described ponors at the feet of mounds on cockpit karst. On tropical karsts, ponors develop at the feet of the mounds even if the karst water table is near the surface. Such ponors were described from Saravak (Wilford and Wall 1965). In this case after a small lowering of the karst water table, parallel to this level, conduits of small gradient (former conduits of the karst water table) make the conduits of the ponor in the mound. The passage is a through cave because it terminates at the opposite side of the mound.

There are several possible ways for the development of a subsurface cavity which later turns into a water conduit. The cavities and shafts may originate in the epikarst zone, independent from the karst water. This may happen through the downward extension of the shaft. In this case the shaft is created by the percolation of the surface watercourse. Its formation happens by way of solution along fractures (Cvijič 1893; Sweeting 1973; Jaskó 1959, 1961) or along bedding planes (Cvijič 1893; Sweeting 1973) or by upward shaft extension (see Sect. 7.3.3.4). Ponor formation by infiltrating water can be well studied on uncovered gypsum. The rainwater infiltrating along the fractures of the gypsum first creates cracks which develop into shafts to which gullies lead (Kósa 1981).

Conduits also originate by collapse. Thus, the ponor Wadi Al Kharruba (Libya) is due to the collapse of cavernous limestone induced by the pressure increase of water accumulated beyond a dam (Kósa and Csernavölgyi 1983). Ponor formations due to similar (but natural) cavity collapse are known to have happened in several instances. On the Ankarana tsingy (Madagascar), a collapse doline on valley floor conducts water flowing on the valley floor, i.e. acts as a ponor.

According to Hamblin (1989), the conduit of the ponor develops from cavities below the karst water table. This is confirmed by direct observations. The pipes (shafts) of katavothra are produced by ascending karst water. Conduit formation below the karst water table is allowed by more rapid flow, which is more common in the proximity of the karst water table (LeGrand and LaMoreaux 1975).

The shafts (or ponor caves) are either vertical or horizontal (Ford and Williams 2007). A vertical conduit more probably forms in epikarst environment at the margin of the overlying cover deposit, while a horizontal conduit is more common along non-karstic rock interbedding (Ford and Williams 2007). The position of the conduit, however, is not only controlled by the bedding of the non-karstic rock. A horizontal conduit also appears at the cover if the conduit derives from cavities below the karst water table or if the ponor conduit is side branch of a cave system and the difference in elevation between the main branch and the ponor is small. Vertical conduits also develop with overlain cover if conduit formation takes place in the epiphreatic zone (see the origin of conduits of katavothron).

7.4.2 *Ponor Formation on Various Covered Karsts*

The covering non-karstic rock is overlain, interbedded (Fig. 5.76) and mantle-like. In the latter case, it is either impermeable (cryptokarst) or permeable (concealed karst).

At the margin of overlain cover, if the karst water table lies deep below the surface, the ponor or its shaft develops in the epikarst zone. This is explained by the development of a vertical shaft extending downwards by water percolation from the surface watercourse (Fig. 7.70a). The valley leading to the ponor evolves into a

Fig. 7.70 (continued) formed in the zone of karst water fluctuation; (**g**) covered karst ponor on the contact of impermeable and permeable rocks; (**h**) ponor formed on evaporites (autogenic cryptokarst and concealed karst). *I* initial stage, *II* mature stage

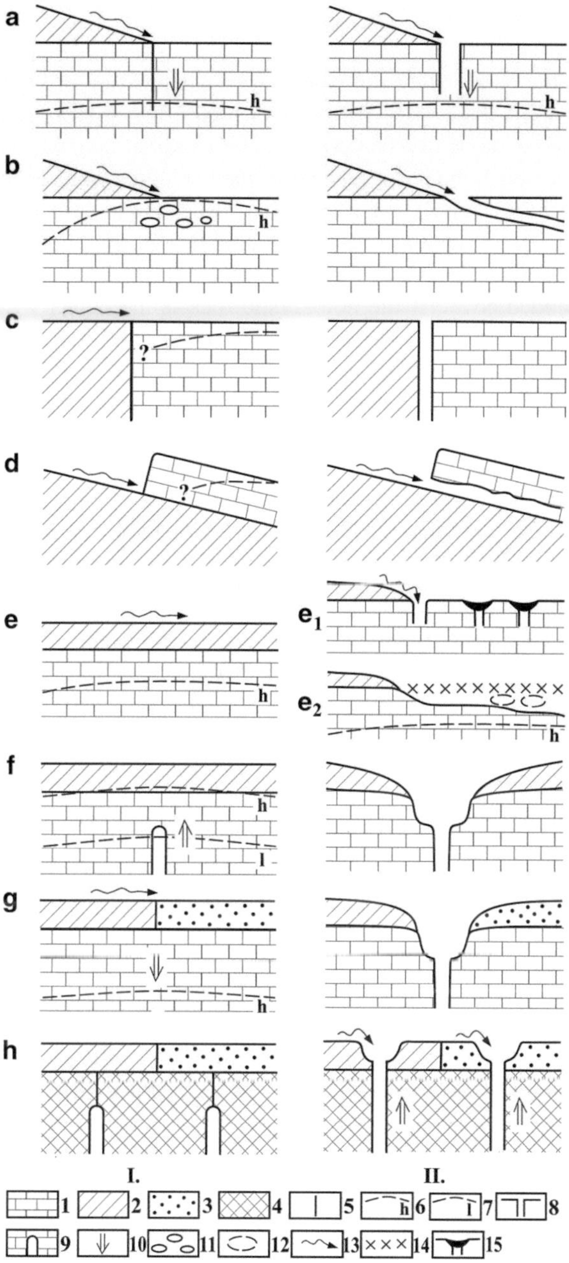

Fig. 7.70 Ponor formation. *1* Limestone, *2* consolidated non-karstic rock, *3* unconsolidated non-karstic rock, *4* evaporite, *5* fracture, *6* high karst water table, *7* low karst water table, *8* shaft, *9* blind shaft, *10* direction of shaft extension, *11* cavity formed below the karst water table, *12* cavity formed below the karst water table and destroyed by erosion, *13* watercourse, *14* former surface, *15* doline developed from ponor. (**a**) Shaft of epikarst origin of the ponor on the margin of overlying rock; (**b**) ponor formed on the margin of overlying non-karstic rock, its shaft developed below the karst water table; (**c**) ponor formed on the margin of interbedded non-karstic rock, with vertical shaft; (**d**) ponor formed on the margin of interbedded non-karstic rock, with horizontal shaft; (**e**) ponor formation (e_1) and opening up of cavity (e_2) below the floor of an epigenetic valley; (**f**) ponor

blind valley. In the blind valley and its environs, a depression of superficial deposit appears on the non-karstic rock. This kind of ponor formation occurs on allogenic cryptokarst. The ponors at overlain cover are either karst marginal or karst interior. The latter are located at the cover patches of the karst interior paleodepressions.

At overlain cover, if the karst water table lies at the karst surface, there is no ponor formation. But if the karst water is sinking slowly, ponors form and the cavities below the karst water table coalesce by surface water inflow (Fig. 7.70b).

With interbedded cover, if the non-karstic rock is in contact with karstic rock along fault (Fig. 5.76c), a vertical conduit (shaft) originates at the contact of rocks by the percolation of waters from the non-karstic terrain (Fig. 7.70c). Blind valley and DSD formation ensue. If the non-karstic rock is in contact with the karstic rock along an oblique surface, horizontal or subhorizontal conduit results (Fig. 7.70d). Ponors evolve into blind valleys and depressions of superficial deposit. This kind of ponor formation is also typical of allogenic cryptokarsts.

With mantle-like impermeable cover, ponor formation may take the following courses:

- There is valley formation on the cover (on cryptokarst evolved from buried karst). If the karst water table lies at the surface, the valley is inherited over the limestone. If valley incision keeps pace with the sinking of the karst water table, no ponor forms. The watercourse of the deepening valley eliminates the cavities which were created under the karst water table but, with the sinking water level, rose above the table. If the lowering of the valley cannot keep pace with the sinking of karst water table (happens at a lower rate), the elevation difference between the valley floor and the karst water table can reach several 100 m. In karst areas, there are numerous epigenetic valleys where this difference is considerable, but there is no ponor in such a valley. This is explained by the lower, but still remarkably rapid, rate of valley incision relative to karst water level sinking. Thus, karst water cavities are destroyed (Fig. 7.70e_2) or open up. The cavities are preserved as residual caves in such valleys (see Sects. 5.8 and 7.6).
- If valley incision takes place at a very low rate, the karst water cavities survive and a ponor forms on the valley rock boundary (Fig. 7.70e_1).
- Vertical conduits (shafts) and ponors form under mantle-like cover where the high karst water table reaches the surface. The conduit develops in the zone of karst water fluctuation (epiphreatic zone). These sites are the katavothra of the karst (Fig. 7.70f).
- Covered karst ponors appear if the cover is a consolidated rock (on transitional cryptokarst), and at the karst windows, caprock dolines emerge. The shafts below caprock dolines are vertical and develop in the epikarst zone through stoping upwards. On evaporites ponor formation is driven by the pipes stoping upwards from great depths (on autogenic cryptokarst, Fig. 7.70h). In this case the solution leading to the formation of pipes is associated with groundwater circulation or the repeated collapses of pipes. The caprock dolines are transformed into karstic ponors. In a similar manner, the upward stoping shafts of the epikarst evolve at

the lateral contact of the impermeable and permeable cover deposits. At such sites too, covered karst ponors come about when the blind shaft opens up to the surface (Fig. 7.70g).

7.5 Origin of Depressions of Superficial Deposit (DSD)

7.5.1 General Characteristics of DSD Formation

DSDs originate by the local denudation of the cover. The denudation of the cover is mostly but not exclusively due to erosion. Erosional evolution also affects other karst landforms, but for DSDs it is predominant, and with progressing evolution, it is becoming ever more significant. In the area of DSDs, surface material transport (if it happened at all) is gradually replaced by the redeposition of the cover deposit into the karst. It is possible if a subsidence doline or a ponor develops and the removal of the cover on the surface becomes non-uniform ever since.

The resulting DSD either expands by karstic or non-karstic processes (erosion, mass movements). Moderate growth by solution (deepening) can primarily take place in the paleodepressions of glaciokarsts, where the cover material is transported in solution. On rock salt deepening by solution can gain prominence. DSDs form by surface solution or solution at greater depths. Solutional subsidence troughs and solution-induced depositional basins are produced if the density of breccia pipes is high (Waltham et al. 2005; Klimchouk 2004).

The formation of karst landforms also contributes to the expansion of DSDs. Thus, they may coalesce with subsidence dolines in their neighbourhood (Figs. 5.85 and 5.92a) leading to their broadening. Broadening is also due to the subsidence dolines on their side slopes (Fig. 5.92b, symbol 3). The subsidence dolines on their floor do not only cause the conveying of their deposits into the karst but also their deepening.

The deepening of the DSD is driven by erosional denudation (either linear or pluvial erosion) if the slope of the floor is sufficiently high. The floor is sloping if a karst landform has formed on it or sediment is transported from its neighbourhood into the existing DSD. In the existing DSD, erosional processes operate if the ponors or subsidence dolines are capable of transporting the cover into the karst. On the denudation map of the cryptokarst and concealed karst floor of the paleouvala of the Hochschwab (Figs. 7.71 and 7.72), it is visible that on the cryptokarst portion linear (gully), erosion operates along the axes of gullies, ravines and valleys associated with the ponor. On the terrains between these landforms – with the exception of a very limited terrain – the floor is being denuded by pluvial erosion. The denudation is of mosaical nature. The removed sediment is redeposited into the gullies and ravines and then into a ponor. On the terrains of concealed karst, areas of limited extension are denuded. Here pluvial erosion is not only driven by gullies but dolines too.

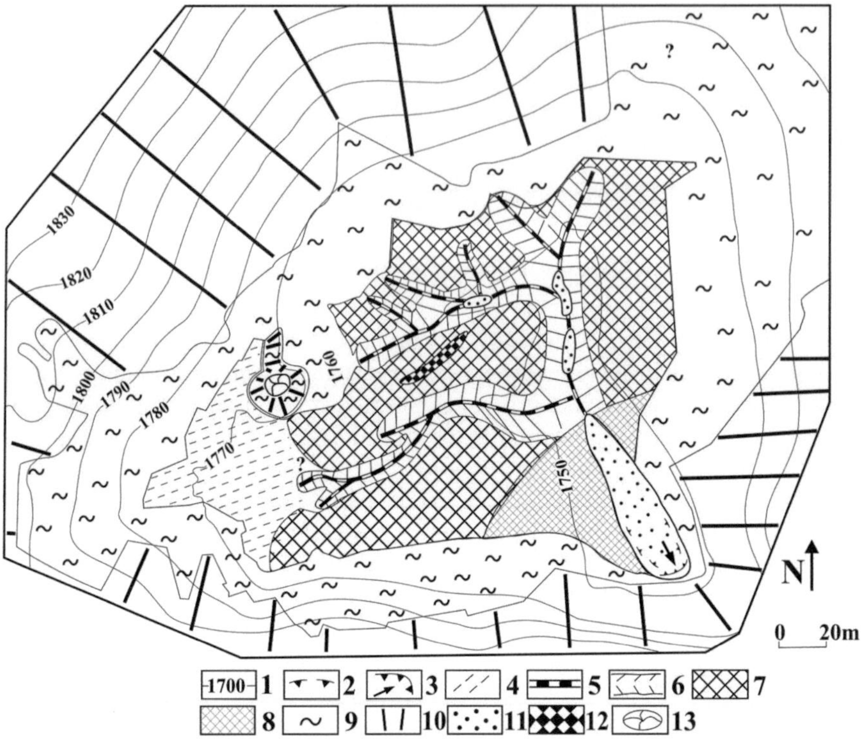

Fig. 7.71 Denudation map of the floor of a paleodepression with cryptokarst (Hochschwab paleouvala). *1* Contour line, *2* infilled ponor, *3* active ponor, *4* concealed karst, *5* linear erosion, *6* slope directly shaped by linear erosion, *7* slope indirectly shaped by linear erosion (slope angles between 0° and 30°), *8* slope shaped by ponor formation, *9* upper accumulation level (accumulation at the termination of the slope of the paleodepression, slope angle 30°–50°, where accumulation and denudation alternate), *10* denuding slope of depression (slope angles above 50°, from where the removed material is transported to the upper accumulation level), *11* lower accumulation level (accumulation in the depressions on the floor of the paleodepression, slope angles between 0° and 10°), *12* terrains without denudation between channels and valleys (slopes angles between 0° and 10°), *13* karstic landform remnant

Pluvial erosion leads to extensive areal denudation of the floor and, therefore, to the uniform deepening of the DSD. Linear erosion, on the one hand, is a preparatory process to pluvial erosion, and on the other hand the valleys resulting from linear erosion broaden by pluvial erosion and join each other. Thus, together with pluvial erosion, they contribute to the areal denudation of the DSD floor.

7.5.2 Origin and Evolution of DSDs

DSDs originate from the coalescence or formation of subsidence dolines, ponor formation and erosional transformation of subsidence dolines and ponors or from katavothra.

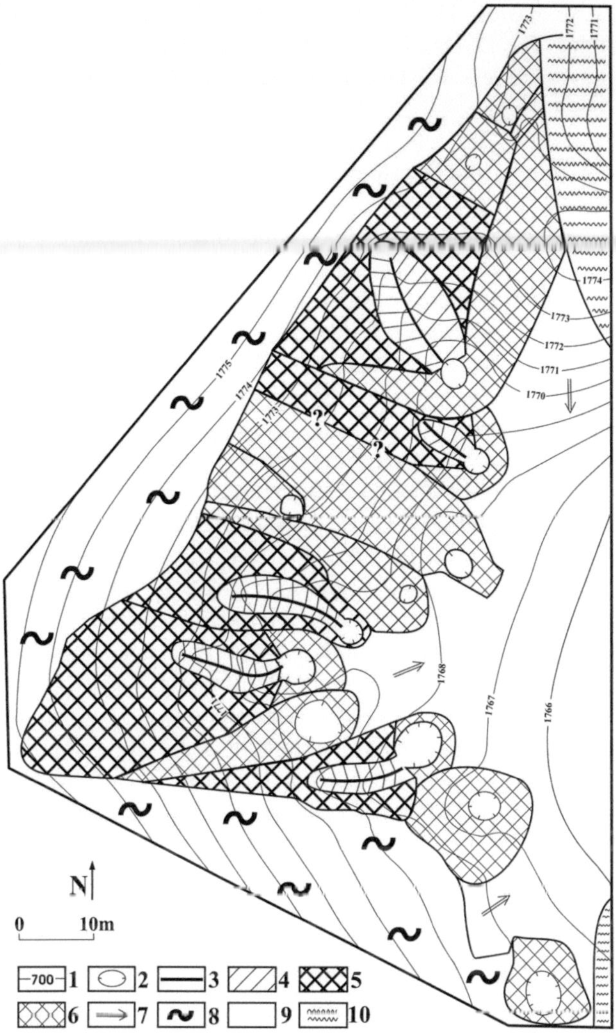

Fig. 7.72 Denudation map of the floor of a paleodepression with concealed karst (Hochschwab paleouvala – modified, Veress 2012b). *1* Contour line, *2* suffosion doline, *3* linear erosion, *4* slope directly shaped by gullies, *5* slope indirectly shaped by linear erosion, *6* slope shaped by doline formation, *7* sediment transport on the surface, *8* upper accumulation level of the paleodepression, *9* concealed karst, *10* cryptokarst

7.5.2.1 Formation by Coalescence

The subsidence dolines close to each other are broadening and, as a consequence, finally coalesce. The broadening of subsidence dolines is caused by the denudation of their slopes through pluvial erosion. The formation of DSDs through the coalescence of subsidence dolines was described by Beck and Sinclair (1986), and they called them large basins. The removed sediment leaves through the non-karstic pipes of the floor. The DSDs of coalescence origin occur on concealed karst, primarily if the cover is a less resistant, non-cohesive rock, such as loess.

7.5.2.2 Formation by the Development of Subsidence Dolines

The origin of this type of DSD is bound to the formation of subsidence doline(s). The process takes place on concealed karsts. Compared to formation by coalescence, it is different in that here it is not the coalescence of existing dolines that is typical but the environs of a doline is denuded. There are two possible ways for this. In one case a ravine is attached to the doline. The ravine and the doline are broadening, mainly by pluvial erosion, and a DSD results, the interior of which is dissected by gullies and subsidence dolines. In the other case, the DSD develops during the development of subsidence dolines. It is favoured by the fact if there is only a patch of superficial deposit on the cover. From the area of the superficial deposit patch, there is no surficial material transport because the rainwaters of the superficial deposit patch seep away, or if they flow out from the area of the cover, they seep away on the bordering limestone. This form is called half-DSD (Fig. 7.73a). From the date when a subsidence doline appears at a given site of the cover (such a site of doline formation can be located where the cover is thinner above an elevation of the bedrock), the sediment transported by rainwater is transported into the karst through the subsidence doline. To the effect of the rainwater flowing towards the subsidence doline, the cover is thinning out and the DSD is deepening. In another site of the DSD, a new doline emerges where the conditions for doline formation are favourable. The evolution of the DSD arrives at a juvenile stage (Fig. 7.73c). With the further denudation of the cover, the DSD is becoming ever deeper, and at the same time, the previously formed subsidence dolines may fill up and are transformed into fossil dolines, resulting in an impermeable cover in their area. The infilling of the dolines is enhanced by the partial or complete loss of catchment of the former doline as a consequence of the emergence of new dolines. The evolution of the DSD reaches its adult stage (Fig. 7.73d). On the floor of the DSD, gullies develop and increase the rate of gully incision (Fig. 7.74). In the area of the DSD, since more and more previously formed doline is filled up, the impermeable character becomes widespread. The mature stage of DSD evolution is reached (Fig. 7.73e, f).

7.5.2.3 Formation by the Development of Ponors

DSD formation governed by ponors is typical of cryptokarsts. On allogenic cryptokarsts, ponor rows form at the termination of the non-karstic rock. Ponors will have blind valleys. Pluvial erosion generates surfaces sloping towards the blind valleys. The area enclosed by the line connecting the highest elevations of the terrains between the blind valleys (divide) constitutes the DSD (Fig. 7.75).

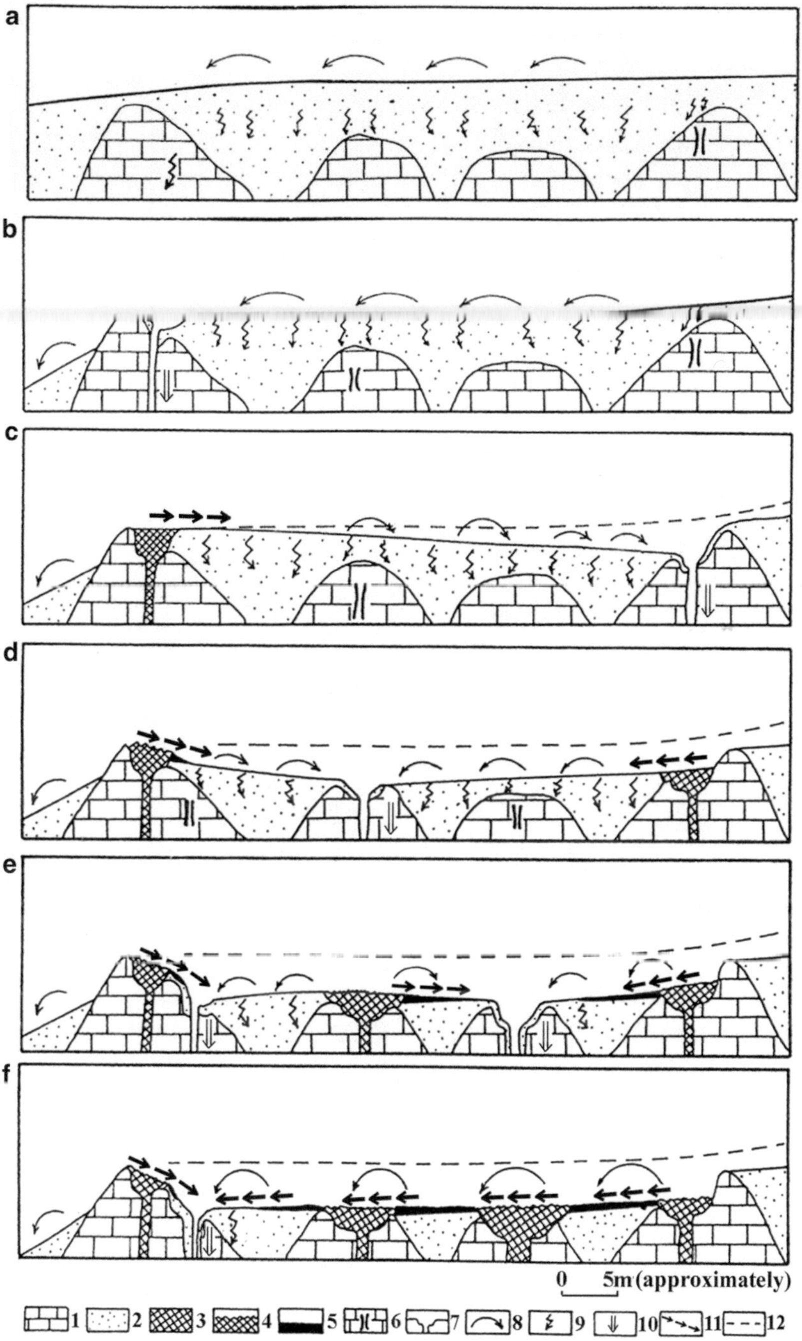

Fig. 7.73 Evolution of a pseudodepression (**a**) half depression, (**b**) embryonic depression, (**c**) young depression, (**d**) adult depression, (**e**, **f**) mature depression, *1* carbonate rock, *2* cover deposit, *3* impermeable fill of subsidence doline, *4* partially truncated sediment fill of subsidence doline, *5* impermeable deposit due to redeposition, *6* karst conduit (blind shaft, shaft), *7* subsidence doline, *8* pluvial erosion, *9* water percolation, *10* material transport at depth, *11* surface runoff, *12* surface at the time of DSD formation

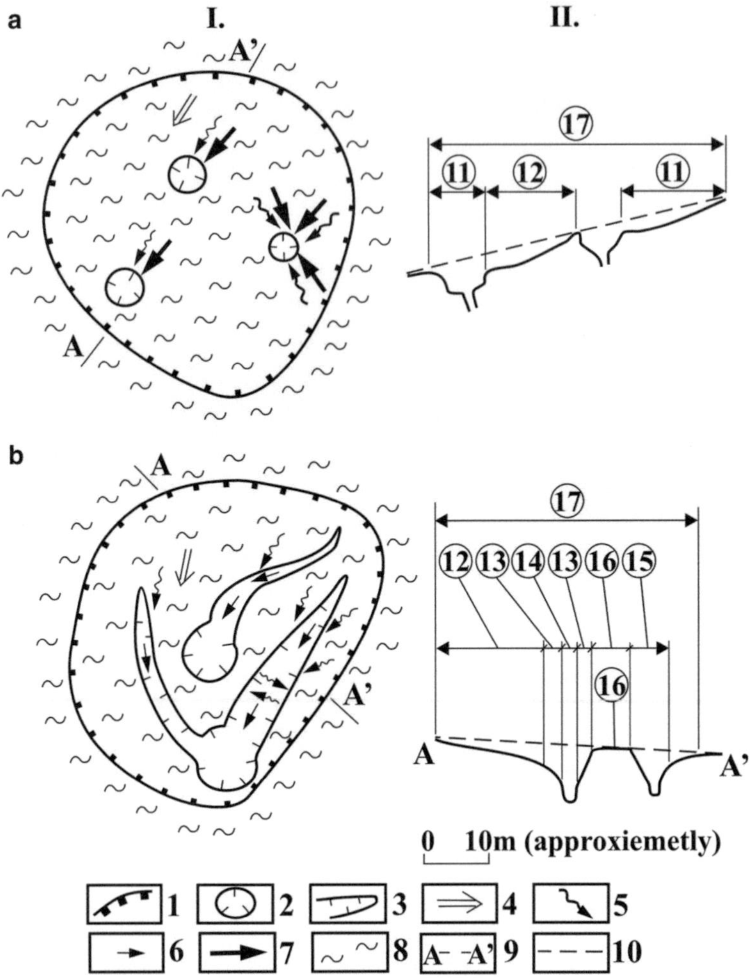

Fig. 7.74 Deepening of DSD through pluvial and linear erosion on concealed karst. (**a**) Subsidence dolines without gullies, depression development by pluvial erosion, (**b**) subsidence dolines with gullies leading to them (linear erosion also contributes to the formation of the depression), *I* groundplan, *II* cross-section, I. *1* depression with margin in cover deposit, *2* subsidence doline, *3* gully, *4* initial surface slope, *5* pluvial erosion, *6* linear erosion, *7* slope generated by pluvial erosion, *8* cover deposit, *9* site of profile, II. *10* initial surface, *11* subsidence doline, *12* surface denuded by pluvial erosion, *13* denudation of the sides of channels and gullies by pluvial erosion, *14* linear erosion, *15* gully, *16* remnant of initial surface, *17* DSD

DSD formation in the area of paleodolines is presented in Fig. 7.76. Ponors and then blind valleys come about. In blind valleys, linear erosion and, on the terrain between them, pluvial erosion drive the transport of the deposits of paleodepressions into ponors.

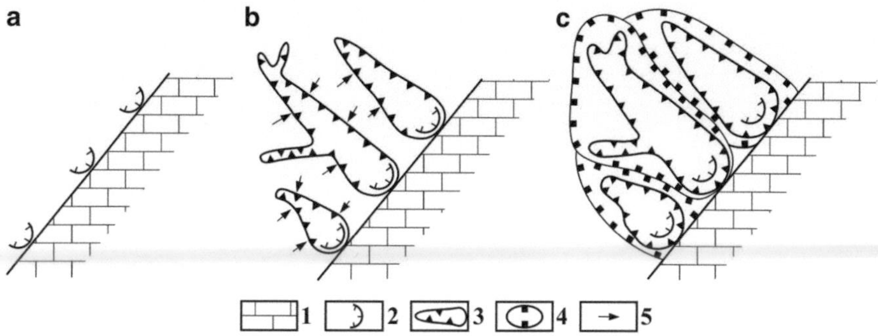

Fig. 7.75 Formation of DSD on allogenic cryptokarst. *1* Limestone, *2* ponor, *3* blind valley, *4* DSD, *5* pluvial erosion. (**a**) Ponor formation on rock boundary; (**b**) blind valley formation, denudation of the terrain between blind valleys by pluvial erosion; (**c**) formation of a divide, the margin of the DSD, on the terrain between blind valleys

7.5.2.4 DSD Formation at Katavothra

This DSD variety forms in two stages. The first is associated with the fluctuation of karst water level. The fluctuation of the karst water level results in the development of one or possibly several subsidence dolines in the middle part of the present DSD in the environs of the katavothron (Fig. 7.60a–c).

In the second stage of its evolution, the subsidence doline is broadening (if several dolines have emerged in the first stage, they coalesce). Pluvial erosion, mass movements on the slope, additional subsidence dolines on the margins and linear erosion all contribute to broadening (Fig. 7.60d–f). The valley that develops to the effect of linear erosion is also regarded part of the DSD.

7.5.3 Material Budget and Transformation of the DSDs

DSD transformation depends on its material budget. DSDs can be of deepening (true DSDs are exhuming) and infilling types. The deepening of a true DSD involves the exhumation of the bearing paleodepression. The DSD is deepening if it receives less material from its environs than it is transported into the karst. The broadening of the DSDs is limited by bedrock morphology or, in the case of those developed at katavothra, the margin of the polje. The diameter of true depressions cannot exceed that of the bearing paleodoline. For pseudodepressions, the side slope of the elevation on the bedrock restricts broadening. If this side slope is gentle, ponors receiving sediment or subsidence dolines form at newer and newer sites, and as a consequence, the DSD is broadening. If the slope of the bedrock is steep, the cover will be so thick in a certain distance from the margin of the DSD that no ponor or subsidence doline can emerge (Fig. 7.77III). The DSD is of infilling type if there is more sediment transported into the depression than the loss from the DSD into the karst.

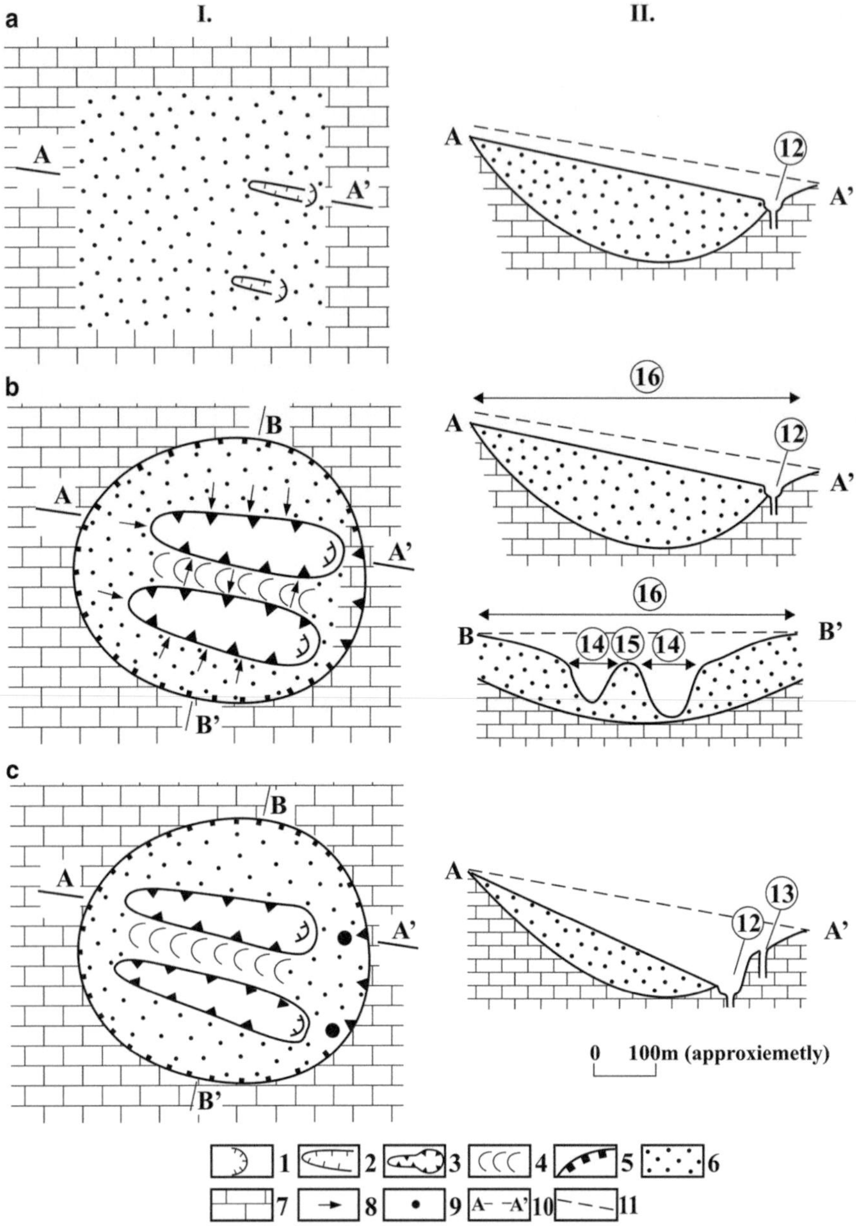

Fig. 7.76 Evolution and development of DSD in the area of a paleodoline. *I* Groundplan, *II* cross-section, in groundplan: *1* ponor, *2* gully, *3* blind valley, *4* interfluvial ridge, *5* DSD margin, *6* superficial deposit, *7* limestone, *8* pluvial erosion, *9* fossil ponor, *10* cross-section. In cross-section: *11* former surface, *12* ponor, *13* fossil ponor, *14* valley, *15* interfluvial ridge, *16* DSD, (**a**) ponor formation, (**b**) blind valley formation, (**c**) DSD formation through the exhumation of a paleodoline. With the denudation of the surface, further ponors develop and the older ponors are truncated

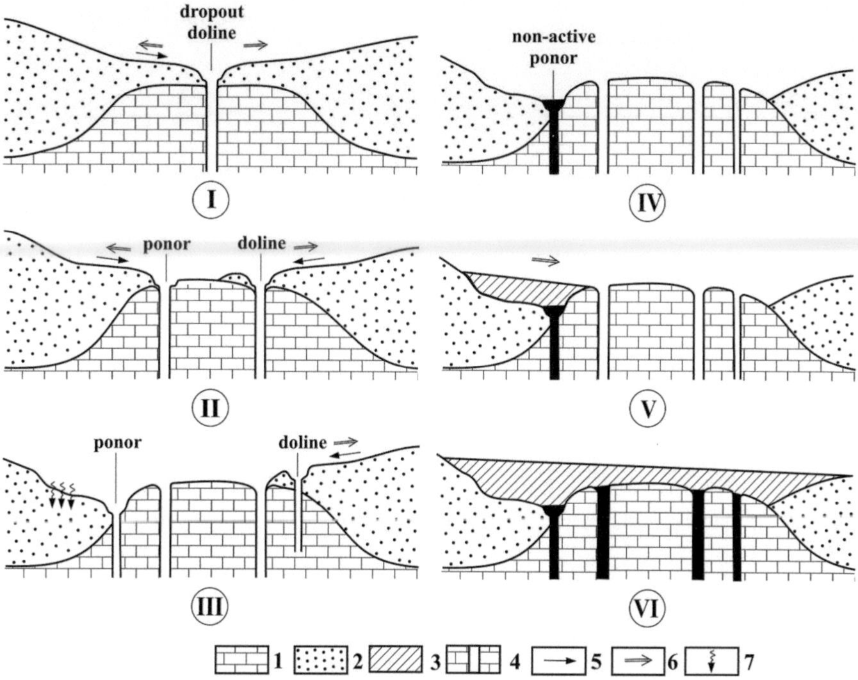

Fig. 7.77 Development and filling of a pseudo DSD (Veress 2012a). *I–III* Development of DSD, *IV–V* the increasing of the DSD stops (the ponor fills), *VI* filled DSD, *1* limestone, *2* superficial deposit, *3* redeposited superficial deposit, *4* shaft, *5* transporting of sediment, *6* the depression increases because of the denudation of its side slopes, *7* deepening of creek floor

If the outermost doline or ponor is filled up (Fig. 7.77IV), the broadening of the DSD ceases and even its infilling with the sediment transported into its interior may start (Fig. 7.79IV–VI), and its entire interior can be filled up. Subsequently, karstification is rejuvenating in the interior (Figs. 7.78 and 7.79).

In Fig. 7.80 the evolution and transformation of DSDs controlled by their material budget eventually leading to their cyclic evolution is presented. Cyclic evolution involves the deepening, broadening and then infilling and burial of the DSD. If karstification resumes in its area and/or outward transport exceeds inward transport, the DSD is rejuvenated. As a consequence of this rejuvenation, the DSD re-appears, or if infilling has already started in its area, it begins to deepen again.

7.6 Formation of Remnant Caves

Remnant caves originate from the opening up of karst water cavities to the surface. They may derive from the cavities of the elevations on covered limestone or from the cavities being close to the flat limestone surface. Because of the exhumation of

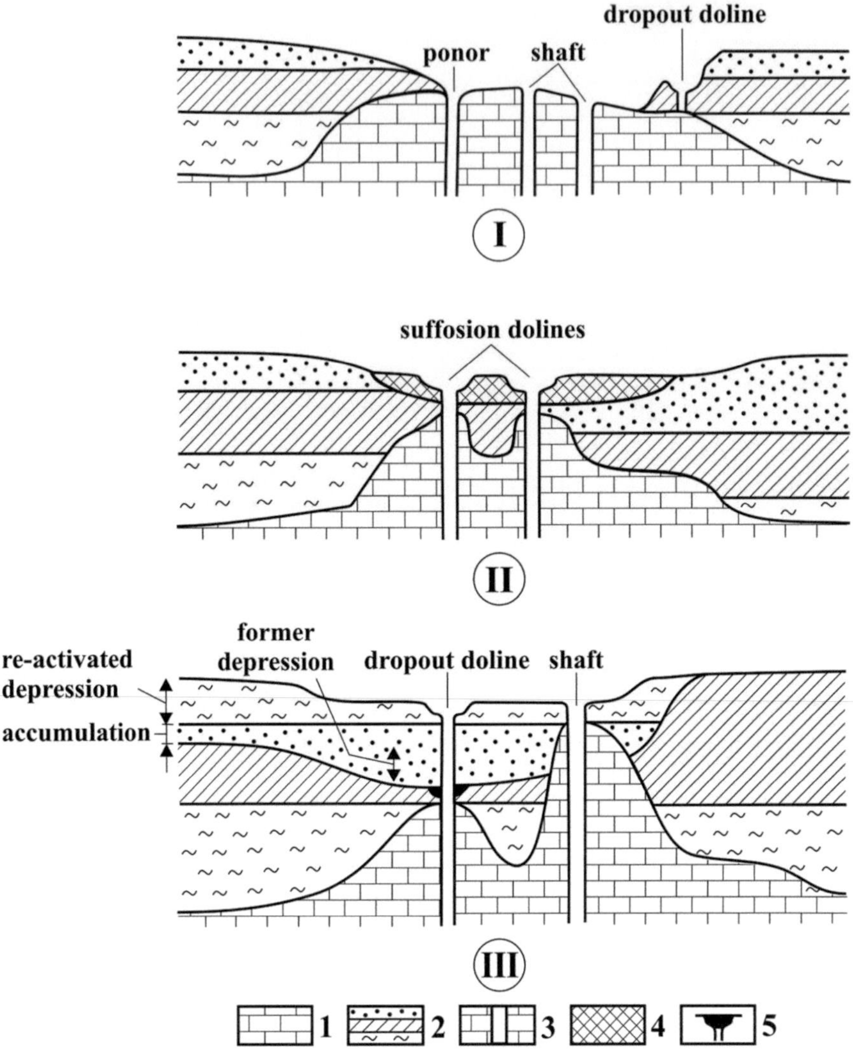

Fig. 7.78 Types of pseudodepression according to material turnover (Veress 2012a). *I* Developing DSD, *II* infilling (senile) DSD, *III* reactivating DSD, *1* limestone, *2* cover deposit, *3* chimney (shaft), *4* cover deposit accumulated in the DSD, *5* sediment fill of doline

the karst elevation or if the flat limestone surface loses its sediment (without valley formation), the cavities open up to the surface if they were open once. Closed cavities are usually opened to the surface by collapses (Fig. 7.81). They form in epigenetic valleys if the karst water table is more than 100 m below the elevation of valley floor (see Sect. 7.5.2). The floor of the deepening valley inherited from the cover reaches the cavities in the rock and opens them up (Fig. 7.82). If the cavities are located at longer distance from the channel, the opening up of the cavities takes place through the denudation of the valley side (Fig. 5.97).

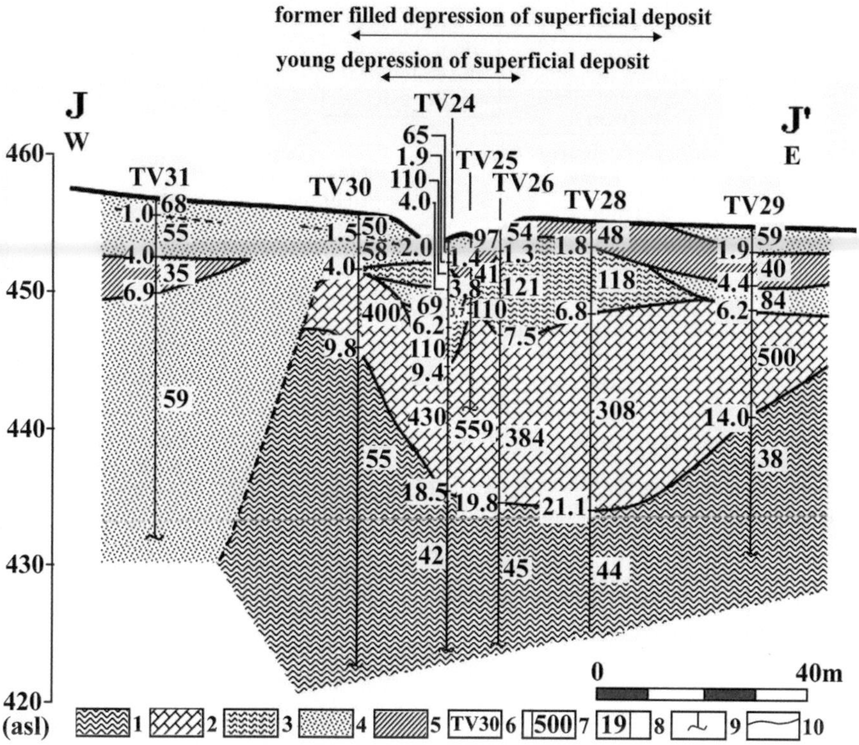

Fig. 7.79 Geoelectric–geological profile J–J′ of the depression D-7 (Tés Plateau) (Veress 2012a). *1* Clay, marl, *2* limestone, *3* loess (sandy or with limestone debris), *4* loess (clayey or silty) or clay with limestone debris, *5* limestone debris (clayey), *6* site of VES measurement, *7* geoelectric resistance of the series (Ohmm), *8* basal depth of the geoelectric series, *9* approximate penetration of the VES measurement, *10* boundary of geoelectric series

7.7 Conclusions

The cover exerts diverse influences on karren formation: it distributes and stores the water arriving onto the bedrock as well as contributes to CO_2 production. The distribution and storage of water (i.e. the duration of solution) primarily depends on particle size. The formation of subsoil karren is governed by water flow in the cover and temporary waterlogging. Irregular karren features are created by solution along bedding planes and fractures in the interior of the rock mass. Vertical karren come into existence if solution penetrates deep into the bedrock. The channels are produced by water flowing on the surface of the bedrock, while grikes are due to the water percolating into joints. The grikes, if not getting narrower downwards, may deepen until the lower boundary of the epikarst or until the karst water table. The deepening and broadening of grikes lead to pinnacle karsts and their varieties such as stone forest, arête and pinnacle karst as well as inselberg karst. Notches develop at the level of the soil but may occur at the level of temporary waterlogging too.

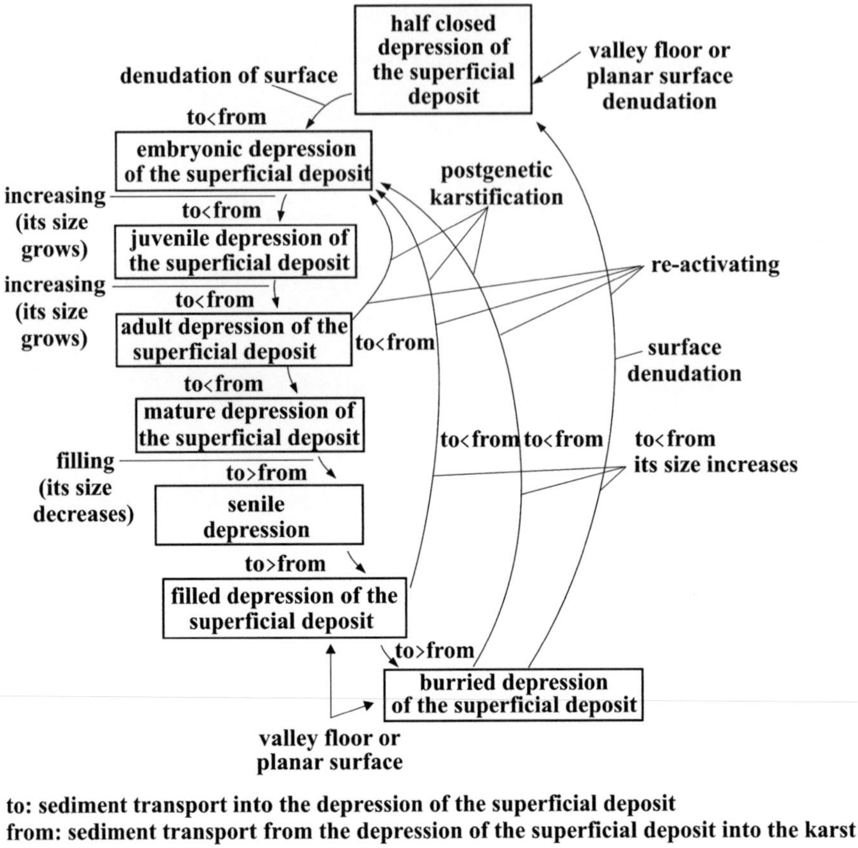

Fig. 7.80 Evolution cycle of pseudodepressions (Veress 2012a)

Other karren (spongy surfaces, bedding plane grikes) are found below the level of temporary inundation.

On rock salt, primarily grikes and channels form. The latter mostly develop from rainwater rills or are inherited from salt breccia. Pseudokarren emerge in fills due to suffosion and collapse.

The C_1 variety caprock dolines originate from the collapse of cavities or from the stoping of chimneys on the ceilings of allogenic caves. C_2 variety dolines are the result of opening up of breccia pipes developed at great depths, while C_3 varieties emerge above chimney (shaft) groups.

There are well-definable conditions necessary for the formation of subsidence dolines: slope inclination (mainly developed if slope angle is lower than 10°–12°), slope height (depending on the elevation of the bearing landforms and the karst water level) and karst water (formed if the karst water level fluctuates or sinks) as well as the properties of the cover. Thus, the greater the chance for doline formation is, the thinner the cover is and the lower its $CaCO_3$ and clay contents are. Both extremely small and large particle sizes favour local solution (shaft development),

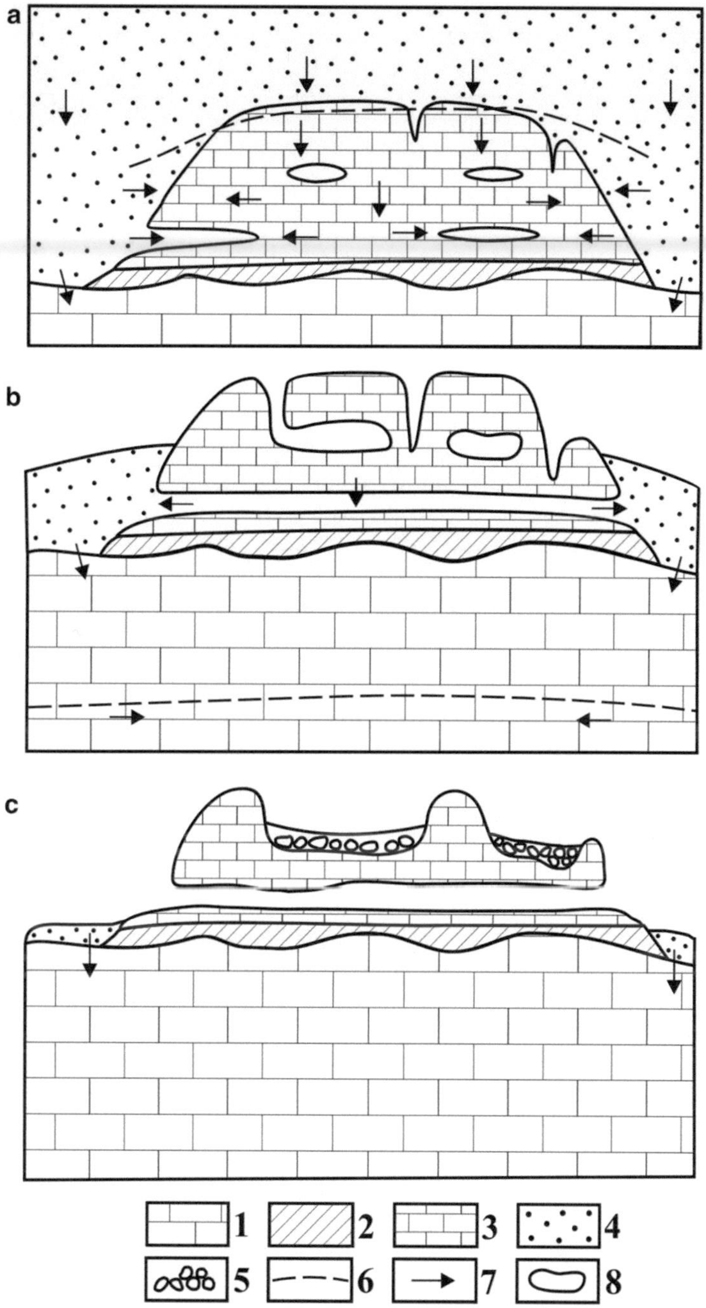

Fig. 7.81 Evolution of karstic cavities of Likas-kő (Veress-Futó 1987). (**a**) Carbonate rock covered and below karst water table; (**b**) partly exhumed and above karst water table; (**c**) karstic cavities are destroyed following further exhumation; *1* Triassic carbonate rock; *2* abraded base conglomerate; *3* Eocene limestone; *4* gravel (Csatka Gravel Formation); *5* collapsed material; *6* karst water table; *7* direction of flow; *8* karstic cavity

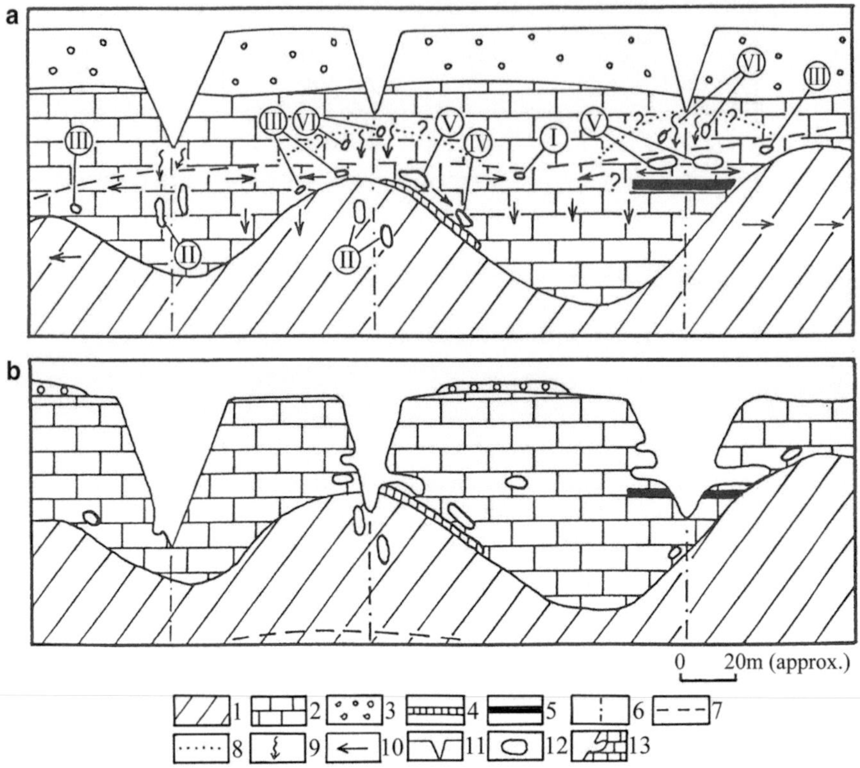

Fig. 7.82 Development of cave remnants (After Veress 2000). (**a**) Valley and cavity formation, (**b**) cavity destruction and exposure, *1* Triassic carbonate rock, *2* Middle Eocene limestone (Szőc Formation), *3* gravel (Csata Gravel Formation), *4* abrasion breccia, *5* marl, *6* fracture and fault, *7* karst water table, *8* local elevation in karst water table from infiltration, *9* infiltrating water from floors of superimposed valleys, *10* karst water flow, *11* superimposed valley, *12* cavity, *13* exposed cavity (cave remnant), *I* cavity development below karst water table, *II* cavity development along fractures and faults, *III* cavity development above dolomite, *IV* cavity development above local impermeable (or partially impermeable) series, *V* cavity development above local impermeable series intensified by infiltration from valley, *VI* cavity development below temporarily elevated karst water table

while medium particle size favours both local solution (shaft development) and continuous solution (karren formation).

With direct inheritance at first, a blind shaft emerges on the bedrock and then shaft forms by collapse. Then this process may be inherited over the cover. The morphology of the shaft and often the morphology and pattern of the depression on the cover is governed by rock structure. Complex shaft systems may be generated. Their blind shafts also extend upwards caused by the impoundment of the water in the shaft. With indirect inheritance, the cover material is transported by collapse, suffosion, solution or material fall or with karst water flow.

Dropout dolines originate through indirect or even direct inheritance if there is a cavity in the cover or the cavity of the bedrock is of large size. For syngenetic suffosion dolines, suffosion and solution is followed by increase in pore volume and then compaction. This process is extending towards the surface. If compaction only affects part of the cover, the portion of the cover above this part subsides, but if the entire cover is compacted, the whole cover sinks. Both processes result in suffosion doline formation. Suffosion dolines, however, also develop through the denudation of the slopes of dropout dolines or non-karstic pipes. With postgenetic suffosion dolines, the loss of the fill of the shaft in the bedrock may be influenced by suffosion, collapse and the melting of a snow plug or ground ice.

On glaciokarsts, syngenetic subsidence dolines form above karren. On mid-latitude karst, they develop above blind shafts developing into shafts above the karst water table. On tropical karsts of fenglin type or in poljes, they occur above shafts and cavities which formed below the karst water table if the water level sinks.

Subsidence pseudokarst depressions are due to non-karstic processes. In their origin several processes may play a part: transport of cover of volcanic rock into gas bubble cavities (by suffosion or collapse) or subsidence of rock masses above mine chambers which extends over the cover.

A condition of ponor formation is the existence of rock boundary (true rock boundary, structural rock boundary or buried rock boundary), lack of surface runoff and the level of karst water below the surface and the formation or opening up of a conduit. Ponors form at the termination of overlain or interbedded cover as well as on a mantle-like non-karstic rock; with overlain cover allogenic water conduction, if the karst water table lies at greater depth or close to the surface, but sinking slowly; and with mantle-like cover on the valley floors of cryptokarst developed from buried karst (at valley rock boundary), on autogenic cryptokarst (at evaporite breccia pipes), on transitional cryptokarst (from caprock dolines of karst windows) and finally where the high karst water reaches the surface of the bedrock.

DSDs develop where the cover material, following local denudation, is transported into the karst through ponors or subsidence dolines. On concealed karsts, the depressions are due to the growth and subsequent coalescence of subsidence dolines. Another way of their formation is that the developing doline is transformed by linear erosion into a more extensive depression (the doline and its ravine extend by pluvial erosion) or that the surface of the environs of the subsidence doline is transformed into a basin by pluvial erosion.

On allogenic cryptokarst, DSDs form from the coalescence of blind valleys. At katavothra first subsidence dolines emerge, and then they broaden by mass movements, by pluvial erosion and by the coalescence of marginal subsidence dolines with the doline broaden into DSD.

The evolution of the DSD depends on its material turnover. Deepening, subsequent infilling and rejuvenation are possible. For true DSDs, width is controlled by the diameter of the depression in the bedrock, while for pseudodepressions, the dip of the bedrock.

Remnant caves are the products of the opening up of cavities formed below the karst water table or of the exhumation of buried cavities. Opening is caused by linear erosion, collapse or the denudation of valley sides.

References

Al-fares W, Bakalowicz M, Guerin R, Dukhan M (2002) Analysis of the karst aquifer structure of the Lamalou area (Herdult, France) with ground penetraradar. J Appl Geophys 51:97–106
Andrejchuk V, Klimchouk A (2002) Mechanism of karst breakdown formation in the gypsum karst of the fore–Ural Region, Russia (from observations in the Kungurskaja Cave). Int J Speleol Thema Issue 31:89–114
Arrington DV, Lindquist RC (1987) Thickly mantled karst of the Interlachen, Florida, area. In: Beck BF, Wilson WL (eds) Karst hydrogeology: engineering and environmental applications. Balkema, Rotterdam, pp 31–39
Axaixiang W, Yezhy S (2008) Granular dynamic theory and its applications. Springer, Dondrecht, 364 p
Balázs D (1984) Exhumált trópusi őskarszt Lapinha vidékén (Minas Gerais, Brazília) (Exhumed tropical paleokarst in the Lapinha area) (Minas Gerais, Brazil). Karszt Barlang II:87–92 (in Hungarian)
Beck BF (1986) A generalized genetic framework for the development of sinkholes and karst in Florida, USA. Environ Geol Water Sci 8:5–18
Beck BF (1991) On calculating the risk of sinkholes collapse. In: Karsting EH, Kasting KM (eds) Appalachian karst, Proceedings of the Appalachian karst symposium, National Speleological Society, Huntsville, pp 231–236
Beck BF (2000) Report on Italian sinkhole conference. KWI Conduit, 8(1) www.karstwaters.org
Beck BF, Sinclair WC (1986) Sinkholes in Florida: an introduction. Florida Sinkhole Research Institute Report 85-86-4, 16 p
Beese AP, Creed MJ (1995) A database for subsidence sinkholes near Cork, Ireland. Dan Geotech Soc Bull 4:19–24
Beggs TF, Ruth BE (1984) Factors affecting the collapse of cavities. In: Beck BF (ed) Sinkholes: their geology, engineering and environmental impact. Balkema, Rotterdam, pp 183–188
Bengtsson TO (1987) The hydrologic effects from intense ground-water pumpage in eastcentral Hillsborough County, Florida. In: Beck BF, Wilson WL (eds) Karst hydrogeology: engineering and environmental applications. Balkema, Rotterdam, pp 109–114
Benson RC, Kaufmann RD (2001) Characterization of a highway sinkhole within the gypsum karst of Michigan. In: Beck BF, Herring JG (eds) Geotechnical and environmental applications of karst geology and hydrology. Balkema, Lisse, pp 103–112
Benson RC, Yuhr LB (1987) Assessment and long term monitoring of localized subsidence using ground penetrating radar. In: Beck BF, Wilson WL (eds) Karst hydrogeology: engineering and environmental applications. Balkema, Rotterdam, pp 161–169
Bergado DT, Selvanayagam AN (1987) Pile foundation problems in Kuala Lumpur Limestone, Malaysia. Q J Eng Geol 20:159–175
Bögli A (1960) Kalklösung und Karrenbildung. Z Geomorphol Suppl 2:4–21
Bögli A (1964) Le Schichttreppenkarst. Un exemple de complexe glaciokarstique. Rev Belg Geogr 1(2):64–82
Bretz JH (1942) Vadose and phreatic features of limestone caverns. J Geol 50:675–811
Bretz JH (1956) Caves of Missouri. Missouri Geological Survey and Water Resources, series 2, 39 p
Brink ABA, Partridge TC (1965) Transvaal karst: some considerations of development and morphology, with special reference to sinkholes and subsidence in the Far West Rand. S Afr Geogr J 47:11–34
Büdel J (1957) Die Doppelten Einebrungsflächen in den feuchten Tropen. Z Geomorphol N F 1:201–288
Bull PA (1980) The antiquity of caves and dolines in the British Isles. Z Geomorphol Suppl 36:217–232
Butzer KW (1976) Geomorphology from the earth. Harper and Row, New York, 463 p

Chen J (1988) Karst collapses in cities and mining areas, China. Environ Geol Water Sci 12:29–35

Chen J, Beck BF (1989) Qualitative modelling of the cover-collapse process. In: Beck BF (ed) Engineering and environmental impacts of sinkholes and karst. Balkema, Rotterdam, pp 98–95

Christiansen EA (1971) Geology of the crater lake collapse structure in southeastern Saskatchewan. Can J Earth Sci 8:1505–1513

Clayton KM (1966) The origin of the landforms of the Malham area. Field Stud J 2:359–384

Cramer H (1941) Die Systematik der Karstdolinen. N Jahrb Miner Geol Paläont 85:293–382

Crawford NC (2001) Environmental problems associated with urban development upon karst, Bowling Green, Kentucky. In: Beck BF, Herring JG (eds) Geotechnical and environmental applications of karst geology and hydrology. Balkema, Lisse, pp 397–424

Currens JC, Paylor RL, Beck EG, Davidson B (2012) A method to determine cover – collapse frequency in the Western Pennyroyal karst of Kentucky. J Cave Karst Stud 74(3):292–299

Currin WE, Barfus BL (1989) Sinkhole distribution and characteristics in Pasco County, Florida. In: Beck BF (ed) Engineering and environmental impacts of sinkholes and karst. Balkema, Rotterdam, pp 97–106

Curtis LF, Courtney FM, Trudgill ST (1976) Soils in the British Isles. Longman, London, 380 p

Cvijič J (1893) 'Das Karstphänomen'. Versuch einer morphologishen Monographie. Geogr Abh von A Penck 5(3):217–329

Day M, Waltham T (2009) The pinnacle karrenfields of Mulu. In: Ginés Á, Knez M, Slabe T, Dreybrodt W (eds) Karst rock features. Karren Sculpturing Zalozba ZRC. Institut za raziskovanje krasa ZRC SAZU, Postojna-Ljubljana. Carsologica 9, pp 423–432

Drumm EC, Kane WF, Yoon CJ (1990) Application of limit plasticity to the stability of sinkholes. Eng Geol 29:213–225

Erhartič B (2012) Geomorfološka dedišeinau Dolini Triglavskih jezer-Geografija Slovenije 23, ZRC SAZU 187p

Foose RM (1953) Ground-water behaviour in the Hershey Valley, Pennsylvania. Bull Geol Soc Am 64:623–645

Foose RM (1968) Surface subsidence and collapse caused by ground water withdrawal in carbonate rock areas. In: Proceedings 23rd international geological congress 12, pp 155–166

Ford DC (1971) Geologic structure and a new explanation of limestone cavern genesis. Trans Cave Res Group G B 13(2):81–94

Ford DC (1984) Karst groundwater activity and landform genesis in modern permafrost regions of Canada. In: LaFleur RG (ed) Groundwater as a geomorphic agent. Allen & Unwin, London, pp 340–350

Ford DC, Williams PW (1989) Karst geomorphology and hydrology. Unwin Hyman, London, 601 p

Ford DC, Williams PW (2007) Karst hydrogeology and geomorphology. Wiley, Chichester, 561 p

Ford DC, Salomon JN, Williams PW (1997) The Lunan Stone forest as a potential world heritage site. In: Song L, Waltham T, Cao N, Wang F (eds) Stone forest: a treasure of natural heritage, Proceedings of the international symposium for Lunan Shilin to apply for World Natural Heritage Status. China Environmental Science Press, Beijing, pp 107–123

Füzesi I (2007) Loess tests carried out in the surroundings of some covered karstic depressions (Tes-Plateau). Carpathian J Earth Environ Sci 2(2):39–44

Guo X (1991) Geological hazards of China and their prevention and control. Geological Publishing House, Beijing, 260 p

Hamblin WK (1989) The earth's dynamic systems. Mcmillan, New York, 410 p

He K, Liu C, Wang S (2003) Karst collapse related to over-pumping and a criterion for its stability. Environ Geol 43:720–724

Hevesi A (2001) A Nyugati-Mecsek felszíni karsztosodásának kérdései (Questions of the surface karstification of the West Mecsek). Karsztfejlődés VI:103–111 (in Hungarian)

Hyatt JA, Wilkes HP, Jacobs PM (1999) Spatial relationships between new and old sinkholes in covered karst, Albany, Georgia, USA. In: Beck BF, Petit AJ, Herring JG (eds) Hydrogeology and engineering geology of sinkholes and karst. Balkema, Rotterdam, pp 37–44

Hyatt JA, Wilson R, Givens JS, Jacobs PM (2001) Topographic, geologic and hydrogeologic controls on dimension and locations of sinkholes in thick covered karst, Lowndes Country, Georgia. In: Beck BF, Herring JG (eds) Geotechnical and environmental applications of karst geology and hydrology. Balkema, Lisse, pp 37–45

Jakucs L (1977) Morphogenetics of karst regions. Adam Hilgar, Bristol, 284 p

Jammal SE (1984) Maturation of the Winter Park sinkhole. In: Beck BF (ed) Sinkholes: their geology, engineering and environmental impact. Balkema, Rotterdam, pp 363–369

Jaskó S (1959) Vízmérések a bakonyi karsztszurdokokban (Water measurements in the gorges of the Bakony karst). Karszt Barlangkutatási Tájékoztatók 4:30–31 (in Hungarian)

Jaskó S (1961) A balatonfelvidéki és északbakonyi patakok vízhozamának a kapcsolata a vízföldtani felépítéssel (The relationship between the discharge of the streams of the Balaton-felvidék and of Northern Bakony and the hydrological structure). Hidr Közl 41:75–85 (in Hungarian)

Jennings JN (1966) Building on dolomites in the Transvaal. Civ Eng S Afr 8:41–62

Jennings JN (1977) Limestone tablet experiments at Cooleman Plain, New South Wales, Australia and their implications. Abhandlungen zur karst und Höhlenkundr; Reine A-Speleologie Heft; Festchrift für Alfred Bögli, pp 526–538

Jennings JN (1985) Karst geomorphology. Basil Blackwell, New York, 293 p

Jones RJ (1965) Aspects of the biological weathering of limestone pavements. Proc Geol Assoc 76:421–434

Kárpát J (1982) Alba Regia-barlang (The Alba Regia cave) – Magyarország barlangtérképei, vol 2. Magyar Karszt és Barlangkutató Társulat, Budapest (in Hungarian)

Kiss K, Zámbó L, Fehér K, Móga J (2007) A lösztakaró karsztosodásban játszott szerepének vizsgálata a Tési-fennsíkon (An investigation of the role of the loess mantle in karstification on the Tési plateau). Karsztfejlődés XII:193–205 (in Hungarian)

Klimchouk A (2004) Evaporite karst. In: Gunn J (ed) Encyclopedia of caves and karst science. Fitzroy Dearborn, New York, pp 343–347

Klimchouk A, Andrejchuk V (2002) Karst breakdown mechanisms from observations in the gypsum caves of the western Ukraine: implications for subsidence hazard assessment. Int J Speleol 31(1/4):55–88

Korzhuev SS (1961) Merzlotnyi karst Srednego Prilen'ya i nekotorye osobennosti yego proyavleniya (The Middle-Lena frozen karst and its characteristics). In: Sokolov NI, Gvozdetskiy NA, Balashov LS (eds) Regionalnoe karstovedenie. Izdatelstvo AN SSSR, Moscow, pp 207–220

Korzhuev SS (1972) Drevniy karst i tsikly karstoobrazovaniya Sibirskoy platformy (A paleokarst and karstification cycles on the Siberian platform). Tr Moskovskogo Obshshestva Ispytateley Prirody XLVII:141–151

Kósa A (1981) Bir Al Ghanam gipsz barlangjai (Líbia) (Gypsum caves of Bir al Ghanam, Lybia). Karszt Barlang I-II:21–26 (in Hungarian)

Kósa A, Csernavölgyi L (1983) Barlangok a líbiai Al Akhdar-hegységben (Caves in the Al Akhadar Mountains of Lybia). Karszt Barlang I-II:35–42 (in Hungarian)

Kruse S, Grasmueck M, Weiss M, Viggiano D (2006) Sinkhole structure imaging in covered Karst terrain. Geophys Res Lett 33:L16405. doi:10.1029/2006GL026975

LaMoreaux PE, Newton JG (1986) Catastrophic subsidence: an environmental hazard, Shelby County, Alabama. Environ Geol Water Sci 8:25–40

Lauriol B, Ford DC, Cinq-Mars J, Morris WA (1997) The chronology of speleothem deposition in Northern Yukon and its relationship to permafrost. Can J Earth Sci 34:902–911

LeGrand HL, LaMoreaux PE (1975) A karszt hidrogeológiája és hidrológiája. In: Burger A, Dubertret A (eds) Karsztterületek Hidrogeológiája. MKBT, Budapest, pp 9–21

Lei M, Jiang X, Yu L (2001) New advances of karst collapse research in China. In: Beck BF, Hering JE (eds) Geotechnical and environmental applications of karst, geology and hydrology. Balkema, Lisse, pp 145–151

Li J, Niu J, Liu Q, Li G (1983) Subsidence and its treatment in karst coal mines in China. In: Proceedings 11th international congress speleology, Beijing, pp 180–181

Lippmann L, Kiss K, Móga J (2008) Az Abaliget-Orfűi karszt karsztos felszínformáinak vizsgálata térinformatikai módszerekkel (Investigation of the karstic phenomenon near Orfű and Abaliget by GIS methods). Karsztfejlődés XIII:151–166 (in Hungarian)

Lungersgauzen GF (1966) Inflyuviy-osobyi geneticheskiy tip materikovykh obrazovaniy (Genetical types of characteristic landform features). Dokl AN SSSR 171(3):690–693

Magdalene S, Alexander EC (1995) Sinkhole distribution in Winona Country, Minnesota, revisited. In: Beck BF (ed) Karst Geohazards. Balkema, Rotterdam, pp 43–51

Mangin A (1997) Some features of the Stone forest of Lunan, Yunnan, China. In: Song L, Waltham T, Cao N, Wang S (eds) Stone forest: a treasure of natural heritage, Proceedings of the international symposium for Lunan Shilin to apply for World Natural Heritage Status. China Environmental Science Press, Beijing, pp 68–70

Mari L, Nagy B, Heiling Z, Nemerkényi Z (2014) Pszeudokarszt a Száraz-Andokban? – Az Ojos del Salado olvadáshoz köthető felszínformái (Pseudokarst in the "Arid Andes"? Landforms of melting origin in the Ojos del Salado area). Karsztfejlődés XIX:231–241 (in Hungarian)

Móga J, Németh R (2005) The morphological research of the basalt and loess covered plateaus in the Bakony Mts. (Transdanubian middle mts.-Hungary). Acta Carsologica 34(2):397–414

Newton JG (1987) Development of sinkholes resulting from man's activities in the eastern United States. US Geological Survey Circular 968, 54 p

Newton JG, Hyde LW (1971) Sinkhole problem in and near Roberts industrial subdivision, Birmingham, Alabama. Geol Surv Ala Circ 68, 42 p

Osmaston HA (1980) Patterns in trees, rivers and rocks in the Mulu Park, Sarawak. Geogr J 146(1):33–50

Osmaston HA, Sweeting MM (1982) Geomorphology of the Gunung Mulu National Park, Sarawak. Sarawak Mus J 51:75–94

Parise M (2011) Surface and subsurface karst geomorphology in the Murge (Apulia, Southern Italy). Acta Carsologia 40(1):79–93

Paton JR (1964) The origin of the limestone hills of Malaya. J Trop Geogr 18:134–139

Penck A (1924) Das unterirdische Karstphänomen. In: Vujevic P (ed) Zbornik Radova Posvecen Jovanu Cvijicu, Belgrade, pp 175–197

Peng J, Cai Y, Yang M, Liang H, Liang F, Song L (2007) Relating aerial erosion, soil erosion and sub-soil erosion to the evolution of Lunan Stone Forest, China. Earth Surf Processes Landf 32:260–268

Plan L, Decker K (2006) Quantitative karst morphology of the Hochschwab plateau Eastern Alps, Austria. Z Geomorphol N F 147.29–54

Pulina M (2005) Le karst et les phenomenes karstiques similaires des regions froides. In: Salomon JN, Pulina M (eds), Les karsts des regions climatiques extremes, vol 14. Karstologia Me´moires, pp 11–100

Renault P (1968) Contribution a l'étude des action mécanique et sédimentologiques dans la spéléogenèse. An Spéléologie 23(3):529–596

Silvestru E (1997) Dolines in the Padiş Plateau (Bihor Mountains, Romania), one peculiar case, many questions. Theor Appl Karstology 10:127–132

Sinclair WC (1982) Sinkhole development resulting from ground-water withdrawal in the Tampa area, Florida. US Geological Survey Water Resources Investigation 81–50, 19 p

Slabe T (1995) Cave rocky relief. Znanstvenaraziskovalni Center Sazu, Llubljana, 128 p

Slabe T (1999) Subcutaneous rock forms. Acta Carsologica 28(2):255–271

Slabe T (2005) Two experimental modelings of karst rock relief in plaster: subcutaneous 'rock teeth' and 'rock peaks' exposed to rain. Z Geomorphol 49(1):107–119

Slabe T (2009) Karren simulation with plaster of Paris models. In: Ginés Á, Knez M, Slabe T, Dreybrodt W (eds) Karst rock features. Karren Sculpturing Zalozba ZRC. Institut za raziskovanje krasa ZRC SAZU, Postojna-Ljubljana. Carsologica 9, pp 47–61

Slabe T, Liu H (2009) Significant subsoil rock forms. In: Ginés Á, Knez M, Slabe T, Dreybrodt W (eds) Karst rock features. Karren Sculpturing Zalozba ZRC. Institut za raziskovanje krasa ZRC SAZU, Postojna-Ljubljana. Carsologica 9, pp 123–137

Song LH (1986) Origination of stone forest in China. Int J Speleol 15(1–4):3–33
Song L, Liang F (2001) Distribution of CO_2 in soil air affected by vegetation in the Shilin National Park. Acta Geol Sin 75(3):288–293
Song L, Liang F (2009) Two important evolution models of Lunanshilin karst. In: Ginés A, Knez M, Slabe T, Dreybrodt W (eds) Karst rock features. Karren Sculpturing Zalozba ZRC. Institut za raziskovanje krasa ZRC SAZU, Postojna-Ljubljana. Carsologica, 9, pp 453–459
Sowers GF (1996) Building on sinkholes. ASCE Press, New York, 202 p
Spooner J (1971) Mufulira interim report. Min J 276:122
Stefanovits P (1981) Talajtan (Soil science). Mezőgazdasági Kiadó, Budapest, 380 p (in Hungarian)
Sweeting MM (1973) Karst landforms. Columbia University Press, New York, 362 p
Szalai S, Veress M, Novák A, Szarka L (2006) Geofizikai vizsgálatok fedett karszton (Homód-árok, Bakony) (Geophysical investigations on covered karsts, Homód-valley, Bakony). Karsztfejlődés XI:155–170 (in Hungarian)
Telbisz T, Mari L, Kohán B, Jelena C (2007) A szerbiai Miroč-hegység töbreinek térinformatikai GPS-es terepi vizsgálata (A spatial informatics GPS study of the dolines of the Serbian Miroč Mountains). Karsztfejlődés XII:71–90 (in Hungarian)
Tharp TM (1999) Mechanics of upward propagation of cover-collapse sinkholes. Eng Geol 52:23–33
Thomas TM (1974) The South Wales interstratal karst. Trans Br Cave Res Assoc 1:131–152
Trudgill ST (1972) The influence of drift and soils on limestone weathering in N. W. Clare Ireland. Proc Univ Bristol Speleol Soc 12:113–118
Trudgill ST (1975) Measurement of erosional weight-loss of rock tables. Br Geomorphol Res Group Tech Bull 17:13–19
Trudgill ST (1976a) Limestone erosion under soil. In: Panos V (ed) Proceedings of the 6th international congress of speleology, vol II. Ba. Academia, Prague, pp 409–422
Trudgill ST (1976b) The erosion of limestones under soil and long term stability of soil – vegetation systems on limestone. Earth Surf Processes 1:31–41
Trudgill ST (1985) Limestone geomorphology. Longman, New York, 196 p
Upchurch SB, Littlefield JR (1987) Evaluations of data for sinkhole-development risk models. In: Beck BF, Wilson WL (eds) Karst hydrogeology: engineering and environmental applications. Balkema, Rotterdam, pp 359–364
Veress M (1982) Adatok a Hárskúti-fennsík morfogenetikájához (Data to the morphogenetics of the Hárskút plateau). Karszt Barlang II:71–82 (in Hungarian)
Veress M (1983) Eltérő magasságú tönkfelszínek karsztosodásának kérdései az Északi-Bakony keleti részén (Questions of karstification of the planation surfaces of different heights on the eastern part of the Northern Bakony Mountains). Folia Mus Historico-Nat Bakonyiensis 2:29–44 (in Hungarian)
Veress M (2000) Covered karst evolution Northern Bakony mountains, W-Hungary. A Bakony Természettud. Kut. Eredményei, 23, Bakonyi Természettudományi Múzeum, Zirc, 167 p
Veress M (2008) A mészkőfekü morfológiájának hatása a fedett karsztosodásra az Északi-Bakonyban (The effect of the morphology of the limestone bedrock on covered karstifiaction in the Northern Bakony Mountains). Karszt Barlang 2004–2005 I–II:33–54 (in Hungarian)
Veress M (2009a) Rinnenkarren. In: Ginés Á, Knez M, Slabe T, Dreybrodt W (eds) Karst rock features, Karren Sculpturing Zalozba ZRC. Institut za raziskovanje krasa ZRC SAZU, Postojna-Ljubljana. Carsologica 9, pp 151–159
Veress M (2009b) Investigation of covered karst form development using geophysical measurements. Z Geomorphol 53(4):469–486
Veress M (2010) Karst environments – karren formation in High Mountains. Springer, Dordrecht, 230 p
Veress M (2011) Adatok a Mecsek-hegység fedett karsztosodásához a Cigány földi mintaterületről vett példák felhasználásával (Some data to the covered karstification of the Mecsek Mountains using examples taken from the Cigány földi research area). Karszt Barlang 2010 I-II:9–30 (in Hungarian)

Veress M (2012a) The development of the depressions of the superficial deposit of the Tési Plateau (Bakony Mountains, Hungary). Karst Dev 2(1):29–44

Veress M (2012b) Glacial erosion and karst evolution (Karren on the surfaces formed by glaciers). In: Veress B, Szigethy J (eds) Horizons in earth science research. Nova Science, New York, pp 1–94

Veress M (2015) Kab mountain: karst under a basalt cap. In: Lóczy D (ed) Landscapes and landforms of Hungary. Springer, Heidelberg, pp 55–62

Veress M, Futó J (1987) Adatok a Hódos-éri Likas-kő morfogenetikájához (Contributions to the morphogenesis of the Likas-kő of the Hódos-ér). Karszt Barlang I-II:9–16 (in Hungarian)

Veress M, Futó J (1990) Fedett, paleokarsztos térszíneken végbement lepusztulás és felhalmozódás kimutatása a Bakony-hegységben (Determination of erosion and accumulation on covered paleokarstic surfaces, Bakony mountains, W. Hungary). Földtani Közlöny 120:55–67 (in Hungarian)

Veress M, Péntek K (1996) Theoretical model of surface karstic processes. Z Geomorphol 40(4):461–476

Veress M, Puskás J (2007) Adalékok az Eleven-Förtési töbörcsoport (Bakony-hegység) karsztosodásához (Contributions to the karstification of the Eleven-Förtes doline group, Bakony Mountains). Karsztfejlődés XII:171–192 (in Hungarian)

Veress M, Lóczy D, Zentai Z, Schläffer R (2008) Kürtőképződés egy madagaszkári sziklás parton (Chimney development on a rocky coast of Madagascar). Karsztfejlődés XIII:135–149 (in Hungarian)

Veress M, Péntek K, Unger Z, Almási I (2012) Development of covered karstic dolines in ground ice environment (Eastern Alps, Austria). Interests of experimental and mathematical modelling. Z Geomorphol 56(2):79–104

Walters RF (1977) Land subsidence in central Kansas related to salt dissolution. Bull Kansas Geol Surv 214:82

Waltham AC (1989) Ground subsidence. Blackie, Glasgow, 202 p

Waltham AC (1997) Pinnacle karst of Gunung Api, Mulu Sarawak. In: Song L, Waltham AC, Cao N, Wang F (eds) Stone forest: a treasure of natural heritage, Proceedings of the international symposium for Lunan Shilip to apply for World Natural Heritage Status. China Environmental Science Press, Beijing, pp 52–55

Waltham AC, Fookes PG (2003) Engineering classification of karst ground conditions. Q J Eng Geol Hydrogeol 36:101–118

Waltham T, Hatherley P (1983) The caves of Leck Fell. Cave Sci 10:245–251

Waltham AC, Smart PL (1988) Civil engineering difficulties in the karst of China. Q J Eng Geol 21:2–6

Waltham T, Bell F, Culshaw M (2005) Sinkholes and subsidence. Springer, Berlin, 382 p

Wassmann TH (1979) Mining subsidence in the East Netherlands. In: Saxena SK (ed) Evaluation and prediction of subsidence. American Society Civil Engineers, New York, pp 283–302

White WB (1988) Geomorphology and hydrology of karst terrains. Oxford University Press, New York, 464 p

White WB, White EL (1992) Sinkholes and sinkhole collapses. In: Majundar SK, Forbes GS, Miller EW, Schmaltz RF (eds) Natural and technological disasters: causes, effects and preventative measures. Academy of Science, Pennsylvania, pp 280–293

Wilford GE, Wall JRD (1965) Karst topography in Sarawak. J Trop Geogr 21:44–70

Williams PW (1966) Limestone pavements with special reference to western Ireland. Trans Inst Br Geogr 40:155–172

Williams PW (2004) Dolines. In: Gunn J (ed) Encyclopedia of caves and karst science. Fitzroy Dearborn, New York, pp 304–310

Williams PW (2009) Arête and pinnacle karst of Mount Kaijende. In: Ginés Á, Knez M, Slabe T, Dreybrodt W (eds) Karst rock features. Karren Sculpturing Zalozba ZRC. Institut za raziskovanje krasa ZRC SAZU, Postojna-Ljubljana. Carsologica 9, pp 433–437

Wilson WL (1995) Sinkhole and buried sinkhole densities and new sinkhole frequencies in karsts of northwest peninsular Florida. In: Beck BF (ed) Karst geohazards. Balkema, Rotterdam, pp 79–91

Xiang S, Chen J, Wilson WL, Beck BF (1988) Sinkholes as a consequence of groundwater development in karst regions. Am Inst Hydrol J 4:160–173

Xu W, Zhao G (1988) Mechanism and prevention of karst collapse near mine areas in China. Environ Geol Water Sci 12:37–42

Yuan D (1987) Environmental and engineering problems of karst geology in China. In: Beck BF, Wilson WL (eds) Karst hydrogeology: engineering and environmental applications. Balkema, Rotterdam, pp 1–11

Yuan D (1991) Karst of China. Geological Publishing House, Beijing, 224. p

Zámbó L (1970) A vörösagyagok és a felszíni karsztosodás kapcsolata az Aggteleki-karszt délnyugati részén (The relationship between red clays and surface karstification at the southwestern part of Aggtelek karst). Földrajzi Közlemények 94(4):281–293 (in Hungarian)

Zhang F, Geng H, Li Y, Yang Y, Ren J, Wang F, Tao H, Li Z (1997) Study on the Lunan stone forest karst. Yunnan Science and Technology Press, Kunming, 155 p

Zseni A (2004) Talaj alatti karrformák (Subsoil karren features). Karsztfejlődés IX:157–175 (in Hungarian)

Zseni A (2009) Subsoil shaping. In: Ginés Á, Knez M, Slabe T, Dreybrodt W (eds) Karst rock features. Karren Sculpturing Zalozba ZRC. Institut za raziskovanje krasa ZRC SAZU, Postojna-Ljubljana. Carsologica 9, pp 103–121

Chapter 8
Evolution of Covered Karst Surfaces

Abstract This chapter presents the evolution of covered karst surfaces. On karsts which develop from bare karst into covered karst, geomorphic evolution (increasing or decreasing extension of the covered karst) depends on the ratio between the inward transport of cover upon the karst and outward transport (removal of material). Here two-phase landform evolution takes place as on recent allogenic covered karst and renewed allogenic covered karst. On other types of covered karst, the cover formed independent from bare karst, and surface evolution has a single phase, such as in the case of horst covered karst, mantled allogenic covered karst, covered karst of glaciokarst and platform covered karst. Surface evolution begins with covered karst stage. In the latter group, a type (horst covered karst) also occurs where surface evolution is differentiated as a consequence of different evolution and different present elevation of karst horsts. Covered karst is continually reproduced where intensive karstification happens. Such type is tropical karst. In an overview of geomorphic evolution, the role bare karst plays in covered karst formation, the phases of bare karst development and the contribution of karstic and non-karstic landforms in landscape evolution are presented. When describing the various paths of evolution, the properties of cover rock are also taken into consideration.

Keywords Covered karst • Surface evolution • Advancing rock boundary • Retreating rock boundary • Infilling of paleodepressions • Exhumation of paleodepressions • DSD development from blind valleys • Transport of cover into the karst • Büdel's model of double planation • Triplex erosion model

8.1 Introduction

The geomorphic evolution of karst surfaces was first studied in the early twentieth century, when Grund (1914) and Cvijič (1918) proposed evolution models. These models applied the principles of W.M. Davis' cycle of erosion to karst. They outlined a general model for the evolution of karst, but did not explain the surface development of different karst types.

In reality, however, the evolution of the individual karst types differs to various extents and shows characteristic properties. This also applies to covered karsts.

M. Veress, *Covered Karsts*, Springer Geology,
DOI 10.1007/978-94-017-7518-2_8

Waltham and Fookes (2003) classified karsts observing engineering aspects (Fig. 8.1). This classification also represents the evolution phases of karst. Two of their classes (juvenile and youthful karsts) are less dominated by karst landforms, and covered karst terrains are of limited extension. For three classes (mature, complex and extreme karsts), karst landforms tend to be larger and larger, and covered karst is becoming more common. (This latter is a novel element compared to previous evolution models.) The above classes can be regarded as phases of the evolution of a tropical karst, but they also show some characteristics of temperate karsts in the early phase of their evolution.

Williams (1985, 1987) outlines tropical karst evolution. Because of intensive dissolution, the surface deepens to the karst water table, and then with the broadening of dolines, corrosion plains and residual hills develop (Williams 1985). He (Williams 1987) presents the formation of stripped terraces and new cockpit generations through the inheritance of watercourses from the cover in the area of intermountain plains (Fig. 8.2).

According to Ford and Williams (2007), however, the dissection of the cover is not always due to incising watercourses but also to the retreat of the cover margin during karstification (Fig. 8.3). Here the spatial configuration of the near-surface rocks decides that in which part of the karst the cover will be removed (see below).

According to Cui et al. (2002), relying on Büdel's (1957) model of peneplanation, the tropical covered karsts in China are due to planation, during which double levelling surfaces form. The lower level is the surface of the bedrock and the upper is that of the cover. The cover is generated during the solution of the limestone, which, although being denuded, is also recharged from the solutional residue of the bedrock. The peneplain of the tropical karst is transformed to the effect of uplift (removal of the cover and dissection of the bedrock), as it happens, for instance, on the Tibet Plateau.

Surface evolution on covered karst is single phased, two phased or continuous. In the case of single-phased evolution, karst formation begins with covered karst conditions. This happens if the karst became rapidly covered (Fig. 8.4). The development of the cover took place either on not yet karstified rock (e.g. during marine sedimentation) or on karstic surface (e.g. if lava covered the karst). In this case covering was independent from karstification. Single-phased surface evolution can take place with either impermeable (lava rock, sandstone) or permeable cover (loess). With impermeable cover, karstic surface evolution is closely associated with fluvial geomorphic evolution or develops from the latter, beginning on buried karst. Single-phased surface evolution is possible on any type of covered karst.

If developed from buried karst, epigenetic valleys, possibly with ponors on their floor, emerge on the non-karstic terrain. The valleys evolve into valleys with doline rows (Jakucs 1977), followed by the complete removal of the cover. This version of surface evolution is not independent, but occurs on several covered karst types (recent allogenic covered karst, horst covered karst). It appears on some portions of the above karst types or characterises the early stage of surface evolution.

In two-phased geomorphic evolution, the cover is extending gradually. Its development is associated with bare karst or karst formation. In the first phase of the

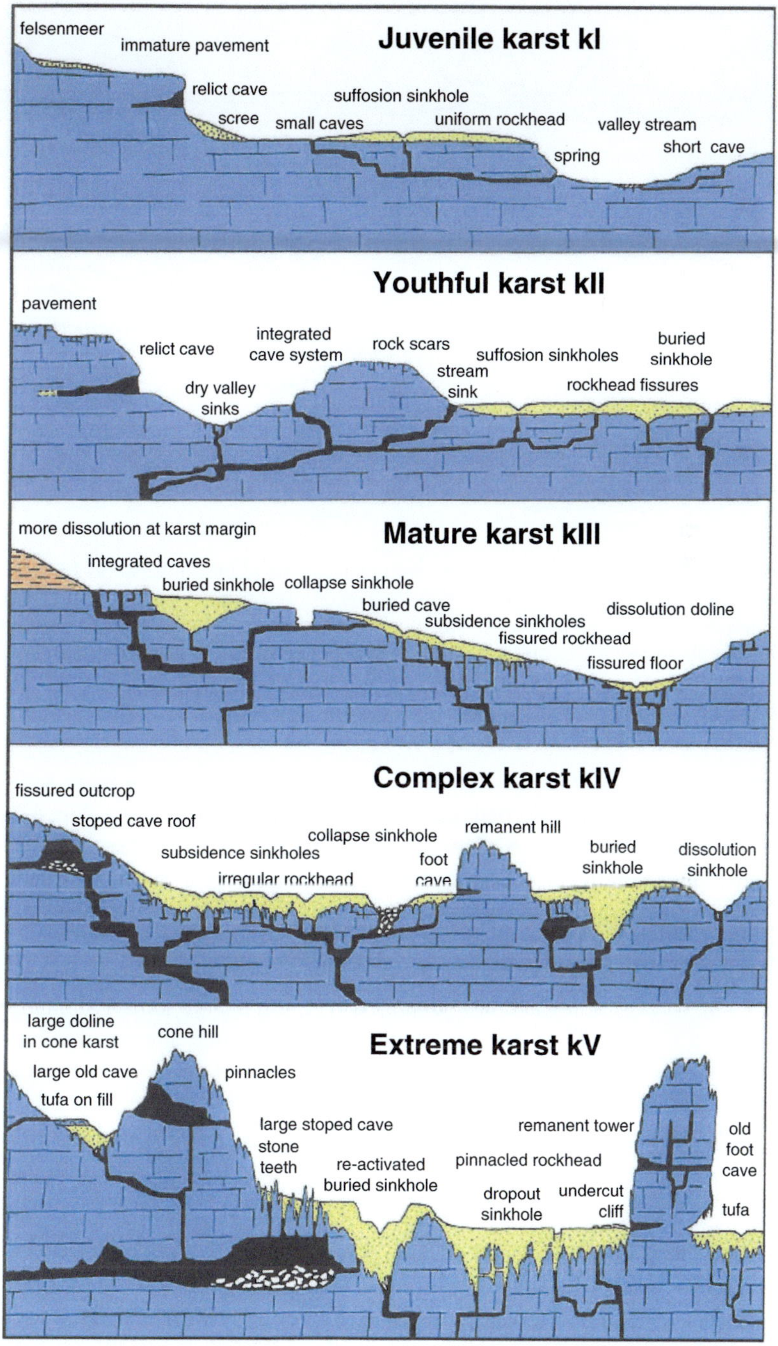

Fig. 8.1 Diagrammatic representation of the five classes in the engineering classification of karst, with various sinkholes and solution dolines forming one of the components of all karst terrains (Waltham and Fookes 2003, 2005)

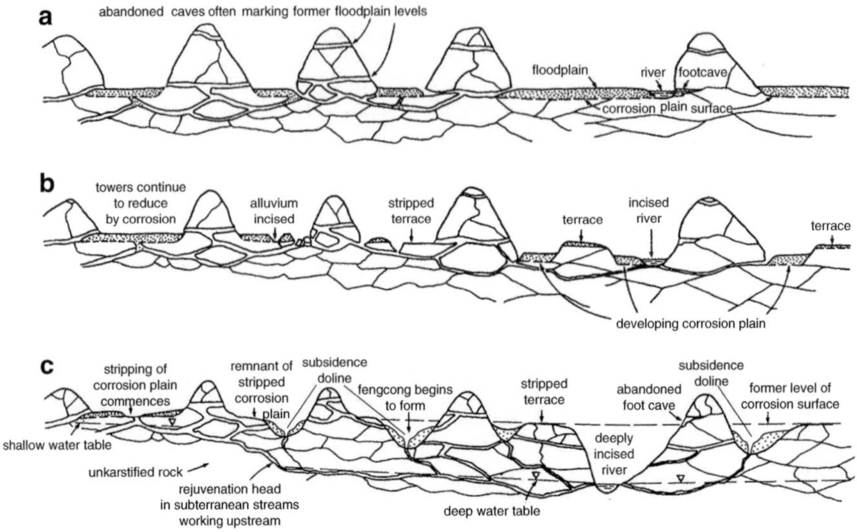

Fig. 8.2 Redevelopment of fengcong karst from rejuvenation of fenglin karst. (**a**) Isolated karst towers on alluvial corrosion plain. (**b**) Shallow incision of river generates lower corrosion plain. (**c**) Deep incision of river result in development of new fengcong. Note that in (**c**) the isolated hills of the previous phase may become the tops of the highest tier of cones in the modern phase (Williams 1987) – modified

two-phased evolution, the non-karstic cover is extending because more sediment accumulates in the area of the bare karst, than it is removed from there (Fig. 8.5a–c). In the second phase of surface evolution, the non-karstic cover is denuded from the karst since more sediment is removed than transported upon the karst (Fig. 8.5d). This type of surface evolution can equally happen with impermeable or permeable cover (Figs. 8.6 and 8.7) and is typical of temperate covered karsts.

Continuous geomorphic evolution takes place if during karstification large amounts of solutional residue are generated (Fig. 8.8). Bare karst is transformed into covered karst, and this state is retained for a long time. Further development (partial or even complete exhumation or burial) depends on the accumulation of solutional residue compared to its removal. This type of surface evolution takes place on tropical and partly on mediterranean karsts.

8.2 Surface Evolution on Tundra and Taiga Covered Karsts

Karstification is hindered by permafrost. Karstic surface development is local (at paleokarst landforms, evaporites) or controlled by valley formation since the watercourses in valleys make the permafrost thaw.

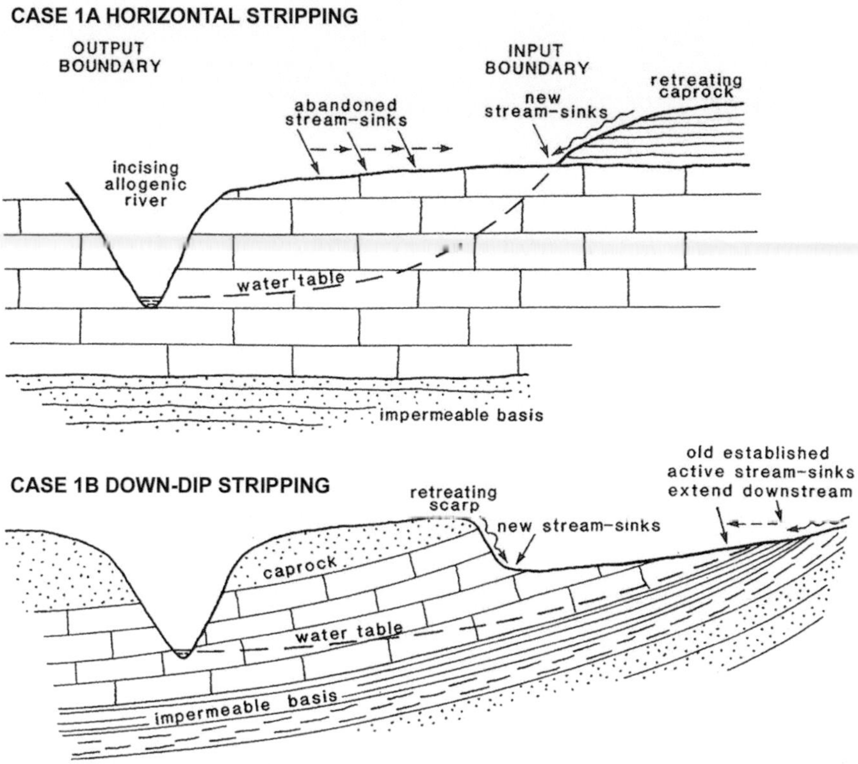

Fig. 8.3 Karstic evolution in uplifted dense carbonate rock protected by cover beds; *case 1a* horizontal beds; *case 1b* strata dip downstream (Ford and Williams 2007)

8.3 Surface Evolution on Temperate Karst

8.3.1 Surface Evolution on Recent Allogenic Covered Karst

The cover redeposited from a non-karstic terrain (cryptokarst or buried karst) accumulates on the uncovered paleokarst surface. In the course of continuous transport, the extension of the cover grows in proportion to the ratio of sediment transport capacity of watercourses to the capacity of ponors as sediment recipients.

If the rate of sediment is higher than ponor capacity, the cover is extending; if its rate is lower, the cover is shrinking. With extending cover, the rock boundary shifts towards the karst interior (advancing rock boundary). During advance the sites of ponor formation are also dislocated towards the karst interior, while the previously formed ponors are buried. Uncovered karst landforms (dolines, uvalas) are filled up to various degrees. In the only partially infilled dolines, the cover can be thin and permeable. Thus, in the foreland of the front of ponor formation, patches of

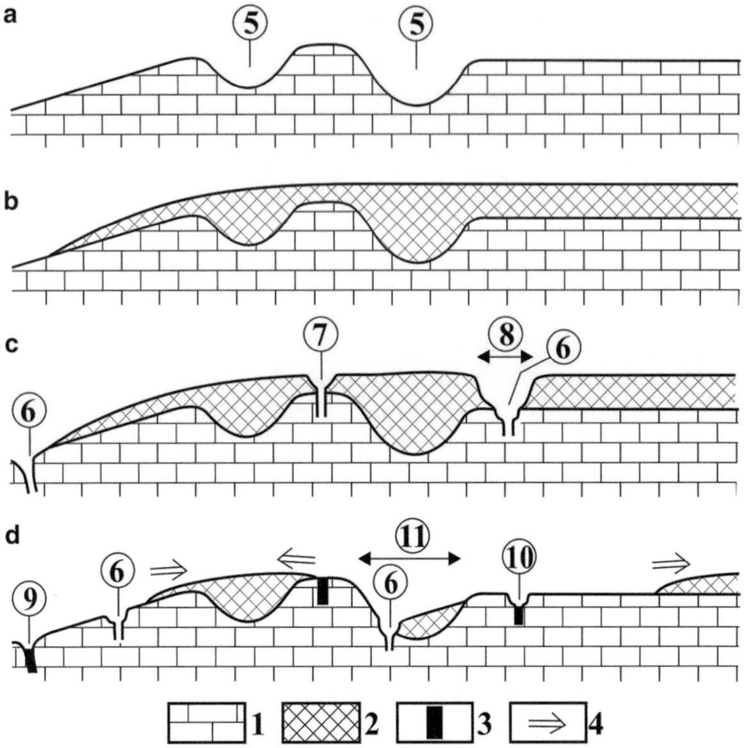

Fig. 8.4 Single-phase surface evolution of covered karst. *1* limestone, *2* non-karstic consolidated rock, *3* karst conduit fill, *4* retreat of rock boundary, *5* solution doline, *6* ponor, *7* caprock doline, *8* valley, *9* inactive ponor, *10* doline (transformed from ponor), *11* true DSD. (**a**) bare karst; (**b**) karst rapidly covered; (**c**) ponors emerge on the margin of non-karstic rock, caprock dolines and ponors in its interior; the cover is dissected into patches, and its extension is reduced; (**d**) the cover patches are shrinking (are only preserved in the former karst depressions, gradually developing bare karst)

concealed karst with subsidence dolines may also emerge. On such a karst, elevated mounds represent the uncovered patches of the covered karst.

If inward transport to the covered karst has a similar rate to that of transport from this area into the karst, the location of the rock boundary does not alter and blind valleys are generated. The site of the rock boundary also remains the same if there are steep slopes (of karst mound or escarpment) between the uncovered karst and the covered karst terrain. Since the end points of blind valleys stay in the same place, their slopes erode heavily and neighbouring blind valleys possibly merge. For this reason, at rock boundaries which remain stable, the chance for the development of DSDs is high. DSD generation involves the dissection of the uniform cover into isolated patches. If less sediment arrives at the uncovered portion of the karst than is transported away from there, the rock boundary shifts towards the opposite margin of the karst (retreating rock boundary). The blind valleys are becoming

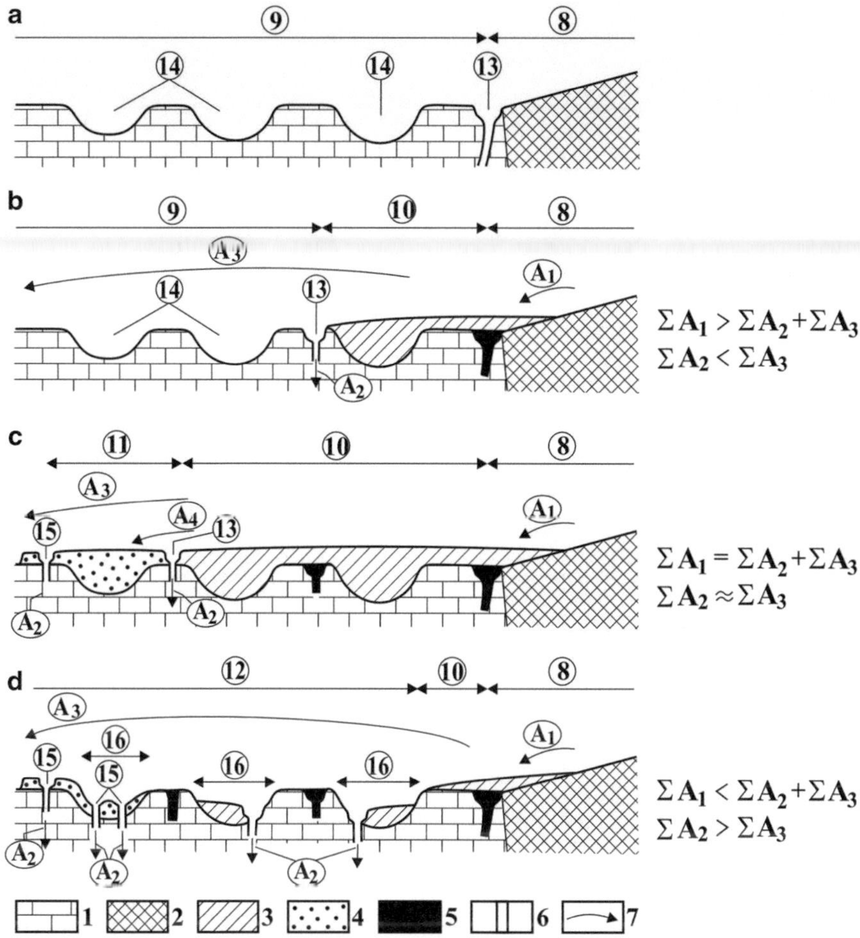

Fig. 8.5 Two-phased surface evolution of covered karst. *1* limestone; *2* consolidated non-karstic rock (impermeable); *3* non-consolidated, impermeable, non-karstic rock; *4* non-impermeable, non-karstic rock; *5* ponor fill; *6* karst conduit; *7* sediment transport; *8* non-karstic terrain (buried karst); *9* bare karst; *10* cryptokarst zone; *11* concealed karst; *12* mixed composite karst; *13* ponor; *14* paleodoline; *15* subsidence doline; *16* true DSD. A_1 sediment transport upon the karst, A_2 sediment transport into the karst, A_3 outward sediment transport from the karst (fluvial), A_4 sediment redeposition within the covered karst. (**a**) bare karst; (**b**) beginning of covered karst evolution (the margin of the cover advances on the karst); (**c**) the bare karst transforms into covered karst (with cryptokarst and concealed karst zone); (**d**) the covered karst transforms into bare karst (the margin of the cover is retreating on the karst; the covered karst patches are shrinking)

shorter, and on their floor, several new ponors emerge, and the site of ponor formation retreats if the karst water table is deep below the surface. The blind valleys develop into valleys with doline rows. The bare karst extends at the expense of the covered karst.

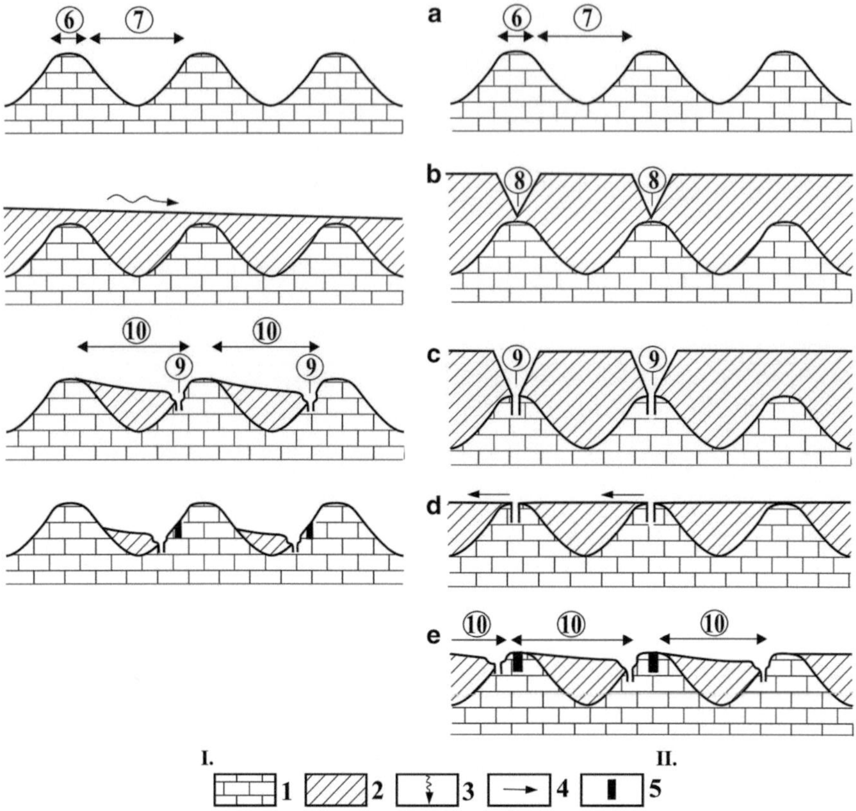

Fig. 8.6 Evolution of temperate buried karst. *I.* the surface slopes perpendicular to the rows of mounds; *II.* the surface slopes in the direction of the rows of mounds; *I.a.* karstic surface develops; *I.b.* it is buried, and then the denudation of the cover deposit begins; *I.c.* ponors develop at true rock boundaries; the cover is transported into the karst; *I.d.* with the lowering of the surface, ponors in lower position and DSDs emerge, and the mounds and paleodolines are exhumed; *II.a.* karstic surface develops; *II.b.* the karst is buried, and valleys form in the cover deposit; *II.c.* ponors develop at valley rock boundaries; *II.d.* valley incision is slowed down, and valley sides in the cover deposit are denuded; *II.e.* ponors develop on the sides of mounds and convey the cover material into the karst, and DSDs emerge. *1* limestone, *2* impermeable cover, *3* direction of transport of cover material, *4* direction of shift of ponor formation, *5* chimney fill, *6* mound, *7* doline, *8* valley, *9* ponor, *10* DSD

The karst can extend over the bordering buried karst terrains. Then ponors emerge on the inheriting valley floors. As a consequence of complete inheritance, the ponors on the valley floors are transformed into dolines, which fill up and turn into covered karst patches.

With shrinking or stable cover, patches of cover emerge on the bare karst, in the area of paleodolines. In the area of covered karst, patch surface evolution is island-like, isolated and independent. Depending on the composition of the cover, the patches are either cryptokarst or concealed karst terrains. The sediment of the patches is transported into the karst through subsidence dolines or ponors.

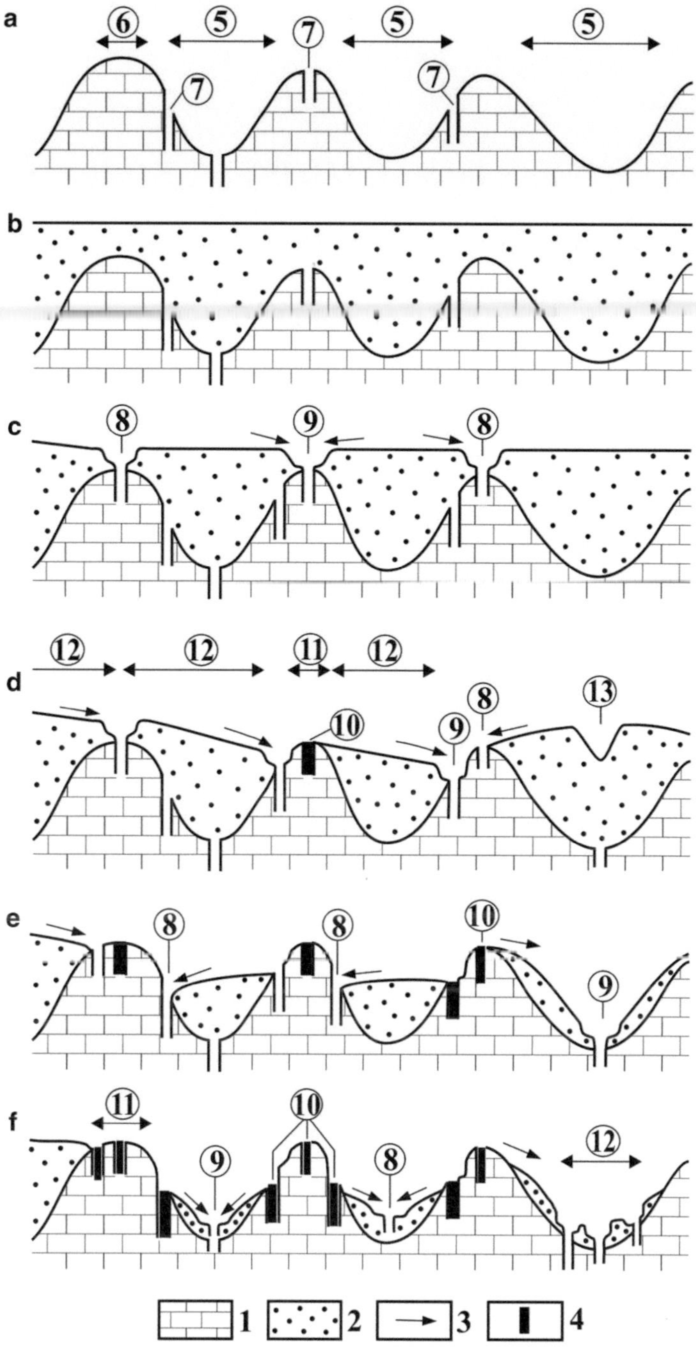

Fig. 8.7 Surface evolution of temperate concealed karst. *1* limestone, *2* permeable cover deposit, *3* direction of transport of cover deposit, *4* chimney fill, *5* doline, *6* mound between dolines, *7* chimney, shaft, *8* syngenetic subsidence doline, *9* postgenetic subsidence doline, *10* destroyed subsidence doline, *11* exhumed mound, *12* DSD, *13* valley. (**a**) bare karst surface; (**b**) burial of karst; (**c**) above the mounds subsidence dolines develop; (**d**) DSDs develop; (**e–f**) paleodolines are exhumed through the transport of the cover into the karst

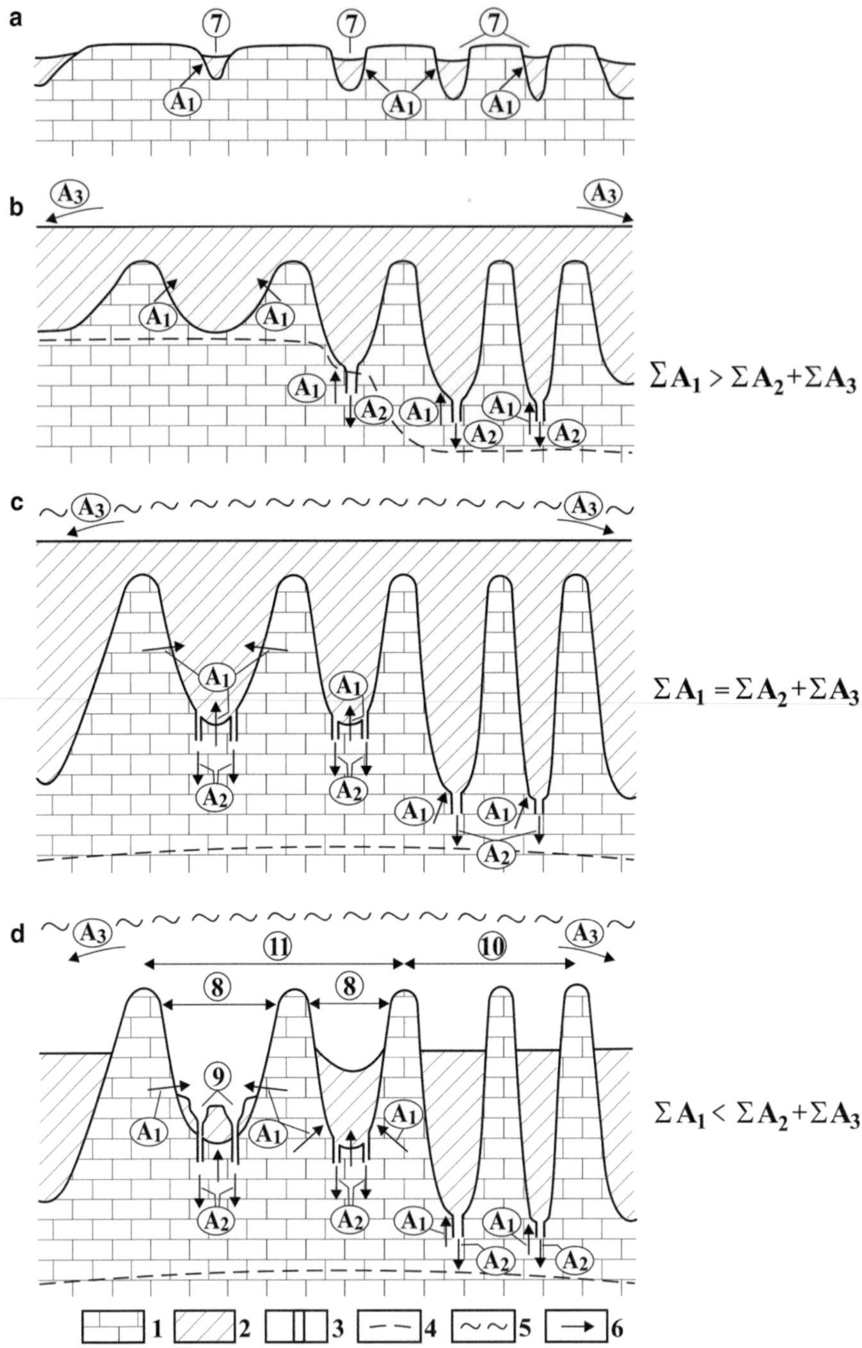

Fig. 8.8 Surface evolution of continuously regenerating covered karst. *1* limestone, *2* solution residue, *3* chimney, shaft, *4* karst water table, *5* former surface, *6* sediment transport, *7* grike, *8* doline or true DSD, *9* subsidence doline, *10* pinnacle karst, *11* inselberg karst (fengcong), A_1 generation of solution residue, A_2 sediment transport into the karst, A_3 sediment transport

DSDs come about, intensively deepening since they do not receive considerable amounts of sediment recharge from their environments. (This situation only changes if sediment transport intensifies from the non-karstic terrain.) With the deepening of the DSD and the exhumation of paleodolines and uvalas, the partially covered karst is transformed into bare karst again.

8.3.2 Surface Evolution on Renewed Allogenic Covered Karst

The uplifted or tilted uplifted karst has lost its sediment source and its certain parts even lost their cover. During exhumation, valleys were inherited over the bedrock, where ponors and then solution dolines emerged.

On the uncovered part of the karst, solution dolines formed both before and after cover formation occur. On the cover patches of the dolines, covered karst formation takes place. On the cryptokarst or concealed karst of the cover, DSDs deepen. The paleodolines are being exhumed.

On portions of the karst which are tilted to slope in opposite direction relative to the overall slope of the karst, cover deposits are transported towards the uncovered zones of the karst. At the edge of the cover, a ponor row with blind valley forms. The cover is dissected into patches, which are preserved in dolines (uvalas) formed before cover development. The paleodolines are also exhumed here (Fig. 8.9).

The section of the karst which acquired a lower position due to tilting does not slope towards the karst interior. Therefore, the watercourses do not flow towards the karst interior but leave the karst. The cover being thick, the valleys cannot be inherited over to the bedrock, or if they are inherited, only in the case if the height difference between the karst and the local base level is in relatively high position. Instead of karstic, erosional surface evolution is typical.

8.3.3 Surface Evolution on Mantled Allogenic Covered Karst

A consolidated rock covers the karstic limestone surface (transitional cryptokarst). Both at the edge of the cover rock and in the interior at karst windows, karstification takes place. At the edge of the cover, ponors with blind valleys emerge, while in the interior, caprock dolines emerge. Caprock dolines transform into ponors which coalesce and thus constitute DSDs. The denudation of the edge of the cover results in its retreat. In the interior of the cover, karstic denudation is a local phenomenon (observed at karst windows). The valleys leading to karst landforms cut through the cover and divide it into patches (Fig. 8.4).

Fig. 8.8 (continued) from the karst (fluvial). (**a**) bare karst with grikes; (**b**) covered karst forms, and the cover thickens; (**c**) dynamic equilibrium of the covered karst conditions; certain parts take pinnacle and other fengcong shape; the vertical dissection of karst increases; (**d**) exhumation, further intensification of vertical dissection of karst; the extension of the covered karst is reduced at the expense of the bare karst

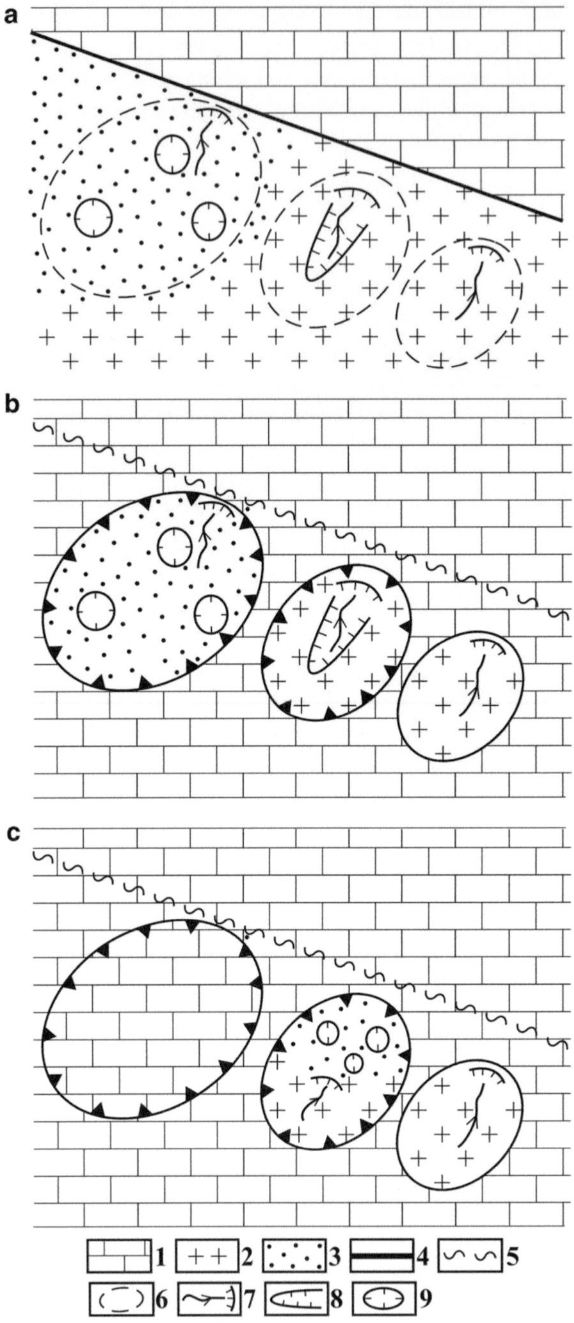

Fig. 8.9 Evolution of an allogenic karst dissected with paleokarst depressions. *1* limestone, *2* impermeable cover, *3* permeable cover, *4* rock boundary, *5* former location of rock boundary, *6* buried paleodepression, *7* ponor, *8* blind valley, *9* subsidence doline, (**a**) uniform cover, (**b**) cover dissected into patches, (**c**) paleodepressions are exhumed

8.3.4 *Surface Evolution on Horst Covered Karst*

In the area of mountains dismembered into blocks, karstification and, thus, geomorphic evolution are mosaical. Next to each other's blocks, groups of blocks of karstic and non-karstic evolution occur. (The karstic evolution of neighbouring blocks can be largely different, too.) The possible varieties of surface evolution are the following:

- The surface of the block is densely dissected by crests and mounds (Fig. 8.10). From the areas between crests and mounds, if the block is tilted, the cover is removed by pluvial erosion. On the thinning cover of these sites, subsidence dolines come about. Beyond some point of time, the cover cannot be transported on the surface (the exhuming crests create barriers to sediment transport), and the sediment is redeposited into the subsidence dolines. Between the crests, DSDs emerge. Denudation and DSD deepening is promoted by the presence of fossil subsidence dolines since in their area the cover is impermeable. This way the cover can be redeposited from their area into the dolines by pluvial erosion.
- On the block surfaces, the density of mounds is lower. In this case subsidence dolines mostly form above buried mounds. Through the coalescence of dolines or by erosional processes (ravine formation, pluvial erosion on the doline slopes), DSDs develop. To the effect of fossil dolines, the removal of the cover is also enhanced in this case. In the above cases, the denudation of the cover will not be complete. In the neighbourhood of the exhuming mound, if the cover is thick, no further subsidence doline emerges. Subsequently, from the environment of the mounds, the cover is removed, at most, by sediment transport on the surface. Karstification only involves the local denudation of the cover.
- Subsidence dolines emerge above the paleodolines buried on the block surface. In the environs of subsidence dolines, the denudation of the cover exposes paleodolines. The karstic denudation of the cover is also local in this case.
- Subsidence dolines come about on the cover accumulated subsequently on the inherited valley floors of the block. Through the redeposition of their cover into the subsidence dolines, details of the valley floor may be transformed into DSDs, and this results in distinctive covered and uncovered sections of the valley floors.
- The block is overlain by impermeable deposits and epigenetic valleys emerge. Valley evolution is enhanced by the cavities formed below the karst water table. Surface evolution on the block is erosional. The percolation of watercourses from the inherited valley floors may moderate the rate of erosional valley evolution.

On the individual blocks, the above described varieties of surface evolution are occasionally present simultaneously. Therefore, the evolution of block surfaces can be jointly influenced by several processes.

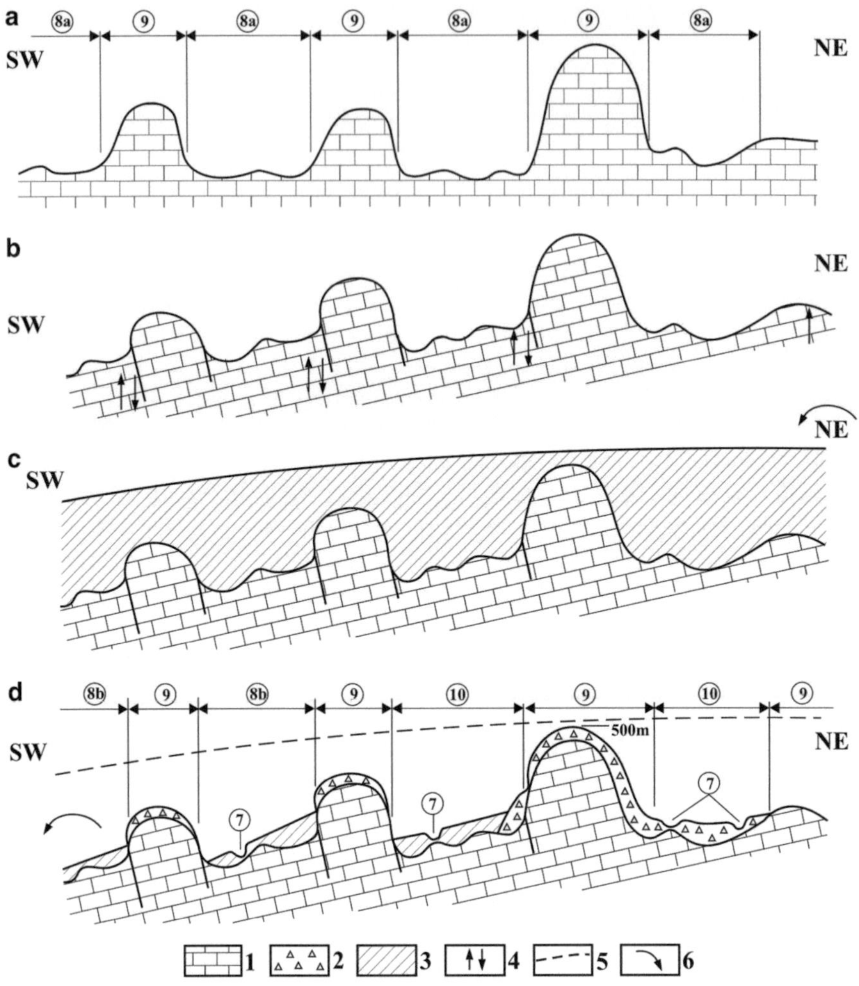

Fig. 8.10 Surface evolution of a tilted block (Mester-Hajag). *1* limestone *2* frost-shattered debris, *3* cover deposit, *4* tilting, *5* former surface with cover deposit, *6* sediment transport, *7* subsidence doline, *8a* surface created by karstification, *8b* gently sloping surface with cover deposit between crests, *9* residual feature of karstification, *10* depression of superficial deposit (DSD), (**a**) karstification, (**b**) tilts, (**c**) burial, (**d**) exhumation (the cover deposit is transported between the mounds), subsidence dolines form

8.4 Surface Evolution on Glaciokarst Covered Karst

On glaciokarsts, covered karst is generated in cirques and troughs. Covered karst formation is primarily concentrated in the area of paleodepressions, which are not entirely covered or filled with the products of glacial accumulation or subsequent mass movements. If burial is partial, the cover is interrupted by uncovered parts or

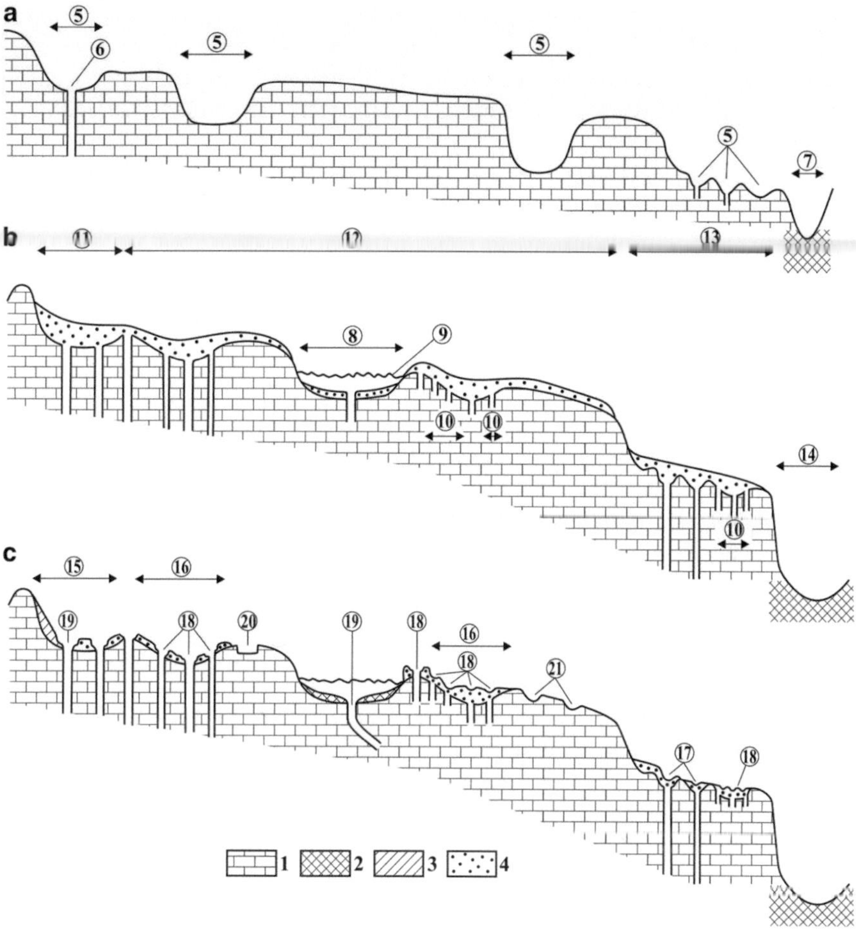

Fig. 8.11 Karstic evolution of glaciated surface. *1* limestone; *2* impermeable non-karstic rock; *3* debris fan built of non-karstic rock; *4* morainic deposit; *5* paleodoline; *6* chimney, shaft; *7* valley; *8* rock basin; *9* lake; *10* karren; *11* cirque; *12* glacial trough; *13* mountain foreland; *14* main glacier valley bordering (part of) the mountains; *15* DSD in cirque; *16* DSD in glacial trough; *17* postgenetic subsidence doline; *18* syngenetic subsidence doline; *19* covered karst ponor; *20* shaft doline; *21* recent solution doline. (**a**) karstification; (**b**) glacial erosion and accumulation; (**c**) recent karstification: the cover deposit is being removed from the thresholds between dolines; in the former dolines DSDs develop, and on terrains without cover deposit, recent karstification takes place (karren and solution dolines come about)

landforms (roches moutonnées). After the ice disappeared, the paleodepressions are regarded (true) DSDs, which may be accumulated or exhumed (see below).

In lack of watercourses and because of the paleodepressions (with counterslopes), surficial sediment transport from the glacial valleys does not happen. The loss of sediments from glacial valleys can take place during transport into the karst, and thus sections of the troughs are exhumed (Fig. 8.11). The entire area of the patchy covered karst is gradually transformed into bare karst. Recent recharge of the cover deposits of glacial valleys can be intensive (by mountain collapses, frost shattering, redeposition of morainic material, etc.). Therefore, individual valley sections are occasionally filled up to a large extent, and the paleokarst and glacial landforms are buried.

The glacially transformed bare karst features develop (extend by karstification) at a moderate rate (mostly wall karren are created). Their slopes are primarily affected by mass movements.

The degree to which paleodepressions are exhumed depends on the ratio between inward and outward transport. Inward transport can be by water or by mass movements, while outward transport mainly takes place by the transport of cover deposits in solution (Trudgill 1985) and through the conduits of ponors and subsidence dolines. Outward transport is possible if the karst is capable of receiving the sediment in the karren or paleokarst forms (shafts, caves) of the epikarst zone. Paleocaves are large enough to function as recipients of particularly large amounts of sediment. The caves, such as those in the Alps (Audra 1994; Audra et al. 2006), occur at highly variable elevations. Therefore, sediment input into the karst also happens at different elevations. The DSDs of glaciokarsts develop in paleodepressions and rock basins. The DSDs of concealed karsts can be non-exhuming, exhuming or infilling. On the floors of non-exhuming DSDs, there are no subsidence dolines, and the cover deposit is permeable and non-karstic. The exhuming DSDs receive small amounts of sediment from the environment. Their deepening is driven by the solution of the cover deposit or the transport of cover deposit into the subsidence dolines. The DSDs where the origin of subsidence dolines is associated with karren under cover are exhuming to a small extent since in this case the karren are only capable to receive small amounts of sediment. The DSDs where the density of subsidence dolines is high and which are formed above paleoshafts are exhuming to medium extent. The DSDs with clayey cover deposits (and ravines on their surfaces) and subsidence dolines above paleoshafts deepen at high rates. The DSDs are infilling if the number of subsidence dolines is low and they have no conduits (there are no cavities in the bedrock).

On cryptokarst, there are both accumulating and exhuming DSDs. The deepening of the latter can be of considerable rate driven by the action of watercourses on the floor. The blind valleys leading to ponors dissect the floor of the DSDs, and during deepening, the rock boundary is shifted and new ponors emerge. If the passage system of the karst is undeveloped, the ponors are clogged and the deepening of the DSDs is stopped. The cover extends in the area of the infilling DSD and the rock boundary is shifted towards the karst interior (similarly to the recent allogenic karst). At the margin of the cover, where the cover is thinning out, subsidence dolines emerge. In the environs of paleoshafts, there is cover deposit accumulation and the paleoshafts are transformed into ponors.

8.4.1 Surface Evolution on Cirque Covered Karst

Exhumation primarily takes place through the dissolution and pluvial erosion of the cover. In the area of cirques, one or more DSDs come about. There are no ponors but postgenetic subsidence dolines are generated.

The recipient capacity of dolines is considerable since they appear above large-size shafts. (Particularly under cirques in higher elevation, the chance of formation of large shafts is high.) Abundant precipitation (in the form of snow which is melting for a long period) favours pluvial erosion. Coarse rock debris, however, reduces the intensity of pluvial erosion. If doline density is high, the floor is denuded at a uniform rate and planated. If doline density is low, the floor is dissected. In both cases, the cirque (i.e. the DSD) is deepening. During the denudation of the cover, if several DSDs form in the cirque, they may coalesce. In the meantime, the extension of the covered karst patches is being reduced, while the uncovered karst sections are extending. If inward transport happens at a high rate, the alluvial fans may extend over the entire cirque floor and the subsidence dolines are buried.

8.4.2 Surface Evolution on Glacial Valley Covered Karst

Covered karst develops in the areas of both the cirque valley and the glacial trough (Fig. 8.11), either restricted to the area of the paleodolines (e.g. in the case of plateau glaciers) or valley glaciers.

In the case of plateau glaciers, paleodolines deepen into the plateau surface. The cover of the paleodolines (DSDs) denudes, and the features formed in the fill deepen as they are not supplied with sediment from their environs.

The covered karst of valley glaciers is either striped (the cover is also present on the valley floors between paleodepressions) or striped and patchy (the cover is only developed in the area of paleodepressions). Striped cover can show transitions towards continuous cover (for pediment glaciers). In the area of paleodepressions, DSDs appear, while on the valley floors between them, subsidence dolines develop. The patchy pattern (and also the striped pattern) develops in paleodepressions (dolines, uvalas) and rock basins (exclusively carved by ice).

In the valleys of valley glaciers (particularly in glacial troughs, from the valley floors between paleodepressions), the cover is removed from ever larger areas. Denudation is differentiated. The cover is denuded from the valley floors between paleodepressions, and it is not transported to the here occurring subsidence dolines but into the paleodepressions instead. The deepening of neighbouring paleodepressions of the same glacial trough can take highly variable rates. In the initial stage, infilling is also possible.

In the paleodepressions where the cover is impermeable if it includes few limestone fragments, ponors (with blind valleys) develop. If there are paleoshafts and paleocaves, the floor is dissected and deepens. In those where there are no paleoshafts on the bedrock and covered karren formation happens, if the cover is rich in limestone debris, the rate of cover denudation and DSD development is low. The floor is deepening at a low rate. Finally, those which receive large amounts of sediment from their environs are filled up. At the termination of the cover, the subsidence dolines and ponors are filled up, and the cover extends in the paleodepressions. Infilling DSDs emerge.

As a consequence, the dissection of the glacial valley is increasing. This does not result from the increasing dissection of the bedrock, but from the fact that some paleodepressions lose sediment, while others are filled up. The loss of cover leads to the extension of uncovered karst terrains.

8.5 Surface Evolution on Tropical Covered Karst

On tropical karst, covered karst formation begins with bare karst or buried karst. After formation it survives over a large area (Fig. 8.8). On tropical karst, pinnacle karst, autogenic karst and mixed allogenic–autogenic karst develop. The development of pinnacle karst has already been partly presented at the description of pinnacle varieties (see Sect. 7.1.3).

On tropical karst, surface evolution begins with grike formation, during which pinnacle terrains and inselberg karst develop from widening grikes. Both evolution types are described by Büdel's (1957) modified model of double planation. According to Paton (1964), the grike walls retreat and inselbergs are produced. The retreat of grike walls starts when the deepening of the grike floor reaches the karst water table or an impermeable bed (Fig. 7.3). Subsequently, the grikes are only broadening. The development of the different varieties of tropical karst depends on the amount of the solution residue (which is a function not only the rate of solution but also the amount insoluble material in the karstic rock) and on the rate of its removal (on the surface or into the karst).

If the accumulation and transport of solution residue is in balance, the extent of exhumation of the karst surface does not change, but its vertical dissection increases (Fig. 8.8a–c). The reason for this is that during the dissolution of the bedrock, the surface of the covered patches is gradually displaced to lower and lower levels. If outward transport exceeds the rate of accumulation of the solution residue, it is not only the dissection of the karst increases, but the ratio between covered and uncovered terrains is modified: the former is reduced, while the latter extend in area (Fig. 8.8d).

8.5.1 Surface Evolution of Pinnacle Terrains

The widening of grikes by solution was evidenced in our laboratory investigations (see Sect. 7.1.3). The relationships between the deepening of grike floors by solution, the denudation of the cover and the exhumation of pinnacles and thus the evolution of pinnacle terrains were presented by Peng et al. (2007) in the 'triplex erosion model'. An important condition of the surface evolution of pinnacle terrains is the presence and thickness of solution residue. If the solution residue is not removed, because of its thickness the rate of karst (covered karst), evolution is reduced and the karst becomes buried.

If denudation keeps pace with the generation of the solution residue from the solution of grike floors, cover thickness remains stable. Surface lowering is differentiated and results in a higher degree of vertical dissection. This means that surface details of grikes will become lower with the exception of inter-grike features (crests, pinnacles), which are exhumed. At the same time, with the broadening of grikes, the pinnacles may be consumed and covered terrains widened. This alternative evolution refers to the initial phase of inselberg karst evolution. On covered karst terrains, subsidence dolines and ponors develop. The cover deposit is transported through these features and through the grikes (cracks) of the bedrock into the karst. The denudation of the cover may also happen by surface material transport. The cover is either redeposited locally (producing stone forest terrain) or transported away. In this case, in addition to or instead of karst landforms, a valley network comes about.

If the rate of outward transport of the cover exceeds that of solution residue generation, the exhumation of the grikes of the pinnacle terrain begins. This also means a reduction in the rate of deepening of grike floors.

8.5.2 Surface Evolution of Autogenic Karst

On the fengcong variety of autogenic karst, dolines or ponors develop in the depressions between inselbergs (Fig. 8.12). Subsidence doline formation deriving from karst water table fluctuations is typical on valley floors or at plain fengcong, where katavothra emerge. Even more common are subsidence dolines of similar origin on the fenglin with intermountain plains, where the karst water table lies close to the surface (Fig. 8.12a).

During the dissolution of limestone, the floors of depressions (solution dolines) are deepening, but sediment accumulates in the depressions (solution residue). If the cover deposit is transported into the karst through the dolines and ponors, DSDs are generated during the sediment transport (Fig. 8.12b). The DSDs are either of concealed karst type or cryptokarst type.

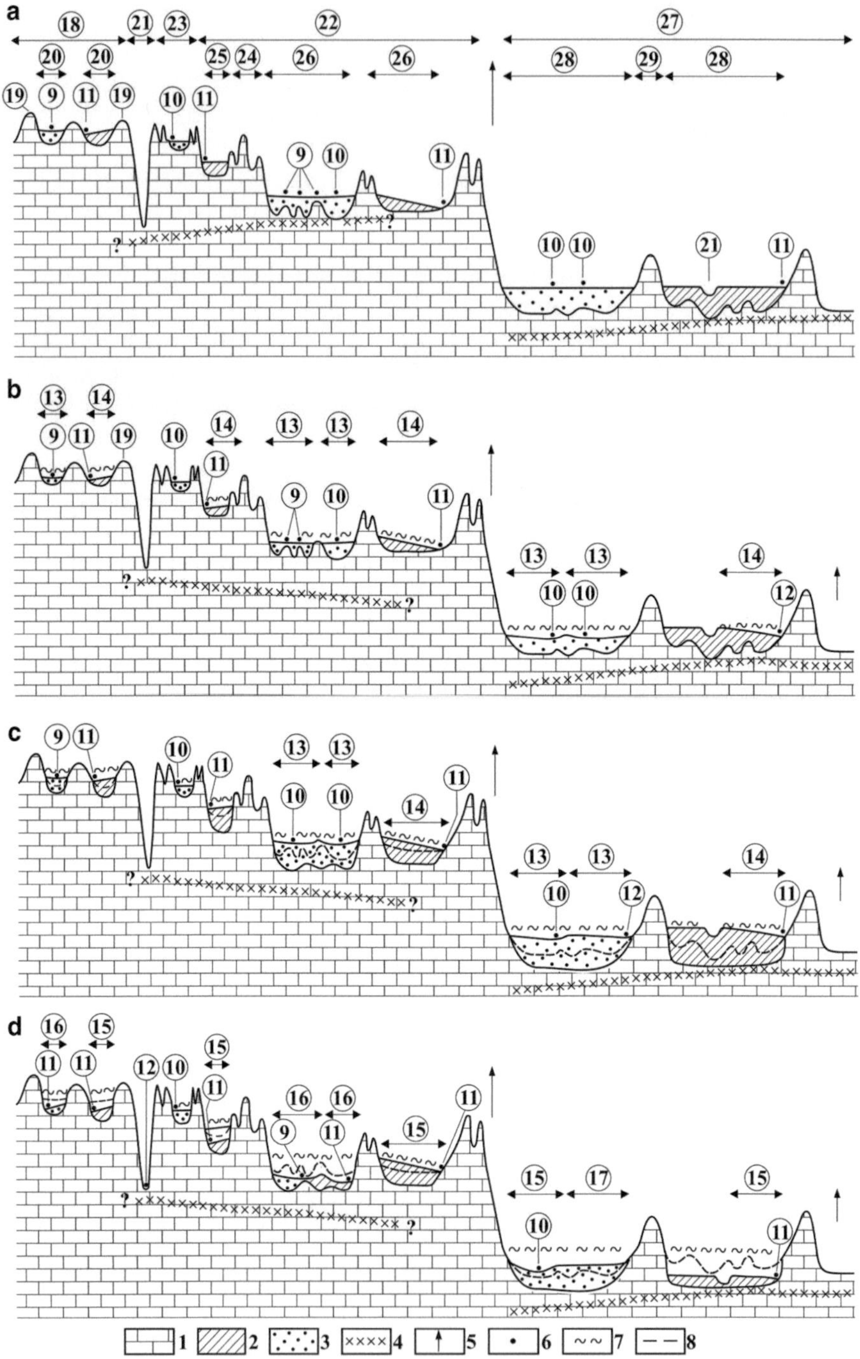

Fig. 8.12 Surface evolution on tropical covered karst. *1* limestone, *2* impermeable cover, *3* permeable cover, *4* karst water, *5* elevation, *6* subsidence doline and ponor, *7* former surface of cover deposit, *8* former limestone surface, *9* syngenetic subsidence doline, *10* postgenetic subsidence doline,

The vertical dissection of inselberg karst increases where in the area of DSDs between inselbergs outward and inward transport is equal (Fig. 8.12c). Since simultaneously with this balance the bedrock deepens and the cover is denuded, the more enduring this balance, the vertical dissection will be more pronounced. Where outward transport is more limited, infilling becomes predominant and the DSDs are filled up (Fig. 8.12c). Particularly in the area of intermountain plains, this situation is common: if the karst water table lies above the bedrock surface (induced by rising base level or sinking karst), subsidence dolines become fossilised. If outward transport exceeds cover accumulation, the DSDs are exhumed (Fig. 8.12d) and the vertical dissection of the karst is enhanced. On the inselberg karst, DSDs of different evolution come about and contribute to the variable morphology of karst type.

On fengcong karst, vertical dissection can reach remarkable dimensions since the karst water table lies under the bedrock surface. For this reason, subsidence dolines and ponors emerge on the karst. The sediment recipient capacity of the karst can be high since there are large caves in the karst. It is not only the vertical dissection of the fengcong that grows, but the bare karst extends at the expense of covered if the rate of outward transport exceeds the generation of solution residue.

On the fenglin type (especially at the intermountain plain variety), the outward transport is limited. Since the karst water table is close to the surface, the cover is less likely to be transported into the karst. It is probable that large-scale transport of the cover is only possible with fluctuating karst water level (sinking) or if at developing intermountain plains, the bedrock did not reach the level of the karst water table. Otherwise continuous removal of the cover and thus increasing vertical dissection are only possible through fluvial transport. The surface transport of the cover is favoured by the fact that intermountain plains are not closed landforms.

Because of the generation of solution residue in variable amounts (particularly on fengcong karst) and because of variable rates of outward transport (particularly on fenglin karst), the extent of vertical dissection in horizontal direction will also be highly variable.

Fig. 8.12 (continued) *11* ponor, *12* katavothron, *13* depression of superficial deposit (its *bottom* is concealed karst), *14* depression of superficial deposit (its *bottom* is cryptokarst), *15* further developed DSD, *16* renewed (transformed) DSD, *17* buried depression, *18* fenglin of high position, *19* conical inselberg, *20* inter-cone depression (tropical doline), *21* valley and river channel, *22* fengcong of high position, *23* depression within mountain group (tropical doline), *24* mountain group of common base, *25* depression between mountain groups (cockpit), *26* flat terrain between mountain groups, *27* lowland fenglin, *28* intermountain lowland, *29* inselberg. (**a**) ponors develop at the margin of the cover deposit of depressions and intermountain lowland, and subsidence dolines emerge on the impermeable cover; (**b**) various DSDs develop through the transport of the cover into the karst; (**c**) solution residue is generated during the dissolution of the limestone, the cover deposit thickens, and the limestone basement of the depressions and intermountain lowlands sinks; (**d**) the vertical dissection of the karst increases: the denudation of the cover is enhanced, new dolines and ponors emerge, the previously developed subsidence dolines are destroyed, new dolines develop, new DSDs come about, and the former DSDs are transformed or infilled

8.5.3 Surface Evolution of Mixed Allogenic–Autogenic Karst

The tropical variety of the mixed allogenic–autogenic karst develops if the fenglin is bordered by non-karstic rock (Sect. 4.6.2.6, Fig. 4.91). Here the fenglin is filled up with sediment from the non-karstic terrain. If the rate of infilling is high (as the marginal ponors are capable of conveying small amounts of sediment into the karst interior), the sites of ponor formation are shifted into the fenglin interior. The intermountain lowlands receive a cover of non-karstic origin, and ponors emerge at the inselbergs and even in the interior of the intermountain lowland. Ponor formation primarily takes place at the inselbergs. Because of the position of karst water table close to the surface, ponor formation is not possible in the interior of the intermountain lowland, while at inselbergs it is possible.

8.6 Surface Evolution on Platform Karst

The strata of the carbonate platform karst are either horizontal or gently dipping. According to Ford and Williams (2007), with horizontal stratification, the margin of the karst cover is retreating from the lower margin of the karst. As a consequence, the cover is preserved on the opposite margin of the karst (Fig. 8.31a). With gently dipping strata, the cover is denuded on the opposite margin of the karst (Fig. 8.31b). At the upper margin of the karst, older ponors occur along the intercalations of non-karstic rocks. At the upper margin of the cover preserved at the lower part of the karst, new ponors emerge and cause the retreat of the cover margin in dip direction.

On evaporites (autogenic cryptokarst), surface evolution is controlled by breccia pipes stoping from great depths. Above them caprock dolines develop to which watercourses and their valleys are attached. The denudation of the cover is caused by these features as well as the watercourses running in outward direction from the karst.

8.7 Conclusions

The surface evolution of covered karsts follows different courses under different climates, but fundamentally three types are identified. It can start from bare karst and buried karst or can regenerate continually. On bare karst the cover either derives from the outside or locally. Here the covered karst stage is developing gradually, and then with the denudation of the cover, the karst is transformed again into bare karst. Such a karst is the recent allogenic covered karst and the renewed allogenic karst. In the case of evolution from buried conditions, the cover developed independent from karstification. During the denudation of the cover deposit, the covered karst is transformed into bare karst. Such a karst is the horst covered karst, the glaciokarst covered karst, the mantled allogenic covered karst, the platform karst as well as the tundra and taiga karst. Continuously existing covered karst is the tropical karst.

With the exception of tropical karsts, the evolution of covered karst ends with the removal of the cover, and the karst is transformed into bare karst. (This does not apply to those karst types which receive sediment recharge from the bordering terrain continuously, also at present.) On glaciokarsts, glacial action proves a connection between former karstification and covered karst formation (glacial erosion transformed the old landforms and supplies the cover material, too). On tropical karsts, covered karst formation is continually present (the covered karst is regenerated), and it is not limited to a single stage of geomorphic evolution. The explanation for this is that limestone solution produces so high amounts of solution residue which is capable of continuously maintaining covered karst conditions. On tropical karst, covered karst is the product of karstification processes.

Cover denudation (and thus the elimination of the covered karst conditions) takes two possible courses: through the surface transport of cover deposits or through the transport of the cover into the karst cavities. In addition to the denudation rate of the cover deposit, the rate of cessation of the covered karst stage is controlled by the development stage of karst cavities and the rate of conveyance of the sediment input by surface landforms (ponors, subsidence dolines).

References

Audra P (1994) Alpine karst speleogenesis: case studies from France (Vercors, Chartreuse, Ile de Crémieu) and Austria (Tennengebirge). Cave Karst Sci 21(3):75–80

Audra P, Bini A, Gabrovšek F, Häuselmann P, Hoblea F, Jeannin PY, Kunaver J, Monbaron M, Šušteršič F, Tognini P, Trimmel H, Wildberger A (2006) Cave genesis in the Alps between the Miocene and today: a review. Z Geomorphol N F 50(2):153–176

Büdel J (1957) Die doppelten Einebrungsflächen in den feuchten Tropen. Z Geomorphol 1:201–288

Cui Z, Li D, Feng J, Liu G, Li H (2002) The covered karst, weathering crust and karst (double-level). Sci China 45:366–378

Cvijič J (1918) L'hydrogeographie souterraine et l'évolution morphologique du karst. Trav Inst Geogr Alp 6:375–426

Ford DC, Williams PW (2007) Karst hydrogeology and geomorphology. Wiley, Chichester, 562 p

Grund A (1914) Der geographische Zyklus im Karst. Geselischaft Erdkunde 52:621–640 (Translated into English in Sweeting 1981)

Jakucs L (1977) Morphogenetics of karst regions. Adam Hilgar, Bristol, 284 p

Paton JR (1964) The origin of the limestone hills of Malaya. J Trop Geogr 18:134–139

Peng J, Cai Y, Yang M, Liang H, Liang F, Song L (2007) Relating aerial erosion, soil erosion and sub-soil erosion to the evolution of Lunan Stone Forest, China. Earth Surf Process Landf 32:260–268

Trudgill ST (1985) Limestone geomorphology. Longman, New York, 196 p

Waltham AC, Fookes PG (2003) Engineering classification of karst ground conditions. Q J Eng Geol Hydrogeol 36:101–118

Waltham AC, Fookes PG (2005) Engineering classification of karst ground conditions. Speleogenesis Evol Karst Aquifers (www.speleogenesis.info) 3(7)
Williams PW (1985) Subcutaneous hydrology and the development of doline and cockpit karst. Z Geomorph N F 29(4):463–482
Williams PW (1987) Geomorphic inheritance and the development of tower karst. Earth Surf Process Landf 12:453–465

Index

M. Veress, *Covered Karsts*, Springer Geology,
DOI 10.1007/978-94-017-7518-2

G

J

K

L

M

N

Q

R

X

Y

Z

MIX
Papier aus verantwortungsvollen Quellen
Paper from responsible sources
FSC® C105338

If you have any concerns about our products,
you can contact us on
ProductSafety@springernature.com

In case Publisher is established outside the EU,
the EU authorized representative is:
Springer Nature Customer Service Center GmbH
Europaplatz 3, 69115 Heidelberg, Germany

Printed by Libri Plureos GmbH
in Hamburg, Germany